People and Wildlife
Conflict or Coexistence?

Human–wildlife conflict is a major issue in conservation. As people encroach into natural habitats, and as conservation efforts restore wildlife to areas where they may have been absent for generations, contact between people and wild animals is growing. Some species, even the beautiful and endangered, can have serious impacts on human lives and livelihoods. Tigers kill people, elephants destroy crops and African wild dogs devastate sheep herds left unattended. Historically, people have responded to these threats by killing wildlife wherever possible, and this has led to the endangerment of many species that are difficult neighbours. The urgent need to conserve such species, however, demands coexistence of people and endangered wildlife. This book presents a variety of solutions to human–wildlife conflicts, including novel and traditional farming practices, offsetting the costs of wildlife damage through hunting and tourism, and the development of local and national policies.

ROSIE WOODROFFE is Associate Professor of Conservation Biology at the University of California, Davis.

SIMON THIRGOOD is Science Leader for Ecology at the Macaulay Institute in Aberdeen, Scotland.

ALAN RABINOWITZ is Director of the Science and Exploration Division for the Wildlife Conservation Society based at the Bronx Zoo in New York.

Conservation Biology

Conservation biology is a flourishing field, but there is still enormous potential for making further use of the science that underpins it. This new series aims to present internationally significant contributions from leading researchers in particularly active areas of conservation biology. It will focus on topics where basic theory is strong and where there are pressing problems for practical conservation. The series will include both single-authored and edited volumes and will adopt a direct and accessible style targeted at interested undergraduates, postgraduates, researchers and university teachers. Books and chapters will be rounded, authoritative accounts of particular areas with the emphasis on review rather than original data papers. The series is the result of collaboration between the Zoological Society of London and Cambridge University Press. The series editors are Professor Morris Gosling, Professor of Animal Behaviour at the University of Newcastle upon Tyne, Professor John Gittleman, Professor of Biology at the University of Virginia, Charlottesville, Dr Rosie Woodroffe of the University of California, Davis and Dr Guy Cowlishaw of the Institute of Zoology, Zoological Society of London. The series ethos is that there are unexploited areas of basic science that can help define conservation biology and bring a radical new agenda to the solution of pressing conservation problems.

Published Titles

1. *Conservation in a Changing World*, edited by Georgina Mace, Andrew Balmford and Joshua Ginsberg 0 521 63270 6 (hardcover), 0 521 63445 8 (paperback)
2. *Behaviour and Conservation*, edited by Morris Gosling and William Sutherland 0 521 66230 3 (hardcover), 0 521 66539 6 (paperback)
3. *Priorities for the Conservation of Mammalian Diversity*, edited by Abigail Entwistle and Nigel Dunstone 0 521 77279 6 (hardcover), 0 521 77536 1 (paperback)
4. *Genetics, Demography and Viability of Fragmented Populations*, edited by Andrew G. Young and Geoffrey M. Clarke 0 521 782074 (hardcover), 0 521 794218 (paperback)
5. *Carnivore Conservation*, edited by John L. Gittleman, Stephan M. Funk, David Macdonald and Robert K. Wayne 0 521 66232 X (hardcover), 0 521 66537 X (paperback)
6. *Conservation of Exploited Species*, edited by John D. Reynolds, Georgina M. Mace, Kent H. Redford and John G. Robinson 0 521 78216 3 (hardcover), 0 521 78733 5 (paperback)
7. *Conserving Bird Biodiversity*, edited by Ken Norris and Deborah J. Pain 0 521 78340 2 (hardcover), 0 521 78949 4 (paperback)
8. *Reproductive Science and Integrated Conservation*, edited by William V. Holt, Amanda R. Pickard, John C. Rodger and David E. Wildt 0 521 81215 1 (hardcover), 0 521 01110 8 (paperback)

People and Wildlife
Conflict or Coexistence?

Edited by

ROSIE WOODROFFE, SIMON THIRGOOD
AND ALAN RABINOWITZ

CAMBRIDGE UNIVERSITY PRESS
Cambridge, New York, Melbourne, Madrid, Cape Town, Singapore, São Paulo, Delhi

Cambridge University Press
The Edinburgh Building, Cambridge CB2 8RU, UK

Published in the United States of America by Cambridge University Press, New York

www.cambridge.org
Information on this title: www.cambridge.org/9780521825054

First published 2005

A catalogue record for this publication is available from the British Library

ISBN 978-0-521-82505-4 hardback
ISBN 978-0-521-53203-7 paperback

Transferred to digital printing 2008

Contents

Contributors

REIDAR ANDERSEN
Norwegian Institute for Nature Research
Tungasletta 2
N-7485 Trondheim
Norway

HENRIK ANDRÉN
Grimsö Wildlife Research Station
Department of Conservation Biology
Swedish University of Agricultural Sciences
SE-730 91 Riddarhyttan
Sweden

CHRISTOF ANGST
KORA
Thunstrasse 31
CH-3074 Muri
Switzerland

VAL ASHER
Turner Endangered Species Fund
1123 Research Drive
Bozeman, MT 59718
USA

EDWARD E. BANGS
US Fish and Wildlife Service
100 N Park, #320
Helena, MT 59601
USA

ELIZABETH H. BRADLEY
School of Forestry
University of Montana
Missoula, MT 59812
USA

URS BREITENMOSER
Institute of Veterinary Virology
University of Bern
Länggasstrasse 122
CH-3074 Bern
Switzerland

CHRISTINE BREITENMOSER-WÜRSTEN
KORA
Thunstrasse 31
CH-3074 Muri
Switzerland

TIM W. CLARK
Yale University School of Forestry and Environmental Studies
205 Prospect Street
New Haven, CT 06511
USA

DAVID COPE
The Macaulay Institute
Craigiebuckler
Aberdeen AB15 8QH
UK

PAUL FERRARO
Andrew Young School of Policy Studies
Georgia State University
33 Gilmer Street Unit 2
Atlanta, GA 30303
USA

HANK FISCHER
National Wildlife Federation
Northern Rockies Natural Resource Center
240 N Higgins, #2
Missoula, MT 59802
USA

JOSEPH A. FONTAINE
US Fish and Wildlife Service
100 N Park, #320
Helena, MT 59601
USA

LAURENCE G. FRANK
Museum of Vertebrate Zoology
University of California
Berkeley, CA 94720
USA

JOHN GOODRICH
Wildlife Conservation Society
2300 Southern Boulevand
Bronx
New York, NY 10460
USA

RAJESH GOPAL
Project Tiger–Government of India
Bikaner House Annexe-5
Sha Jahan Road
New Delhi-110 011
India

IVAR HERFINDAL
Biology Institute
Norwegian University of Science and Technology
N-7491 Trondheim
Norway

STEPHEN HERRERO
Faculty of Environmental Design
University of Calgary
Calgary, Alberta
T2T 2Y2
Canada

CATHERINE M. HILL
Department of Anthropology
School of Social Sciences and Law
Oxford Brookes University
Gipsy Lane Campus
Oxford OX3 0HP
UK

JON M. HUTTON
Fauna and Flora International
Great Eastern House
Tenison Road
Cambridge CBI 2TT
UK

JOHN JACKSON
World Conservation Force
Suite 1045, 3900 N Causeway Boulevand
Metarie, LA 70002
USA

MICHAEL D. JIMENEZ
US Fish and Wildlife Service
100 N Park, #320
Helena, MT 59601
USA

K. ULLAS KARANTH
Wildlife Conservation Society
Bronx
New York, NY 10460
USA

TIMOTHY J. KNICKERBOCKER
Central College
812 University
Pella, IA 50219
USA

UNNI STØBET LANDE
Biology Institute
Norwegian University of Science and Technology
N-7491 Trondheim
Norway

JEAN-MARC LANDRY
KORA
Thunstrasse 31
CH-3074 Muri
Switzerland

NIGEL LEADER-WILLIAMS
Durrell Institute of Conservation and Ecology
Department of Anthropology
University of Kent
Canterbury CT2 7NS
UK

DALE LEWIS
Wildlife Conservation Society
Plot #8471 Haile Salasie Road
Long Acres
P/Bag E891 Post Net #397
Manda Hill, Lusaka
Zambia

JOHN D. C. LINNELL
Norwegian Institute for Nature Research
Tungasletta 2
N-7485 Trondheim
Norway

BORIS LITVINOV
Tiger Response Team
Inspection Tiger
Ministry of Natural Resources of the Russian Federation
Terney, Primorye
Russia

CURT M. MACK
Nez Perce Tribe
1000 Mission
McCall, ID 83638
USA

FRANCINE MADDEN
2001 12th Street NW, Suite 317
Washington, DC 20009
USA

LAUREN MCCAIN
Department of Political Science
University of Colorado
Campus Box 333
Boulder, CO 80309
USA

THOMAS J. MEIER
US Fish and Wildlife Service
100 N Park, #320
Helena, MT 59601
USA

BRIAN J. MILLER
Denver Zoological Foundation
2900 East 23rd Avenue
Denver, CO 80205
USA

DALE MIQUELLE
Wildlife Conservation Society
2300 Southern Boulevand
Bronx
New York, NY 10460
USA

LISA NAUGHTON-TREVES
Department of Geography
University of Wisconsin
550 North Park Street
Madison, WI 53706
USA

CARTER C. NIEMEYER
US Fish and Wildlife Service
1387 Vinnel Way, Rm 368
Boise, ID 83709
USA

IGOR NIKOLAEV
Institute of Biology and Soils
Far Eastern Branch of the Russian Academy of Sciences
Vladivostok, Primorye
Russia

ERLEND BIRKELAND NILSEN
Biology Institute
Norwegian University of Science and Technology
N-7491 Trondheim
Norway

PHILIP NYHUS
Environmental Studies Program
Colby College
Waterville, ME 04901
USA

JOHN K. OAKLEAF
US Fish and Wildlife Service
P.O. Box 856
Alpine, AZ 85920
USA

JOHN ODDEN
Norwegian Institute for Nature Research
Tungasletta 2
N-7485 Trondheim
Norway

MORDECAI O. OGADA
Mpala Research Centre
P.O. Box 555
Nanyuki
Kenya

FERREL V. OSBORN
Elephant Pepper Development Trust
18 Rowland Square
Milton Park
Harare
Zimbabwe

STEVEN A. OSOFSKY
Wildlife Conservation Society
2300 Southern Boulevard
Bronx
New York, NY 10540
USA

HOWARD QUIGLEY
Wildlife Conservation Society
3610 Broadwater Street
Suite 111
Bozeman, MT 59718
USA

ALAN RABINOWITZ
Wildlife Conservation Society
2300 Southern Boulevard
Bronx
New York, NY 10540
USA

RICHARD P. READING
Denver Zoological Foundation
2900 East 23rd Avenue
Denver, CO 80205
USA

STEPHEN REDPATH
Centre for Ecology and Hydrology
Banchory AB31 4BW
UK

MARCUS ROWCLIFFE
Institute of Zoology
London NW1 4RY
UK

KETIL SKOGEN
Norwegian Institute for Nature Research
Fakkelgården
N-2624 Lillehammer
Norway

EVGENY SMIRNOV
Sikhote-Alin State Zapovednik
Terney, Primorye
Russia

DOUGLAS W. SMITH
National Park Service
P.O. Box 168
Yellowstone National Park, WY 82190
USA

EVGENY SUVOROV
Sikhote-Alin State Zapovednik
Terney, Primorye
Russia

JON E. SWENSON
Departmernt of Biology and Nature Conservation
Agricultural University of Norway
Box 5014
N-1432 Ås
Norway

SIMON THIRGOOD
The Macaulay Institute
Craigiebuckler
Aberdeen AB15 8QH
UK

CHRIS THOULESS
Namibia Tourism Development Programme
PO Box 25781
Windhoek
Namibia

ADRIAN TREVES
Wildlife Conservation Society
2300 Southern Boulevard
Bronx
New York, NY 10540
USA

JULIET VICKERY
British Trust for Ornithology
The Nunnery
Thetford
IP24 2PU
UK

JOHN WAITHAKA
African Conservation Centre
Box 62844
Nairobi
Kenya

MATTHEW J. WALPOLE
Fauna and Flora International
Great Eastern House
Tenison Road
Cambridge CB1 2TT
UK

JEAN-MARC WEBER
KORA
Thunstrasse 31
CH-3074 Muri
Switzerland

DAVID WESTERN
African Conservation Centre
Box 62844
Nairobi
Kenya

ROSIE WOODROFFE
Department of Wildlife, Fish and Conservation Biology
University of California
1 Shields Avenue
Davis, CA 95616
USA

Foreword

In the origin myths of many human cultures, a central theme is the distinguishing of humans from the rest of nature (Lévi-Strauss 1964). According to these histories, before humans as humans existed, people lived in a way indistinguishable from other animals – depending on wild species and eating them raw. After the origin, after emerging, after being created, after The Fall, humans are distinguishable from the rest of nature through the acquisition of culture and by means of it. People cook their food, people alter their landscape, people cultivate crops and raise domestic animals. In the words of the Judaeo-Christian Bible: 'the Lord God sent him forth from the Garden of Eden, to till the ground . . . ' (Genesis 3: 23). And by contrast, wild animals define humanity. The relationship with wild species is fundamental to our self-identity. At its core, this book is about that relationship.

We humans value nature and wild species in many contexts and situations. Wild species are of cultural and social importance. They are valued as resources. We value their very existence. On the other hand, humans often and increasingly come into conflict with wildlife. We humans channel more and more of the world's resources to support our own kind. We now channel more than 40% of the terrestrial net primary productivity, which is the sustenance of all animals and decomposers, to our own ends (Vitousek *et al.* 1986). Forty-five per cent of Asia is under cropland and permanent pasture, 36% of Africa, 30% of North America, 35% of South America, 47% of Europe, 25% of the former Soviet Union. These figures do not include managed forests, and urban and suburban sprawl. And in the seas, humans are pre-empting 25–35% of the total production of continental shelf ecosystems (World Resources Institute 1994). As wild species lose habitat, individual animals necessarily come into conflict with human beings.

Some species thrive in this human dominated world. Some we tolerate and even enjoy. Songbirds are part of our landscape. Grey squirrels (at least in North America) bother few people, with the exception of those with bird feeders. White-tailed deer seem part of the ex-urban landscape, but with their population growth probably are too much of a good thing. Other adaptable species we seek to extirpate. Rats, cockroaches, termites, we call 'pests', and their irrepressibility supports a whole industry of pest control. These animals affect or are perceived to affect our lives and livelihoods.

This book is really about that subset of species that are valued but also negatively affect our lives and livelihoods. Jaguars are cultural icons throughout South America, but they also are major predators of cattle. Baboons exhibit social shenanigans that keep ecotourists enthralled, but they also raid crops. Elephants elicit inordinate attention from conservationists, but they are a threat to human life and limb. Pigs, goats and donkeys are valued by animal rights advocates, but they tear up our parks and reserves. This book explores the conundrum when animal species that are valued in one context or by some people are in conflict with human needs and aspirations in another or by others.

Resolving the conundrum has absorbed considerable intellectual energy. In Europe from the Middle Ages, and to some extent in the New World, one approach was to turn to the legal system (Evans Pritchard 1906). Pigs were put on trial for killing children. Cats were put to death for excessive caterwauling (Darnton 1985). Shakespeare in *Merchant of Venice* (Act iv, scene 1) notes a wolf 'hanged for human slaughter, even from the gallows . . . .' Judicial proceedings were initiated against domestic and wild animals. Those animals found guilty and in human custody were usually put to death. A more modern approach was to argue the case in the court of public opinion seeking to have a species declared a pest or in need of protection. Once a species was considered a pest, it could be subject to lethal control, and this book catalogues the species on which we have focussed such attention. Alternatively, once a species was considered to be endangered or threatened, then it could be protected.

Many of the examples in this book avoid this dichotomous choice. Most accept that choosing unequivocally between humans and wildlife is rarely a choice. Both humans and wildlife must be accommodated. Some advocate the spatial separation of wildlife and humans, and the creation of heterogeneous landscapes. For many large-bodied animals, for predators, and for species whose existence depends on undisturbed habitats, this clearly will be necessary for their long-term survival. Other chapters suggest that coexistence in the same place is possible if the conflict can be mitigated.

Some argue that conflict can be mitigated through the scientific process, by which information made available to actors can allow more informed and rational choices. Still others urge a realigning of the economic incentives so that the benefits of coexistence outweigh the costs. Clearly both approaches have been successful on occasion. It is also clear, however, that there are many cases where scientific and financial arguments have no validity. For instance, no amount of scientific information or economic analyses can seemingly sway opponents of prairie dogs and wolves. Here the issue is less about 'human–wildlife conflict' and more about 'human–human conflict'. Here different people have very different perceptions and values about wildlife, and resolution must reconcile different interest groups. But all of the chapters suggest that understanding the issues is a first step towards reconciling the conflict. Ultimately, if the human society is both to continue to conserve wild species and to sustain the lives and livelihoods of people that come into conflict with wildlife, then we must find ways to mitigate and resolve human–wildlife. If we fail, then we will either not meet our fundamental ethical obligation to steward the world's species and natural systems, or not meet the obligation to sustain our fellow human beings. Such a failure is unthinkable.

John G. Robinson

Acknowledgements

This book is very much a collaborative effort, and we would like to take this opportunity to thank the 62 chapter authors and co-authors whose collective wisdom undoubtedly outweighs that of the editors. The authors not only contributed their own chapters, they also provided, along with Nick Hanley and Claudio Sillero-Zubiri, insightful reviews of other authors' contributions. We certainly look forward to working with all of them in the future. We would also like to thank John Robinson for taking the time out of a hectic life to write an evocative and inspired foreword to this volume.

This book sprang from a symposium held at the Zoological Society of London in December 2002. We would like to thank the Wildlife Conservation Society, the Michael Cline Family Foundation and the Zoological Society of London for their support for the meeting, and Deborah Body and Georgina Mace for their help in organizing it. We also wish to thank Liz Bennett, Glyn Davies, James Deutsch, Sarah Durant, John Harwood, David Macdonald, John Robinson and Kim Rollins for their various valuable contributions to the meeting. RW also wishes to thank the many contributors to the symposium 'People and predators', held at the 8th International Theriological Congress, South Africa, in 2001, and ST would also like to acknowledge the participants of the symposium 'Birds of prey in a changing environment', held in Edinburgh in December 2000. These two meetings and the ideas that they generated encouraged us to organize the London symposium.

At Cambridge University Press, we would like to thank Alan Crowden, Maria Murphy, Carol Miller and Anna Hodson for their editorial help. This book has appeared a mere two years after the London meeting, primarily as a result of their efficiency.

This book developed over a number of years, and we would especially like to thank our friends and colleagues whose insights and experience helped to shape our opinions on this difficult conservation issue, particularly Josh Ginsberg, Laurence Frank, Steve Redpath, Karen Laurenson,

Claudio Sillero-Zubiri and John Robinson. We also thank our employers, the University of California, Davis (RW), the University of Stirling, Frankfurt Zoological Society and the Macaulay Institute, Aberdeen (ST) and the Wildlife Conservation Society (AR), for their support in preparing this book.

Rosie Woodroffe
Davis, California, USA
Simon Thirgood
Serengeti National Park, Tanzania
Alan Rabinowitz
New York, USA

1

The impact of human–wildlife conflict on natural systems

ROSIE WOODROFFE, SIMON THIRGOOD AND ALAN RABINOWITZ

INTRODUCTION

This book is concerned with resolving conflicts that occur between people and threatened wildlife. Wildlife are often subject to control if they are perceived to harm the livelihoods, lives or lifestyles of people. Many wildlife species can thrive despite such control: our continued need for mouse- and cockroach traps is testament to the resilience of some species in the face of extensive lethal control. While a panoply of invertebrate (especially insect) pests, and adaptable vertebrates such as coyotes (*Canis latrans*), ground squirrels (e.g. *Spermophilus californicus*) and red-billed quelea (*Quelea quelea*) continue to out-wit pest control experts, other species are not so well equipped to resist the effects of lethal control, Many have become seriously endangered as a result. This raises a serious challenge: what do we do when a highly endangered animal genuinely causes serious damage to human lives or livelihoods? How can we reconcile the need to conserve the species with the need to protect the rights and property of people who share its environment? Resolving such conflicts will be crucial to the success of conservation development plans that require coexistence of people with wildlife. For many sensitive species, effective conservation will be near-impossible to achieve unless such conflicts can be resolved or at least mitigated.

The scope and structure of this book

In this book, we seek resolutions to the most widespread and serious conflicts involving people and threatened wildlife: crop raiding, livestock depredation, predation on managed wildlife (such as farmed or otherwise managed game species) and, least common but most emotive, killing of people. We term this phenomenon *human–wildlife conflict*. These conflicts

People and Wildlife: Conflict or Coexistence? eds. Rosie Woodroffe, Simon Thirgood and Alan Rabinowitz. Published by Cambridge University Press.

involve a taxonomically diverse array of wild species, but many of the solutions may be generally applicable. To preserve this generality, we have omitted some less common forms of conflict, such as control of wildlife to limit the spread of infectious disease. We have focussed on the best-studied systems, which leads to an inevitable bias towards studies of large vertebrates. Nevertheless, this is not a book about *vertebrate pest management*. An extensive literature on vertebrate pest management has developed over the past 30 years, particularly in the USA, which has recently been summarized by Conover (2002). Our book differs in that we seek solutions that will result in an improvement of the conservation status of wildlife that come into conflict with people. Our perspective is the management of species of conservation concern, and we consider the management of common, successful species only insofar as this contains lessons for more threatened species. However, we are aware that some of the approaches that we discuss may have local application to the management of common species. We have also chosen to focus entirely on terrestrial ecosystems. Conflicts between people and wildlife do exist in the marine and freshwater environments, such as the debate over the role of marine mammals in preventing the recovery of commercial fisheries (Yodsis 2001), but the solutions to such issues are likely to be very different from those in terrestrial systems.

This book falls into three parts. The first section (Chapters 1 and 2) sets the scene by reviewing the impact of lethal control on wild populations of threatened species (this chapter), and the impacts of threatened species on human lives and livelihoods (Thirgood *et al.*, Chapter 2). Our second section (Chapters 3 to 10) reviews various approaches to resolving conflicts between people and wildlife, including technical measures to mitigate wildlife impacts (e.g. guard dogs, electric fencing: Breitenmoser *et al.*, Chapter 4), economic incentives that may offset the costs of wildlife impacts (e.g. ecotourism: Walpole and Thouless, Chapter 8) and policy approaches (e.g. zoning: Linnell *et al.*, Chapter 10). The third section (Chapters 11 to 23) presents a broad array of case studies which discuss specific attempts to resolve conflicts between people and threatened wildlife. Finally we present (in Chapter 24) our conclusions and our hopes for the future.

LETHAL CONTROL

Where wildlife cause – or are perceived to cause – serious damage to human livelihoods, a common response has been to kill them. We choose to term this practice 'lethal control' because an alternative word, 'persecution', implies that such control is unjust or unwarranted. To the contrary, wild animals – even beautiful, charismatic and highly endangered wild animals – can and do kill people, destroy their crops and kill their livestock (Thirgood *et al.*, Chapter 2). If we fail to appreciate this crucial fact, we can never fully understand the causes of conflicts, and will never successfully resolve them.

People have been subjecting wildlife to lethal control for centuries if not millenia. As early as AD 800, the Emperor Charlemagne employed a cadre of professional wolf-hunters tasked with ridding the Holy Roman Empire of this menace (Boitani 1995). Some societies have taboos against killing particular species, even if they cause serious damage (e.g. Menon *et al.* 1998), but this is rare. Cultural factors strongly influence people's willingness to tolerate wildlife damage (Woodroffe 2000), but in many cases a primary limitation on the level of lethal control has been people's ability to capture and kill wildlife. Deliberate killing of wild animals perceived as pests has taken place on all of the inhabited continents, as well as in the sea and fresh water, and involves threatened species as diverse as orang utans (*Pongo pygmaeus*: Rijksen and Meijaard 1999), snow leopards (*Uncia uncia*: Ahmad 1994), peregrine falcons (*Falco peregrinus*: Thirgood *et al.* 2000b), prairie dogs (*Cynomys* spp.: Miller *et al.* 1996) and fur seals (Wickens 1996).

Wildlife perceived as 'problem animals' are killed both legally and illegally, by private individuals, informally organized communities, bounty hunters, and local and national governments. In developed countries the most common methods are shooting, trapping and poisoning, but traditional methods are also used. For example, in East Africa large carnivores are not infrequently killed with spears (Frank *et al.*, Chapter 18), as are African elephants (*Loxodonta africana*), and chimpanzees (*Pan troglodytes*: Ghiglieri 1984; Moss 2001). Innovative (if sometimes grisly) new methods have also been devised. In India, for example, farmers may deliberately modify power lines to electrocute crop-raiding Asian elephants (*Elephas maximus*), or pack explosives into jackfruit baits (Menon *et al.* 1998). Novel methods may also be highly selective; for example protective collars fitted to livestock ensure that coyotes are killed only when they bite the throat of a sheep and pierce the collar's reservoirs of 1080 poison (Burns *et al.* 1996).

HISTORICAL IMPACTS OF LETHAL CONTROL

Species extinctions

Lethal control has led to the extinctions of several species. The Guadelupe caracara (*Polyborus lutosus*), a raptor species confined to the island of Guadelupe off the Pacific coast of Mexico, was reported to kill juvenile goats and was shot and poisoned by local people for this reason

(Greenaway 1967). While the last few individuals were killed by collectors, lethal control is believed to have been the principal factor leading to the species' extinction in 1900 (Fuller 2000). Likewise, conflict with people over sheep depredation led to the extinction of two carnivorous mammals, the thylacine or marsupial wolf (*Thylacinus cynocephalus.*, in 1930), restricted to Tasmania, and the Falkland Island wolf (*Dusicyon australis,* in 1876: IUCN 2002). The Carolina parakeet (*Conuropsis carolinensis*) was killed as a pest of fruit crops, and this is believed to have been a primary cause of the species' extinction in 1904 (IUCN 2002). Reports from the time describe how, once one parakeet was shot, others would hover and scream above the carcass, making it easy to destroy entire flocks (Greenaway 1967).

Range collapses

Only a handful of species have been completely extirpated through human persecution, but many species have experienced massive contractions of their geographic ranges. Some of the most impressive range collapses occurred in North America, perhaps because the 'pioneer spirit' of European settlers pitted well-armed and highly motivated people against wildlife with very little experience of lethal control. In 1900, colonies of prairie dogs – not, in fact, dogs but burrowing squirrels – are estimated to have covered 410 000 km^2 of North America's short grass prairies. However, the farming industry perceived them as vermin which could compete with livestock for forage, and they were subjected to a massive government-sponsored poisoning campaign. By 1960, prairie dogs' geographic range had collapsed to less than 2% of their former distribution, and this range was still further reduced by the end of the twentieth century (Reading *et al.*, Chapter 13). Likewise, wolves (*Canis lupus* and *C. rufus*) were formerly distributed throughout the USA south of Canada, but, following a concerted (and, once again, government-sponsored) attempt to eradicate a species perceived as a threat to livestock, by 1960 they were confined to northeastern Minnesota and Isle Royale National Park in Lake Superior (*C. lupus*) and to a small area on the Texas–Louisiana border (*C. rufus*). Hence, wolves were extirpated from nearly 8 000 000 km^2 of their former range in North America alone (Fig. 1.1a). African wild dogs (*Lycaon pictus*) were eradicated from 25 of the 39 countries they formerly occupied (Fig. 1.1b), not only because they were considered a threat to livestock but also because they were thought to suppress densities of 'game' species inside protected areas (Bere 1955). Similar range collapses have affected most of the larger mammalian carnivores and are almost too numerous to mention. Both lions (*Panthera leo*) and cheetahs (*Acinonyx jubatus*) were all but eradicated from Asia by the early twentieth century, and today occupy greatly reduced distributions in

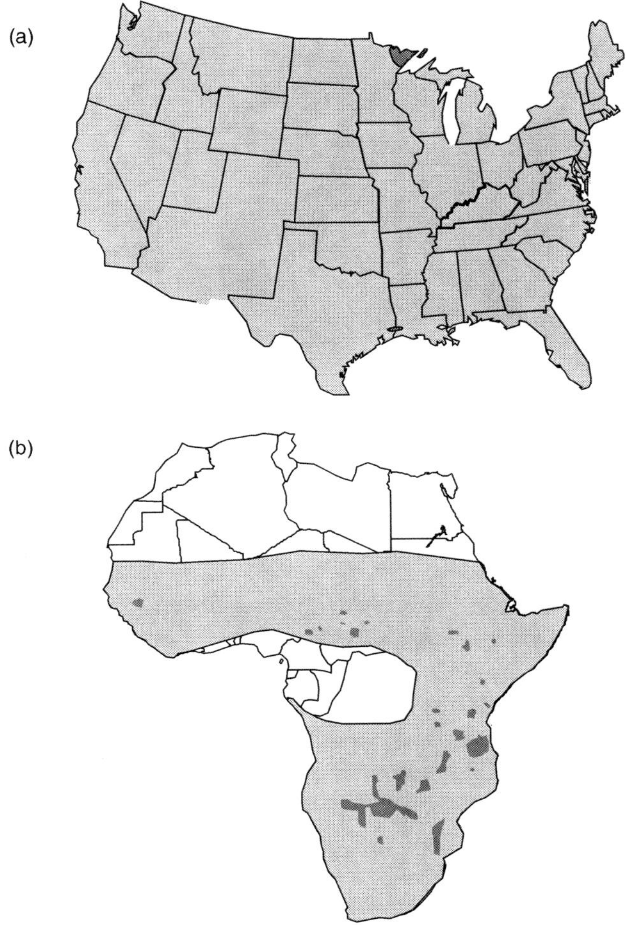

Figure 1.1. Range collapses of wildlife species in conflict with people. Maps compare historic distributions with more recent distributions. (a) Grey wolves in North America: 1700 (light grey) vs. 1970 (dark grey), based on data from Thiel and Ream (1992); (b) African wild dogs: ~1800 (light grey) vs. 1997 (dark grey), based on data from Fanshawe *et al.* (1997); (c) hen harriers in Britain (1825–1975; (for colours see key), based on data from Watson (1977).

Africa (Nowell and Jackson 1996). Brown bears (*Ursus arctos*), lynx (*Lynx lynx*) and wolves had disappeared from most of western Europe by the end of the nineteenth century (Woodroffe 2001a). Jaguars (*Panthera onca*) have shown a similar range contraction in Central and South America (Sanderson *et al.* 2002), as have dingoes (*Canis familiaris dingo*) in Australia (Glen and Short 2000).

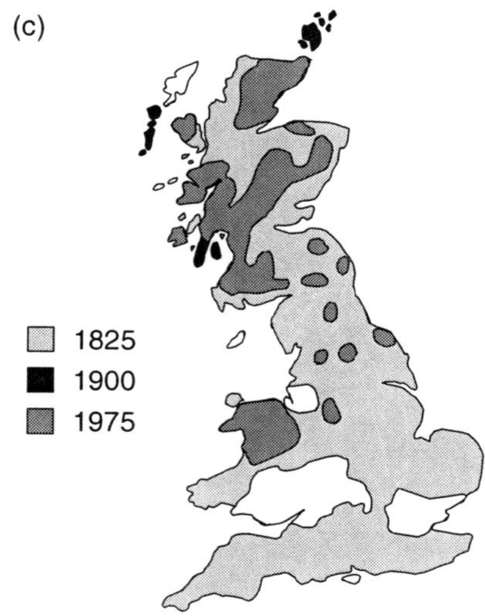

Figure 1.1. (*cont.*)

Avian predators have shown similar patterns of range collapse. Birds of prey come into conflict with hunters and farmers because of their predation on livestock and on game species, and lethal control has had a major impact on the status and distribution of numerous species throughout the world (Newton 1979, 1998; Manosa 2002; Vinuela and Arroyo 2002). This has been particularly well documented in Scotland where conflicts have been intense, and where there is a strong tradition of ornithology to assess the impacts of sustained killing of raptors (Galbraith *et al.* 2003). A good example is the hen harrier (*Circus cyaneus*) which comes into conflict with grouse (*Lagopus lagopus*) hunters and landowners because of the impact of its predation on grouse harvests (Thirgood *et al.* 2000a, b, c; Thirgood and Redpath, Chapter 12). Hen harriers were historically widespread in Scotland, but sustained killing during the nineteenth century eradicated them from the mainland leaving a few remnant populations in the Western Isles (Watson 1977) (Fig. 1.1c). Harriers recolonized the mainland during the 1940s, but have not reoccupied their original geographic range.

Similar patterns of range collapse have been recorded for Scottish populations of golden eagles (*Aquila chrysaetos*) and white-tailed eagles (*Haliaeetus albicilla*), although in these cases conflicts were perceived to be with sheep-farmers as much as grouse-hunters (Watson 1977). Range contractions have also occurred in other European countries where game-bird shooting is a significant form of land use. Of particular concern is the situation in Spain, where deliberate killing of raptors involves extremely endangered species such as Bonelli's eagles (*Hieraetus fasciatus*) and Spanish Imperial eagles (*Aquila adalberti*: Vinuela and Villafuerte 2003).

CURRENT POPULATION IMPACTS OF LETHAL CONTROL

Large-scale population suppression

Where lethal control has not caused local extinction, it may still cause local declines or suppress populations. Multiple studies have shown that carnivore populations are limited by human interventions. For example, wolves living in human-dominated landscapes are almost invariably limited by people. Where wolf-hunting is legal, this accounts for much of the offtake, but human activities can still limit local populations even where wolves are legally protected (Boitani 1992; Fuller 1989; Bangs *et al.*, Chapter 21). Likewise, lion, red fox (*Vulpes vulpes*), dingo and European badger (*Meles meles*) populations may be suppressed below carrying capacity by control efforts (Allen and Gonzalez 1998; Heydon and Reynolds 2000; Le Fevre *et al.* 2003, Frank *et al.*, Chapter 18).

Estimates of the impacts of lethal control on elephant populations are few in number. In southern India, 17% of female elephant mortality was due to either shooting or electrocution carried out by people in defence of their crops (Sukumar 1989). While demographic modelling suggested that this level of control was just sustainable, small increases in mortality were projected to cause declines. Among males (which have tusks), retributive killing (8–17% of deaths) was apparently dwarfed by poaching (48–57%) as a cause of mortality; this combined offtake caused marked decline of the male population (Sukumar 1989). Unconfirmed reports suggest even more severe impacts of lethal control on Asian elephants in Sri Lanka (Sri Lanka Wildlife Conservation Society 2000).

As in Asia, estimating the impact of retributive killing on African elephant populations is difficult, in part because studying elephant demography requires expensive long-term studies, and in part because data are complicated by the issue of ivory poaching. However, the available data indicate that problem animal control is an important cause of mortality. Information presented to the Convention on International Trade in Endangered Species (CITES) in 1997 suggests that problem animal control may be as serious a mortality cause as ivory poaching. Of 1224 elephant deaths recorded in Botswana in 1989–96, 230 were due to problem animal control and 259 were poached (CITES 1997). Likewise, in Kenya 467 elephants were recorded killed in problem animal control during 1993–8, compared with 355 poached (R. Hoare pers. comm.). Figures for Namibia suggest a somewhat smaller impact of retributive killing, with 148 poached during 1990–6 and 29–50 killed as problem animals (CITES 1997). While these data do not provide quantitative estimates of the impact of retributive killing on local elephant populations, they do suggest that if the current level of ivory poaching is a threat to Africa elephants, problem animal control is an equally serious threat (see also Western and Waithaka, Chapter 22). Killing of crop-raiding elephants is a widespread phenomenon in Africa (e.g. Dudley *et al.* 1992; Tchamba 1996), but its impact on regional elephant populations appears largely unknown.

Almost no quantitative data are available on the impact of retributive killing on crop-raiding primates, in part because few population studies have been carried out where primates are in direct conflict with people.

Local effects: source–sink dynamics

Because populations are connected with one another through animal movement, even localized lethal control can influence populations over wide areas. Frank *et al.* (Chapter 18) show how lethal control of livestock-killing lions on a single 180-km^2 ranch generated a sink affecting the lion population over at least 2000 km^2. Mace and Waller (1998) described a similar impact of localized mortality on a regional grizzly bear (*Ursus arctos*) population. On a still larger scale, Etheridge *et al.* (1997; see also Thirgood and Redpath, Chapter 12) showed that illegal killing of hen harriers on moorland managed for commercial hunting of red grouse transformed an entire habitat type into an extensive population sink, sustained only by immigration from moorland not managed in this way. Of course, the more widely individuals of a species range (or disperse), the greater the spatial scale across which source–sink effects may operate.

Source–sink dynamics can have especially damaging effects when they involve nominally protected populations. Where people use lands adjacent to reserves, their activities can threaten wildlife nominally protected by the park. One of the best examples of such edge effects comes from Algonquin Park, Canada, where the wolf population was driven into decline by persistent killing of animals that ranged beyond park boundaries (Forbes and Theberge 1996). Banning the killing of wolves close to the park border dramatically reduced overall mortality in the park (Forbes and Theberge 1996). Likewise, most of the African elephant mortality recorded in a population under study in Amboseli National Park, Kenya, was due to spearing by neighbouring cattle-farmers (Moss 2001). Edge-related mortality of this kind is extremely common among large carnivores, occurring in all species for which data are available (Woodroffe and Ginsberg 1998, 2000). Mortality is particularly severe where reserves are surrounded by high densities of people (Harcourt *et al.* 2001).

Such localized edge mortality need not necessarily contribute to overall population decline. For example, the Amboseli elephant population was increasing at 2.2% per year despite regular offtake due to conflict with neighbouring cattle-farmers. However, if edge mortality is high enough, it can cause population decline or even local extinction. Woodroffe and Ginsberg (1998) showed that these effects are a globally important cause of extinction among nominally protected carnivore populations. Wide-ranging species such as grizzly bears and African wild dogs are particularly sensitive to extinction through these effects, and persist only in the very largest national parks and reserves. This is almost certainly because they are especially likely to wander over the borders of protected areas and into contact with people.

Indirect effects

Even where lethal control has relatively small direct impacts on population density, there is a possibility that its effects might be magnified by social factors. For example, killing of seven male chimpanzees (from a community of about 80) in a crop-raiding incident profoundly affected the social structure of a group under study in the Taï Forest, Côte D'Ivoire (Boesch and Boesch-Achermann 2000). This social disruption was believed to have reduced the group's ability to counter leopard attacks, which subsequently led to high mortality. Likewise, Courchamp and Macdonald (2001) argued that quite small reductions in pack size of African wild dogs (as might occur through lethal control) could dramatically affect the group's ability to hunt and raise young, thus having disproportionately large impacts on population density.

Behavioural responses to lethal control

Lethal control has very clear population impacts, but it may also have indirect effects through its impact on behaviour. Local extinction may occur because all the animals in an area are killed, but it can also occur because all the animals move elsewhere. Collapse of African elephant populations in areas of high human density (Hoare and Du Toit 1999) probably occurs through such behaviour. Elephants will move away from areas where conspecifics are killed (Whyte 1993), and this may disrupt natural seasonal migrations (Tchamba *et al.* 1995), with unknown consequences for population viability. One result of such movements is that elephants may become compressed in relatively small protected areas (e.g. Wittemyer 2001), a situation that is probably not sustainable in the long term. Fortunately, savanna elephants also become aware quite rapidly when formerly dangerous areas become safe, and will recolonize areas surprisingly readily if habitat connectivity has been preserved (e.g. Thouless 1995).

Animals' ability to recognize areas that are dangerous may, of course, greatly reduce human impact on their populations. Frank *et al.* (Chapter 18) suggest that lions may avoid areas where they face high mortality risks. This behavioural response may help to explain why lions have historically been highly extinction-prone outside protected areas, yet persist well, and at high densities, in quite small reserves (Woodroffe 2001a).

IMPACTS OF HUMAN–WILDLIFE CONFLICT ON ECOSYSTEM FUNCTION AND HABITAT DESTRUCTION

Trophic cascades

The outcome of conflicts between people and wildlife may extend beyond populations, to affect entire ecosystems. Many 'conflict' species (e.g. elephants, large carnivores) are also keystone species whose removal affects the structure of entire ecosystems. Extirpation of grey wolves and grizzly bears from parts of the northern Rocky Mountains has been shown to influence, through its impacts on ungulate density and behaviour, habitat suitability for neotropical migrant birds (Berger *et al.* 2001), and restoration of grey wolves has affected many facets of this montane ecosystem (Smith *et al.* 2003).

Perhaps the best example of a trophic cascade triggered by human–wildlife conflict involves prairie dogs. Prairie dog colonies constitute a unique grassland habitat which support a remarkably biodiverse community (Kotliar *et al.* in press). Systematic attempts to eradicate prairie dogs from very large areas will have adversely affected all members of this community, but its most high-profile (and expensive) impact was the extinction in the wild of the black-footed ferret (*Mustela nigripes*), a highly specialized species that is an obligate predator of prairie dogs (Miller *et al.* 1996). Black-footed ferrets in the last wild population to go extinct (prior to intensive recovery efforts and multiple reintroductions) were very few in number and ultimately killed by infectious disease; hence initial recovery issues focussed primarily on these issues (Seal *et al.* 1989). However, it was deliberate destruction of the ferrets' habitat and prey base that drove their decline, and which continues to dog recovery efforts (Miller *et al.* 1996).

Sea otters (*Enhydra lutris*) are another strongly interacting species affected by human–wildlife conflict. For many years, conflict with lucrative shellfish fisheries on parts of the California coast prompted local laws to prevent sea otter recovery in designated 'no otter zones' (US Fish and Wildlife Service 2003c). Sea otters' role in structuring marine communities is very well established (Estes *et al.* 1996); hence this management decision could have had a very marked effect on many aspects of California's coastal ecosystems, beyond its influence on fisheries.

Elephants also have the ability to profoundly affect the structure of the ecosystems they inhabit (reviewed in Sinclair 1995); hence control of problem elephants could influence entire ecological communities. The extent to which this actually occurs probably depends upon habitat type, as well as the (largely unknown) impact of lethal control on elephant population densities. Most human–elephant conflict involves agricultural areas; here natural habitat is presumably in the course of being converted to cultivation and hence elephant impacts on ecosystem structure are probably dwarfed by human impacts. However, if conflict areas are sinks, influencing elephant populations ranging over a larger area, intense problem animal control could affect a wider region. Equally, ecosystem impacts of elephant depletion might be observed where intense conflict occurs between elephants and pastoralists in dryland areas unsuitable for agriculture (such conflicts are mentioned in Moss 2001).

Habitat destruction

At worst, human–wildlife conflict may directly drive habitat destruction. A case of this was described by Ottichilo (2001), who noted that traditional calving grounds of wildebeest (*Connochaetes taurinus*) in southern Kenya had been preferentially converted to wheat farms by local Masai pastoralists. This land conversion was a major reason for an 81% decline in the local wildebeest population, and presumably also affected a diverse array of other wildlife species. Masai chose to convert these particular areas in a deliberate attempt to reduce contact between their cattle and calving wildebeest, because wildebeest carry malignant catarrhal fever, a viral infection which does not harm wildebeest but can cause high mortality of cattle. Transmission is particularly likely during wildebeest calving (Talbot 1963). To avoid the risks of such infection, Masai preferentially allowed wildebeest calving grounds to be used for wheat-farming (Ottichilo 2001). Such destruction of habitat as a way of reducing human–wildlife conflict may be more common than is currently realized.

CONCLUSIONS

Conflict between people and wildlife has had profound impacts on wild ecosystems, and continues to do so. These impacts reach beyond the suppression or local extinction of populations in conflict with people, to the structuring of ecosystems, the destruction of habitat, and even the extinction of associated species. Legal protection has comparatively little impact on this problem, because people will kill 'problem' wildlife illegally if they are sufficiently motivated. Even large protected areas may be inadequate, because animals that range beyond reserve boundaries are still vulnerable to lethal control. Conflict with people is the most serious threat faced by many vulnerable species, and the ecosystem consequences of losing these species is only beginning to be recognized (e.g. Ray *et al.* in press).

These wide-ranging conservation impacts of human–wildlife conflict point to an urgent need for effective solutions to reduce killing of problem wildlife. Protected areas provide only a limited solution, because lethal control can still affect species occupying all but the very largest reserves (Woodroffe and Ginsberg 1998). Moreover, while protected area networks are being expanded in some areas, many new reserves are 'multiple-use areas' occupied by people as well as wildlife (World Parks Congress 2003). In the majority of cases, therefore, effective conservation will demand some means of peaceful coexistence of people and wildlife outside protected areas. If such means cannot be found, further extinctions, with their associated consequences for regional biodiversity, cannot be avoided.

2

The impact of human–wildlife conflict on human lives and livelihoods

SIMON THIRGOOD, ROSIE WOODROFFE AND ALAN RABINOWITZ

INTRODUCTION

Human populations interact with wildlife in numerous ways. Our species has directly exploited wild animals for food and furs for millennia and more recently for sporting or cultural reasons. Humans have greatly modified habitats and landscapes through agriculture and other extractive industries with far-reaching and typically negative impacts on wildlife populations. We have also translocated species around the globe, either deliberately or accidentally, with major consequences for native fauna. From the human perspective, our interactions with wildlife are often positive – we gain material benefit from harvesting species for food or other animal products. In other situations, however, human interactions with wildlife are negative. Wild animals may eat our livestock and damage our crops, they may compete with us as hunters for wild prey populations, and they may even injure or kill us. In the previous chapter, we outlined the negative consequences for wildlife that come into conflict with people, focussing on the global and local extirpations and range contractions suffered by a variety of predators and crop-raiders. Here in this second scene-setting chapter, we summarize the negative impacts of threatened wildlife on people. Focussing again on predators and crop-raiders we assess the costs to stakeholders of living with wildlife. We focus on the *direct costs* to stakeholders of living with wildlife such as loss of human life, livestock, wildlife resources and crops and try to calculate these in financial terms. Wildlife may also impose *indirect costs* in terms of time and money spent in preventing wildlife damage and *opportunity costs* in terms of the income foregone from those activities that are precluded by the presence of wildlife (Norton-Griffiths and Southey 1995; Emerton 2001). This latter area has received much less attention to date and we draw attention to the need for inclusive studies.

People and Wildlife: Conflict or Coexistence? eds. Rosie Woodroffe, Simon Thirgood and Alan Rabinowitz. Published by Cambridge University Press.

HUMAN FATALITIES AND INJURIES

One of the most serious causes of human–wildlife conflict is the fear of being killed by a large carnivore or mega-herbivore. The fear of carnivores is deeply rooted in the human psyche and has been interpreted as an instinctive anti-predator response (Kruuk 2002; Quammen 2003). Attacks on humans by cats, bears and wolves have been reported in both historical and contemporary circumstances throughout the world. By nature most reports are anecdotal and rigorous verification is often difficult. However, a substantial body of evidence suggests that hundreds of people are killed annually by wildlife on a global basis (Kruuk 2002; Quigley and Herrero, Chapter 3).

Cats

Some of the most famous accounts of carnivores killing humans involve big cats, particularly lions (*Panthera leo*) and tigers (*P. tigris*). Notorious early-twentieth-century man-eaters include two lions in Tsavo, Kenya, who killed over 100 workers building the Mombasa-to-Nairobi railway, and eight tigers in India who were alleged to have killed 1000 people (Kruuk 2002). Human predation on such scales no longer occurs because the cats are shot before their depredations become very extensive. However, in some parts of Asia and Africa, human mortality to big cats is still a common occurrence. Up to 100 people are killed by tigers each year in the Sundarbans of India and Bangladesh (Sanyal 1987). Lethal attacks on humans elsewhere in Asia are rarer but 22 people were killed by tigers in the period 1985–2001 around the Kanha Reserve in Madhya Pradesh (Karanth and Gopal, Chapter 23) and 28 people were killed by lions during 1978–91 around the Gir Forest in Gujarat (Saberwal *et al.* 1994). Lion attacks on humans are a recurrent problem in Africa although they are often restricted to the borders of protected areas. In Tanzania, however, lion attacks are widespread particularly throughout the southern and eastern regions of the country (Whitman and Packer in press). Between 1989 and 2004, government reports recorded 175 fatalities from four regions representing about 34% (299 745 km^2) of mainland Tanzania. It is believed that these records are an underestimate of human mortality as many attacks are not reported. Lethal attacks by cats occur much less frequently in the developed world with, for example, only 10 people killed by cougars in North America in the twentieth century (Beier 1991).

Bears

Attacks on humans by bears have been frequently reported (Herrero 1985; Quigley and Herrero, Chapter 3). About 50 people were killed by brown bears (*Ursus arctos*) and 23 people killed by black bears (*U. americanus*) in North America during 1900–80. The majority of deaths caused by brown bears were in parks and were predominantly by females defending cubs. In contrast, black bear attacks were primarily outside parks with predation or scavenging the motive. Within Europe, brown bears are thought to be more aggressive in the eastern countries, with, for example, 12 people killed and 94 injured during eight years in Romania, whilst in Scandinavia there has only been one unprovoked bear-inflicted human fatality in the twentieth century (Swenson *et al.* 1996). Polar bears (*U. maritimus*) also attack people, with six fatalities in Canada during 1965–85 (Fleck and Herrero 1989) and 50 attacks on Svalbard during 1973–86 (Gjertz and Person 1987). More frequent attacks have been reported from India where sloth bears (*Melursus ursinus*) killed 48 people and injured 687 others in Madhya Pradesh during 1989–94 (Rajpurohit and Krausman 2000). Elsewhere in India attacks by bears are less frequent but still totalled more than 200 incidents in five states during the same five-year period.

Wolves

Attacks by grey wolves (*Canis lupus*) fall into three types: attacks by rabid wolves; defensive attacks; and predatory attacks (Linnell *et al.* 2002). The majority of attacks involve rabid wolves and are reported throughout the world. Defensive attacks typically occur in the context of wolves biting shepherds guarding livestock but human deaths have not been reported. Predatory attacks are rare but have occurred in historical and contemporary contexts. Records indicate that wolves killed hundreds of people in Europe during the eighteenth and nineteenth centuries (Kruuk 2002). In a single parish in Estonia, 48 children were killed in 1808–53. Predatory attacks from the twentieth century are less common, but include five children in Poland (1937), four children in Spain (1957–74) and 36 children in Russia (1944–53). Kruuk (2002) also describes contemporary accounts of wolves killing children and adults in Belarus in 1995–6. Attacks by wolves on children were commonplace in nineteenth-century India and there are well-documented reports from Uttar Pradesh and Madhya Pradesh of several hundred human fatalities in the past 20 years (Linnell *et al.* 2002). In contrast, there are no documented cases of people being killed in predatory attacks by wolves in North America during the twentieth century (Linnell *et al.* 2002).

Mega-herbivores

Similar numbers of people are killed globally by mega-herbivores as by predators. Asian elephants (*Elephas maximus*) kill 100 to 200 people annually in India (Veeramani 1996). In Bihar, south India, there were 242 human deaths and injuries caused by elephants from 1989 to 1994 (Rajpurohit 1999). Records from Rajaji and Corbett National Park in north India indicate that 13 people were killed by elephants during 1993–9 (Williams *et al.* 2001). African elephants (*Loxodonta africana*) also kill people although quantitative data are scarce. A summary of trends in human–elephant conflict in Kenya is given by Western and Waithaka (Chapter 22). Elephants killed 221 people in Kenya during 1990–7 compared to 250 people killed by predators over the same period. Hippos (*Hippopotamus amphibius*) are also known to kill people in Africa but quantitative data are not available (see Duke University (2005) for case studies).

In summary, wild animals kill hundreds if not thousands of people globally each year. This figure must be placed in the context of a global human population of over 6 billion. Whilst losses of human lives to wildlife are tiny in a global context, and trivial in comparison to human deaths caused by disease, famine or war, they may be critical in determining the tolerance of local communities to wildlife. In this context, the fear of wildlife itself, as opposed to actual attacks, may be sufficient to result in pre-emptive killing.

TRANSMISSION OF DISEASE

Whilst beyond the main focus of this book, the transmission of disease from wildlife to humans or livestock can be a major cause of conflict. Wild carnivores can act as reservoirs of zoonotic diseases of humans with rabies as the classic example. Rabies is a viral disease which attacks the central nervous system. The symptoms may be paralysis followed by death, or alternatively, animals may develop the 'rage' which results in biting other animals, including humans, and transmitting the disease. Rabies was lethal to humans until a vaccine was developed in 1885 and in some countries it remains a major killer. Rabies is responsible for ~50 000 deaths globally each year with 10 000 deaths occurring annually in India alone (WHO 1998). These figures are two orders of magnitude more than the numbers of people killed by wildlife. Rabies reservoirs occur in a range of wild carnivores – red fox (*Vulpes vulpes*) in Europe, striped skunks (*Mephitis mephitis*) and racoons (*Procyon lotor*) in North America, and yellow mongoose (*Cynictis penicullata*), bat-eared fox (*Otocyon megalotis*) and black-backed jackal (*Canis mesomelas*) in southern Africa (Aubert 1993; Jenkins *et al.* 1998; Bingham *et al.* 1999; Courtin *et al.* 2000). It should be stressed, however, that even where wild carnivores are the reservoir that sustains rabies infection in an area, domestic dogs (*Canis familiaris*) are responsible for 98% of transmission to humans (Bishop *et al.* 2002). Rabies can also have a very significant impact on livestock production. In Latin America the annual cost to the livestock industry has been estimated at $400 million (Macdonald 1980). Livestock losses to rabies may also be very significant on a local scale. Rabies is estimated to cost $7 per annum per household to agro-pastoralists in the Bale Mountains in Ethiopia where average annual income is $120 and similar losses have also been documented around the Serengeti in Tanzania (K. Laurenson pers. comm.). Other human pathogens for which carnivores are the reservoir include the tapeworm *Echinococcus multilocularis* which is transmitted by canids (Eckert *et al.* 2000) and the protozoon *Toxoplasma gondii* which is transmitted by felids (Jones *et al.* 2001). Wildlife are also implicated in the transmission of other generalist pathogens to livestock including the microparasites *Brucella abortus* and *Mycobacterium bovis* which cause the economically important diseases brucellosis (cattle and bison (*Bison bison*) in the USA) and bovine tuberculosis (cattle and badgers (*Meles meles*) in the UK) (Hudson *et al.* 2002).

PREDATION ON LIVESTOCK

Livestock depredation, particularly by large carnivores, is probably the most common cause of human–wildlife conflict on a global basis. A survey of carnivore researchers revealed that livestock depredation was the most frequently cited (40%) reason for conflicts involving carnivores (Sillero-Zubiri and Laurenson 2001). Several factors may contribute to this situation. First, domestic livestock have retained little anti-predator behaviour and are therefore easy prey for wild carnivores. Second, livestock may compete with wild herbivores for grazing and reduce the abundance of wild prey for carnivores. Finally, particularly in the developed world, there have been changes in livestock husbandry, especially in areas where carnivores were greatly reduced in density. In much of Europe and North America livestock are no longer guarded by people or dogs and are thus easy prey for recolonizing carnivores (Breitenmoser *et al.*, Chapter 4; Linnell *et al.*, Chapter 10; Swenson and Andren, Chapter 20).

Carnivore predation on livestock

Despite the frequency of conflict between carnivores and livestock-herders, there have been surprisingly few studies that have quantified the extent of livestock predation by carnivores and its economic impact in the context of other losses. This is particularly notable in the developing world where a review of the literature reveals only eight studies in Africa and Asia that quantify livestock loss to large carnivores. The six African examples encompass a range of management systems (nomadic pastoralist, communal lands, commercial ranches), the full suite of large carnivores (lion, leopard (*Panthera pardus*), striped hyaena (*Hyaena hyaena*), African wild dog (*Lycaon pictus*)) and report from < 1% to 10% annual losses of cattle, sheep and goats to carnivores (Kruuk 1980; Frank 1998; Frank *et al.*, Chapter 18; Mizutani 1999; Rasmussen 1999; Butler 2000; Woodroffe *et al.* 2005). Where data are available these losses appear to be lower than losses to disease which are often 10%–20% per annum. However, livestock predation is often unevenly distributed and, where it does occur, can have a significant economic impact on individual farmers. For example, Woodroffe *et al.* (2005) found that, in an average attack, African wild dogs killed 3.2 sheep or goats, which equates to 22% of average per capita annual income in Kenya. The two published studies in Asia from pastoralist communities in the Himalayas report comparable losses to snow leopards (*Panthera uncia*) and grey wolves and the associated financial loss appears to be very significant with up to 50% of annual income lost to depredation (Oli *et al.* 1994; Mishra 1997). The studies described above have typically quantified the direct economic costs of predation and have not considered the indirect costs of living with predators such as time spent guarding animals or building bomas. In one exception, Frank (1998) quantified the additional costs to commercial ranchers in Laikipia, Kenya, of living with lions and found that ranchers needed just 3% fewer herders if there were no predators because active guarding was also necessary to deter livestock theft.

Carnivore depredation on livestock in Europe and the Americas has, in contrast, received considerable research attention. The problem is widespread but appears to be particularly acute in areas where carnivores have returned after being temporarily absent (Blanco *et al* 1992; Quigley and Cranshaw 1992; Cozza *et al.* 1996; Kaczensky 1996; Breitenmoser 1998; Treves *et al.* 2002). The conflict between Eurasian lynx (*Lynx lynx*) and sheep-farming is particularly well studied and serves as a good example of the problem. Lynx were formerly widespread in forested areas but by 1950 had been eradicated from much of continental Europe with the exception of the Carpathians, Balkans and part of the Alps (Breitenmoser 1998). Lynx were also persecuted in Scandinavia and were reduced to low numbers by the early 1900s (Swenson and Andren, Chapter 20). Management policies more sympathetic to carnivores combined with reintroductions have led to a recovery of lynx throughout the Alps and Scandinavia (Breitenmoser *et al.* 2000). The great majority of these lynx live outside protected areas in multiple-use landscapes where they come into conflict with sheep-farmers. The conflict is exacerbated by modern sheep-husbandry techniques that do not involve guarding. Comprehensive studies on lynx predation have been conducted in the Swiss Alps, French Jura and in Norway. In the Alps and Jura, sheep losses on summer pastures are low on a regional scale at about 0.25% of the available flock totalling around 100 sheep killed per anum in each region (Breitenmoser and Haller 1993; Stahl *et al.* 2001a). Predation is concentrated in 'hotspots' with about 50% of attacks in fewer than 5% of farms (Stahl *et al.* 2002). This has led to a management policy of targeted removal of 'problem individuals'; however, this is only a short-term remedy (Stahl *et al.* 2001a). In contrast, the conflict between lynx and sheep in Norway is an order of magnitude more severe in terms of total numbers of sheep killed. Some 2.5 million sheep graze unattended in Norwegian forests and about 8500 (0.34%) are killed by lynx each year (Linnell *et al.* 2001a). Lynx predation on sheep occurs throughout sheep range and male lynx have higher predation rates than females leading to the identification of a 'problem sex' rather than problem individuals (Odden *et al.* 2002). Free-ranging sheep in Norway and Sweden are also killed by wolverines (*Gulo gulo*), brown bears and grey wolves and these carnivores also come into conflict with the Sami herders of semi-domestic reindeer (*Rangifer tarandus*) (Sagør *et al* 1997; Landa *et al.* 1999; Pedersen *et al.* 1999). Swenson and Andren (Chapter 20) review these conflicts in detail and describe the mitigation schemes adopted by these two countries.

Comparable studies have also been conducted on the conflict between sheep-farmers and coyotes (*Canis latrans*) in the western USA. Commercial sheep-farming has declined in the USA primarily because of low economic returns and coyote predation is cited as a major causative factor (Knowlton *et al.* 1999). Generalizations about the magnitude and economic impact of predation can be misleading because of the varied nature of sheep-farming and the environmental circumstances surrounding individual operations. Long-term studies on an experimental farm in California indicate that 2% of breeding ewes and 3% of lambs were killed by coyotes each year (Blejwas *et al.* 2002). Adult breeding coyotes were responsible for most predation (Sacks *et al.* 1999a) and there was no relationship between non-selective control and subsequent predation rates (Conner *et al.* 1998). Selective removal of breeding pairs of coyotes during the lambing season reduced predation rates for up to three months (Blejwas *et al.* 2002). The resolution of coyote predation on livestock in the USA remains controversial for sheep-farmers, resource managers and the general public with different sections of society attaching different values to coyotes and to farm economics (Knowlton *et al.* 1999). Similar complex issues surround predation by grey wolves on livestock – both sheep and cattle – in the western USA and are reviewed by Bangs *et al.* (Chapter 21).

Raptor predation on livestock

Lamb predation by large eagles is perceived to be a problem throughout the world. Species such as the wedge-tailed eagle (*Aquila audax*) in Australia, black eagle (*Aquila verreauxii*) in South Africa, golden eagle (*Aquila chrysaetos*) in the USA and white-tailed eagle (*Haliaeetus albicilla*) in the UK have been killed in great numbers to protect sheep. Has such persecution been based on any evidence of eagles killing livestock? Numerous studies have assessed eagle diet and estimated their impact on sheep production (Leopold and Wolfe 1970; Brooker and Ridpath 1980; Bergo 1987; Matchett and O'Gara 1987; Phillips and Blom 1988; Davies 1999, 2000; Marquiss *et al.* 2002). Estimated losses to eagles are typically small (0–3% per year) compared to lamb mortality (20–30% per year). Necropsies conducted on lamb carcasses yield reliable data on causes of death and can also indicate whether lambs killed by predators were viable. Davies (1999) compiled data from 44 studies worldwide involving 30 000 necropsies. Most sheep farms experience low predation losses – on average predators removed 4.9% of lambs compared to other mortality of 13.7%. Avian predation was less frequent than mammalian predation and losses to corvids exceeded those to eagles.

The complexities of eagle predation on livestock are well illustrated by a study of black eagles in South Africa (Davies 1994, 1999, 2000). Black eagles occupy a controversial position – do they harm the farmer by preying on lambs or do they benefit him by limiting the abundance of rock hyrax (*Procavia capensis*), which compete with sheep for forage? Studies of black eagle diet and predation rates showed that lambs comprise less that 1% of prey remains and prey items delivered to nests. Resident pairs of black eagles killed about 0.3 lambs per farm per year at an economic cost of $7 per eagle per year. In contrast, black eagles removed about 11% of hyraxes per year which was equivalent to about a third of the annual recruitment. A simulation model suggested that the amount of grazing consumed by the additional hyraxes in the absence of black eagle predation on the average farm was valued at $700 per eagle. This benefit of having eagles on the farm outweighs the cost of occasional lamb predation by eagles by two orders of magnitude.

In summary, data on livestock losses to predators are generally poor, particularly in the developing world. Where good data do exist, livestock losses to predators are typically less than 10% per year. These losses to predators are usually less than losses to diseases, but the financial impact of predation can nonetheless be high in relation to income. A further important point is that livestock predation is often patchily distributed and losses to individual farmers may be severe. The perception of the risk of attack may be disassociated from the true risk of attack because the consequences of an attack are severe (Joffe 2003). Taken together, the direct and indirect costs of livestock predation have been a key factor in driving the global predator decline described by Woodroffe *et al.* (Chapter 1).

PREDATION ON GAME

The question of whether vertebrate predators can limit prey populations has long been controversial. The scientific consensus for much of the twentieth century, following the influential studies of Errington (1946), was that predators had no effect on vertebrate prey populations as they only took those individuals that would have died from other causes – the so-called 'doomed surplus'. This view has been questioned in numerous studies from 1980 onwards and a new paradigm has emerged which suggests that predation may have profound limiting effects on vertebrate prey populations (Marcstrom *et al.* 1988; Gasaway *et al.* 1992; Krebs *et al.* 1995; Tapper *et al.* 1996; Korpimaki and Norrdahl 1998; Thirgood *et al.* 2000a, b, c). Conflicts between predators and human hunters emerge when the prey concerned is of economic or cultural value as game species. Such conflicts have historically been resolved by lethal control of predators and this has had a major impact on the abundance and distribution of many predatory species, particularly in Europe and North America (Newton 1979; Reynolds and Tapper 1996; Woodroffe *et al.*, Chapter 1).

Carnivore predation on game

The controversy over predator limitation of economically valuable prey populations is typified by the long-standing and acrimonious debate in North America about the impact of grey wolves and brown bears on moose (*Alces alces*) (Boutin 1990). Perhaps most revealing is the eight-year experimental study of Gasoway *et al.* (1992) in Alaska. This suggested that predation by large carnivores was responsible for limiting moose populations to 148 animals per 1000 km^2. When wolves and bears were culled, the same moose populations increased to 663 animals per 1000 km^2. Moose harvest by hunters showed similar trends, with hunting yields with predator control of 20–130 moose per 1000 km^2, compared to 0–18 moose per 1000 km^2 with no predator control. However, as with most complex predator–prey systems, the impact of predators on moose populations is strongly influenced by the presence of alternative prey, in this case caribou (*Rangifer tarandus*). Comparably detailed studies on the Isle Royale in Michigan also suggest that wolf predation regulates moose populations (Peterson 1999). These and other studies have led to controversial culls of wolves in Alaska to protect moose and caribou harvests.

The debate over the impact of predation by large carnivores on ungulate harvests is matched by that concerning predation by smaller carnivores on small-game harvests (Reynolds and Tapper 1996; Newton 1998). In this latter case, however, there is good evidence that generalist mammalian predators can reduce both breeding success and breeding density of game birds and lagomorphs. The classic study is that of Marcstrom *et al.* (1988) who experimentally removed red foxes and pine martens (*Martes martes*) from islands in the Gulf of Bothnia in Sweden. The study was very rigorous with replication and switching of experimental treatments after four years. During years with predator control, spring densities of mountain hares (*Lepus timidus*) were 2–3 times higher and densities of capercaillie (*Tetrao urogallus*) and willow grouse (*Lagopus lagopus*) 1.7 times higher than in years without predator control. This study was repeated in the agricultural landscapes of southern England with the added complication that the prey species concerned, the grey partridge (*Perdix perdix*), is declining rapidly across its range in Europe. Tapper *et al.* (1996) experimentally controlled foxes and corvids on large plots of farmland for three years and then reversed treatments for a further three years. Predator control increased spring densities of partridges by 2.5 times and autumn densities (and thus harvestable surpluses) by 3.5 times. Taken together these and other less rigorous studies are used to justify the traditional practice of mammalian predator control in game-bird management in the UK (Reynolds and Tapper 1996; Sillero-Zubiri *et al.* 2004). Similar debate surrounds the issue of mammalian predator impact on waterfowl populations in the Prairie Pothole region of North America (Greenwood *et al.* 1995).

Raptor predation on game

Generalist mammalian predators of game birds and waterfowl are typically fairly common and robust to lethal control (Treves and Naughton-Treves, Chapter 6). Putting welfare and animal-rights issues aside, there is rarely a major concern about their conservation status. The same is not true for generalist raptors that come into conflict with hunters because of their perceived impact on populations of game birds and other small game. This is an emerging conflict across Europe and has resulted in widespread killing of raptors with significant consequences for the population status of several species (Thirgood *et al.* 2000b; Vinuela and Arroyo 2002). Raptors are killed by hunters because they are perceived to reduce game-bird harvests. But is this belief justified?

Relatively few studies have assessed the impact of raptor predation on game birds. Those that have are almost exclusively in northern Europe where predator–prey communities are typically simpler than in southern Europe thus precluding their extrapolation to more diverse ecosystems (Valkama *et al.* 2005). Nonetheless, where detailed studies have been conducted there is evidence that raptor predation can limit game-bird populations and reduce harvests. The most comprehensive study was conducted on the impact of hen harrier (*Circus cyaneus*) and peregrine (*Falco peregrinus*) predation on red grouse (*Lagopus lagopus scoticus*) populations in Scotland and is described by Thirgood and Redpath (Chapter 12). This study demonstrated that raptor predation can reduce post-breeding densities of red grouse by 50% (Thirgood *et al.* 2000b). A quasi-experimental addition of raptor predation to a grouse population prevented a cyclic increase in grouse density and resulted in the cessation of a harvest valued at $150 000 per year (Thirgood *et al.* 2000c). Harrier–grouse dynamics were complex, however, and were strongly influenced by alternative prey and land management (Redpath and Thirgood 1999). The harrier–grouse case has been widely publicized across Europe and has raised the issue of raptor–game-bird conflicts elsewhere. But how typical is the Scottish study?

A number of studies in Fennoscandia suggest that goshawks (*Accipiter gentilis*) are significant predators of hazel grouse (*Bonasa bonasia*) and willow grouse, and in the presence of alternative prey may result in low density and stable grouse populations (Linden and Wikman 1983; Tornberg 2001). Hunters in these countries appear to accept this competition from natural predators and there is little persecution of raptors. Goshawks are also responsible for high mortality in both wild and released pheasants (*Phasianus colchicus*) in Sweden and mitigation techniques include the translocation of individuals from areas of conflict with game rearing (Kenward 1977, 1981). There is also some evidence that goshawk predation on red-legged partridge (*Alectoris rufa*) populations in Spain may be additive to other causes of mortality and thus limit populations (Manosa 1994). Reif *et al.* (2001) concluded that common buzzards (*Buteo buteo*) reduce the breeding success of small game in Finland, in particular during the decline phase of the field vole (*Microtus agrestis*) cycle when they shift to alternative prey and their breeding densities are still high. Similarly, predation by gyrfalcon (*Falco rusticolus*) on ptarmigan (*Lagopus mutus*) in Iceland may accelerate the decline, accentuate the amplitude and prolong the low phase of the ptarmigan cycle (Nielsen 1999). Finally, there is correlative evidence that harrier predation may influence grey partridge populations in France (Bro *et al.* 2001).

In summary, establishing the impact of avian and mammalian predators on economically important game species is not trivial, requiring long-term and/or experimental studies. A number of studies have, however, demonstrated that generalist predators may, under certain circumstances, have a major impact on prey population dynamics and thus on harvesting. The response of hunters, particularly in Europe where prey is commercially valuable, has been to adopt the precautionary principle and remove predators before damage to game populations occurs. This lethal control has had major impacts on the population status of several species of predatory mammals and birds. These management practices are increasingly unacceptable in a socio-political context and research is needed to find practical non-lethal alternatives.

CROP-RAIDING

In many parts of the world, crop-raiding by wild animals gives rise to significant conflict between local communities and wildlife conservation. Although certainly not a new phenomenon, crop-raiding has recently received considerable attention from conservation biologists, particularly in relation to African elephants (Hoare 2000; Osborn and Parker 2003; Osborn and Hill, Chapter 4; Naughton-Treves and Treves, Chapter 16). Conflicts between elephants and local communities are widespread in Africa and are a major concern for both elephant conservation and rural development. Human–elephant conflict occurs throughout elephant range, both in the forests of west and central Africa (Barnes 1996; Lahm 1996) and the savannas of east and southern Africa (Hoare and du Toit 1999; Sitati *et al.* 2003). The issue of human–elephant conflict has become increasingly significant as human populations expand into elephant habitat and as elephant populations locally recover from poaching. Although elephants are not the most economically important crop pests in Africa, they can cause severe localized damage (Naughton-Treves 1998; Naughton-Treves and Treves, Chapter 16). Unlike primate crop-raiders (Hill 2000), elephants also kill several hundred rural Africans each year. As a result, elephants are widely perceived as a major threat to rural lives and livelihoods and inspire great animosity and fear among people who live in elephant range (Naughton-Treves *et al.* 1999).

The economic significance of elephant crop-raiding must be judged in the context of overall losses to crop pests in Africa. Although hard data are lacking, annual crop losses to vertebrates, invertebrates, weeds and pathogens in African subsistence agriculture are in region of 50% (Yudelman *et al.* 1998). The most comprehensive study that has quantified damage from elephants and other vertebrate pests in Africa is that of Naughton-Treves *et al.* (1998; Chapter 16) around the 'hard edge' of Kibale National Park in Uganda. This study showed that there was considerable spatial variation in crop damage with 90% of the damage within 160 m of the forest edge. Within this band 5% of crops were lost annually to vertebrates, equivalent to \$6 per farmer in a region where GDP is \$200. The five most important wildlife crop-raiders were olive baboons (*Papio cynocephalus*), bush pigs (*Potamochoeros* spp.), red-tailed monkeys (*Cercopithecus ascanius*), chimpanzees (*Pan troglodytes*) and elephants – and livestock caused more damage than elephants. Elephant damage was highly localized but catastrophic when it did occur, destroying half of average annual income. The average losses described above are relatively meaningless to affected farmers who fear elephants more than any other crop-raiders.

How typical are these data from Kibale? Although there have been many studies on human–elephant conflict the database on patterns of crop damage is poor and burdened by ill-defined methods that limit comparisons between sites. Naughton-Treves *et al.* (1999) reviewed 16 studies that quantified crop damage by elephants. Average losses ranged from 0.2% (Niger) to 61% (Gabon) of planted fields and costs per affected farmer ranged from \$60 (Uganda) to \$500 (Cameroon). Naughton-Treves *et al.* (1999) also reviewed 25 studies that ranked wildlife pests in Africa. Elephants were relatively low in the ranking and, across all these studies, less than 10% of crop losses were blamed on elephants. However, elephants also impose additional indirect costs that are difficult to quantify but may outweigh the direct costs of crop-raiding. These indirect costs include restrictions on movements, loss of sleep, reduced school attendance and increased exposure to malaria due to guarding fields from elephants. The combination of direct and indirect costs combine to make elephant crop damage one of the most significant causes of human–wildlife conflict in Africa.

Human–elephant conflict is now thought to be one of the biggest threats to elephant conservation in Africa, particularly over the 80% of elephant range that occurs outside protected areas (Hoare 2000). Mitigation requires a detailed understanding of spatial and temporal variation in the occurrence of conflict. Conflict usually takes place at night and crop-raiding in particular is often strongly seasonal (Hoare 1999). Predicting the spatial occurrence of human–elephant conflict has been more problematic. The level of problem elephant activity in 15 000 km^2 of northern Zimbabwe showed huge variation and could not be explained by elephant density, proximity to protected areas, human density or area of human settlement (Hoare 1999). This suggested that the occurrence of conflict was a feature of the unpredictable behaviour of the male elephants that were responsible for the majority of crop-raiding incidents. In contrast, a finer-scale study in 1000 km^2 of unprotected elephant range adjacent to the Masai Mara National Reserve in Kenya suggested that crop-raiding was clustered into distinct conflict zones and could be predicted on the basis of the area under cultivation and proximity to settlements (Sitati *et al.* 2003). A number of conflict-mitigation approaches have been developed, ranging from traditional scaring with noise and fire to high-technology repellents and fencing (Osborn 2002; Osborn and Parker 2003; Osborn and Hill, Chapter 5).

CONCLUSIONS

A number of clear conclusions have emerged from this review. First, and most importantly, wildlife – including threatened species – can cause significant loss of human lives and livelihoods. Ignoring or belittling this fact does not help in the conservation of these species, as retributive killing of problem animals can cause serious population decline. Second, damage caused by wildlife – whether predation on livestock and game or damage to crops – is often patchily distributed with most losses accruing to a minority of stakeholders. However when losses do occur, they can be serious, even catastrophic, in economic terms. This reinforces negative attitudes among stakeholders, because whilst the true risk of attack may be quite low, the potential cost of attack can be very high. Third, this potential cost is reinforced when the animals concerned are large, dangerous and can injure or kill people. In these situations, local people, particularly in developing countries, feel powerless to protect their livestock or crops because of the risk of being killed. Addressing these issues and the resultant killing of problem animals is the focus of the rest of this book.

ACKNOWLEDGEMENTS

ST was based at the Centre for Conservation Science at Stirling University whilst writing this chapter and at the Frankfurt Zoological Society Africa Regional Office in the Serengeti whilst revising it. Critical reviews by Karen Laurenson and Claudio Sillero-Zubiri greatly improved the chapter.

3

Characterization and prevention of attacks on humans

HOWARD QUIGLEY AND STEPHEN HERRERO

INTRODUCTION

Interactions between wildlife and humans come in many forms, from indirect observation, like the remote detection of smells or signs, or sightings (mutual and singular), to direct contact. In this chapter, we address the latter form, direct contact. Specifically, we deal with perhaps the most negative and dramatic of these interactions (from the human perspective), attacks on humans by wildlife. And, from such actions may follow perhaps the most unacceptable result, serious human injury or the loss of human life. Outside of rare or unwitting contact, most direct contacts between wildlife and humans can be viewed as negative for the individual wildlife involved.

Attacks on humans, with the animal intending to repel or even kill, fall in the extreme end of the direct contact category (Thirgood *et al.*, Chapter 2). Attacks on humans by wildlife are not new. It is important to note that humans have been preyed upon from the earliest forms of our genus *Homo* and even earlier forms of hominids (Kruuk 2002). Although we have become increasingly accomplished in our ability to prey upon and defend ourselves against other animals, early hominids were highly vulnerable to a wide variety of predators and competitors (Kruuk 2002; Miller 2002). Even today, some of our primate relatives have some of the most intricate and developed forms of predator avoidance known (Miller 2002). However, the early and increasingly more elaborate development of tools separated us from other primates. Human society grew on the foundation of these tools and the ability to deal with competitors and threats, real or perceived.

Modern human society has continued to dominate the landscape of most of the world, and human impact and movements go far beyond the edges of cities and into wild habitats. This incursion of humans into wild

People and Wildlife: Conflict or Coexistence? eds. Rosie Woodroffe, Simon Thirgood and Alan Rabinowitz. Published by Cambridge University Press.

habitats, plus habituation of some wildlife species, has resulted in increased potential for human–wildlife encounters, including wildlife attacks on humans. Although there has been a substantial and recent increase in our appreciation and understanding of wildlife, especially in the past 30 years after the Western environmental movement of the 1970s, attacks on humans and their property tend to elicit strong, often negative, responses. With injuries or loss of life, there may be a subsequent, retaliatory attack on the offending wildlife species (e.g. Anonymous 2002a) that causes harm. Individuals may be destroyed and, in some cases, whole wildlife subpopulations eliminated (Woodroffe *et al.*, Chapter 1). With the human population of the earth exceeding 6 billion and growing at an estimated rate of 1.2% per year, or about 80 million (US Census Bureau 2002), human–wildlife encounters will continue to increase. Attacks on humans are perhaps the least understood of these encounters, but the most interesting and emotionally connected to people. This lack of understanding, coupled with the intense interest that attacks elicit, makes the situation ripe for reactions that will not only cause human injury or death but will also damage wildlife populations. There is a need for new approaches to wildlife attacks on humans, based on understanding (e.g. Baron 2003; Quammen 2003). Reducing attacks has the potential not only to reduce injury and loss of lives in human populations, but to conserve wildlife populations, promote good will toward wildlife, minimize economic loss and improve quality of life for humans.

However, despite the need to understand attacks and circumstances associated with attacks, little objective information exists about many attacks and there are few compilations of the information that does exist. Moreover, there is a need for a global perspective. Local successes in reducing attacks and conserving wildlife can bring gratification and can inspire others. But the sharing of such information and the standardization of information-gathering can bring a greater good out of an integrated approach.

Our objective is to describe attacks on humans, provide examples and ideas about reducing these attacks or the damage they cause, and, finally, make a case for the overall need to reduce such attacks on humans. We endeavour to characterize attacks on humans by examining a wide variety of situations in such a way that we develop an amalgamated view of the situation that can be referenced for future management and research approaches. In the subsequent discussion about deterrents, we also aim to provide a platform for the development of methodologies to reduce conflicts from attacks, and to illuminate creative approaches to the reduction of attacks. Finally, we make the case for the responsibility we have in human society to aggressively search for solutions as humans continue to develop the earth.

Our scope of study is limited in two important ways. The first is simply the limitation of available and accessible information. The second limitation is self-imposed. We focus on terrestrial carnivores in this chapter. We avoided other taxa, except where they offer good examples for comparisons of numbers of attacks or specific characteristics. Thus, specifically, we examine the topic as it relates to humans and large terrestrial members of the Order Carnivora. These large carnivores are not only a challenge for protecting human welfare; their persecution has had grave effects on their own survival (Woodroffe *et al.*, Chapter 1).

CHARACTERIZATION

Each attack and combined group of attacks will have characteristics of animal behaviour within a context that will be important to that understanding. Additionally, an important part of that foundation of understanding will be an understanding of the ecology and behaviour of the species involved. We attempt two types of characterizations. First, we look at the environmental circumstances associated with the attack. Second, we attempt to document, summarize and characterize actual attack characteristics.

We placed all attacks into the two broad categories of provoked or unprovoked (Table 3.1). A provoked attack can take many forms. The basic definition we used is when a person(s) enters an animal's personal space or purposely tries to touch, injure or kill the animal and the animal attacks, or the person(s) had human food or garbage attractants that brought the animal nearby, again, within the animal's personal space. This personal space can be defined as the area around an animal where the animal reacts to the presence of the person. This space and the distance at which this threat is initiated can change with the conditions of the situation. That is, an animal may be particularly aggressive due to the presence of offspring that require defending; or an animal may be more determined to obtain food due to a high level of hunger, thus becoming less tolerant and more aggressive once food is detected.

One of the difficulties of using the word 'provoked' to describe this type of attack is that it commonly evokes the image of a person prodding an animal, or an image of a cornered and/or wounded animal. These types of attacks do occur and they often are the source of magazine stories and hunting tales (e.g. Grey 1922). However, a provoked attack can also be brought about without purposive aggression by the person(s), as when a photographer approaches too close to an animal, provoking the animal to attack. For example, both grizzly bears (Herrero 1985: 75–6) and polar bears (Stirling *et al.* 1977) are known to charge long distances towards people. We

Table 3.1. *A summary of characterization information from attacks on humans by 15 species of large carnivores*

Species	Attack characteristics[a]	Correlates/causes of unprovoked attacks	Reference
Tiger	P–UP; UP attacks by surprise usually, high incidence of fatality, attack at head/shoulders from rear or side, humans consumed, can be occasional or dedicated man-eaters.	*Nepal*: increasing tiger population, protected area, injured and old animals, prey density. *Russia*: injured, sick animals. *India*: density of people in forest, individual preference for humans as prey.	McDougal *et al.* 2001 Miquelle *et al.* Chapter 19 Hendrichs 1975
Leopard	P–UP; can become common predator of domestic stock and pets while avoiding humans, can be occasional or dedicated man-eaters, attacks commonly associated with night-time.	*Africa*: human vulnerability.	
Jaguar	P; attacks rare.		Rabinowitz 2000
Cougar	P–UP; UP attacks often non-fatal, attack from front or rear, attack at head/shoulders, attacker often driven away, will consume humans if undisturbed, direct 'near attacks' documented with threatening behaviour.	*USA*: human vulnerability. *USA/Canada*: human vulnerability, increased number of people in cougar habitat, aberrant individual behaviour, size of person.	Young and Goldman 1946 Beier 1991, 2004; Torres 1997; Danz 1999; Deurbrouck and Miller 2001
Lion	P–UP; UP attacks by African lion usually fatal but Asiatic lions much less so, attacks sudden and by surprise.	*Africa*: decrease in native prey, drought, increased number of people in lion areas, size of person, human vulnerability. *Asia*: drought, human vulnerability	Treves and Naughton-Treves 1999; Skuja 2002 Saberwal *et al.* 1994
Brown bear	P–UP; accounts for highest number of bear attacks, attacks often near water or concentrated sources of bear foods, including garbage, attacks usually in day	*North America*: increased number of people in grizzly bear habitat, defence of personal space, defence of young, concentrated bear foods.	Herrero 1985; Herrero and Higgins 1999; Herrero and Higgins in press Swenson *et al.* 1996

	or crepuscular hours, threatening behaviour common in close encounters, but rarely followed by attack, high incidence of females with cubs in attacks, low fatality of attacks, high rate of serious injury, consumption rare.	*Europe/Asia*: high density of bears in some regions, high human activity in bear habitats.	
American black bear	P–UP; threatening behaviour common in close encounters, but attacks rare, attacks commonly associate with minor injuries, but sometimes serious or rarely fatal, high numbers associated with campground bears until recently, fatal attacks mostly predatory and involve adult male bears.	High human activity in black bear habitats, human vulnerability, concentrated bear foods and human presence.	Gunther 1994; Herrero and Higgins 1999; Herrero and Higgins in press
Asiatic black bear	P; attacks uncommon and probably defensive, serious injuries rare, attacks focus on head.	Development in bear habitat, increased use of bear habitat by people, defence of personal space.	Chestin 1999; Hazumi 1999; Sathyakumar 1999; Wang 1999; Yamazaki 2002
Sun bear	P; attacks occasional and associated with surprise encounters in which they stand and appear larger, can inflict serious wounds.	People in bear habitat, agricultural endeavours like oil palm or coconut plantations.	Servheen 1999; D. Garshelis pers. comm.
Sloth bear	P–UP; very unpredictable but often attacks when encountering humans, attacks mostly appear defensive, but some predatory behaviour associated, attacks focus on head, arms and legs, low fatality of attacks, rare consumption rate.	Interface of agriculture with bear habitat, defence of personal space, human vulnerability	Garshelis *et al.* 1999; Rajpurohit and Krausman 2000; Chauhan *et al.* 2002

Table 3.1. *(cont.)*

Species	Attack characteristics[a]	Correlates/causes of unprovoked attacks	Reference
Polar bear	P–UP; UP attacks considered predatory, male bears more commonly attack, humans consumed.	Defence of young, concentrated bear food, human vulnerability.	Herrero and Fleck 1990; Gjertz and Scheie 1998
Spectacled bear	P; attacks only associated with hunting or surprise encounters.	Defence of personal space.	Orejuela and Jorgenson 1999; Peyton 1999
Giant panda	P; attacks on people rare, highly unlikely to attack people, shy but defensive in some situations.	Repeated approaches invoke aggression.	Schaller *et al.* 1985; Reid and Gong 1999
Grey wolf	P–UP; large percentage of attacks associated with rabies which include direct advance on victim during daylight hours and multiple bites, defensive attacks related to dens, trapped animals, or domestic dogs, some predatory characteristics.	*Europe, Canada*: rabies, habituation/food conditioned.	Kruuk 2002; Linnell *et al.* 2002; McNay 2002
Spotted hyaena	UP; attacks focussed on head, resting or sleeping people vulnerable, during daylight hours.	Human vulnerability.	Clarke 1969; Treves and Naughton-Treves 1999; Kruuk 2002

[a] P, provoked attack; UP, unprovoked attack.

classify all of these as provoked attacks. In many cases, it is also enlightening to describe such attacks as defensive. In all of these cases, the animal is defending itself or some attractant or possession. The offensive action – the attack – is initiated by the animal due to a perceived threat.

We define unprovoked attacks as those where the animal approached and attacked with the principal attraction being the person(s), not people's food or other attractants. Unprovoked attacks may be predatory, a consequence of disease, or involve an animal wanting right of way. Predatory attacks are self-explanatory and, simply put, the object of the attack, the person, is attacked as prey. These attacks are associated with predatory behaviours such as stalking, pursuit or testing of the targeted person. Disease can also be a factor in unprovoked attacks and render an animal aggressive much like a wound can. We classify these situations as unprovoked, but caution that attacks mediated by disease could warrant a separate classification. However, for our purposes, attacks in which the animal has a proven or suspected debilitating disease are best classified as unprovoked. Lastly, right-of-way attacks can be further described as a form of interference competition between carnivores and people. In such instances, the animal and the person are intent on using the same space and the animal attacks when it is not given the right of way and the person is not able to scare off the attacking animal. In these cases, it is not clear that the animal intended an attack and they usually are associated with threatening vocalizations and postures but there is no associated attractant such as food or garbage.

TAXONOMIC REVIEW OF ATTACKS ON PEOPLE

Although we propose that all attacks can be placed in the categories of provoked or unprovoked, effective preventative actions will be dependent on a more thorough understanding of attacks. Taking each taxonomic group in order, and then offering a summary, we provide a set of descriptions that will allow comparisons and contrasts we feel are important in reference to taxonomic evaluations and the circumstances of attacks.

Cats

Felids are often described as the most highly specialized terrestrial predators.

Tiger

The tiger is perhaps the most famous for its attacks on people and gained extensive coverage through such writers as Corbett (1946). McDougal (1987) correlates the high prevalence of tiger attacks with areas that have growing or stable populations of tigers being forced into marginal habitat where people are more commonly encountered. These conclusions seemed reasonable for the time, but would require some monitoring over time to validate and test. In these situations with high numbers of attacks, McDougal speculates tigers are forced to override a natural tendency to avoid humans, and they begin attacking them opportunistically as prey, or they become 'dedicated man-eaters' (McDougal 1987: 446). In later summaries (McDougal 1999; McDougal *et al.* 2001), he also correlates tiger attacks in Nepal with disturbed areas (i.e. logging activities) and animals stressed by lack of natural prey or injuries.

It is clear that there is a proportion of tiger attacks in which people are taken as prey, thus categorized as unprovoked attacks. The Sundarbans in eastern India is a consistent focus of tiger attack reports (Hendrichs 1975; Chaudhary and Chakrabarti 1979). Hendrichs (1975) concluded that about one-third of the Sundarbans tiger population was aggressive toward people, but that only 3% were dedicated man-eaters. Attacks by the other two-thirds of the population are only brief and the victim only mauled, and if killed, not carried away or eaten. For those tigers that take humans opportunistically or as dedicated man-eaters, the attack occurs from the back or side with a bite to the neck; the body is dragged to a place where it is fed on starting 'from the belly and buttocks' (Hendrichs 1975: 191). McDougal *et al.* (2001) also reports two unprovoked attacks in which the tigers were surprised and killed their victims, but did not feed on the carcass. McDougal (1999) reports initial feeding from the head and torso and, in agreement with Hendrichs, says that almost all attacks occur during the daylight hours. The largest portion of these reported attacks by tigers appear unprovoked and the attack is quick and only known at the time of contact by the victim. There are common records of tigers relinquishing their victims when approached by groups of people to reclaim the body (e.g. Chaudhary and Chakrabarti 1979; McDougal *et al.* 2001). In these cases, vocalizations are sometimes given by the tigers as threats.

Cougar

As with tigers, there are also a number of accounts of cougar attacks on people that give summary data (e.g. Beier 1991; Torres 1997; Danz 1999), but there is also extensive detailed coverage of attack specifics in both scientific and popular literature (e.g. Beier 1991 and Mueller 1985, respectively). First, generally, there has been an increase in cougar attacks that can be explained through increasing human and cougar populations in the western USA, with the accompanying increase in encounter rate. Hornocker (1992), however, speculates that this may be partly due to the selective influence cougar-hunting may have by selecting out those less aggressive individuals and favouring those more prone to attack. Still, relative to the tiger, the number of cougar attacks and fatalities are few, averaging, for example, fewer than one fatality per year over the last decade in the USA and Canada (Chester 2004). Beier (1991) found most cougar attacks were on children under 17 years of age and almost all attacks on adults were on people travelling alone. Typically, the attacks came from the back or side of the victim, with the likely first point of contact being the head and shoulders, and the direct cause of death (if fatal) was a bite, or bites, to the head or neck (e.g. Deurbrouck and Miller 2001: 110). If left undisturbed, cougars appear to feed on humans starting with the face and torso, and they will cover the body between feeding bouts.

African lion

African lions, much like tigers, have a few noted records of man-eating that dominate the somewhat fabled reputation for attacking humans. The most noted of these has been the pair of lions that killed scores of workers building the trans-Africa railroad (Patterson 1907; Kerbis-Peterhams 1999). However, although dedicated man-eaters are not common, it appears that African lions will attack humans opportunistically (Guggisberg 1961). These unprovoked attacks are not well documented and objective accounts are difficult to obtain (Nowell and Jackson 1996). Skuja (2002) surveyed people in seven rural villages in northern Tanzania and documented 14 separate attacks over a period of 26 months, resulting in 13 deaths. At least six of these were children under the age of 16; all but two of the attacks occurred in the evening hours. Skuja correlates attacks with the increase of immigration to the area and the decrease of wild prey due to drought. Little detail is available about these attacks, but they appear to be sudden and almost always fatal. This contrasts with reports of attacks by Asiatic lions (Saberwal *et al.* 1994) in which only 14.5% of the attacks from 1977 to 1991 resulted in mortality. In both areas, drought is also implicated as a factor increasing attacks. Treves and Naughton-Treves (1999), in a review of government reports in Uganda, found lions were responsible for at least 275 attacks on people over a 72-year period; 75% of these attacks were fatal. Only 14% of these were attacks by wounded animals and considered provoked; the balance were considered unprovoked.

Leopard

We obtained several accounts directly from witnesses of leopards preying on pets and domestic stock in which leopards specifically seemed to avoid people available as prey in the same situation. However, there are also credible accounts of leopards as unprovoked, dedicated man-eaters (Inverarity 1894; Corbett 1948), and as occasional predators on humans in unprovoked attacks (Moss 1903; Treves and Naughton-Treves 1999). Daniel (1996) believes that leopard attacks in India peaked in number during the late nineteenth century and early twentieth century, a time of great human expansion. His review of attacks in India indicates most leopard attacks happen at night inside villages or very near to villages. There are several accounts of leopards entering houses to attack people in their sleep.

Jaguar

Finally, among the felids, the only other species of note for attacks on humans is the jaguar. The information is somewhat contradictory, but seems to support the statement by Rabinowitz (2000) that among the big cat species, 'the jaguar is the least known to attack man unprovoked and is virtually undocumented as a man-eater' (Rabinowitz 2000: 201). Attacks on people appear to be rare and often described as involving children and new settlements, but little detail is available beyond this.

Bears

All adult bears have the size and strength to seriously injure or kill people. Fortunately, this seldom occurs because most bears are secretive and avoid people. However, because attacks do occur, this becomes a special challenge to human safety and bear conservation. Bears use intraspecific aggression to allocate resources within a population and many of the same signals used between bears are also directed toward people. Thus, knowledge of the details of agonistic behaviour among bears has some application to human safety (Herrero 1985). During agonistic encounters between bears, serious injury or death occasionally occurs, as it does with people, albeit at somewhat higher frequency.

While carnivores by evolutionary origin, bears are functionally omnivorous and much of their diet, except for that of the largely carnivorous polar bear, is composed of calorically concentrated vegetative foods such as berries, nuts, the underground storage organs of plants, and lush green vegetation. Meat or fish is eaten when it can be caught or taken away from other predators such as wolves or even people. It is near these foods or near water that bear–human encounters often occur (Herrero 1985). When nutritious, easily digestible foods are not available, as during winter for many species, bears hibernate, thus dramatically decreasing the danger of bear attack during the time they are in their dens unless they are provoked. In subtropical or tropical climates, bears may not hibernate. Polar bears in the southern arctic have a metabolic rest during summer after the ice has melted (Stirling and Guravich 1988). Because of their omnivorous nature, most bears can be attracted by peoples' crops, food or garbage. This brings a potentially dangerous animal into close proximity to people and has been associated with serious bear-inflicted injuries in North America (Herrero 1970, 1985, 1989; Gunther 1994; Gniadek and Kendall 1998; Herrero and Higgins 1999, 2003).

Bears of even one of the most aggressive and powerful species, the brown bear, will usually flee when their personal space is entered by a person (Herrero 1985: 13). At times, they display their stress by making aggressive gestures or sounds in these situations, such as by clacking their upper and lower sets of teeth together, growling, blowing, running at a person, or by swatting the ground with a paw or paws (Herrero 1985). Rarely, such sudden encounters and displays are followed by attack. We classify this type of attack as provoked and defensive. When attack does occur by brown bears under these circumstances, it is of short duration (i.e. a minute or two), although very serious injury may be inflicted (Herrero 1985). The bear's motive when its personal space is entered and attack occurs does not appear to be predation, because victims are seldom killed and consumed (or even stayed with), but more to stop a perceived threat (Herrero 1985).

Bears of several species also make rare, unprovoked, usually predatory, attacks on people. In these circumstances, the bear seldom acts as if stressed or gives much warning that attack is about to occur, beyond simply following a person, then running at the person and attacking. In contrast to provoked, defensive attacks, predatory attacks may last for as long as the person can stand off the bear, sometimes lasting for hours (Herrero 1985; Herrero and Higgins 1995). A bear that kills a person under these circumstances feeds on its prey much like it would an ungulate, including sometimes covering the carcass with debris to keep away scavengers if it leaves for some period of time. Fortunately, such incidents are rare, otherwise people would not choose to coexist with bears. Old age and disease of bears are occasionally associated with unprovoked attacks, but most have occurred without these conditions (Herrero 1985; Herrero and Higgins 1999). Rabies appears to be rare in bears and has only been reported for American black bears. Between 1954 and 1992, only seven of almost 50 000 animals confirmed to have rabies in the province of Ontario, Canada, were American black bears; however, 'most rabid (American) black bears in Ontario are reported to be aggressive toward humans' (R. Walroth *et al.* pers. comm.).

Brown bear

The brown bear accounts for more serious or fatal injuries to people than any other bear species. The scientific literature regarding brown (grizzly) bear attacks on people in North America has been recently and briefly summarized (Herrero and Higgins 1999). In Yellowstone National Park, grizzly- and black-bear-inflicted injuries decreased dramatically from the 1930s to the 1990s (Gunther 1994). This decrease was associated with making people's food and garbage less and less available to bears. Food and garbage availability brings bears into close association with people and thus increases the danger of attack. Other circumstances associated with brown-bear-inflicted injuries in North America were wounding, defence of carcasses, predatory attack, harassment by dogs or photographers, and starving or sick bears (Herrero 1985). An overview of brown bear attacks, emphasizing those that were fatal or resulted in serious injury, cannot help exaggerating the danger of individual encounters with this species. Each year in North America there are probably millions of interactions between brown bears and people. Most are totally benign. A minute fraction of these results in human injury. Extensive and high-quality information is available regarding how to minimize the risk of bear attack in North America (Herrero 1985; Safety in Bear Country Society 2001). We discuss this later in this chapter.

Polar bear

Polar bears are the most predatory of modern bears. Serious and fatal attacks on people by polar bears have been reported and analysed for the Canadian arctic (Fleck and Herrero 1988) and for Svalbard (Gjertz and Persen 1987; Gjertz and Scheie 1998). All fatal attacks were unprovoked and predatory. Male bears, especially subadults, were disproportionately involved in actual or attempted predation. In a few cases female polar bears appeared to have been provoked and attacked in defence of young; however, the majority of polar bear attacks were unprovoked and predatory. Several fatal attacks occurred in a context where polar bears were attracted to human settlement by garbage, other food or carcasses, and then attacked as predators. Some predatory attacks had no attractants beyond the person involved.

Sloth bear

Sloth bears are known for their aggressiveness toward other animals, including people (Garshelis *et al.* 1999). While 'they seem to avoid human contact' (Garshelis *et al.* 1999), they often are found in landscapes near dense human populations or in agricultural areas, so encounters are relatively frequent. Phillips (1984: as cited in Garshelis *et al.* 1999) reported that sloth bears were second only to rogue elephants as the most feared animal among jungle villagers of Sri Lanka. While two of the three people killed by sloth bears in the North Bilaspur Division were consumed (Chauhan *et al.* 2002), this may have been scavenging. However, there is reason to believe predatory attacks occur. Garshelis (pers. comm.) interviewed a person in Nepal who was attacked by a sloth bear reportedly for 'hours' – the person kept fighting the bear, and it kept fighting him, occasionally leaving then returning. This behaviour is consistent with a predatory attack. The majority of injurious encounters with sloth bears seem to have been defensive and provoked by people silently coming nearby a bear or family group and the bear's personal space being entered. However, provoked, defensive attacks by brown bears seldom result in death since the bear appears to want to only remove a threat (Herrero 1985). In this context, the 48 reported sloth bear fatalities seem an anomaly if the attacks were provoked and defensive. The attacks appear to focus on the head area and leg and arm muscles, much like brown bear attacks in North America (Herrero 1985; Rajpurohit and Krausman 2000). Strategies for protecting the head in the event of attack should be particularly valuable because of the ease of seriously damaging the human face (Herrero 1985).

American black bear

Comparison of American black bear and brown bear injury rates for Alberta and British Columbia give substance to the colloquial knowledge that in North America the brown bear is much more dangerous (Herrero and Higgins 1999, 2003). During most close range encounters with American black bears, the bear either flees or uses agonistic displays to indicate stress. People recognize this language and usually back away, thus defusing this situation. Attack very seldom follows. During the period from the 1930s to the 1970s, American black bears inflicted thousands of minor and a few major or fatal injuries (Herrero 1985; Herrero and Higgins 1995). The many minor injuries were associated with American black bears begging for handouts and foraging in campsites and at roadsides, thus being drawn into very close proximity to people. Since the 1970s–1980s, the cleaning up of campsites and the prohibition of roadside feeding have dramatically decreased injury associated with these circumstances (Herrero 1985; Gunther 1994).

Other bears

Finally, attacks on people by Asiatic black bears, sun bears, pandas and spectacled bears appear to be uncommon (Table 3.1).

Grey wolf

Stories of wolves' ferocity and aggressiveness have been portrayed for centuries, in the Bible, from the hand of Aristotle, to Red Riding Hood and Peter and the Wolf, to White Fang (e.g. London 1913; Boitani 1995;

Kruuk 2002; Linnell *et al.* 2002). There has also been an abundance of popular literature produced lately about wolves (e.g. Casey and Clark 1996). Such history and writing make objective analysis of the situation even more imperative. Recent summaries (e.g. McNay 2002) have greatly advanced the science and objectivity in analysing wolf attacks. Linnell *et al.* (2002) surveyed worldwide and concluded that wolf attack records were highly fragmented and of highly variable quality. However, they were able to verify that wolves do attack people throughout wolf range and that there are literally hundreds of records over the past two centuries. McNay (2002) reviewed 80 cases of wolf–human encounters from Alaska and Canada over the twentieth century and concluded that wolf attacks, while still infrequent, were on the increase in the final three decades of the century. Still, Linnell *et al.* (2002) stated that 'people do not appear as regular items of wolf prey' (p. 36) and the risk is impossible to quantify because it is so low.

There are some discernable patterns and characteristics in wolf attacks. Linnell *et al.* (2002) place attacks into three categories: attacks by rabid wolves, predatory attacks, and defensive attacks. In our definitions (see section 'Characterization', above), the first two types are contained within our definition for unprovoked and the latter is contained within our definition of provoked. Wolves are a good example of the need to consider a separate category for disease-influenced attacks. Linnell *et al.* (2002) state: 'the most important factor explaining the incidence of present day, and probably most historic, wolf attacks is the presence of rabies' (p. 36). They documented incidents where multiple people were bitten in a single attack. These attacks, in which the typical injury was a bite or multiple bites, almost always resulted in the death of the victim in the era before the development of a treatment (in the 1890s), or if treatment was not available.

Unprovoked attacks on people by wolves have mixed characteristics and descriptions. McNay (2002) reduced 80 encounters between people and wolves to 18 non-rabies, unprovoked aggressions between 1969 and 2000. He further divided the attacks into habituated and non-habituated individuals and found that 11 were habituated wolves in park (protected area) settings or industrial sites, five of which were food-conditioned in their approach behaviour (see also Strickland 1999). The remaining attacks were inconclusive for patterns, ranging from protection of a kill to a leap toward the victim's face with no obvious motivation (McNay 2002). Linnell *et al.* (2002) also noted the importance habituation may play in human–wolf interactions. Wolves appear to lose fear of humans as they associate them with food, much as in the relationship between bears and humans (Herrero 1985, 1989). One of the perplexing characteristics of unprovoked attacks by wolves is the dramatic difference between the lack of attacks in North

America and the consistent level of attacks in Eurasia. Kruuk (2002) relates several specific incidents in Europe in which it would appear that wolf attacks on humans are opportunistic and predatory (unprovoked). Evidence from North America indicates no such level of encounter between wolves and people (McNay 2002).

Wolf attacks on humans occur principally during the day or the evening hours and they are initiated with little warning except the direct approach of the wolf toward the victim (Pollard 1966; McNay 2002). Linnell *et al.* (2002) point out that there is an important age difference in the victims of rabies-influenced attacks versus non-rabies attacks: 'the vast majority' of victims of rabid wolves are adults (and mostly men) and the victims of predatory, unprovoked attacks are mostly (90%) children under the age of 18 (p. 37). They speculate this is a reflection of the randomness of rabies-influenced attacks, opposed to the selective predation on more vulnerable prey, respectively. There are few descriptions of pre-attack behaviour in wolves except in popular publications. McNay (2002) describes growling associated with some unprovoked attacks, and tooth-baring is commonly described in popular literature (e.g. Pollard 1966). McNay (2002) also described 'prey testing' behaviour related to a small number of human–wolf encounters and that some encounters never reached the level of attack, even when threat behaviour was involved.

Spotted hyaena

Of the hyaenas, the striped and spotted hyaenas are both known to attack people (Nowack 1991). The spotted hyaena is the most important in this context due to the number of actual and potential attacks on adults and children. Due partly to the lack of official reporting of wildlife attacks on people in Africa, it is difficult to measure the extent of the situation. Clarke (1969: 89) states that 'hundreds of Africans throughout Africa' have been attacked by spotted hyaenas. Treves and Naughton-Treves (1999), however, found records of only four attacks by spotted hyaenas on humans in Uganda over a 70-year period beginning in 1923. Balestra (1962) reported a case in Malawi where man-eating appeared to develop in spotted hyaenas responsible for killing and eating 27 people over a period of five years (see also Kruuk 2002). Recent reports from Malawi (Lonely Planet 2004) indicate attacks continue in that country. Another news report (World Wildlife Fund 2004) indicates there have been 52 hyaena attacks resulting in 35 deaths in a recent 12-month period in Mozambique. Woodroffe (pers. comm.) reports a small number of hyaena attacks (possibly associated with drought) in the Laikipia region of Kenya, and Thirgood (pers. comm.) reports occasional hyaena attacks (associated with rabies) in Serengeti District of Tanzania. The details of attacks by spotted hyaenas are lacking and therefore are difficult about which to generalize. However, many of the victims were children and commonly the attack was to the head, many times directly to the face, while the person was resting or sleeping (Kruuk 2002). Also, as Kruuk also points out, attacks can come during the daylight hours on solitary adults. There appears little, if any, warning before hyaena attacks except the direct approach of the attacker(s). We found no records indicating provoked attacks in hyaenas.

To summarize this section, when viewed across species and families, there exists wide variation in the characteristics of attacks. Certainly, it appears from the data that children are more vulnerable and that, overall, most attacks occur during the daylight hours. Most of the attacks seem to be a surprise to the person being attacked and they come from the side or from behind; this seems to be less true of the hyaena and wolf attacks, and almost never true with bear attacks. It is also interesting to note the geographic variation in the presence of attack behaviour in large carnivores (Kruuk 2002). For example, there are some regions of the USA where no cougar attacks on people have been recorded, while nearly half the fatal attacks documented by Beier (1991) occurred on Vancouver Island. In another example, hundreds of unprovoked attacks by wolves on people have been recorded in Eurasia, while attacks in North America are relatively very rare. The presence of rabies is certainly part but not all of the explanation in this latter example. But the larger question of geographic variation in attacks leads to a variety of questions that centre on learned behaviour and the potentially 'unnatural selection' of certain behaviours in subpopulations of large carnivores. As a last note, we encourage a greater attention to reporting attacks by carnivores. Only through the examination of detailed accounts of attacks on people will we be able to fully understand the behaviours and the contexts in which they occur, and thus mitigate, reduce and prevent them in the future.

SOLUTIONS

We already know we have the ability to eliminate these species and eliminate the threat of attacks. However, the elimination of species or a population of a particular species is not the goal and is ethically questionable in most cases (Treves and Naughton-Treves, Chapter 6). Deterrents and prevention are sought as more acceptable solutions. It is useful to approach attack prevention on a continuum, with an actual attack on one extreme; on the other extreme would be the circumstances surrounding the potential of an encounter with a large carnivore and, thus, for it to attack. In this latter end of the continuum, carnivores are present and the actions of the person(s) can reduce the chances of the carnivore encounter. We begin this discussion with the latter and move through attack prevention toward the actual attack of an animal and the actions that can be performed to reduce injury in such cases.

Attack prevention

Living, working or travelling in areas with large carnivores takes awareness and preparedness. As Herrero (1985: 123) states for bears, 'Your best weapon to minimize the risk of a bear attack is your brain.' This holds true for other large carnivores. Entering or living in carnivore country has certain accompanying responsibilities. Remember that often when an attack occurs – whether provoked or unprovoked – the animal is pursued and killed. Reducing the potential for attacks can save injury to oneself, and save the life of the wild carnivore involved.

Often, when travelling or working in carnivore range, the most important thing one can do to reduce the potential for attack is to advertise one's presence. This advertisement can come in the form of noise or obvious motion and it is based on the assumption that the species involved exhibits natural avoidance behaviour toward humans. This is not always the case (e.g. habituated man-eaters as found in some tiger populations). However, this is a reasonable assumption for most species. In more developed parts of the world, much of this type of travel is related to recreation. However, in developing countries, travel in carnivore range is more likely to be related to work. Still, the same approach applies: noise and commotion can advertise to animals that a human is in the area, thus evoking avoidance or flight. Noise generators can be passive, such as bells hung on backpacks, or they can be more active, such as periodically calling or shouting, or even using horns or whistles. Whether travelling in carnivore range for work or for recreation, the potential for attack is much greater if one is travelling alone, or with only one other person.

In some cases, increasing safety for travelling or working in carnivore range may be brought about through strategies that decrease the potential of being viewed as prey. Although perhaps not as often in open habitats like the open arctic and the African savannas, people can be observed or even stalked by the potential carnivore attacker and not be aware of it. It is one thing to advertise your presence to induce avoidance; it is another to assume you might be seen by the potential predator, and to decrease your attractiveness. Decreasing your attraction is equivalent to decreasing your appearance as vulnerable prey. Actions to reduce one's attraction and vulnerability would be to travel with small children off the ground, such as carrying them in a backpack. A backpack or heavy clothing may also increase one's silhouette to appear larger and less vulnerable to attack. In parts of the Sundarbans of India and Bangladesh, some success appears to have been found by using backward-facing masks. The masks ward off attacks, it is believed, by appearing to extract the element of surprise for the attacker. Like rear-facing masks, use of electrified dummies to condition tigers has also had mixed results.

Just as travelling or working in carnivore range brings responsibilities, living in carnivore range brings the same. Whether out of necessity or choice, having one's home in or near carnivore habitat means learning the habits of the animals and reducing the potential a particular carnivore species will come near and, thus, increase the potential for attack. Again, the basic approach is to reduce the attractiveness and increase the repulsing nature of the area around which people live. For many of the solitary large carnivores, this means reducing the amount of cover near the house or living quarters. These species are normally loath to expose themselves in the open. This is especially important when small children are using the area because they increase the attractiveness of the area to large predators. Because many of the species are most active in the crepuscular hours or at night, the presence of a light, natural (i.e. fire) or artificial, may reduce movement of the animals around the immediate area. The intensity and persistence of carnivore movement around the area may be accentuated by the actual placement or site of the buildings (or living area) in the overall landscape. That is, for example, in northern latitudes, if it lies in an important ungulate wintering area, or in southern latitudes, it lies in an important dry-season watering area for game, these prey can be attractants to carnivores. Thus, in these cases, modification of the immediate area around houses can become more important. In some cases, the placement of structures for human work or living quarters may be set and established for years. But, where there is the potential for planning, the landscape can be analysed to identify habitat, prey densities and travel corridors where carnivores are more likely to be present. The proper placement of human development to avoid these areas has the potential to reduce carnivore–human conflict, carnivore attacks and long-term costs of carnivore conflicts and management.

Perhaps the single most important action that can be undertaken to reduce the potential for attack is to reduce or eliminate food as an attractant. For bears, this can be garbage, food in storage or preparation areas, or any place where there is a concentration of food or food smells. For big cats, food attractants may be pets, domestic stock or even habituated wildlife such as deer near areas of human use. Even wolves and hyaenas can be drawn to areas by the presence of prepared foods or the presence of vulnerable animals. Proper food and garbage storage is essential. Residents in carnivore range should also maintain pets and domestic stock in areas of low vulnerability. Physical exclusion of carnivores can also be effective. In Africa, this may mean rigorous maintenance of boma fencing to physically exclude lions, hyaenas and leopards. Fencing can also be effective for excluding bears. Whatever the action, preventative action is essential; once an animal has obtained food, the development of habituated behaviour and attraction to human food is an important factor in attacks.

Equally important in the developed world is the feeding of wildlife. Direct feeding of carnivores should be strictly avoided and enforced with laws such as those in most national parks in North America. The feeding of wildlife in general is a common activity in developed countries. This occurs in urban, suburban and rural areas and it involves active feeding of a variety of wildlife species, including birds, ungulates and even carnivores. Of concern is also the more passive feeding of wildlife like deer and elk that may enter areas around buildings to forage on ornamental shrubs and trees, as well as food gardens. This is a fruitful area for government agency involvement. Agencies can establish and enforce laws against feeding of wildlife. These laws and regulations can be very effective in reducing conflicts, property damage and attacks (Harms 1980). At least as important, however, can be education and assistance programmes that inform the public and engage landowners and other residents in planning for an environment that enhances the wildlife–human interaction and reduces conflicts (e.g. VanDruff *et al.* 1994; Decker and Chase 1997; Schusler and Decker 2002).

What to do if attacked

What if preventative actions fail and a person suddenly feels their body being violated by teeth or claws or both? Most people's instinct in these cases is to fight for their lives trying to use their strength or any available or improvised weapon. If the attack is unprovoked and predatory or disease-motivated, instinct yields the action most likely to minimize risk of serious injury. People have warded off or even killed bears during predatory attacks by using a knife, rocks or even fists (Herrero 1985: p. 122, 63). In the event of a close-range (0–6 m) aggressive encounter with a brown (grizzly) or American black bear, aerosol sprays containing capsicum derivatives have been shown effective in deterring attack (Herrero and Higgins 1998). Perhaps because cougars are often smaller than North American bears, there are many examples of people successfully fighting off attacks. One example is the case of a woman attacked on Vancouver Island, Canada (Etling 2001: 85–6). She was walking to her campsite when she was confronted by an aggressive cat. When the cat attacked her, she screamed and fought back, the cat backed off and her screams brought others to her assistance. Children had been playing in the area just before the incident and may have been the original attractant for the cat. According to both Beier (1991) and Torres (1997), fighting back does influence the severity and outcome of attacks on humans by cougars. Fighting back is always the appropriate response to predatory attacks. If several people are present, they should act together to try to deter the attacking animal using every possible means to intimidate or injure the animal.

Ironically there are times when fighting an attacking carnivore may increase, not decrease, the chance of serious or fatal injury. This is best understood with regard to the brown bear. Most, but not all, attacks by this species are provoked when the bear's personal space is entered by a person. The extent of injury appears to have been less when people adopted passive resistance, lying on their stomachs, face down, on the ground with their hands behind their necks (Herrero 1985: 24–30). Sometimes fighting back deterred brown bears involved in unprovoked attacks but the odds seem to favour minimizing injury by passive resistance. One of the reasons for this is that a brown bear is usually much stronger than a single person. As mentioned, unprovoked attacks by brown bears are often of short duration, lasting only several minutes, however serious injury may be inflicted. Whether extent of injury would be minimized by passive resistance during provoked attacks by other species is not yet demonstrated. It is important to better understand this to be able to offer advice to minimize the extent of human injury during provoked attack by large carnivores.

How does one recognize whether an attack is provoked or unprovoked? This decision is obviously important and should be based on a person's assessment of the situation and the attacking animal's behaviour. Prior to unprovoked predatory attack people are typically approached as prey; such approaches are often silent and a person may see or hear the predator, or only feel it, as it attacks. The animal may, however, be uncertain regarding the person, approaching to find out more. If the approaching animal has been given right of way and yet still approaches, now is the time to let the animal know you are capable of hurting it if it continues to approach. Shout, throw things, use any deterrents you many have. Convey the message you are not easy prey.

Provoked attacks result when the animal has been stressed as a result of the person entering the animal's personal space. Bears have many different sounds and actions to indicate their stress and to further assess the situation. They may stand on their hind legs to better sense the situation; they may blow, snort, growl, strike the ground with one or both paws, or run at a person and stop nearby (Herrero 1985). These sounds and actions indicate that a bear is uneasy with having the person nearby. The person should back away, watching the bear and stopping, standing their ground if the bear continues to approach. The situation usually eases when the person is outside the bear's personal space. Again, the extent to which this is true with regard to bear species other than American black bear and brown bear, or with respect to other large carnivores, is unclear and requires study. For cougar confrontations, it is recommended that people stand their ground or back away, lift children off the ground, make oneself look as big as possible, and fight back if the attack ensues (Torres 1997; Anonymous 2000).

Attacks in context

As we attempt to orient wildlife–human conflicts in the context of global conservation (Woodroffe *et al.*, Chapter 1; Thirgood *et al.*, Chapter 2), wildlife attacks on humans offer us an intense and dramatic model of the challenge we face in understanding and reducing conflict. First, we must not only understand the species of concern (i.e. their behaviour, distribution, movements, etc.), but the environmental influences on their behaviour. Second – and basic to this understanding – we must develop protocols for recording and reporting incidents. An overwhelming impression from our analysis is that people have dealt with the subject in a manner that is commonly less than objective. As a result, the base of information on attacks is scant and offers less guidance in providing solutions than it would if reporting were objective and detailed. And, third, reducing attacks on humans is altogether possible. But this is only the case if we continue to elevate the attention we apply to solutions and continue to learn from these applications. As such, we will advance our techniques and get closer to the goal of a stable, less conflicted coexistence of humans and carnivores.

This coexistence must be based on an ethic accepting the premise that human existence and quality of life is enhanced by the presence of wildlife, including large carnivores. The human population of the earth continues to grow exponentially and human development will continue to engulf wildlife habitat. Human contact with carnivores will inevitably increase. Perhaps the biggest question about this relationship is not so much 'Why do they attack and kill people?' but 'Why don't they kill more people?' Certainly, wolves, bears, lions and hyaenas have the power and the opportunity. Is this avoidance and abstinence a result of intense selection against predation on humans because people eliminated the individual problem animals? Or is it something else? Whatever the reason, it gives us an advantage and it helps as we continue to understand our responsibility to conserve nature, including species that attack us. We know that all species have ecological roles to play, and we continue to increase our understanding of the potentially keystone roles that large carnivores can play in community organization. We also understand their vulnerability; their life-history characteristics of low reproductive potential and low densities make them especially susceptible to extinction. With increasing understanding, initiatives based on an inclusive ethic for carnivores and other wildlife, plus the resources and attention to the challenge, attacks on people can continue to be rare while these animals enhance our lives and we become better stewards of our natural world.

Non-lethal techniques for reducing depredation

URS BREITENMOSER, CHRISTOF ANGST, JEAN-MARC LANDRY, CHRISTINE BREITENMOSER-WÜRSTEN, JOHN D.C. LINNELL AND JEAN-MARC WEBER

INTRODUCTION

Ever since humans domesticated the first animals several thousand years ago, there have been conflicts with large carnivores attacking livestock (Kruuk 2002). Every year, thousands of cattle, sheep, goats, poultry or farmed fish are killed by wild carnivores worldwide (Thirgood *et al.*, Chapter 2) (Table 4.1). The farmers, in turn, kill the predators. Lethal control of stock-raiders is common in all cultures and has a devastating impact on many populations of large carnivores (Woodroffe *et al.*, Chapter 1). Retaliatory killing was the most important reason for the historic eradication of large carnivores in large areas (Breitenmoser 1998). In addition to killing the predators, herdsmen have tried to protect their livestock, mainly because lethal control alone rarely reduced depredation to an acceptable level. For a traditional society, the investment in terms of labour and resources for the protection of livestock was high (Kruuk 2002). Rural cultures have consequently adopted a combination of non-lethal measures, lethal control and – strongly varying between cultures – an acceptance of losses. The application of non-lethal techniques was mainly a matter of technology and of cost–benefit considerations. From the perspective of modern society, there are two more reasons to propagate preventive measures: conservation (lethal control threatens many carnivore populations), and ethical arguments (moral reservation against the killing of predators and against livestock being exposed to pain and suffering). In this chapter, we review both traditional and modern methods that have been applied to prevent depredation on livestock, and focus less on the welfare of individuals than on the conservation of populations. To make a difference at the population level, non-lethal techniques must be not only effective, but also applicable and acceptable and cost-effective for farmers on a large scale.

People and Wildlife: Conflict or Coexistence? eds. Rosie Woodroffe, Simon Thirgood and Alan Rabinowitz. Published by Cambridge University Press.

Table 4.1. *Main diet and conservation status of mammalian carnivore species preying occasionally or regularly upon livestock*

Family, species and common name	Region	Status[a]	Wild prey (WP) and livestock (L)	Significance of livestock[b]	References
Canidae					
Canis aureus Golden jackal	Asia, Europe	Not listed	WP: rodents, lagomorphs, carrion; L: cattle (calves), small stock	Low	Yom-Tov *et al.* 1995
C. latrans Coyote	N.America	Not listed	WP: lagomorphs, rodents; L: small stock, cattle	Low	Sacks and Neale 2002 Windberg *et al.* 1997
C. lupus Grey wolf	Asia, Europe, N.America	Lower risk–vulnerable	WP: ungulates; L: small stock, cattle,	Low–high	Boitani 2000
C. l. dingo Dingo	Australasia	Not listed	WP: wallabies, rabbits; L: small stock, cattle	Low–medium	Corbett 1995
C. mesomelas Black-backed jackal	Africa	Not listed	WP: rodents, carrion; L: small stock	Low	Bothma 1971
C. rufus Red wolf	N.America	Critically endangered	WP: deer, rodents, lagomorphs; L: small stock, cattle	Low	Macdonald and Sillero–Zubiri 2004
Lycaon pictus African wild dog	Africa	Endangered	WP: ungulates; L: cattle, small stock	Low	Woodroffe *et al.* 1997

Ursidae					
Tremarctos ornatus Spectacled bear	S.America	Vulnerable	WP: fruits; L: cattle (calves)	Low	Goldstein 1992
Ursus americanus American black bear	N.America	Not listed	WP: fruits, berries; L: livestock, honey	Low	Jonker *et al.* 1998
U. arctos Brown bear	Asia, Europe, N.America	Not listed	WP: berries, fish, rodents; L: livestock, honey	Low–medium	Swenson *et al.* 2000
U. thibetanus Asiatic black bear	Asia	Vulnerable–critically endangered	WP: fruits, berries; L: small stock, cattle	Low	Servheen *et al.* 1999
Mustelidae					
Gulo gulo Wolverine	Asia, Europe, N.America	Vulnerable	WP: lagomorphs, rodents, carrion; L: sheep, reindeer	Low–medium	Landa *et al.* 2000c
Lutra lutra European otter	Europe	Vulnerable	WP: fish; L: fish (fish farms)	Low	Bodner 1998
Mellivora capensis Honey badger	Africa	Not listed	WP: small vertebrates, honey; L: honey	Low	Mills and Hes 1997
Hyaenidae					
Crocuta crocuta Spotted hyaena	Africa	Lower risk	WP: ungulates, carrion; L: small stock, cattle	Low	Kruuk 1972, Mills 1990

Table 4.1. (*cont.*)

Family, species and common name	Region	Status[a]	Wild prey (WP) and livestock (L)	Significance of livestock[b]	References
Hyaena brunnea Brown hyaena	Africa	Lower risk	WP: carrion, insects; L: small stock, cattle (calves)	Low	Mills 1990
H. hyaena Striped hyaena	Africa, Asia	Lower risk	WP: carrion; L: small stock	Low	Kruuk 1980
Felidae					
Acinonyx jubatus Cheetah	Africa, Asia	Africa: vulnerable–endangered; Asia: critically endangered	WP: ungulates; L: small stock, cattle (calves)	Low–high	Nowell and Jackson 1996, Marker 2000a
Caracal caracal Caracal	Africa, Asia	Not listed	WP: rodents, ungulates, lagomorphs; L: small stock	Low	Nowell and Jackson 1996
Puma concolor Puma	N. and S. America	Near threatened–critically endangered	WP: ungulates; L: cattle (calves), small stock	Low	Hansen 1992, Nowell and Jackson 1996
F. temmincki Asiatic golden cat	Asia	Vulnerable	WP: rodents, ungulates; L: small stock	Low	Nowell and Jackson 1996
Lynx lynx Eurasian lynx	Europe	Near threatened	WP: ungulates; L: small stock	Low	Breitenmoser *et al.* 2000

L. rufus Bobcat	N.America	Not listed	WP: lagomorphs; L: small stock	Low	Nowell and Jackson 1996
Panthera leo African/ Asiatic lion	Africa, Asia	Africa: vulnerable; Asia: critically endangered	WP: ungulates; L: cattle	Low–medium	Kruuk 1980, Saberwal *et al.* 1994
P. onca Jaguar	S. America	Near threatened	WP: ungulates; L: cattle	Low	Nowell and Jackson 1996
P. pardus Leopard	Africa, Asia	Africa: not listed–critically endangered; Asia: endangered–critically endangered	WP: ungulates; L: small stock	Low	Nowell and Jackson 1996
P. tigris Tiger	Asia	Endangered–critically endangered	WP: ungulates; L: cattle	Low–medium	Miquelle *et al.* 1999b
Uncia uncia Snow leopard	Asia	Endangered	WP: ungulates; L: small stock	Low–high	Oli 1991

[a] Conservation assessment according to the IUCN Red List 2002 (Red List: Table 4.3).
[b] Importance of domestic prey for the survival of the predator.

This may not be the same for all regions, as it depends on the technical and economic standard of a country, along with topographical and habitat factors and on the relative importance of carnivore conservation to the local society. We must consider such socio-cultural and economic differences when assessing the usefulness of individual techniques. Finally, we will discuss the implementation of non-lethal techniques in the wider context of carnivore conservation.

REVIEW AND ASSESSMENT OF PREVENTIVE MEASURES

The number of traditional and modern non-lethal techniques to reduce depredation is almost immeasurable. Variations on a common theme (shepherds, dogs and fences) have evolved into locally adapted solutions. Ideas based on new technology and biological understanding have been proposed and tested over and over again. Most of the experience has never been published in readily available journals. Table 4.2 summarizes the non-lethal techniques and the technical and descriptive literature, and Table 4.3 lists relevant URLs.

Zootechnical methods and livestock husbandry

Guarding livestock has been the natural response to depredation losses since the beginning of domestication. Where shepherds are present, losses are generally lower than in free-ranging herds (Kaczensky 1996; Linnell *et al.* 1996). Guided grazing allows the avoidance of depredation hotspots, penning stock at night, and quick and flexible responses to predator attacks. Shepherds are most effective in concert with herding and guarding dogs (Rigg 2001). In addition, this method allows the rapid detection of sickness and the disposal of carcasses that might attract scavengers. In developed countries, where large carnivores have been eradicated, farmers have developed husbandry systems where livestock range freely on mountain pastures and in forests, which expose the livestock to very high rates of depredation when the predators return (Linnell *et al.*, 1996; Breitenmoser 1998). Zootechnical options in a free-grazing system are limited. In regions with seasonal transhumance (seasonal movement of livestock between different regions), controlled reproduction is essential, so that no calving and lambing occurs on the exposed summer pastures, or during movement. Nevertheless, wherever large carnivores recover, shepherding must be re-established. This is, however, often limited in practice through high labour costs and lack of experienced shepherds. Consequently, the solution to depredation is sought in new techniques (Table 4.2). In the USA, 39% of the cattle, 88% of the sheep and 63% of the goat operations applied non-lethal prevention methods in 1999, summing up

Table 4.2. *References to non-lethal techniques used for carnivore damage prevention*

Method	References	Remarks on efficiency (E), manpower requirement (M) and cost
Technical measures		
Fences		
Natural fences	Charudutt 1997; Fitzwater 1972; Jhala 2000; Kruuk 1980[a]; Mizutani 1993	[a]2–10% losses due to depredation, of which only 10% in bomas.
Electric fences	Andelt 1996; Angst *et al.* 2002[a]; Bangs and Shivik 2001[b]; Bourne 2002; Dorrance and Bourne 1980[c]; Fitzwater 1972; Gates *et al.* 1978[c]; Gutleb 2001[d]; Huygens and Hayashi 1999[d]; Knowlton *et al.* 1999[e]; LeFranc 1987; Levin 2002[f]; Linhart *et al.* 1982[g]; Linnell *et al.* 1996[f, h]; Mertens *et al.* 2002[c]; Nass and Theade 1988; Thompson 1979; Timm and Connolly 2001	[a]E high for game in enclosures. [b]Electric fences economic for night corrals. [c]Costs $713–1125/km. [d]E high against bears at beehives. [e]Coyotes dig under fence; maintenance labour-intensive. [f]Cats difficult to exclude, good results with captive lynx. [g]E high against coyotes, costs $0.7–2.8/m. [h]High costs for construction and maintenance (keep vegetation low), obstruction of wildlife.
Conventional netting fences	DeCalesta and Cropsey 1978[a,b]; Linnell *et al.* 1996[c]; Thompson 1978[b], 1979[b], 1984[d]	[a]Cost $1543/km. [b]Properly built and maintained netwire fences E high against coyotes. [c]Inefficient for bears, cats difficult to exclude. [d]Dingoes significantly excluded from main sheep area.
Fladry	Maughan (Table 4.3)[a]; Musiani and Visalbergh 2001[b]; Okarma and Jedrzejewski 1997[c]; Rilling *et al.* 2002[d] Volpi *et al.* 2002[b]; WGF report (Table 4.3)[e]	[a]M: four persons mount 4 miles in 4 days. Fladry is short-term deterrent, but not permanent solution. [b]Wolves excluded from food source in captivity. [c]Wild wolves captured with fladries and nets. [d]Only short time effect on captive wolves. They crossed the fladries after a few hours. [e]Jureano Mts wolf pack crossed fladry after several months and killed a calf.

Table 4.2. *(cont.)*

Method	References	Remarks on efficiency (E), manpower requirement (M) and cost
Repellents		
Visual repellents	Boggess *et al.* 1980[a]; Bomford and O'Brien 1990; Green 1994 *et al.* 1994; Koehler *et al.* 1990; Linhart *et al.* 1984[b]; Robel *et al.* 1981[c]	[a]Placing lights over corrals. [b]Strobe lights with sirens prevented depredation for 53–91 nights. [c]Electronic guard frightening device ($250): NWRC, Table 4.3). Lights over corrals reduced depredation.
Acoustic repellents	Andelt 1996[a]; Bomford and O'Brien 1990; Koehler *et al.* 1990; LeFranc *et al.* 1987; Linhart *et al.* 1992[a] and Pfeifer and Goos 1982[b]; Robel *et al.* 1981[c]; Shivik and Martin 2001[d]	[a]Electronic frightening device (strobe light and siren) reduced sheep losses on high summer range by 60%. [b]Propane gas exploders easy to use and cheap; reduced coyote depredation for 1 to 180 days. [c]Bells over night corrals reduced depredation. [4]Radio-activated guard prolongs time-span for habituation.
Projectiles	Linnell *et al.* 1996; Smith *et al.* 2000b[a]	[a]Rubber bullet shotguns and 12-gauge plastic slugs immediate positive results with bears. Use only by trained persons on short distance; risk to injure or kill the predator.
Electric collar	Andelt *et al.* 1999b; Asher *et al.* 2001; Linhart *et al.* 1976[a]; Shivik and Martin 2001[b]; Shivik *et al.* 2002[b]	[a]Successfully tested in captive coyotes attacking sheep. [b]High costs; not tested under field conditions.
Conditioned taste aversion (CTA)	Burns 1980[a], 1983[a]; Connolly 1995[b]; Conover and Kessler 1994[a,c]; Conover *et al.* 1979[a]; Dorrance and Roy 1978[a]; Ellins and Catalano 1980[d]; Ellins *et al.* 1977[d]; Forthman 2000[d]; Gustavson *et al.* 1974, 1982[d]; Linnell 2000[a,e]; Linnell *et al.* 1996[f]	[a]No positive effect under field conditions. [b]Denver Wildlife Research Center stopped CTA tests with LiCl because no reduction of depredation. [c]After 10 years testing CTA (54% of farmers considered it successful), only one used CTA. [d]Positive results. [e]Logistic problems in application. CTA prevents consumption, but not attacking or killing. [f]Ambiguous results regarding E.

Chemical repellents	Burns *et al.* 1984[a]; Hatfield and Walker 1994[a]; Landa and Tømmerås 1996, 1997[b]; Landa *et al.* 1998b[b]; Lehner 1987[a]; Linnell *et al.* 1996[c]	[a]Chemical repellents appear to be ineffective against coyotes but effective to some extent against [b]wolverine and [3]bears. [b]Certain E on wolverine and bears. [c]Capasaicin (bearspray) showed best results with bears.
Protection collars with non-toxic repellents	Burns *et al.* 1984[a]; Burns 1996[b]	[a]No significant effect with different chemicals in sheep neck collars. [b]Vichos collars did not deter coyotes.
Protection collars	Angst *et al.* 2002[a]; Hill and Simper 2002[b]; Linnell *et al.* 1996[a]; Steinset *et al.* 1996 [a]; King Collar (Table 4.3)[c]	[a]No significant effect against lynx and wolverine. Costs €20–30 per collar. [b]Studies with King Collars against coyotes in the USA. [c]Efficient against jackal, but not against caracals. 170 000 King Collars (polyethylene collar) in use against jackals in South Africa. Costs $1 per collar.
Zootechnical measures		
Flock/herd management		
Shepherding	Andelt 1996[a]; Angst *et al.* 2002[a]; Coppinger and Schneider 1995[b]; Jackson *et al.* 1994[a]; Jorgensen 1979[a]; Kaczensky 1996[a]; Kruuk 1980[a]; Landry 1998[a,b]; Linnell *et al.* 1996[a,b]; Nass *et al.* 1984[a]; Odden *et al.* 2002; Poulle *et al.* 2000; Tigner and Larson 1977[a]; Wick 1995[a]	[1]Fewer losses with shepherds present. [b]Combination with LGD and night-time corrals recommended.
Hotspots avoidance	Angst *et al.* 2000, 2002[a]; Ciucci and Boitani 1998[a]; Fritts 1982[a]; Jackson *et al.* 1994[a]; Kaczensky 1996[a]; Muzitani 1993; Pearson and Caroline 1981; Rabinowitz 1986[a]; Robel *et al.* 1981[a]; Stahl *et al.* 2002[a]	[a]Carnivores kill livestock mostly in specific areas (steep, rocky areas, forest land, areas with tall grass and shrubs). Avoiding such areas reduced depredation.
Control of calving and lambing	Angst *et al.* 2002[a]; Dorrance 1982[a]; Stahl *et al.* 2001a[a]; Robel *et al.* 1981[a]	[a]Newborn livestock is highly vulnerable to most medium-size and large carnivores.

Table 4.2. *(cont.)*

Method	References	Remarks on efficiency (E), manpower requirement (M) and cost
Confining livestock at night (night pens)	Andelt 1996[a]; Angst *et al.* 2002; Dorrance and Roy 1976; Jackson *et al.* 1994[b]; Knowlton *et al.* 1999[a]; Kruuk 1980; Linnell *et al.* 1996[b]; Nass *et al.* 1984; Robel *et al.* 1981	[a]Crowding at night requires enhanced health care.[b]Night pens must be predator-proof, otherwise risk of excessive killing.
Disposal of livestock carcasses	Andelt 1996[a]; Fritts 1982; Jones and Woolf 1983[a]; Linnell *et al.* 1996[a,b]; Robel *et al.* 1981[a]; Todd and Keith 1976	[a]Disposing of carcasses (disease) reduced losses to predators.[b]Removing kills (predation) provokes additional kills.
Guarding animals		
Livestock guarding dogs (LGDs)	Andelt 1999a, b[a]; Andelt and Hopper 2000[a]; Bergman *et al.* 1998; Black and Green 1985; Clemence 1992; R. Coppinger, 1992; Coppinger and Coppinger 1982, 1994, 1995, 1998, 2001; Coppinger *et al.* 1983[c,f], 1985, 1987, 1988[a]; Coppinger and Schneider 1995; Green and Woodruff 1988, 1990[b]; Hansen and Bakken 1999; Hansen and Smith 2001; Landry 1998[b], 2001[e]; Lorenz 1985[b]; Lorenz and Coppinger 1986[b]; Lorenz *et al.* 1986[g]; Olsen 1985; Pitt 1988[f]; Rigg 2001; Rousselot and Pitt 1999[c]; Scott and Fuller 1965[c]; Serpell and Jagoe 1995; Vandel *et al.* 2001; Warbington 2000; [d]Wick 1998	[a]Efficient against all predators, work best in combination with shepherds. [b]Training (socialization) and taking care of the pups very time-consuming. [c]Socialization not easy; many LGDs do not correctly bond to flock. Best E in age of 2–3 years. [d]LGDs often disturb (play) flock when young; may even injure or kill lambs. [e]LGDs may frighten hikers. [f]LGDs may roam and hunt wildlife. [g]Premature death of LGDs.
Donkey	Andelt 1995[a]; Braithwait 1996[b]; Bourne 1994[c]; Landry 2000[d]; Linnell *et al.* 1996; Marker 2000a[g]; Tapscot 1997[h]; Donkey (Table 4.3)	[a]Good E against small predators, but not against packs or large carnivores. [b]Unneutered jacks can be aggressive towards sheep. [c]May disturb ewes when lambing or trample lambs. [d]Not all individuals are protective; test

		donkey with a dog before use. [g]Using donkeys younger than 1 year is not recommended. [h]Some plants are toxic for donkeys.
Llama	Andelt 1995; Angst *et al.* 2002; Bangs and Shivik 2001[b]; Braithwait 1996[a]; Franklin and Powell 1993[c]; Markham 1995; Meadows and Knowlton 2000; Warbington 2000	[a]Some E against small and medium predators, [b]but not against packs or large carnivores. [c]Using llamas younger than 1 year is not recommended.
Other species	Braithwait 1996; Marker-Kraus *et al.* 1996[a]; Pitt 1988	[a]Some anecdotal trials with zebras, cattle, baboons, stallions, horned oxen, mules, cows with calves and multi-species grazing.
Physiological measures and translocation		
Sterilization	Bromley and Gese 2001a, b; Conner 1995; Knowlton *et al.* 1999; Sacks 1996; Saunders *et al.* 2002[a]; Till and Knowlton 1983[b]	[a]Sterilization feasible with red fox.[b]Removing pups reduced coyote depredation.
Immuno-contraception	Balser 1964[1]; DeLiberto *et al.* 1998[b]	[a]Reproduction inhibited with diethylstilbesterol. [b]Immunocontraception using porcine zona pellucida (PZP) promising method to control coyote fertility.
Translocation	Funston 2001[a]; Linnell *et al.* 1997[a,b]	[a]Low to no effect. [b]Homing behaviour; most individuals continued depredation; reduced survival or reproduction after translocation; expensive.

Table 4.3. *Uniform resource locators (URL) to websites with additional information on non-lethal techniques*

Keyword	URL
Donkey	http://www.gov.on.ca/OMAFRA/english/livestock/sheep/facts/donkey2.htm
	http://www.agr.state.tx.us/pesticide/brochures/pes_donkeys.htm
Fladry	http://www.defenders.org/wildlife/wolf/idaho/fieldreports/fladry.html
Honey badger	http://www.badgers.org.uk/badgerpages/honey-badger-11.html
King Collar	http://www.kora.unibe.ch/en/proj/cdpnews/cdpnews007.htm King
Livestock guarding animals	http://www.awionline.org/farm/alt-farming3.html guards
Maughan	http://www.forwolves.org/ralph/fladry-jureano2.htm
National Wildlife Research Centre	http://www.aphis.usda.gov/ws/nwrc/index.html
Red List	http://www.redlist.org
Red wolf	http://www.epa.gov/docs/fedrgstr/EPA-SPECIES/1995/April/Day-13/pr-220.html
	http://southeast.fws.gov/pubs/alwolf.pdf
WGF report	www.defenders.org/wildlife/wolf/wolfupdate/fieldreports/fieldreport11.html

to almost $8 million in investment by the livestock owners. The National Wildlife Research Centre spent an additional $7.5 million on non-lethal management techniques (Shivik 2001).

Fencing

An anti-predator fence must keep out animals that not only can jump but also climb or dig. Different forms of protective fences have been developed worldwide, using natural materials or artefacts as thorn bushes, logs, bricks, wire and synthetics. Depending on the situation fencing can be used to (1) protect livestock in small night-time pens, (2) provide predator-proof grazing areas, or (3) exclude carnivores from entire regions. The type of fence varies according to the goals and materials available.

Fences made of natural material (wood, thornbush, earth, rocks, ditches filled with water) are mainly used for night-time pens (Linnell *et al.* 1996; Charudutt 1997). A famous example is the African boma (thornbush corral: Kruuk 1980; Ogada *et al.*, 2003). Conventional wire-netting fences are used worldwide in different forms, depending on the terrain and the target predator (Linnell *et al.* 1996). Tests in Canada and the USA have tried to exclude dogs, coyotes (*Canis latrans*), black bears (*Ursus americanus*), grizzly bears (*U. arctos*) and polar bears (*U. maritimus*), but conclusive assessments have been limited (Linnell *et al.* 1996). Properly built and maintained net-wire fences were effective in tests with captive coyotes (Thompson 1979). DeCalesta and Cropsey (1978) successfully implemented Thompson's (1979) recommendations in the field. Grizzly bears and black bears made holes in chainlink or woven fences, dug underneath, or climbed over (Clarkson and Marley 1995). The world's longest wire-netting fence is in Australia, a 5300-km fence meant to exclude dingoes (*Canis lupus dingo*) from the main sheep areas of southeastern Australia. The fence is not completely dingo-proof, but limits immigration and hence assists lethal control (Thompson 1984, 1986). Considering the jumping and climbing ability of felids, netting fences without additional impediments have limited effect against them. Experiences from the Lakipia case study have shown that wire fences are less effective than brush fences at excluding lions and hyaenas (Ogada *et al.* 2003; Frank *et al.*, Chapter 18).

Electric fences, causing a shock of several thousand volts, can improve existing wire-netting fences. One to three electric wires are added above or outside the fence. Such constructions prevented coyotes from entering pastures (Andelt 1996), and Eurasian lynx (*Lynx lynx*) from climbing into game farms (Angst *et al.* 2002). Plain electric wire fences with alternating live and earth wires must be adjusted in height, the number of and spacing between wires and voltage according to the carnivore that is to be excluded (details in Linnell *et al.* 1996). Multi-strand electric fences provide a cost-effective method for preventing coyote depredation (DeCalesta and Cropsey 1978; Dorrance and Bourne 1980; Linhart *et al.* 1982; Nass and Theade 1988). Electric fences are being widely and successfully used to protect beehives, livestock, garbage dumps and campsites from bears (Wade 1982; LeFranc *et al.* 1987; Clarkson and Marley 1995; Kaczensky 1996; Huygens and Hayashi 1999; Gutleb 2001; Levin 2002). Few studies address the efficiency of electric fences against felids, which seem to be difficult to exclude (Linnell *et al.* 1996). In South Africa, additional strands of electric wire atop jackal (*Canis mesomelas*) fences worked against caracals (*Caracal caracal*), and even leopards (*Panthera pardus*) (Bowland *et al.* 1993). Levin (2002) excluded Eurasian lynx in captivity from food resources using multi-strand wire fences. Solar-powered electric-fences are advantageous for remote areas, but like all electric fence systems, they need regular maintenance, which sometimes is a considerable problem limiting their efficacy especially if large ungulates like moose (*Alces alces*) or elephants (*Loxodonta africana*) occasionally destroy them (Knickerbocker and Waithaka, Chapter 14).

Fladries, brightly coloured flags sewn on ropes (Fladry: Table 4.3), have been used for hunting wolves (*Canis lupus*) in Europe for centuries. Most wolves fear fladries and rarely cross such barriers. Okarma and Jedrzejewski (1997) captured wild wolves, and Musiani and Visalberghi (2001) kept captive wolves away from food resource by means of fladries, although captive wolves crossed the fladries after a short time (Rilling *et al.* 2002). The practical value of this technique in preventing attacks is not yet known, but fladry is cheap, highly mobile, and easy to use even in difficult terrain (Musiani *et al.* 2003).

In summary, fences can successfully protect livestock. They are most useful and cost-effective on small pastures and as night-time corrals. Fencing off larger areas is not only less efficient and more expensive, but also problematic, as long fences fragment the habitat, and wire-nettings sometimes become deadly traps for terrestrial wildlife.

Livestock-guarding animals

Livestock-guarding animals (LGAs) are able to detect an approaching predator and to interrupt the attack sequence (Smith *et al.* 2000a) (see also Table 4.3). LGAs live constantly within the flocks (Coppinger *et al.* 1983). They are either potential prey (herbivores) acting in self-defence or potential competitors (carnivores) to the attacking predators. The attention given to the herd is the key to the success of all LGAs (Coppinger and Coppinger 1998).

Livestock-guarding dog breeds (LGDs) appeared 5000 years ago in Asia (Olsen 1985). Transhumance and human migration have led to some 50 breeds across the Palaearctic (overview in Landry 1998). LGDs can protect any domestic species (R. Coppinger 1992), also in combination with other preventive systems in different habitats (Landry 2001). Their weight averages 35–65 kg (Landry 1998). Smaller dogs are sometimes used (e.g. in Navajo country: Black and Green, 1985; or in Africa: Coppinger and Coppinger 1998) but they might not be able to fight off a large predator and they act more as an alarm dog. In Africa (e.g. Kenya: Ogada *et al.* 2003), some dogs do not really defend the herd, but they bark at approaching predators and alarm the herders.

Juvenile LGDs must be socialized with the domestic species they will protect during a limited time-window of their development (Scott and Fuller 1965). The methodology of socialization was described by Lorenz (1985), Lorenz and Coppinger (1986) and Green and Woodruff (1990). Play behaviour – often simulating stalking and attack – occurs from 3 to 14 months and occasionally later (Landry 2001), but it should disappear from the adult dogs' behavioural repertoire (Coppinger and Schneider 1995) if a dog is properly trained to not disturb the flock. Subsequently, the LGD behaves as a member of the flock, in contrast to the herding dog, which shows a predatory behaviour towards livestock (Clemence 1992). LGDs chasing domestic or wild animals may be the result of bad selection or bad training and these individuals should be removed (Landry 2001).

LGDs are selected for their ability to respond (e.g. by barking) to any strange event and hence to disturb the predator. Although the dog may fight with predators such as wolves, it is not necessary to select it for aggressiveness (Black and Green 1985). LGDs are most effective in concert with shepherds. Several LGDs are essential against cooperatively hunting predators, such as wolves, and in rugged or wooded terrain (Landry 1998). Typically, three to seven dogs can protect a herd of 1000 head of livestock. LGDs should judge the severity of a situation and show an appropriate reaction, especially in areas where there are tourists and their pet dogs (Landry 2001). The correct use of LGDs may provide effective livestock protection, although never at 100% (review in Rigg 2001).

Dogs need special food, compared to the stock they protect. The advantage of other LGAs is that they are herbivores, live longer and usually do not frighten hikers (Andelt 1995; Landry 2000). Experiences with zebras (*Equus* sp.), horse stallions, horned oxen, mules, baboons (*Papio* sp.), cows with calves, and multi-species grazing (e.g. mixed sheep and cattle herds) remain anecdotal (Pitt 1988; Braithwait 1996; Marker *et al.* 1996), but donkeys and llamas (*Lama glama*) have certain significance as LGAs. Both are social animals and integrate best into a sheep herd when kept without conspecifics (Andelt 1995). Socialization is simple (Franklin and Powell 1993; Andelt 1995; Marker 2000a) and donkeys and llamas can be introduced into a herd at any age. They react to disturbance within the flock and chase away any intruder (Andelt 1995). Sheep tend to gather around their protector instead of fleeing (Andelt 1995; Landry 2000). Donkeys and llamas work best in small, penned herds where it is easy to maintain an overview (Bourne 1994; Andelt 1995). They have proved effective against small predators like red foxes (*Vulpes vulpes*) or coyotes (Franklin and Powell 1993; Raveneau and Daveze 1994; Andelt 1995). Donkeys worked against jackals, caracals, and cheetahs (*Acinonyx jubatus*) in Namibia (Marker 2000a). Donkeys and llamas are however potential prey of larger carnivores themselves. They feared pumas (*Puma concolor*: Andelt, 1995), and llamas were killed by wolves (Bangs and Shivik 2001). The llama's practical efficiency therefore remains controversial (Linnell *et al.* 1996; Meadows and Knowlton 2000).

Deterrents and repellents

Numerous deterrents and repellents have been tested against many different carnivores, but only a few have produced practical results (Table 4.2). We distinguish two conceptual groups (Smith *et al.* 2000b): frightening devices and aversive devices.

Frightening devices are those that immediately disrupt a predator's action, such as flashes, sirens, projectiles, explosives, and aggressive chemicals (Mason *et al.* 2001). It is difficult to assess their efficiency because of the general lack of experimental controls (Bomfort and O'Brien 1990), but they can be useful tools for a limited period in small areas (e.g. Angst *et al.* 2002). The effect can be strengthened through varying position, appearance, duration or frequency of the frightening stimuli, but they should be used in emergency situations only, to avoid habituation of the predator. The latest technology is a radio-activated guard (RAG), where the radio-tagged predator triggers the device on approach (Shivik and Martin 2001). Effectiveness of frightening devices probably depends on the availability of alternative prey. This must be considered, especially if the alternative is not game but the neighbours' sheep. Regardless of their actual effectiveness, initial results indicate that they are popular with livestock producers, and therefore may help reduce some of the social aspects of the carnivore–livestock conflicts (Bangs, in press).

The second group consists of aversive devices, that activate a learned negative association after a link between an unwanted behaviour (e.g. attacking sheep) and a negative stimulation, through repeated negative experience. Gustavson *et al.* (1974) suggested that conditioned taste aversion (CTA) might be an effective prevention tool, but many pen and field tests with chemicals (mostly LiCl) produced conflicting and inconclusive results (Burns 1983; Conover and Kessler 1994; Linnell 2000). Linhart *et al.* (1976) and Andelt *et al.* (1999) demonstrated that commercial dog-training collars worked for conditioning coyotes in captivity, and Shivik *et al.* (2002) expanded this concept to wild wolves. However, this technique still faces many practical problems regarding its application in the wild, and it is only applicable on an individual level.

Physiological measures and translocations

Anti-fertility agents have been proposed as a mean to limit coyote populations (Balser 1964) and to reduce livestock losses, because non-reproductive coyotes would kill fewer livestock. However, anti-fertility agents have failed to control reproduction of coyote (Balser 1964; Stellflug *et al.* 1984) or red fox (Allen 1982) in the field, mainly because of limited bait consumption. Recently, the interest in reproductive inhibition using immunocontraception has been revived (DeLiberto *et al.* 1998), mainly to control red foxes (Saunders *et al.* 2002). A similar concept is the sterilization of territorial, breeding coyotes in areas of chronic depredation. The principle is based on research by Till and Knowlton (1983) which indicated that depredations can be reduced by removing coyote pups, assuming (1) that breeders are the principal killers of livestock, and (2) that they will continue to maintain territories (and keep conspecifics away) while sterile (Zemlicka 1995; Sacks 1996). However, Scrivner *et al.* (1985), Conner (1995) and Sacks (1996) argued that the presence of lambs, rather than pups, determined coyote predation patterns. Even if the measures work, it is hard to see how they can be applied over large scales (practicality and economics), or for carnivore species of conservation concern. A sterilized individual is in effect dead from the point of view of population viability, and also occupies space that a reproductive individual could occupy. Translocation of individual carnivores has been a standard management tool for decades to mitigate depredation (and other conflicts with large carnivores) mostly in North America and southern Africa (Treves and Naughton-Treves, Chapter 6). This method suggests that only a few specialists are stock-raiders, and that these individuals, when displaced into areas with reduced conflict potential, will not return. The success rate of translocations is relatively low. The removed problem animals have often (1) showed homing behaviour, (2) reduced survival, or (3) resumed their conflict behaviour at the release site (review in Linnell *et al.* 1997). Moreover, the concept of specialists is controversial. Linnell *et al.* (1999) hypothesized that most large carnivores at least occasionally kill livestock. Stahl *et al.* (2001a) and Angst *et al.* (2002) found that in certain 'hotspots', any lynx would kill sheep, and that the removal of the problem animal did not resolve the problem.

Indirect measures

The application of non-lethal techniques alone will never solve the conflict between livestock breeders and carnivores. Indirect measures are intended (1) to increase wild prey availability, and (2) to mitigate the economic consequences of depredation through compensation. Both approaches must be combined with carnivore damage prevention into a holistic, integrated concept (see below).

APPLICATION OF NON-LETHAL TECHNIQUES

In Europe, most livestock problems are supposedly caused by wolves, Eurasian lynx, brown bears (*Ursus arctos*) and wolverines (*Gulo gulo*). For instance, approximately €2 million is paid annually for livestock losses to predators in Italy (Ciucci and Boitani, 2000). In Norway, lynx killed about 19 000 sheep between 1990 and 1995 (Breitenmoser *et al.* 2000) and wolves in the French Alps have been responsible for the death of about

8650 sheep (at ~€1.65 million compensation) since their return to France in 1992 (Poulle *et al.* 2000; Dahier 2000, 2002). Compensation programmes have been launched to cover predator damages in about half of the European countries (Boitani 2000; Breitenmoser *et al.* 2000; Landa *et al.* 2000c; Swenson *et al.* 2000; Swenson and Andrén, Chapter 20), but livestock protection remains a more effective solution to alleviate conflicts between predators and humans. Shepherding, guarding dogs and fencing have been constantly used in European countries where large predators have never disappeared. In areas where they were exterminated, the knowledge of these techniques has been lost, and, as a consequence, people need to (re)learn how to coexist with the large carnivores. Farmers are encouraged through financial incentives to implement traditional and new livestock protection techniques, for example in countries such as France or Switzerland (Poulle *et al.* 2000; Weber 2000). Even smaller carnivores that are potentially harmful to livestock, like red foxes (Hewson 1984) or river otters (*Lutra lutra*: Bodner 1998), may be deterred from killing sheep or farmed fish, respectively, by non-lethal techniques (pers. obs.; Bodner 1998). However, despite financial support as well as demonstrated effectiveness (review in Rigg 2001; Angst *et al.* 2002), implementation of non-lethal preventive measures is not always popular. In Norway and Sweden, livestock owners still prefer to control the numbers of wolverines through hunting them instead of herding sheep and semi-domestic reindeer (*Rangifer tarandus*) or translocating wolverines as is done in Finland (Landa *et al.* 2000c; Swenson and Andrén, Chapter 20).

Coyote depredation on livestock is an important economic problem for farmers in North America (Green *et al.*, 1984). Although depredation on sheep varies spatially and temporally, it is assumed that coyotes kill between 1% and 2.5% of adult sheep and between 4% and 9% of lambs in the Western USA (Andelt 1992). Whether reintroduced in the Rocky Mountains or recovering naturally in the Great Lakes region, wolves have also an impact on livestock production in the USA. In total, 148 cattle and 356 sheep were killed by wolves in the Rocky Mountains from 1987 to January 2001 (Bangs and Shivik 2001), and the injury or death of 377 domestic animals caused by wolves was documented in Wisconsin between 1976 and 2000 (Treves *et al.* 2002). A wide spectrum of non-lethal techniques has been developed throughout the continent to reduce coyote and wolf damage. Guarding dogs have generally been successful in protecting herds (Rigg 2001). Besides increasing efforts to relocate depredating wolves, more non-conventional prevention measures such as conditional taste aversion, shock collars, sterilization (only for coyotes: Bromley and Gese 2001a, b) or frightening devices are currently being tested. Preliminary results seem relatively encouraging (Bangs and Shivik 2001; Rigg 2001). In North America, bear depredation may be locally important, but the use of LGDs and/or electric fences to protect livestock, and on some occasions beehives, as well as displacement of cattle closer to human residences, were rated effective by most producers (Jonker *et al.* 1998; Andelt 1999a). In contrast, removal or harassment of nuisance bears and use of repellents such as human urine or kerosene did not solve the problems (Jonker *et al.* 1998). Guarding dogs also proved effective to protect cattle and sheep against puma (Andelt 1999a), a species known to take its toll of domestic animals locally (Nowell and Jackson 1996).

Large predators are problematic for livestock throughout Africa, not only along the border of protected areas (Nowell and Jackson 1996) but also well outside reserves (Marker 2000a). In the late 1970s, East African pastoral tribes lost between 2% and 10% of their livestock per year, mostly to lions (*Panthera leo*), spotted hyaenas (*Crocuta crocuta*) and black-backed jackals (Kruuk 1980, 2002). Cheetahs, wild dogs (*Lycaon pictus*) and striped hyaenas (*Hyaena hyaena*) were also involved to a lesser extent (Kruuk 1980). Livestock was herded in daytime and fenced in traditional enclosures (bomas) during the night. On open-range cattle farms in Namibia, leopards and cheetahs each killed annually an average of 320 cattle and 375 sheep between 1986 and 1991 (Nowell and Jackson 1996). However, the use of donkeys to protect livestock, especially calving herds, reduced losses considerably as long as guidelines for successful implementation were properly followed (Marker, 2000a). An LGD programme was also launched and it apparently has brought some promising results so far (Marker 2000b).

Livestock predation is also a serious issue in Asia. Acute depletion of wild prey in most of their range forces large predators to feed extensively on domestic animals, and indeed depredation on livestock caused by Asiatic lions, leopards, snow leopards (*Uncia uncia*) or tigers (*Panthera tigris*) has increased inexorably (Seidensticker *et al.* 1990; Miller and Jackson 1994; Saberwal *et al.* 1994; Miquelle *et al.* 1999b). In some areas, wolves are also highly dependent on domestic animals (Kumar 2001). To protect livestock, the pastoral communities usually use night-time vigils and guarding dogs, and build thorn corrals (Jhala 2000). The effectiveness of guarding dogs remains, however, a controversial question (Jhala 2000; Kumar 2001). In India, compensation for livestock losses is paid under specific conditions (Nowell and Jackson 1996). However, Oli (1991) considered that damage compensation has the best potential to reduce conflicts between local people and snow leopards.

INCORPORATION OF NON-LETHAL TECHNIQUES INTO A WIDER CONSERVATION CONCEPT

Non-lethal techniques to prevent livestock depredation are studied, propagated and eventually implemented worldwide today to promote carnivore conservation. This happens in Europe and the USA, where large carnivore populations are recovering as a consequence of protective laws, wild prey increase and a consolidation in agriculture, leaving more space for wild animals, and also in developing countries, where the growing human population and domestic stock increases the pressure on natural habitat and carnivore populations. Although the need for non-lethal techniques is obvious in both (and all intermediary) situations, the socio-cultural and economic conditions for their implementations differ considerably. What may be operational in one place may not work somewhere else. Here, we want to explore the integration of non-lethal techniques into a more general conservation concept.

The strongholds of many carnivore populations are protected areas, which may even act as a source for neighbouring areas. The surroundings are multiple-use landscapes where carnivores and wild prey coexist with humans and their livestock herds. Such landscapes are used for a variety of agricultural activities, hunting and recreation. These are the conflict areas where human-caused mortality of the predators is high. Nevertheless, the extended multiple-use landscape is important for the survival of the population, as few protected areas are large enough to host viable populations of large carnivore species (Woodroffe and Ginsberg 1998).

The abundance of a carnivore in the multiple-use landscape depends on: (1) the availability of wild prey, (2) the accessibility of livestock, which together define the ecological carrying capacity, and (3) the tolerance of humans towards the predator ('social carrying capacity'). Compared to the protected area as a reference, the carrying capacity regarding wild prey is lower in the multiple-use landscape, but the carrying capacity of wild and domestic prey combined – and hence the potential abundance of the predator – is higher, because livestock normally reaches very high densities compared to wildlife. The social carrying capacity however is limiting, as people usually do not tolerate predator densities as high, or higher, than those within the protected area.

If we want to adjust the use of non-lethal techniques to basic conservation goals, the crucial questions are: (1) how important is domestic prey, and (2) how significant is retaliatory killing for the survival of the carnivore population? We can imagine three different scenarios. First, in the case where availability of wild prey is high and livestock are rarely killed,

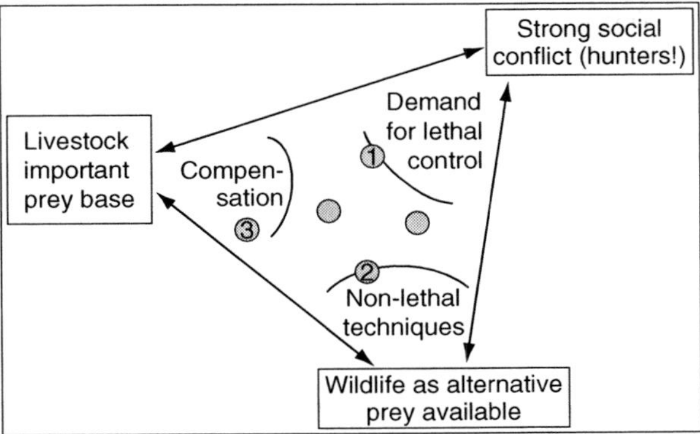

Figure 4.1. The 'depredation triangle' illustrates the cornerstones of the coexistence of large carnivores in a landscape used for livestock husbandry and other human activities such as hunting. The application of non-lethal techniques to prevent depredation will only be beneficial for the conservation of the population if wildlife as an alternative prey is available and if depredation is the main cause of the conflict. In practice, a balance between prevention, compensation and lethal control will be the likely solution for projects (grey dots). The numbers 1–3 refer to the approximate positions of the three examples given in the text.

depredation may not be frequent enough to encourage the habitual use of preventive measures (position 1 in Fig. 4.1). This is, for instance, the situation in the Swiss Alps, where lynx occasionally kill free-ranging sheep, but mainly prey upon roe deer (*Capreolus capreolus*) and chamois (*Rupicapra rupicapra*) (Angst *et al.* 2000). Due to the rareness of these attacks and the fact that government compensates for losses, the incentive for sheep-breeders to apply preventive measures is weak. Habitual stock-raiders are shot, a measure more useful for managing the controversy than for substantially reducing depredation (Angst *et al.* 2002). But as most lynx prey entirely on wildlife, the occasional removal of problem animals does not harm the population.

Second, where wildlife and livestock are both readily available, but domestic stock is preferentially killed, the introduction of preventive measures, forcing the predator to switch to wild prey, would support the conservation of the carnivore population (position 2 in Fig. 4.1). An example is the return of the wolf into the Alps of France, Italy and Switzerland. Although wild ungulates are abundant, the wolves preferentially kill the widespread, free-ranging sheep, which are an easy prey. The attacks on sheep herds provoked a major conflict and wolves were illegally killed. The implementation of preventive measures such as shepherds, LGDs, electric fences, etc. reduced the attacks considerably in the French Alps (Poulle *et al.* 2000), but does not hinder the further expansion of the wolf population, as an unexploited supply of wild prey is available.

Third, if wildlife is sparse and livestock readily available, predators inevitably prey on livestock. This was the case in parts of nineteenth-century Europe, when the wild ungulates had been almost eradicated, but the large carnivores still persisted (Breitenmoser 1998), and it seems today to be the situation for many carnivores, especially in northern Africa, the Middle East and Central Asia (see above), where wild prey have declined severely as a consequence of over-hunting, poaching and over-grazing by livestock (Karami 1992). It is also the case in northern Fennoscandia where semi-domestic reindeer are the only available ungulate prey for lynx and wolverine populations (Pedersen *et al.* 1999). The carnivore populations may suffer considerable losses from retaliatory killing, but the implementation of preventive measures would cut off the predators from their crucial food resource, still resulting in a decrease of the population. In such a situation, non-lethal techniques are not a real alternative to lethal control. The only solution would be to provide compensation that would motivate herdsmen to accept losses without retaliation, and to start a wild prey recovery programme where appropriate (position 3 in Fig. 4.1). In Turkmenistan's Kopetdag range, where a strong decrease of wild prey over the past 10 years has forced leopards to switch to domestic prey, the World Wide Fund for Nature (WWF) has established a compensation sheep herd, from which the local herdsmen can choose an equivalent in sheep for each cow, horse, sheep or goat lost to leopards (Lukarevsky 2002). The advantage of the compensation flock is not only that sheep remain a potential food resource for leopards, but also that the assessment procedure provides an additional monitoring of leopard presence.

In practice, compensation, non-lethal techniques and lethal control will not be exclusive alternatives, but rather concurrently applied management tools, and we must find the optimum balance for their application (Fig. 4.1), considering a wide scope of parameters defining the coexistence of predators and people (see also Treves and Naughton-Treves, Chapter 6). Depredation is often not the only cause of the human–carnivore conflict, and even successful implementation of prevention may not totally solve the conflict. In the example of the lynx in the Swiss Alps, attacks on sheep herds were prominently covered by the media (and hence taken seriously by the state authorities), but the conflict with hunters, who accused lynx of destroying local roe deer populations, was more threatening to the lynx population (Breitenmoser and Breitenmoser-Würsten 2001). Wherever we face a strong social conflict, the request for lethal control is strong (Fig. 4.1), and even the successful implementation of preventive measures may not result in a considerable decrease of illegal killing of predators. The combination of, and the balance between, non-lethal techniques (prevention), compensation of losses, and lethal control must be adjusted according to an assessment of any given situation, considering not only depredation, but also wild prey availability and underlying socio-cultural conflicts.

CONCLUSIONS

Traditionally, herdsmen across the world have applied preventive measures because lethal control of the predators alone was not sufficient to limit the losses of livestock. Today, the (re-)introduction of old and new non-lethal techniques aims to mitigate the eternal conflict between predators and herders in order to minimize the killing of carnivores. The motivation is often, especially in Europe and North America, ethical scruples about killing animals. Sophisticated technical or physiological methods have been developed and tested, sometimes intended to change the individual predator's behaviour. Such techniques will have very limited significance for the conservation of threatened taxa or populations, especially in developing countries. In order to be conservation-effective, non-lethal methods must be acceptable (in accordance with local traditions) and applicable on a large scale (cheap and easy to use). Furthermore, the implementation and application of non-lethal techniques must be considered in the context of the conservation goals and all other management actions. Preventive measures may not always be the cheapest, and sometimes not even the most effective way of conserving a carnivore population. Non-lethal techniques, however, offer the opportunity to involve herders actively into carnivore management, and they can – in combination with other conservation measures such as compensation of losses and removal of habitual stock raiders – help to mitigate the conflict and hence to promote the coexistence of carnivores and people.

ACKNOWLEDGEMENTS

The work of C. Angst, U. Breitenmoser, Ch. Breitenmoser-Würsten, J. M. Landry and J. M. Weber was supported by the Swiss Agency of Environment, Forest and Landcape and the Federal Office for Agriculture; the work of J. Linnell by the Norwegian Directorate for Nature Management.

5

Techniques to reduce crop loss: human and technical dimensions in Africa

FERREL V. OSBORN AND CATHERINE M. HILL

INTRODUCTION

Crop-raiding by wildlife is neither a new phenomenon nor a rare one. Farmers throughout the world are faced with trying to reduce or eradicate the impact of crop damage by wildlife to their standing crops. Insects, birds, rodents and ungulates are perhaps the most common sources of such conflict and there is a growing literature that details various strategies for reducing crop losses (Fiedler 1988; Giles 1989; Adesina *et al.* 1994); describes factors that increase farmer tolerance to losses (Decker and Purdy 1988; Messmer 2000); and discusses the introduction of systems for compensating farmers for losses incurred (Rollins and Briggs 1996; Nyhus *et al.* 2003). For example, wild ungulates and rodents cause an estimated $60-million-worth of damage to forest plantations annually and bird damage to agricultural crops is a multi-million-dollar problem in the USA (Dolbeer *et al.* 1994). Perhaps not surprisingly, then, rodents, invertebrates and birds receive the greatest attention within the literature on pest management. However, other animals – some of them threatened species – may also pose a considerable threat to farmers' livelihoods. Two types of mammal pests, elephants and primates, exemplify the complexities of managing intelligent and potentially dangerous crop pests, which are also of conservation concern. The problem may be chronic or sporadic, predictable or sometimes unpredictable (at least at the level of the individual); whatever the pattern, crop damage threatens the livelihood security of farmers living near wildlife.

In Africa, animals such as wild pigs (*Potamochoerus* spp.), elephants (*Loxodonta africana*), hippopotamus (*Hippopotamus amphibius*) and certain species of primate are commonly cited as problem animals where they live in proximity to agricultural areas (Shafi and Khokhar 1986; Conelly 1987;

People and Wildlife: Conflict or Coexistence? eds. Rosie Woodroffe, Simon Thirgood and Alan Rabinowitz. Published by Cambridge University Press.

Boulton *et al.* 1996; Kharel 1997). This contributes significantly to farmers' negative attitudes towards wildlife and conservation (Conover and Decker 1991). However, where revenue from wildlife is distributed to rural communities, there is some evidence that such negative perceptions toward animals that crop-raid can change (Archabald and Naughton-Treves 2001). Rural communities frequently have an influential political voice and crop-raiding often becomes a 'flash point' for a range of local issues, such as access to land and resources (Barnes 1996). Understanding the reasons why this conflict persists and why most methods to reduce crop loss have failed is central to developing effective strategies for reducing the impact of this constraint on farmers' livelihoods.

The zone between wildlife areas and agricultural land is a complex and dynamic place. Bell (1984a) has described the human–animal interface as 'a rolling zone of attrition' in which wildlife has largely been eliminated, but utilized in the process. Areas available to large mammal species in Africa have been reduced as the human population has increased over recent decades. Consequently, large mammals have become confined to either state-protected areas and/or contiguous multi-use zones where restricted hunting is allowed. The unfenced boundaries of these zones allow movements of wild animals in and out of these areas to forage, and also allow humans to enter wild areas to gather plants or hunt animals (Naughton-Treves 1997). Historically, local people were able to hunt crop pests as a form of compensation. These activities are now largely illegal yet farmers who lose crops often feel that they should have rights to hunt in protected areas as recompense for the damage caused by wild animals.

Most wildlife, and particularly protected species, are generally perceived by rural people as property of the State. State institutions that manage protected areas are, therefore, considered responsible for control of 'their' animals. Associated with this attitude is rural peoples' expectation that wildlife must be contained within protected areas, and that the primary function of game guards and wildlife authorities is to keep wildlife away from areas occupied by people. Wildlife authorities in Africa are generally financially and logistically ill equipped and unable to keep people and wildlife apart to this degree, and hence are often blamed for losses to crops and property.

Further, ethical considerations have become increasingly included in the decision-making process regarding 'pest management'. Endangered animals that are also pests present a particularly troublesome problem as certain sectors of society may want a species to be preserved, while communities affected by the damage want them eliminated. This can further exacerbate human–wildlife conflict where protected species are in conflict with cultivators because 'not only are farmers more restricted in the ways they are able to respond to crop raiding, but their methods of crop protection may also suddenly come under scrutiny from outsiders, and this can further fuel the tension between rural populations and conservation agencies' (Hill, in press).

THE BASIS OF CONFLICT

Elephants and primates consume all human food plants and those fed on by domestic herbivores because cultivars typically have greater nutritional value and lower toxicity than do wild plants. In dietary terms, elephants and primates therefore compete with people directly for their food plants, and indirectly by using the same resources as livestock (Parker and Graham 1989). There are many ecological similarities between elephants, primates and people. Graham (1973) notes that in the case of the conflict between elephants and humans, 'we have one of the best examples of two superficially dissimilar animals sharing common biological needs and therefore competing vigorously whenever they come into contact'.

Primates and elephants pose particular problems to farmers and are often cited as major crop pests (Mascarenhas 1971; Horrocks and Baulu 1994; Strum 1994). They are highly social and intelligent animals and their excellent communication and cognitive skills, combined with dietary and behavioural flexibility, make them extremely adaptable and effective crop-raiders (Else 1991; Barnes 1996; O'Connell-Rodwell *et al.* 2000). The manual dexterity and extreme agility of many primate species, and the large size, great strength and nocturnal/crepuscular activity patterns of elephants mean that primates and elephants present an especially formidable challenge to cultivators.

The environmental impacts of crop loss should also be taken into consideration. Farmers often burn large fires at the edges of fields to discourage crop-raiders, wasting fuel wood. Additionally, farmers clear forest remnants near fields because they believe these islands act as a refuge for crop-raiding animals such as primates and elephants. There is also a significant impact of non-specific defences such as poisons, snares and traps. Poorer yields as a consequence of crop damage by wildlife can result in the need to cultivate progressively larger areas to meet basic nutritional requirements (Osborn and Welford 1997). Hence, crop damage can contribute both directly and indirectly to increased land clearance and habitat change.

PREDICTORS OF RISK

Various factors contribute to a farmer's risk of experiencing crop damage. The position of fields in relation to wildlife refugia, the type of crops grown and the time crops ripen all influence primate raiding patterns (Naughton-Treves 1996; Hill 1997, 2000). Fruiting of wild food resources influences when and why certain farmers are raided (White *et al.* 1993; Naughton-Treves *et al.* 1998). Additional factors include the behavioural ecology of the pest species, their migration patterns and their need to access certain resources such as water and minerals (Sukumar 1991; Hoare 1999; Hill 2000). Not surprisingly studies note considerable temporal and spatial variation in crop-raiding.

There also may be a range of influences on elephant raiding behaviour. For example, in Zimbabwe, the seasonal availability and quality of ephemeral food items (grass and forbs in southern Africa) appear to act as a trigger for elephants to begin raiding (Osborn 1998). Elephants may also raid when they have been attracted to fruiting trees in the vicinity of fields (White *et al.* 1993) or when preferred tree species have been depleted from protected areas (Osborn 2002b). Hoare (1999) examined the possibility of predicting crop-raiding by elephants on a landscape level and concluded that crop-damage incidents represent opportunistic feeding forays by a specific segment of the male elephant population, and the intensity thereof is dependent on the behaviour and ecology of these individuals. In rainforest habitats, the secondary forests associated with agriculture are the preferred habitat for elephants, thus increasing the possibility of direct conflict (Dunn 1991; Barnes 1996). Wildlife pests tend to raid crops opportunistically and enter fields that are poorly defended; thus crop loss is frequently negatively correlated with farmers' vigilance (Osborn 1998). For example, in Gabon, over one-third of farmers whose crops were destroyed by elephants had done nothing to deter them (Lahm 1996). At the risk of stating the obvious, being aware of the presence of crop pests may be the key to improving the effectiveness of control methods.

CONTROLLING THE DAMAGE

To 'control' problem wildlife simply means to reduce the impact of the wild animal on the contested resource to within tolerable limits (Monaghan and Wood-Gush 1990). Wildlife damage control is, essentially, the art of trying to reduce the impact of a particular species' natural habits on an item that humans value (Dolbeer *et al.* 1994). Animal damage control must be a means to accomplish an objective and never an end in itself (Hawthorne 1980). If the objective is to 'reduce the conflict', then raising tolerance of the damage is as important as reducing the damage itself.

In the past, efforts were made to destroy 'pest' species entirely, but this often proved to be more difficult than it first appeared (Treves and

Naughton-Treves, Chapter 6). For example, thousands of elephants were shot in the forests of Uganda because they fed on tree species that were valued as timber (Laws *et al.* 1975). Even under sustained hunting, elephants could not be entirely eliminated from the forests. Smaller and faster-reproducing creatures are nearly impossible to destroy completely (e.g. *Quelea* birds). Eventually, a cost–benefit analysis must dictate the value of a sustained assault on pests. The outcome of such analyses is often acceptance of the impact of crop-raiders, rather than elimination of the problem. For many vulnerable species, the ultimate outcome of sustained lethal control is eventual extinction from the sustained reduction of numbers (Woodroffe *et al.*, Chapter 1).

Different types of animals pose different problems to farmers when they raid crops. The species involved can determine not just the types of crops at risk but also the plant part or stage in the crop's development. This is a reflection of dietary preferences, body size, dexterity and food processing capabilities of different wildlife species (Hill, in press). Similarly, animals' activity patterns and ranging behaviour influence the types of crops damaged, and particularly daily and seasonal patterns of crop damage, which can have a significant impact on the degree to which a farmer's coping strategies are effective or not (Sukumar 1990).

In some cases, a pest animal is legally protected or simply cannot be non-lethally repelled from fields. In developed countries a multifaceted approach is often adopted where farmers are compensated for their loss, attempts are made to repel pest species and people are made aware of the positive values of living with problem animals (e.g. Bangs *et al.* Chapter 21). These points highlight not only the technical problems with controlling predation, but the social costs as well.

Theoretical research into the foraging strategies of insects, birds and mammals has helped wildlife managers to minimize or eliminate crop-raiding in a number of situations (Monaghan and Wood-Gush 1990). Understanding how crop-raiding behaviour fits into overall foraging strategies and the ecology of pest species populations will help managers develop methods to resolve this conflict. The technique selected to limit crop damage is often specific to the species involved. Thus, the first step in developing a strategy for controlling pests' impact is a general understanding of the ecology of the target species. Once this has been achieved, two broad methods of control are available to wildlife managers: lethal action against the pest species, and non-lethal action.

LETHAL CONTROL METHODS

Control hunting was a common feature of crop protection strategies set up and administered by Game Control Departments in African colonies. More recently, there has been a move by government authorities to reduce the numbers of wild animals killed in crop protection, and lethal control is now usually used for animals that are common, such as certain rodents, small carnivores and some bird species, with much more restricted use against larger-bodied and more threatened species. The response by some wildlife management authorities to crop-raiding is to send personnel to the location to assess the extent of the damage and then attempt to kill one or more from the problem group. Centralized Problem Animal Control (PAC) units have been established across Africa and they are tasked with reacting to reports of crop-raiding among communities. For elephants, it was generally believed by wildlife authorities that shooting one at night while raiding is the best way to 'teach' the other elephants to stay away. While this method is still practised to some degree throughout much of the elephant's range in Africa, most wildlife managers think it is generally not effective as a long-term solution and is a drain on a valuable resource (see also Treves and Naughton-Treves, Chapter 6). In many situations the elephant(s) responsible for the majority of the damage cannot be identified, and a token animal is killed. These PAC units are constrained by logistical issues such as lack of transport, or by being unable to reach remote settlements during the wet seasons.

For primates, hunting is occasionally reported as a means of reducing local populations and thus reducing their potential impacts on local farms (FitzGibbon *et al.* 1995), though the degree to which this is a successful means of reducing crop losses from these animals is unclear in many cases. Naughton-Treves (2001) reports that hunting reduced the impact of bush pigs on farms and even on neighbouring farms where the farmers themselves do not hunt. However, in this study it was not clear whether hunting provided an effective method of control against primate raiders, even though 15% of the people interviewed admitted to hunting, trapping or using poison baits against primates. Where farmers opt to use lethal control methods that are not species-specific, such as traps or poisoned bait, measures targeted at common species can, and indeed do, affect protected species. A well-documented example is the impact of snares on the Budongo Forest Reserve chimpanzee population, where a significant proportion of the resident chimpanzees show evidence of permanent limb damage from snares (Waller and Reynolds 2001).

In Barbados, trapping for biomedical research has been used to control introduced primate populations that are in conflict with farmers. Control trapping maintains vervet monkey populations at a level where losses to primate crop raiding are tolerated. Additionally, it generates income which helps offset the negative feelings local people hold against these animals (Horrocks and Baulu 1994). In Japan, large numbers of macaques are shot each year as part of a government control programme to protect orange groves and other agricultural crops (Sprague 2002).

While large-scale control shooting or trapping can provide an effective means of reducing conflict between primates and farmers under particular circumstances, it can be a costly venture and needs to be continued and regulated over a period of time, not least because primate socio-ecology suggests that when prime resource-rich habitats are vacated, as would be the case in control shooting, removed individuals are quickly replaced. Unless incoming animals have not yet acquired crop-raiding behaviour, and are prevented from doing so by assiduous crop-protection activities, then farmers are soon faced with the same problem all over again (Else 1991; Strum 1994). Primates learn well in a social context: thus, new animals are likely to learn through observation of, and association with, animals that already practise crop-raiding behaviours. Primates and elephants also quickly habituate to 'empty' threats such as scarecrows, drum beating and shouting (Thouless 1994; Nyhus *et al.* 2000). For example, in some cases persistent elephant bulls are not deterred by gunfire, even when one of the group is shot (R. Martin pers. comm.). Hoare (2001) has speculated that there is a 'problem component' where regardless of how many problem elephants are removed others will always replace them. Osborn (1998) suggested that crop-raiding is a learned behaviour, taught to young bulls by older bulls within male groups. Killing the leaders or the propagators of these behaviours can reduce conflict temporarily; however, aversively conditioning these leaders has also proved effective in changing the behaviour of crop-raiding groups (Osborn 2002a).

In summary, there is little published evidence that lethal control reduces the impact of crop-raiding unless all of the pest animals are removed (see also Treves and Naughton-Treves, Chapter 6). Killing appears to be a temporary measure prior to other animals taking up the activity again or new individuals recolonizing the vacated home range.

NON-LETHAL CONTROL METHODS

Guarding and scaring

Non-lethal methods of controlling animals perhaps requires an even more detailed understanding of the behaviour and requirements of pest animals than does lethal control. Most studies have focussed on people guarding against these animals by patrolling fields and chasing away any animals observed in the fields (Maples *et al.* 1976; Strum 1994; Hill 1997; Osborn 1998). The most common way that rural farmers across Africa attempt to chase mammal crop pests out of fields is by shouting and banging objects to make loud noises. This is known as the 'drive them away' defence. Farmers use a range of noise-makers, such as beating drums and tins, 'cracking' whips in addition to yelling and whistling to chase away wild animals. These noises are usually accompanied by fires, either located on the boundaries of fields or as burning sticks that farmers carry with them. Farmers may also throw rocks, burning sticks and, occasionally, spears. For example, in Uganda, people use dogs, spears, bows and arrows, and bells to help scare away raiding animals, and sometimes work cooperatively, chasing away intruders from their neighbours' fields as well as their own (Hill 1997).

The degree to which these methods are effective is often unknown and difficult to quantify. Several studies have investigated farmers' perceptions of the effectiveness of different methods, but few have carried out a detailed, independent assessment of the effectiveness of traditional crop-protection strategies such as guarding, throwing rocks and using noise-makers. However, Naughton-Treves (2001) suggests that, in Uganda, intensive guarding is at least partially successful because primates avoid farms that are intensively guarded. Guarding appears most effective when people actively threaten primates through use of spears and bows and arrows, or by throwing missiles or using slingshots (King and Lee 1987; Priston 2001). Indeed, many farmers regard guarding as a necessary crop-protection activity, especially against diurnal primates (van Oosten 2000; C. M. Hill unpubl. data). King and Lee (1987) suggest that a guard known to the monkeys as dangerous is likely to be more effective in their guarding than those that are either unknown or known not to carry through a threat. This fits with the various studies that report that men are more successful than women and children as guards and that particularly armed men are most successful. Baboons retreat more readily when approached by men on farms as compared with women or children, and people carrying bows and arrows or spears are generally avoided and cannot get as close to the animals as unarmed people (C. M. Hill unpubl. data).

Guarding fields by day for primates and/or by night for elephants has considerable social implications for families living with wildlife across Africa (see also Thirgood *et al.*, Chapter 2). Children and adolescents are particularly affected, either by losing sleep or by the danger associated with walking to school early in the morning when elephants are still in the fields. People who guard fields are also at an increased risk of being injured by wild animals, and of contracting malaria, as they must sit outside through the night when *Anopheles* mosquitoes are active. Guarding also represents a considerable commitment of people's time and may contribute to seasonal bottlenecks in household labour requirements (Hill 2000), with children being removed from school to assist in seasonal guarding for example

(L. Naughton-Treves pers. comm.). Female-headed households may also be particularly vulnerable where there are social and cultural rules that prevent women from carrying out particular tasks.

Fences and barriers

Systems that exclude pest animals involve placing barriers that prevent access by pest species into areas or structures. Fencing is the most widely used means to keep animals out of agricultural or pastoral lands, or within a sanctuary. Depending on the species, fence design and materials can be very expensive and must be constantly maintained. Animals such as wild pigs and most carnivores can dig under most fences, thus barriers targeting these species must be partially buried, thereby adding to the construction costs. Primates quickly learn how to negotiate electric fences with impunity and walls and ditched areas are ineffective because of primates' agility and climbing skills (Strum 1994). However, electrified fences are used successfully at several chimpanzee sanctuaries in Africa to retain animals within large enclosures (A. Kabasawa pers. comm.). Areas have to be kept clear of vegetation on either side of the fence to prevent the animals using overhanging branches to move across the fence. This suggests that such fencing might be a more useful option for restraining larger, less agile animals such as great apes than for baboons (*Papio* spp.) and guenons (*Cercopithecus* spp.) that are able to jump considerable distances. Strong, non-electrified fences have also been used successfully to restrict elephant movements in many parts of Africa and Asia. These fences are usually built with wooden or steel poles or railway sleepers driven vertically into the ground. Heavy-gauge wire or cable is strung between the poles and drawn tight.

Electric fencing and trenches have met with success in protecting small areas or cash crops from elephants (Blair *et al.* 1979; Taylor 1999). However, the cost of materials, installation and maintenance make these methods impractical for large-scale application in developing countries unless funded by international aid agencies. It is generally recognized that this approach is unsustainable due to theft of materials and improper maintenance. Thouless and Sakwa (1995) conclude that elephants can overcome modifications to fences, implying that design, construction and voltage do not determine a fence's effectiveness. Consequently, an expensive 'arms race' can develop between managers and elephants which are able to adapt quickly to new fence features designed to electrify the parts of the fence that elephants destroy.

Repellent systems

Other non-lethal methods include systems that repel or deter animals. It is necessary to make a distinction between a repellent (something that wards off or repulses) and a deterrent (something that hinders or discourages). Hunt (1983) offered two definitions in the context of capsicum spray research on bears:

repellents are activated by humans and should immediately turn an animal away in a close approach or attack. A deterrent should prevent undesirable behaviours by turning an animal away before a conflict occurs. Deterrents need not be monitored or manually activated by humans.

There are three basic types of repellents, all of which attempt to frighten an animal from an area to be protected: visual, acoustic and chemical repellents.

Visual repellents, i.e. 'scarecrows' of various types, are the simplest form of repellent, and are used mainly against birds. Scarecrows using either dolls or dead primates appears to meet with some initial success, but animals soon habituate to them (van Oosten 2000; Saj *et al.* 2001). There is no evidence to indicate that visual repellents alone have any significant impact on crop-raiding by elephants or primates.

Acoustic repellents are commonly used for birds and include propane cannons that produce loud explosions at timed intervals. Recorded alarm or distress calls of birds broadcast over a speaker system can also frighten birds (Bomford and O'Brien 1990). Kangwana (1993) used playback recordings of Maasai cattle noise to repel elephants in Amboseli National Park, Kenya. O'Connell-Rodwell *et al.* (2000) found that the playback of elephant sounds had no deterrent effect on elephants tested around waterholes in Namibia. However, advancements in technology and the reduction of the cost of these units may present opportunities for wider application of acoustic deterrents in the future. The problems with using elephant vocalizations as a repellent are: (1) most are of very low frequency and thus require expensive equipment to record and play back; (2) a large repertoire of recordings would probably have to be used to avoid habituation; and (3) the potential exists for disrupting these animals' communication and social systems.

Long-lasting, passive chemical repellents have been used extensively for the control of crop damage by insects, birds and mammals. A variety of non-lethal repellents specific to birds have been used to protect important agricultural crops (Avery 1989; Mason 1989; Nolte *et al.* 1993). For example, apples poisoned with methiocarb are effective as a repellent to blackbirds by eliciting conditioned (i.e. learned) avoidance of treated foods via post-ingestational malaise (Conover 1984). When 'Avitrol', a chemical widely used for bird pests, is ingested, an affected bird emits distress cries while flying in erratic patterns before dying. This behaviour frightens the rest of the flock away from the targeted fields for some time (Dolbeer *et al.* 1994). However, tests with HATE C4 (an effective deer repellent) sprayed onto crops had no effect on raiding by either elephants (Bell 1984) or primates (Strum 1994).

Forthman-Quick (1986) investigated the possibility of inducing taste aversion in crop-raiding primates. She injected baboons with an emetic, lithium chloride (LiCl), inducing maize avoidance for up to three months after a single application. However, primates taste salt at the concentrations necessary to induce vomiting so for this research the animals were trapped and injected with LiCl. While Forthman-Quick demonstrated that taste aversion might be an effective strategy to use against primates, she suggested other substances such as ethyl estradiol could be tested as this substance has apparently been used with other animals successfully and can be administered safely at low doses (Forthman-Quick 1986; Forthman-Quick and Demment 1988). In another study in Japan, primates were given novel foods over several trials and then injected with cyclophosphamide. On release back into the wild the experimental animals avoided novel foods associated with vomiting, and in addition some animals were seen to avoid the other control foods (Matsuzawa *et al.* 1983). Consequently taste aversion might be worthy of further investigation with respect to problem primates, though any such investigations would require very careful design to ensure that the animals were responding to the appropriate stimuli.

Capsicum repellents

One of the most promising chemical repellents for a number of pest species is a concentrate made from fruits of the *Capsicum* genus. This compound has been used successfully to repel attacking bears and to aversively condition habituated problem bears (Hunt 1983, 1985; Smith 1984). Capsicum powder has been tested on refuse dumps in Amboseli National Park (C. Moss pers. comm.) and applied to fence-posts to stop elephant damage in the Laikipia district; neither trial was reported to have yielded conclusive results (V. Booth pers. comm.). Neither of these tests attempted to atomize or distribute the capsicum in the air, but more recent work has developed a procedure for delivering capsicum to crop-raiding elephants in an aerosol vapour cloud that renders capsicum an effective deterrent (Osborn 2002a). Elephants have been repelled from fields by burning a mixture of capsicum powder and elephant dung. Additionally, a mixture of grease and capsicum, applied to string fences, reduced the regularity of incursions by elephants into fields (Osborn and Parker 2002).

Strum (1994) and Chalise (2001) have suggested that use of capsicum directly on crops may have some effect on baboons, macaques and langurs because it reduces crop palatability. Food items and plants were sprayed with a capsicum solution made by boiling fresh chillies in water, as part of an attempt by personnel of the Colobus Trust in Kenya to reduce the local conflict between people, baboons and vervet monkeys; however, the primates were observed to eat the sprayed food items (J. Anderson pers. comm.). In Cameroon, farmers used capsicum in ashes spread on primate paths but there is no evidence to show whether it is effective or not (van Oosten 2000). Therefore, evidence for capsicum derivatives being an effective deterrent against primates is mixed but the concentration of the solutions used and the mode of delivery are likely to affect the success of any such trials. To understand how these repellents work, it is vital to understand what constitutes a 'threat' to the target species/animals.

TRANSLOCATION

The removal of a problem animal through translocation is widely used with large carnivores such as leopards and lions in Africa, tigers in Asia and bears in North America (Treves and Naughton-Treves, Chapter 6). A number of conservation organizations have suggested translocation as an option for crop-raiding elephants, but this technique is expensive, and often the animals return to the areas from which they were removed (Lahiri-Choudhury 1993). In Kenya, the wildlife authority made a decision that translocation was a central part of the country's conflict-reduction efforts, but currently no information has been published that indicates that this is an effective strategy.

Translocation for primates, aside from being a costly venture in terms of time and money, and highly stressful for the animals concerned, is unlikely to be successful, except under very specific conditions. Any primate troupe being translocated must be moved to an area that not only has suitable habitat for the animals concerned, but is also empty of cultivators (otherwise the problem of crop-raiding is just transported to a new area and there is the additional risk that the incoming animals will introduce this feeding strategy to any naive animals in the area). In addition, there must be adequate natural resources such as food, water and sleeping sites available to accommodate the incoming group as well as any groups already resident at the site.

Many translocation efforts require the recapture of animals, making the process prohibitively expensive. Though some translocation schemes have apparently met with success (Strum and Southwick 1986), they are never without significant expense (e.g. $500 per animal: Forthman-Quick 1986). Consequently, translocation might be appropriate only as a last resort under extraordinary conditions, e.g. where the animals concerned are from a highly endangered species, and there is a suitable site available for moving the animals to.

THE WAY FORWARD

No single management option will stop all problem animal-conflict situations. Centralized interventions suffer from logistical problems, and 'traditional' methods are generally ineffective. Most interventions aimed at reducing crop loss come from organizations outside of the affected community, which include government wildlife departments and external development organizations. Farmers expect the conflict to be resolved and, when it is not, often turn against the agencies involved. Donor-funded technical solutions are often not sustainable because of the high maintenance costs which external agencies are reluctant to provide. Consequently, methods need to be within the financial and technological capacities of the people implementing them, if they are to provide long-term solutions (Kangwana 1995).

The central theme that emerges from examination of intervention failures is the need to decentralize responsibility for crop protection to the farmers. This represents a considerable shift in thinking for farmers who have historically depended on centralized PAC units to reduce this conflict. This shift may only be achieved if farmers are then presented with locally appropriate schemes that can provide a tangible reduction in crop loss. The problem with decentralizing the responsibility for crop pests is when this is not matched by devolving the authority to tackle the problem. For example, if the state retains all the authority, such as the laws that ban the killing of wildlife, then decentralizing the responsibility merely aggravates the problem.

CONCLUSION

Primates and elephants continue to threaten farmers' income and food security across many African sites, despite considerable research and resources that have been devoted to resolving this problem. Wildlife managers can kill the pests, or utilize a range of non-lethal techniques to reduce crop loss. Conflict persists in many wildlife-rich areas due to a combination of four factors: namely, deficiencies in one-off technical solutions, lack of farmer vigilance and cooperation, habituation of animals to any one method and the high human and social costs of living with wildlife. As there are no one-off non-lethal technical solutions, we argue here that the problem can be addressed only through an 'adaptive management' approach. It is essential to bring farmers into the process of solving conflict by taking responsibility for the problems of crop pests. Such an approach is likely to be more successful, and more sustainable in the long term, than programmes that are dependant on external funding and technical support.

ACKNOWLEDGEMENTS

The authors wish to thank the Wildlife Conservation Society for funding this research and Simon Anstey, Phyllis Lee, Lisa Naughton-Treves, Guy Parker, Shirley Strum and Lucy Welford for comments on this chapter.

6

Evaluating lethal control in the management of human–wildlife conflict

ADRIAN TREVES AND LISA NAUGHTON-TREVES

INTRODUCTION

Throughout human history, agriculturists have used an array of techniques (irrigation, cultivation, fertilizer, herbicides, pesticides, fences, etc.) to give domesticated species a competitive edge over wild plants and animals. Often the cheapest and most practical strategy came down to killing the competition – especially large vertebrates. Government agencies traditionally responded to agriculturalists' needs without concern for wildlife survival. In fact, the original mission of many wildlife management agencies was not to protect wildlife, but rather to kill all wild animals that threatened human safety or agricultural development (Graham 1973). Because of their slow reproductive rates and low density, large vertebrates proved relatively easy to eliminate, especially as people added poison, guns and bounty payments to their arsenal. Thus in the name of economic progress wolves were extirpated from most of the USA in a few decades (Young and Goldman 1944). Similarly, colonial officers 'liberated' vast tracts of fertile land in Africa from elephants, leopards and other threatening species (Naughton-Treves 1999; Treves & Naughton-Treves 1999). Elsewhere in the world, formal and informal lethal control programmes have driven the decline and even the extinction of several wildlife species (Breitenmoser 1998; Naughton-Treves 1999; Wilcove 1999; Woodroffe *et al.*, Chapter 1).

Environmentalists today look back on these militaristic, morally charged campaigns in horror. Their calls to restore and protect wildlife are inspired by an increased appreciation of non-materialist values of wildlife. Now wildlife managers must respond to two seemingly contradictory mandates. Part of the public (mainly urbanites) demands wildlife be protected from people, and part of the public (mainly agriculturalists and livestock producers) demands people be protected from wildlife.

People and Wildlife: Conflict or Coexistence? eds. Rosie Woodroffe, Simon Thirgood and Alan Rabinowitz. Published by Cambridge University Press.

In this chapter, we consider the role of lethal control in fostering coexistence between people and wildlife. Despite the devastating history of many lethal control programmes, removal may have a legitimate role in wildlife conservation. First, well-managed lethal control has the potential to reduce threats to human lives and livelihoods without entailing serious extinction risks. Second, removing wildlife may placate local citizens and deter them from illicit killing of wildlife. Similarly, if the removal strategy channels benefits to local citizens (e.g. they obtain meat or hunting revenue) it may build local support for conservation efforts. Third, the elimination of some problem wildlife may select for conspecifcs that avoid humans and their property, thereby exerting directional selection for a wilder population of that species (Jorgensen *et al.* 1978; Treves 2002; R. Woodroffe and L. G. Frank unpubl. data). However, all of these conjectures must be rigorously evaluated lest lethal control do more harm than good. Indeed, lethal control programmes must be undertaken with care given the technical challenges surrounding the number and type of animals killed, as well as political and moral issues concerning who is allowed to kill animals and how.

Here we evaluate different forms of lethal control and their effects on long-term coexistence of wildlife and people. If lethal control is to foster coexistence of people and wildlife, it must reduce the impact of wildlife on people or raise public tolerance for damage without a significant reduction in the viability of wildlife populations. Thus, we consider three criteria for evaluating lethal control:

(1) Effectiveness in reducing future threats to human lives and livelihoods.
(2) Impact on the viability of wildlife populations.
(3) Public acceptance and stakeholder participation.

For simplicity we focus primarily on large mammals (>2 kg) but we extract general principles for the management of conflict with other taxa. We also consider translocation as a control method that is intended to be more humane but nonetheless leads to animals being lost from a population.

DEFINING TYPES OF REMOVAL PROGRAMMES

In the broadest sense, lethal control could include human activities that only incidentally diminish wildlife populations, such as habitat conversion, pollution or invasive species. Although these incidental sources of mortality ultimately constitute the greatest threat to the planet's wildlife, we choose to focus on deliberate efforts to reduce or remove wildlife to protect human lives and livelihoods. These can be substantial. For example, between 1996 and 2002 the US agency responsible for control of wildlife damage killed 15 260 640 wild vertebrates or 2.18 million animals in the average year (US Department of Agriculture 2005).

Deliberate removal programmes vary according to the proportion of animals removed and the selectivity used to remove individuals. Here we array the programmes into four overlapping classes and briefly discuss the most common motivations behind their use with examples of their application.

Eradication campaigns aim to extirpate problem wildlife from entire regions by all means available. Depending on the intensity of effort and the resilience of the target species, extirpation campaigns may produce local, regional or global extinctions. Powerful factors may motivate people to eradicate wildlife. For example, the elimination of bison from the American plains was fuelled by profit motives and the desire to subjugate native Americans, as much as the desire to protect agriculture (Isenberg 2000). Similarly, the value of ivory and skins promoted wildlife removal in the name of 'protecting natives' and 'opening agricultural land' in British-held East Africa (Beard 1963; Naughton-Treves 1999; Treves and Naughton-Treves 1999). Large carnivores have often been singled out for eradication due to the perceived and real risks they present to livestock and people (Woodroffe 2000) and their symbolic association with 'untamed wilderness' (Lopez 1978). To this day, colonists at the agricultural frontier in the Peruvian Amazon eliminate carnivores as *cazeria sanitaria* ('hunting to clean the forest'), in their eyes a first step toward economic progress (Naughton-Treves *et al.* 2003a). Today there are few country-wide deliberate eradication campaigns. Existing eradication programmes are generally more localized and might form part of a broader policy of coexistence in other areas (Linnell *et al.*, Chapter 10).

Culling programmes aim to reduce subpopulations of problem wildlife around sites of anticipated conflict (Blackwell *et al.* 2000; Hoare 2001; Cope *et al.*, Chapter 11) under the assumption that reducing wildlife populations will reduce conflicts. Culling encompasses the killing of wildlife in a specific area (but not its entire range), prior to or in the absence of specific, recent complaints about wildlife. Typically, the methods, actors and locations of culling are prescribed. Examples of culling programmes include the aerial shooting of coyotes prior to the release of sheep into summer grazing areas in the western USA (Wagner and Conover 1999) and the proactive removal of European badgers to avoid transmission of tuberculosis to cattle in the UK (case study in Box 6.1).

Government-sponsored culling engages trained agents to kill wildlife in specified areas, but private citizens also cull on private land. A good example

Box 6.1. Lethal control of European badgers in the UK

Bovine tuberculosis (TB) is a serious disease of cattle which can be transmitted to people. During the 1930s over 2500 people died each year from this disease in Great Britain alone. While TB is no longer a major threat to human health in Britain, over 1000 cattle outbreaks were confirmed in 2002, entailing substantial economic losses to farmers and government.

European badgers were first implicated in transmitting TB to cattle in the early 1970s: badger control, carried out by government staff, has been a mainstay of British TB policy ever since. This approach remains highly controversial: TB incidence in cattle has risen steadily since the mid-1980s, leading welfare lobbyists to argue that badger control is ineffective and should be discontinued, and farming groups to argue that control has not been sufficiently vigorous and should be extended (Woodroffe *et al.* 2002). In fact, the effectiveness of badger control in controlling cattle TB is not yet fully known (Krebs *et al.* 1997).

In 1998 the British government – guided by a committee of independent scientists – started a large-scale field experiment to evaluate badger population control as a component of TB policy. The trial was carried out in areas with the highest cattle TB risks and involved three treatments: proactive culling (reduction of badger densities to very low levels across wide areas), reactive culling (control targeted at particular badger social groups only when the farms they occupied experienced TB outbreaks), and badger population monitoring with no culling. An alternate reactive strategy, which attempted to identify and remove only infected badgers, was rejected because diagnostic tests could only identify 41% of truly infected badgers (Woodroffe *et al.* 1999). Each treatment was replicated 10 times in 100-km^2 trial areas, to cover a total of 3000 km^2. The trial was designed on such a large scale because, although TB is a serious economic problem, in statistical terms outbreaks occur rather rarely, requiring that a large number of farms be included to provide statistical power sufficient to measure the effects of culling with the required precision.

Proactive culling has a *population impact*: badger densities are markedly reduced in proactive areas (Le Fevre *et al.* 2003). The viability of the national badger population will not be influenced by the trial, but there would be major regional impacts were this approach to be adopted as national policy. Opinion surveys indicate that proactive culling of badgers is unlikely to prove *publicly acceptable* (White and Whiting 2000), and government ministers have already stated that they are unlikely to accept widespread culling as a future policy; the treatment is included in the trial largely for the epidemiological data it provides (Woodroffe *et al.* 2002).

Differences between treatments in the incidence of cattle TB provide a measure of the *effectiveness* of culling. Unexpectedly, reactive culling has been associated with a significant increase in cattle TB (Donnelly *et al.* 2003); hence this was dismissed as a future policy. This finding suggests that past culling policies may have been equally ineffective at controlling cattle TB, and also casts doubt on claims (Eves 1999) that near-elimination of badgers effectively reduces cattle TB, which are based on comparison of elimination areas with areas where localised culling of badgers occurred (and where cattle TB may thus have been inflated) (Donnelly *et al.* 2003).

This research programme illustrates the scale (3000 km^2) and cost (about \$12 million annually) of study needed to evaluate the impact of lethal control on conflicts that, while important, may in fact occur comparatively rarely.

is the widespread practice of mammalian predator control on private land managed for game-bird hunting in many European countries (Reynolds and Tapper 1996). Local populations of red fox, stoat, weasel and other relatively common small carnivores are reduced through trapping, snaring and shooting by professional gamekeepers.

Public hunts – in contrast to eradication – often include regulations governing the actors, location, timing or method of killing animals, in addition to limits on the number and type of animals that can be killed. In contrast to government-sponsored culling programmes, private citizens pay or volunteer to remove wildlife usually without reference to the location of past conflicts (Sagør *et al.* 1997; Sunde *et al.* 1998). Among the motivations for public hunting, some promote them as conflict mitigation strategies under the assumption that reducing populations of problem wildlife will reduce threats to human safety, economy or recreation. Public hunting is also promoted as a way to build a constituency for unpopular species by giving them value as game, food, fibre, etc. (Hamilton 1981; Linnell *et al.* 2001a; Leader-Williams and Hutton, Chapter 9, Cope *et al.*, Chapter 11).

Selective removal of wildlife is aimed specifically at the individuals suspected to have damaged property. Hence the location, methods and target are specified narrowly – which usually means that only trained authorities kill wildlife. Selective removal differs from culling in targeting fewer individuals and being reactive rather than pre-emptive: no animals are killed unless damage has occurred. The assumption underlying selective removal is that conflicts will diminish when problem individual animals are removed. Selective control is most often employed by governments to manage problems with rare or endangered wildlife. Examples include the removal of wolves radio-located near livestock kills (Bangs *et al.*, Chapter 21), and the issuance of kill permits to private livestock producers in areas that have had 20 or more depredations by lynx (Angst 2001). Selective removal can include translocation from sites of past human–wildlife conflict (Jorgensen *et al.* 1978; Linnell *et al.* 1997; Hoare 2001; Bradley 2004; Bradley *et al.* in press).

EFFECTIVENESS OF LETHAL CONTROL IN REDUCING FUTURE THREATS TO HUMAN LIVES AND LIVELIHOODS

The main justification for lethal control, and removal in all its forms, is conflict prevention and the underlying assumption is that conflict declines when wild animals are removed. There is no doubt that eradication campaigns can drastically reduce losses – at least those caused by the targeted species. The absence of wolf predation on Scottish sheep illustrates how eradication benefited agriculture. However, eradication can have unpredictable consequences. Reducing the density of top predators may cascade through ecosystems with meso-predators increasing in density, which can have unpredictable consequences for prey populations, conflict rates and the services ecosystems provide to humans (Reynolds and Tapper 1996; Estes *et al.* 1998; Terborgh *et al.* 2002). Well-studied examples of this occurring in conflict situations include the increased predation on wildfowl by skunks in the Prairie Pothole region of Canada after the eradication of red fox and coyotes (Greenwood *et al.* 1995), increased predation on rabbits by mongooses in southern Spain following the removal of the Iberian lynx (Palomares *et al.* 1995), and increasing levels of bush pig and baboon crop-raiding associated with the widespread removal of lions and leopards across Uganda (Naughton-Treves 1999). In short, eradication of one predator species may have the opposite result from that intended if a smaller predator at higher density takes its place. For example, the extirpation of wolves from all but a few tiny areas of the USA probably reduced conflicts overall for cattle and other large livestock, but conflicts with coyotes and other smaller carnivores remained frequent or have increased for smaller livestock such as sheep (Newby and Brown 1958; Taylor *et al.* 1979; Pearson and Caroline 1981). In sum, eradication of one species of problem wildlife can have unpredictable effects in the long term but certainly will reduce that species' threats to human life and livelihood. However, eradication of any problem species is clearly in conflict with efforts to promote coexistence of people and wildlife.

The effectiveness of culling programmes, public hunts and selective removal methods is far less clear because they have rarely been properly evaluated. This is particularly the case for large mammals because of the extensive spatial and temporal scales required for meaningful comparisons between different control techniques (Box 6.1). Less rigorous before-and-after comparisons indicate that the removal of large mammals has a mixed record of success in preventing future conflicts (Allen and Sparkes 2001; Hoare 2001; Osborn and Parker 2003). In Table 6.1, we review a handful of systematic studies from North America and Europe that assessed various removal programmes in preventing economic losses to carnivores.

The data in Table 6.1 suggest that removal of carnivores tends to achieve only temporary reduction in conflict if immigrants can rapidly fill the vacancies left after removals. This is consistent with findings for non-carnivores (e.g. moles: Edwards *et al.* 1999; elephants: Hoare 2001; Osborn and Parker 2003). Wolf translocation operations of the US Fish and Wildlife Service illustrate this point (Bangs *et al.* 1998; Bradley 2004; Bradley *et al.* in

Table 6.1. *Systematic studies of removal to prevent human–carnivore conflict*

Carnivore, country	Source	Type of removal	Conclusions
European badgers, Ireland	Eves 1999	Eradication	Tuberculosis in cattle declined more in a badger eradication area than in a surrounding 'reference' area where more selective lethal control was enacted (see Box 6.1 and below).
Brown bears, Norway	Sagør *et al.* 1997	Public hunt	No detectable reduction in sheep losses the following year because of recolonization and renewed depredations.
Wolverines, Norway	Landa *et al.* 1999	Public hunt	Lamb losses declined for one year, ewe losses did not change. Recruitment and recolonization led to renewed depredations.
Cougars, USA	Evans 1983	Culling	No effect on endangered prey species survival following removal of cougars.
Coyotes, USA	Conner *et al.* 1998	Culling	Non-selective removal in and around one farm did not reduce sheep losses in subsequent years.
Red foxes, UK	Reynolds *et al.* 1993	Culling	Reduced game-bird predation for less than one year because removed individuals were replaced in the same season.
Wolves, Canada	Bjorge and Gunson 1985	Culling/Selective removal	Cattle losses declined for two years, followed by recolonization and renewed depredations.
European badgers, Britain	Donnelly *et al.* 2003	Selective removal	Tuberculosis in cattle increased in nine areas where targeted lethal control of badgers occurred, relative to matched areas without lethal control.
Wolves, Canada	Tompa 1983	Selective removal	Following complaints of harassment, lethal control prevented further conflict for more than a year in 13.5% of cases. Following livestock loss, lethal control prevented further conflicts for more than one year in 34% of cases.
Wolves, USA	Bradley 2004; Bradley *et al.* in press	Selective removal	Removal of depredating packs increased the interval between successive depredations by 270 days on average. However, remaining individuals or translocated packs depredated again within a year after 23 (68%) of the 34 removals. When entire packs were removed, recolonizing wolves usually (six of seven cases) also caused depredations (after 99 to 383 days).
Wolves, USA	Fritts *et al.* 1992	Selective removal	Where wolves were removed 34% of farms had another depredation within one year whereas 23% of farms with verified depredations yet without wolf removal had another depredation within one year.

press; Bangs *et al.*, Chapter 21). In the Greater Yellowstone Area, 10 wolf packs were entirely removed following depredations. Recolonization of vacant habitat occurred in seven (70%) of these instances. Six new wolf packs recolonized within one year of the previous pack's removal and one recolonized five years later. Six recolonizing packs killed livestock, five of which preyed upon livestock in the ranches that had been previously affected (Bradley 2004; Bradley *et al.* in press). Data from Wisconsin are virtually identical for one cattle farm suffering chronic losses that had at least three wolf packs removed by translocation (Wisconsin Department of Natural Resources, unpubl. data). Such recolonization might be a good sign for population viability because it reflects resilient recovery following removals but recolonization does not bode well for prevention of conflicts. Recurrence of conflict despite wildlife removal has led some to conclude that the problem lies with husbandry as much as the wildlife (Stahl and Vandel 2001; Bradley 2004; Wydeven *et al.* 2004). Whether these findings hold for non-carnivores and other forms of property remains to be seen.

Table 6.1 may overestimate the effectiveness of culling programmes, public hunts and selective removal in preventing future damage. Without experimental controls, we cannot distinguish property loss followed by removal of wildlife from an isolated incident of property loss that would not have been repeated regardless of control action. Studies of radio-collared carnivores and longitudinal data on livestock operations both indicate that isolated incidents without repeat are common (Jorgensen 1979; Tompa 1983; Angst 2001; Stahl and Vandel 2001; Treves *et al.* 2002; Oakleaf *et al.* 2003; Bradley 2004; Wydeven *et al.* 2004; Bradley *et al.* in press). For example, Tompa (1983) described 49 cases of livestock predation by wolves where lethal control was denied for various reasons. In 39 (80%) of these cases, problems with wolves did not persist beyond one year. Some such cases may reflect unreported removal by the property owners, but others may reflect isolated events by transient animals or single incidents by resident animals triggered by brief changes in the relative availability of wild food.

Our review underscores the need for more experimental studies to understand the true effectiveness of control operations as part of species conservation programmes (Box 6.1). Properly designed experiments are badly needed to evaluate removal operations for large mammals in particular, because worldwide these are among the most endangered species and come into conflict with people commonly (Treves and Karanth 2003). To date such experiments have largely been restricted to small carnivores and predatory birds (Reynolds and Tapper 1996).

Population control of small carnivores is a common technique in game management in Europe and to a lesser extent in North America (Reynolds *et al.* 1993; Reynolds and Tapper 1996). We feel this deserves attention here because people often claim ownership of wild game and therefore predators trigger human–wildlife conflict by our definition when they prey on the contested wild game (e.g. Thirgood and Redpath, Chapter 12, Miquelle *et al.*, Chapter 19). The effectiveness of removal of predators as a game management tool has received considerable attention. The classic study is that of Marcstrom and colleagues (1988) who experimentally removed red fox and pine martens from islands in the Gulf of Bothnia, Sweden. The study was rigorously replicated and included switching of experimental treatments. During years with predator removal, densities of capercaillie and willow grouse were 1.7 times higher than in years without predator removal. A similar study was conducted in farmland in southern England (Tapper *et al.* 1996). In this case, experimental removal of red foxes and corvids increased spring densities of grey partridge 2.5 times and autumn densities 3.5 times. Cote and Sutherland (1997) reviewed 20 published studies of such predator removal programmes in a meta-analysis of their effectiveness to enhance or protect game-bird populations. Removing predators (either mammalian or avian) had a large positive effect on hatching success of the target bird species, with removal areas showing higher hatching success, on average, than 75% of the control areas. Similarly, predator removal significantly increased post-breeding population sizes (autumn densities) of the target bird species. The effect of predator removal on breeding population size (spring densities) was not significant, however, with studies differing widely in their reported effects. Predator removal may therefore fulfil a requirement of game management, which is to enhance harvestable post-breeding populations, but it is much less consistent in achieving an aim of conservation managers, which is to maintain or enhance breeding population size. Indeed these different objectives of game managers and conservation managers go some way to explain the different attitudes and interpretations of predator control programmes (Newton 1998).

An intriguing question is whether the greater success described above of removal of small predators to protect wild game-bird populations relative to the poorer record for removal of large predators to protect property (Table 6.1) tells us something important about the control of wildlife damage or whether it simply reflects different measures of success. With livestock, a handful of depredation events may be considered failure because no one measures success of predator control by how many cattle survive each year, which is precisely how they measure success in terms of game-bird management. Programmes that seek to protect wild prey may achieve some success with each predator removed because virtually all individual predators pose a threat (to their wild prey or the game claimed as property of humans). In short, there is little need for selective removal when all targets of removal pose a threat. However, programmes to protect livestock typically confront a minority of the predator population (see below), hence selective techniques are needed to generate the same incremental improvement in prey survival. Inaccurate removal of non-culprits may even yield unpredictable effects if culprits benefit from the removal of non-culprits (e.g. opening territorial vacancies).

Public hunts are least well represented in Table 6.1. As an ancillary motivation for public hunts, conflict prevention has not been studied as well. However, the recurrent justification of hunting as a conflict-reduction strategy merits more attention (Howard 1988; Linnell *et al.* 2001a; Knight 2003).

Without careful research on public hunts, it will be hard for managers to overcome several challenges that face public hunts as conflict-prevention strategies. For one, hunters are often unable or unwilling to target those individuals likely to participate in conflicts (Faraizl and Stiver 1996; Jackson and Nowell 1996; Sunde *et al.* 1998). For example, safari hunters participating in Zimbabwe's CAMPFIRE programme (see Leader-Williams and Hutton, Chapter 9) prefer to hunt mature bull elephants amidst wild habitat than to shoot younger animals amidst maize fields. As a result, licenses to hunt crop-raiding elephants are offered at a discount (Murombedzi 1992). Second, private hunters may be less well trained or use less effective killing methods than professional wildlife removal agents, resulting in higher frequencies of injured animals; injured animals can cause more problems than healthy ones, at least in some carnivores (Rabinowitz 1986; Linnell *et al.* 1999). Third, hunting may itself precipitate conflict by increasing the likelihood of encounters between wildlife and people or their valuable hunting dogs (Aune 1991; Treves and Naughton-Treves 1999; Treves *et al.* 2002; Wydeven *et al.* 2004). Finally, some carnivore ecologists speculate that heavily hunted populations generate more conflict, because their age structure shifts towards younger, inexperienced predators, which may turn to predictable but risky foods like livestock (Haber 1996; Conner *et al.* 1998). The potential problems described above should be taken into account in designing public hunts to prevent human–wildlife conflicts. Systematic applied research on this topic will be a welcome contribution.

Our examination of culling programmes and public hunts above prompts us to ask whether the short-lived effects of removal operations on large carnivores (Table 6.1) could be due to occasional or frequent removal of the wrong animals. The literature is unanimous that the majority of individual carnivores in a population will not kill livestock despite having access. This is well illustrated by work at the Hopland Sheep Research Station in

California (Conner *et al.* 1998; Knowlton *et al.* 1999; Sacks *et al.* 1999a, b; Blejwas *et al.* 2002). Breeding pairs of coyotes are responsible for the vast majority of incidents of predation on sheep and they kill sheep only within or on the periphery of their territories. Removal of one or both members of a breeding pair or destruction of litters reduced or eliminated livestock losses. Although new coyotes did eventually immigrate into the vacant territory, selectively killing only the breeding coyotes during the lambing season was an effective way of reducing lamb losses. Killing other coyotes was not. Studies of radio-collared pumas, wolves, lynx and grizzly bears corroborate that not all carnivores with access to domestic animals will prey on them. Some individual carnivores avoid humans and their property, others remain nearby without causing problems, and a subset causes damage (Andelt and Gipson 1979; Jorgensen 1979; Suminski 1982; Mace and Waller 1996; Bangs and Shivik 2001; Stahl and Vandel 2001; Treves *et al.* 2002; Bradley 2004; Wydeven *et al.* 2004). Similar analysis of crop-raiders and other sorts of problem wildlife would be valuable.

Non-target animals often fall victim to control operations. We estimated the proportion of non-target carnivores killed to prevent conflicts, from a handful of studies that used different methods (Table 6.2). Depending on the study, between 30% and 81.3% of the carnivores killed in control operations bore no evidence of involvement in conflicts. We caution against uncritical use of these data, because absence of evidence for an individual animal's involvement in conflict cannot fully exonerate it. The estimates in Table 6.2 therefore probably reflect maximum error estimates in some of the studies. Nevertheless, rigorous research by Sacks and colleagues (1999a) suggests the numbers are not unduly inflated (see also Box 6.1). Table 6.2 suggests removal of animals around the damaged property shortly after the damage is inflicted is the most accurate technique for the taxa examined, compared with removal later and far from the damage location (see also Bjorge and Gunson 1985).

There may be several reasons why non-target animals are killed in control operations. Habitual culprits are often hard to capture as their experiences with humans make them wary of human scent and devices (Corbett 1954; Turnbull-Kemp 1967; Conner *et al.* 1998; Sacks *et al.* 1999a, b). Yet one of the most common and effective removal techniques is to shoot, trap or poison animals that return to the damaged resource. For example, ranchers in Kenya shoot suspected culprit lions by concealing themselves in blinds at the sites of fresh livestock kills (Frank *et al.*, Chapter 18), and a similar approach is sometimes taken to shoot or trap livestock-killing wolves in the USA (Minnesota Department of Natural Resources 2001; Wisconsin Department of Natural Resources 2002; Bangs *et al.*, Chapter 21). These

Table 6.2. *Accuracy of lethal control in carnivore depredation management*

Carnivore species	Control method	Probable culprits among those removed	Source of evidence
Black and grizzly bears *Ursus americanus*, *U. arctos*	Trapping	30% ($n = 60$)	Horstman and Gunson (1982): estimated from necropsy, cessation of depredations and evidence at kill sites.
Coyote *Canis latrans*	Trapping	11–64% ($n = 113$)	Gipson (1975): estimated from necropsy of stomach contents; percentage varies with type of agricultural loss.
Coyote *Canis latrans*	Trapping, snaring, shooting, explosives, denning	55–71% ($n = 42$)	Sacks *et al.* (1999a: Table 1): estimated from age class of killed animals; upper bound assumes pups cause no depredations, lower bound assumes neither pups nor yearlings cause depredations for all methods of control.
Coyote *Canis latrans*	Aerial shooting	45% ($n = 11$)	Connolly and O'Gara (1987): estimated from the proportion of individuals killed that lacked a marker dye experimentally introduced into sheep.
African lion *Panthera leo*	Shooting over a livestock kill	70% ($n = 20$)	Frank *et al.*, Chapter 18: estimated from the proportion of lions shot that were too young to kill for themselves.
European badger *Meles meles*	Cage trapping and shooting	18.7% ($n = 18, 141$)	Krebs *et al.* (1997): the proportion of TB-infected badgers that were killed on farms which had experienced TB outbreaks in cattle.

approaches can be selective for the culprit(s), but the selectivity would be expected to decline under several conditions. First, the bait should not be left out too long lest non-target animals be attracted (Ratnaswamy *et al.* 1997). Trappers and other control agents sometimes leave out carcasses or attractive baits for days or even weeks (Jorgensen *et al.* 1978; Horstman and Gunson 1982; Shivik and Gruver 2002; Nemtzov 2003). Second, non-target mortality will increase when using baits if wildlife have become habituated to scavenging from human refuse. The improper disposal of garbage and carcasses in the area is believed to create problem animals (Andelt and Gipson 1979; Jorgensen 1979; Mech *et al.* 2000; Rajpurohit and Krausman 2000). Non-target mortality is expected to increase in species without strict territorial defence of space or resources. Among carnivores, individuals that played no part in killing wild prey and other species may be drawn to kills even in territorial animals (Frank *et al.* 2003; Shivik *et al.* 2003; Smith *et al.* 2003). Finally, non-target mortality may increase when several related taxa are difficult to distinguish from sign at kills, or when culprits are transients.

In summary, the short-lived effectiveness of culling programmes and selective lethal control seems to reflect recolonization of territories left vacant after removal, and high rates of removal of non-target animals.

IMPACT OF LETHAL CONTROL ON THE VIABILITY OF WILDLIFE POPULATIONS

While eradication campaigns are specifically designed to cause local extinction, other forms of removal are expected to have less of an impact on population viability. It is beyond the scope of this chapter to provide a thorough evaluation of wildlife population dynamics under different removal programmes. From our review of the literature, it appears the culling programmes and selective removals are both assumed to have little or no impact on wildlife populations. This is rarely examined systematically however (but see Blackwell *et al.* 2000; Cope *et al.*, Chapter 11).

Culling by government agents/programmes would seem to offer control over the number of wildlife removed. With prudent management and a careful balance between human needs and wildlife habitats, expert culling programmes have the potential to reduce wildlife densities in high-conflict areas without causing regional extinction. However, the government must receive incentives that promote wildlife population sustainability, rather than gain from their destruction. For example, the value of leopard skins and elephant ivory to the Ugandan colonial authorities of the twentieth century promoted widespread, energetic culling far beyond the needs of agricultural protection (Naughton-Treves 1999; Treves and Naughton-Treves 1999). Also, governments must control other sources of wildlife mortality lest government culling be additive with private, illicit killing and together undermine wildlife population persistence. In some cases where private individuals are involved in culling, the culling is intentionally or unintentionally extended to protected species (Goldstein 1991; Gonzalez-Fernandez 1995; Nemtzov 2003). For example, consider the killing and nest destruction of hen harriers and peregrine falcons on moorland managed for red grouse hunting in the UK (Thirgood *et al.* 2000b); Thirgood and Redpath, Chapter 12). The illegal killing of raptors to reduce their impact on game-bird populations is one of the main factors limiting raptor distribution and abundance throughout Europe (Valkama *et al.* 2004).

Public hunts are usually monitored better than culling or selective removal programmes although doubts have been raised about several issues. Setting hunting quotas based on previous years' harvests or previous years' conflict rates may not correlate well with population levels (Bennett 1998; Sunde *et al.* 1998; Landa *et al.* 1999; Logan and Sweanor 2001). The governance of hunting on communal lands may promote corruption and greed which may foster unsustainable hunting levels (Du Toit 2002; Virtanen 2003). Legal but poorly regulated hunting in multiple-use areas of the tropics can quickly extirpate large-bodied vertebrates from wide areas (Naughton-Treves *et al.* 2003a). Finally, the additive effects of public hunting and private removal of agricultural pests may elevate human causes of wildlife mortality to unsustainable levels unbeknownst to managers (Jorgensen *et al.* 1978). On the other hand, public hunts can contribute to the management of wildlife populations both directly via mortality and indirectly by generating revenue or scientific information to help manage wildlife populations (Stowell and Willging 1992; Faraizl and Stiver 1996; Nelson 1997; Andersone and Ozolins 2000).

Selective removal would appear to have a smaller population impact than any of the preceding removal strategies. First, the minimum number of animals is removed from the population. For example, India has preserved the last Asian lions and the largest population of tigers – despite a human population approaching 1 billion – by enforcing the protection of reserves and using lethal control only when problem wildlife become habitual threats to human life and property (Karanth 2002; Karanth and Madhusudan 2002). Second, culprits are eliminated from the gene pool while non-culprits are left in place to reproduce and spread their (learned or innate) avoidance of humans and their property, which could reduce future conflicts and removals (Jorgensen *et al.* 1978; Treves 2002; R. Woodroffe and L. G. Frank unpubl. data).

PUBLIC ACCEPTANCE OF LETHAL CONTROL

Making global conclusions about public opinion regarding lethal control is an overwhelming task given the diverse forms of human–wildlife conflict and the dramatic differences between and within societies regarding the acceptability of killing animals. Moreover, attitudes toward wild animals (and killing them) are value-laden and formed early in life (Kellert 1991). With these caveats in mind, we do discern some general trends in public acceptance of lethal control.

One way to understand varying acceptance to lethal control is to array people's values along a continuum with strong 'wildlife protection' on one end and strong 'wildlife use' on the other (Manfredo and Dayer 2004). People whose values lie at the protection end of the continuum believe that wildlife have rights similar to humans and generally oppose lethal control, as an unethical and cruel endeavour (Berg 1998). At the other end of the spectrum are people who believe wildlife should be used for human benefit, and embrace hunting (Howard 1988). Protectionist values tend to be found in urban populations more than rural while the opposite is true for use values (Kellert 1991; Wells *et al.* 1999). This divergence produces the fundamental tension for lethal control programmes: the rural citizens who are themselves more likely to experience conflict with wildlife are more likely to welcome lethal control, while urban populations with lower vulnerability but contributing more tax revenue tend to object to it (Manfredo *et al.* 1998; Reiter *et al.* 1999; Naughton-Treves *et al.* 2003b). Individual exceptions abound in all studies. Moreover, the specifics of the conflict will shape public approval for management. Table 6.3 describes how attributes of the wildlife and the context of the conflict can shape attitudes toward lethal control.

Campaigns to eradicate regional populations of native wildlife have largely been discontinued because of changing public attitudes to wildlife conservation and animal welfare (but see Reading *et al.*, Chapter 13). Few citizens want to see the complete eradication of a species, even if they are suffering losses, whether they are ranchers in Wisconsin, USA or western Ugandan farmers (Hill 1998; Naughton-Treves *et al.* 2003b).

Public acceptance of culling and hunting as a management tool varies greatly between cultures. Some advocates of management by hunting and culling argue that theirs is an efficient and humane population-control technique (Shelton 1973; Howard 1988). Yet tolerance for hunting and culling is generally diminishing in developed countries such as the USA and UK (Suminski 1982; Evans 1983; Harbo and Dean 1983; Shaw *et al.* 1988; Bennett 1998; Cope *et al.*, Chapter 11).

Table 6.3. *Factors shaping approval for lethal control*

Factor shaping acceptance	Acceptance less likely	Acceptance more likely
Where animal is killed	Public land	Private land
How animal is killed	Poison, snares, kill traps	Sharp-shooters, live traps followed by euthanasia
Type of threat by animal	Nuisance	Attacks human
Who kills animal	Commercial hunter	Government agent
Cost of damage by animal	Low	High
Perceived attributes of animal		
Aggressiveness	Peaceful (dove, cranes)	Fierce, cunning (coyote, fox)
Intelligence	Low (rodent)	High (chimpanzee)
Appearance	Beautiful (swan), neotenous (big eyes, round head, cute), human-like (monkey)	Ugly (vulture), alien (snake)
Abundance	Scarce	Superabundant
Sociality	Strong bonds (elephant)	Loners (leopard)
Reproductive status	Lactating female	Lone male
Health status	Prime or young in age	Ill, injured, decrepit

Attitudes to government culling vary greatly, and in part reflect public attitudes towards government itself. In developing nations, rural, small-scale agriculturalists and other stakeholders may welcome government interventions to control wildlife even if conflict rates remain the same (Bell 1984a; Hoare 2001). This may promote coexistence of wildlife and people if satisfaction with government culling reduces illicit killing of wildlife. By contrast, in some affluent communities of the USA, people have objected to the use of their taxes to remove wildlife, particularly if removal was perceived to have been carried out in response to political pressure by industrial interests (Torres *et al.* 1996; Fox 2001). Similarly, public attitudes to private culling vary tremendously and often reflect attitudes to game management or sport hunting itself. For example, the UK-based Royal Society for the Protection of Birds (RSPB) is Europe's largest non-governmental conservation organization with more than 1 million members. In contrast, the Game Conservancy Trust (GCT), a UK non-governmental conservation organization having strong associations with game management and sport hunting has approximately 24 000 members. Whilst it is certainly an oversimplification to suggest that these membership figures reflect the attitudes of the UK general public to protection versus use of wildlife, one can say that the use of lethal control in game management is not widely supported.

Public hunts are argued to improve tolerance for potentially dangerous wildlife like carnivores because hunters gain a sense of ownership of the wildlife (Linnell *et al.* 2001a). For example, the public in Wisconsin, USA tolerate large populations of species designated as game, such as white-tailed deer and black bears, despite millions of dollars in annual property damage and occasional human injury and death (Stowell and Willging 1992; Nelson 1997). By contrast, tolerance for wolves, a non-game species that causes less damage and does not threaten people, is far lower (Naughton-Treves *et al.* 2003b). Presumably, such tolerance reflects a sense of ownership and self-determination; hunters may accept wildlife on their land if they can use them or participate in their management, but not those species strictly protected by the government. Hunter tolerance is important because it is often hunters who are armed in remote areas and can therefore subvert wildlife protections afforded by law without great fear of prosecution.

Different types of hunting face differing levels of public support. For example, approximately 75% of rural Wisconsin residents are hunters (Naughton-Treves *et al.* 2003b) and a recent ballot initiative made hunting, fishing and trapping a state constitutional right. But proposals for crane-hunting to control crop damage, and wolf-hunting to control livestock damage produced public outcry. In short, public hunts may be acceptable but the methods must be considered carefully if one goal is public acceptance of removal as a control strategy. Promoting public hunts may also be unrealistic for certain flagship or totem species. For example, Native American groups in Wisconsin oppose any form of removal of wolves because the species has symbolic significance to them. There are similar feelings among Japanese hunters about killing monkeys, serow and bears (Knight 2003).

Although selective removal of culprits may seem to be the form of lethal control most likely to receive public acceptance, some animal-welfare groups remain opposed to any lethal control. Animal-welfare advocates argue that livestock husbandry is as important as predator removal but receives far less attention (Berg 1998). For example, the US Department of Agriculture's refusal to protect sheep operations with non-lethal techniques before initiating lethal control of coyotes led local authorities in Marin County, California to seek a private contractor willing to use non-lethal strategies (Fox 2001). Although an unusual case, this illustrates how a non-responsive agency can have its lethal control programme terminated regardless of effectiveness. At the opposite extreme, some groups remain opposed to problem wildlife conservation efforts no matter how responsive and accurate the removal programme. Although the US Fish and Wildlife Service has removed problem wolves from the Greater Yellowstone Area, USA, in a highly selective manner and losses to wolves have been lower than expected (Thompson 1993; Bangs *et al.* 1998; Bangs *et al.*, Chapter 21), some key stakeholders in the livestock industry remain implacably hostile to the wolf population and continue to seek wolf extirpation. In some cases, selective lethal control can frustrate both agriculturists and conservationists because it is reactive rather than preventive; both property and wildlife are lost. However, from a pragmatic standpoint, refinements to lethal control may meet less resistance from rural or agricultural interests and wildlife managers than efforts to change to non-lethal techniques. From this perspective, improvements in the accuracy of selective lethal control or improved hunting regulations represent a compromise position that may be increasingly attractive to wildlife managers in coming years (Box 6.2).

CONCLUSIONS

Arguing that lethal control is a legitimate strategy to promote wildlife conservation is difficult given the historical record of militaristic campaigns across the globe to eradicate species in the name of progress. Efforts to poison, shoot, trap or otherwise eliminate all inconvenient or threatening species were often as much about territorial conquest and the subjugation of nature (and indigenous people) as about protecting property such as crops and livestock. This grim history demands a conservative approach to lethal control today. Our review of the benefits and risks of contemporary lethal control programmes suggests that lethal control is a legitimate part of wildlife management and as such can play a role in conservation. The more difficult questions lie in who should be allowed to kill which animals and under what circumstances. We rest our review on two arguments:

(1) When highly endangered species kill livestock or take human lives, the best form of lethal control is highly accurate, selective removal of 'problem' animals by formally appointed and trained agents. Although killing a problem animal may temporarily placate local complainants, it does nothing to instil ownership or a sense of responsibility for the species among rural citizens who will probably continue to resent the presence of 'the government's animals' on their land in

Box 6.2. Improving the accuracy of selective removal

The main impediments to precise removal of culprits are technical and economic ones. Culprits are difficult to identify because humans can rarely observe conflicts or distinguish culprits from their conspecifics (Osborn and Parker 2003). Necropsy has been the main source of indirect evidence to implicate carnivores in livestock depredations (Gipson 1975; Andelt and Gipson 1979; Horstman and Gunson 1982). Post-mortem evidence remains important in evaluating selective lethal control (see also Box 6.1), but it is obviously too late to exonerate a non-culprit.

There have been two classes of approaches to improving the accuracy of selective lethal control. One is to develop new methods or refine existing methods for selectively killing suspect animals. The other is to assess the evidence against individual suspects before they are killed.

Improving selectivity

For years, researchers have explored the use of toxic chemicals to eliminate culprits at the site of conflict (Ratnaswamy *et al.* 1997; Mason *et al.* 2001). Unintended mortality of non-target animals and learned avoidance have haunted such endeavours. However, the development of livestock-protection devices loaded with toxic chemicals may selectively target problem carnivores, if these behave in predictable ways. For example, Burns and colleagues (Burns *et al.* 1991, 1996) demonstrated that coyotes were killed quickly and effectively when they delivered their stereotypical throat bite to sheep wearing collars loaded with toxin. Highly specific devices like these do not completely eliminate unintended damage to non-target wildlife because of leakage from punctured, defective or mishandled collars and scavenging of the carnivores that succumbed to the toxin. However, these side effects seem minimal. On the negative side, it is not clear whether neophobic animals commonly avoid the collars, or if such devices can work against other taxa than coyotes. Also the considerable cost of such devices almost certainly precludes their use in less-developed countries without considerable external donor support.

Highly focussed studies of the behaviour of problem animals and their conspecifics have revealed ways to improve selectivity of lethal control. Sacks and colleagues (Sacks *et al.* 1999a) contrasted the age classes of coyotes killed by various methods. Based on their work and that of colleagues (cited in the text), they concluded that shooting breeding pairs near dens during the pup-rearing season would be the most selective method of lethal control. Highly specific recommendations like these underscore the need for well-informed, trained professionals conducting selective lethal control.

Implicating and exonerating culprits via indirect evidence

Culprits may be identifiable by scent or other trace evidence. Scent dogs are used routinely to track mountain lions for hunting as well as depredation control. Proponents argue that these hounds are extremely discriminating and will bypass the trail of non-target animals (Suminski 1982; A. A. Smith *et al.* 2001). The technique may hold promise, particularly for wildlife at low population densities and for wildlife authorities with limited resources. Alternatively, captured wildlife may retain evidence of feeding on the human property. Remains and odours on the body of the wild animal may be useful in discriminating those having damaged human property from those that have not. Currently, rapid DNA fingerprinting assays are unavailable but this is likely to change given the advantages such tests would provide in human law enforcement. By contrast, examination of stomach contents and other tell-tale remains collected from live-trapped animals is well within current technical capabilities of developing countries – without killing the subject first.

Improving selective lethal control holds promise but it is questionable if this intensive level of individual wildlife management is economical except in situations that involve very valuable animals, such as endangered species.

the absence of substantial economic benefits from the animal. In short, selective removal of problem animals by government agents may be necessary to protect wildlife from extinction via widespread, illicit retaliation. The accuracy of removal becomes critical for populations near extinction so that non-culprits are left in place to reproduce (Box 6.2). Such may be the case presently for managing the roughly 300 lions persisting in India (Chellam and Johnsingh 1993), and for wild dogs recovering in Kenya (R. Woodroffe unpubl. data). This type of selective removal *was* important for protecting the recovering population of wolves in Wisconsin (Treves *et al.* 2002). However, as the population of an endangered species recovers and expands its range, more flexible, participatory types of control can be implemented.

(2) Public hunts are more participatory and cost-effective than selective removal by government agents when 'problem' species are numerous and widespread. But two cautionary lessons emerge from our review: (a) public hunts are more effective in improving public tolerance of the species (and support for the agency charged with its management) than in preventing damage caused by the species, and (b) regulated harvests may alienate urban constituents who place higher value on non-consumptive use of wildlife.

Of course there are myriad options lying between highly selective removal programmes and large-scale public hunts. Judging which form of removal to promote is challenging given the many ecological and socio-political factors in play. Social and ecological science expertise will be needed. Detailed demographic, ecological or forensic analyses may be required to judge the effectiveness and impact on wildlife populations of one control programme over another. Approval for management and tolerance for conflicts must be surveyed. The relationship between the control method and illicit killing by stakeholders must also be considered and quantified. All of these data and the technical skills to analyse them and make appropriate management recommendations will be lacking in many conflict situations. Even if armed with rich scientific data, policy-makers must judge broad public approval for alternative removal programmes. Without such approval, wildlife managers may lose full, flexible control.

No single prescription will be appropriate for all conflict situations. Instead, the entire constellation of political, economic and aesthetic demands of affected human populations should dictate local and regional solutions (Treves and Karanth 2003). Therefore design of a control programme requires stakeholder input, and consideration of the material and non-material values of the wildlife, stakeholders' perceptions of government intervention, views of the human role in nature, and rarity of the species in question.

Given uncertainty about stochastic causes of mortality in most large animal populations, we suspect that erring on the side of caution is the best way to maintain wildlife population viability for certain species. The prospects for coexistence of wildlife and people have improved in many parts of the world where wildlife eradication campaigns have been replaced with efforts to promote coexistence. Achieving this coexistence will entail technological innovation, including developing better non-lethal deterrent methods, more accurate identification of problem animals and conflict sites, and improved monitoring of the impacts of control programmes. It will also require negotiation to reach a compromise between people who demand the removal of all inconvenient or threatening species and those who demand protection for every wild animal.

ACKNOWLEDGEMENTS

We thank Rosie Woodroffe, Simon Thirgood, Luke Hunter and one anonymous reviewer for their detailed and helpful contributions to this chapter. AT was supported by the Wildlife Conservation Society during manuscript preparation. LN-T was supported by the Center for Applied Biodiversity Science–Conservation International and the University of Wisconsin–Madison. All opinions expressed in the chapter are only the authors' and we take responsibility for all errors.

7

Bearing the costs of human–wildlife conflict: the challenges of compensation schemes

PHILIP J. NYHUS, STEVEN A. OSOFSKY, PAUL FERRARO, FRANCINE MADDEN AND HANK FISCHER

INTRODUCTION

As the cases in this volume vividly illustrate, human conflict with wildlife is a significant – and growing – conservation problem around the world. The risk of wildlife damage to crops, livestock and human lives provides incentives for rural residents to kill wildlife and to reduce the quantity and quality of habitat on private and communal lands.

Recognition among conservationists that the cost of conserving large and sometimes dangerous animals is often borne disproportionately by farmers and others living closest to wildlife has spawned strategies to reduce this imbalance. One popular response is to compensate rural residents for the costs of wildlife damage. By spreading the economic burden and moderating the financial risks to people who coexist with wildlife, conservationists hope to reduce the negative consequences of human–wildlife conflict.

Few systematic efforts have been made to evaluate the efficacy of these programmes or the best way to implement and manage these schemes for endangered species (Sillero-Zubiri and Laurenson 2001). In this chapter, we build on our recent study (Nyhus *et al.* 2003) which asked whether compensation programmes really help endangered species in conflict with humans. We surveyed 23 international experts in large mammal conservation to learn about common pitfalls associated with running a compensation programme and the resources that managers need to succeed. Here, we also draw on additional published studies and reviews to explore the role of compensation in resolving conflicts between people and wildlife. We analyse the prospects and challenges of using these schemes in both developed and developing countries as part of a comprehensive suite of approaches to mitigate the effects of human–wildlife conflict on the long-term survival of

People and Wildlife: Conflict or Coexistence? eds. Rosie Woodroffe, Simon Thirgood and Alan Rabinowitz. Published by Cambridge University Press.

endangered species. We also introduce the idea of performance payments and other alternatives to traditional compensation schemes.

COMPENSATION IN PERSPECTIVE

In their most common form, compensation schemes reimburse individuals or their families who have experienced wildlife damage to crops, livestock or property, or who have been injured, killed or physically threatened by wildlife. A farmer who experiences wildlife damage may receive compensation in the form of cash or in-kind assistance. Compensation can range from more than fair market value to just a fraction of the value of the lost crops or livestock.

Compensation programmes typically target single species or small groups of species. Payment for damage by large or predatory protected species is common. What or who is eligible for compensation may be narrowly defined. For example, compensation for damage by specific large predators may be limited to livestock owners following specified animal husbandry guidelines. Some programmes may target single species damaging specific crops, others may pay for any damage resulting from any protected species or from any species if the damage occurs in a prescribed area (Cozza *et al.* 1996; de Klemm 1996). Eligibility for compensation may depend on where an attack occurs, such as inside or outside a protected area, or upon officials' assessment of the danger to farmers of driving animals away from their crops or livestock.

State-sponsored efforts to manage human–wildlife conflict are not new. Historically, governments have used economic incentives to reduce conflict by supporting bounties to exterminate problem animals. The wolf (*Canis lupus*) in North America (Bangs *et al.*, Chapter 21) is just one example where bounties contributed to the successful eradication of an animal from much of its historic range. Many compensation programmes have been initiated after management and conservation efforts increased the size of diminished wildlife populations, and so many programmes are relatively recent in origin (Wagner *et al.* 1997).

Legal protection of endangered species can restrict the time-honoured practice of lethal control, yet may not provide precedent for state-sponsored compensation for wildlife damage. As de Klemm (1996) suggests, the issue boils down to the elimination of the right of farmers to protect themselves from damage from legally protected animals. The inability of farmers to kill certain species sets up conditions favourable for compensation, even if legal protection of wildlife has not made governments liable for damage.

In fact, in the USA, courts have historically viewed wildlife as *res nullius* (having no owner), thus limiting liability of the state (de Klemm 1996). The

Box 7.1. Compensation and wolf recovery in the USA

Hank Fischer[1]

Prospects for restoring wolves to Yellowstone National Park were grim in 1984. But when William Penn Mott became Director of the US National Park Service, he gave me brilliant advice. 'The single most important action conservation groups could take to advance Yellowstone wolf restoration would be to develop a fund to compensate ranchers for any livestock losses caused by wolves,' he said. 'Economics makes ranchers hate the wolf. Pay them for their losses and you'll buy tolerance and take away their only legitimate reason to oppose wolf recovery.'

Defenders of Wildlife made its first compensation payment in 1987 and by 1992 established a permanent fund to pay for verified livestock losses in the northern Rockies and later the Southwest. This was the first private compensation programme for wolves established in North America.

Here's how the programme works. If a rancher believes a wolf has killed his livestock, he contacts the appropriate government agency. A trained biologist, usually on the scene within 48 hours, investigates to determine whether wolves were responsible. If the investigator verifies that wolves killed the livestock, Defenders is notified. Defenders strives to get cheques to ranchers within two to four weeks of receiving verification of a loss. When ranchers were concerned about unverified losses, Defenders established a category for 'probable' losses, for which it compensates livestock producers at half the market value.

From its inception through 2002, the Defenders programme paid over $270 000 to nearly 225 ranchers to compensate for 327 cows, 678 sheep and 34 other animals killed by wolves. Some feel this is a huge sum to pay for wolves, others feel it is a tiny price. But wolf experts agree that shifting economic responsibility for wolves away from ranchers and toward wolf supporters has created broader public acceptance for wolf recovery and reintroductions. According to US Fish and Wildlife Service wolf recovery coordinator Ed Bangs, 'The livestock compensation programme has made wolf recovery more tolerable to livestock producers and has made wolf recovery more easily attainable.'

[1] Hank Fischer developed Defenders of Wildlife's compensation programmes and managed them for nearly 15 years.

US federal government does not provide direct compensation for wildlife damage (Bangs and Shivik, 2001), leaving states to fill the void. Where government programmes are unavailable, non-governmental organizations have occasionally spearheaded compensation programmes, particularly for endangered species (see Box. 7.1).

ECONOMICS OF COMPENSATION

A major benefit attributed to compensation programmes is that they may increase tolerance of wildlife and promote more positive attitudes and support for conservation among people who live closest to endangered and dangerous animals (Wagner *et al.* 1997). When carried out effectively, compensation programmes raise awareness about community concerns and shift economic responsibility to a broader public. Compensation may result in a landowner giving a wild animal additional chances or result in discussions of how to prevent conflict in the first place. Conservation education and moral persuasion may be more effective in the presence of compensation. In the absence of effective compensation programmes, revenge killing or poaching may be more likely.

However, there is little quantitative evidence to support these claims. Although compensation programmes are conceptually appealing, their effectiveness in reducing local efforts to eradicate nuisance wildlife is largely unknown (for an exception see Box 7.1).

In a world of perfect information, establishing an effective compensation programme would not be difficult. However, the world of human–wildlife interactions is characterized by imperfect information. Thus, compensation proponents must struggle to verify that wildlife damage has occurred, to estimate its cost, to deliver payments quickly and efficiently over time and space, and to ensure that incentives created by the compensation programme do not make achieving the ultimate objective – maintaining or increasing endangered wildlife populations – more difficult. These tasks, discussed below, are the core elements of any compensation scheme.

Verifying damage

Verification is a vexing problem for many programmes because evidence of the wildlife that can cause damage, such as predator spoor, can disappear quickly. Even when investigated immediately, the true cause of damage may not be easy to ascertain. For example, it may be difficult to distinguish damage caused by different animals (Van Eerden 1990; Olsen 1991; Hötte and Bereznuk 2001). To address this uncertainty, there is a need for sound ecological research to quantify animal-inflicted damage and economic losses in relation to other sources of damage (e.g. Thirgood *et al.*, 2000b), and for techniques to differentiate among different 'problem' animals, such as new DNA analyses that can even forensically differentiate individual animals (Ernest and Boyce 2000).

The problem of unverified losses remains one of the most critical challenges for many programmes. Research in Wyoming, USA, using radio-collared livestock found that for every verified livestock loss from grizzly bears (*Ursus arctos*), the equivalent of another two-thirds of an animal is never found, potentially resulting in under-compensation (Wyoming Game and Fish Department, unpubl. data 1996). The case of Norway provides an extreme example of this problem: in 2001, only 5% to 10% of compensated sheep losses were actually documented kills by wild carnivores (Linnell and Brøseth 2003), a scenario which may represent overcompensation.

Endangered wildlife sometimes takes more than its share of the blame. For example, predation may not be the largest source of mortality among livestock (Kruuk 2002). Sutton *et al.* (2002) found that although elephants (*Loxodonta africana*) and other wildlife cause substantial crop damage in Namibia, domestic livestock damage to crops is much higher than that of any other animal (see also Naughton-Treves and Treves, Chapter 16). A survey respondent in Switzerland noted that sheep-farmers frequently claim that sheep that die or go missing have been killed by wolves and lynx (*Lynx lynx*). Yet, when researchers studied this problem, they found that the same numbers of sheep are lost in areas with and without large carnivores. These farmers, of course, have an incentive to attribute losses to carnivores. Unless they have private insurance, which most do not, they are only compensated for losses due to large carnivores – not lightning, falling rocks, or unidentified domestic dogs. These incentives lead to conflict between farmers and game wardens, who have to verify the cause of death, and can strain relationships with researchers, who may be summoned to make an expert judgement when verification is difficult (U. Breitenmoser pers. comm.). Outside verification is essential because self-reporting of wildlife damage may result in overestimates of damage by compensation seekers (Sekhar 1998) or fraudulent claims.

Although the problem of overstating losses is difficult, practitioners can draw from the economic research on the design of contracts that encourage people with private information to tell the truth (Wu and Babcock 1996; Moxey *et al.* 1999). Accurate verification takes time, but delaying verification or having such high standards that few claims are verified can lead to hostility from farmers. In India, a respondent familiar with one state-sponsored compensation programme reported that officials may take months to verify a wildlife attack, and unless it is a case of multiple killings, losses may not be verified at all. As a result, this respondent concluded, 'the actual payment of compensation is rather corrupt and whimsical'.

Making payments

Timely payment can help victims to get over their anger, and may reduce their incentives to retaliate against the animals that caused the damage, or to complain to neighbours or the media. Programmes self-reported to be effective (Nyhus *et al.* 2003) typically compensate quickly following verification, and strive to provide a transparent and fair claims process. A transparent process can be vital to avoid abuse, such as when a few individuals manipulate the system (Sillero-Zubiri and Laurenson 2001), and insufficient information about a programme can lead to frustration among participants (Wagner *et al.* 1997). Programmes self-described as successful (Nyhus *et al.* 2003) also commonly separate the verification arm of the compensation scheme from the arm providing the actual payment.

Determining loss and compensation values can be a challenge because the value of livestock (or crops) may vary with age, size or reproductive status. For example, farmers may receive compensation for a young animal killed by a carnivore, but resent not receiving compensation for the value the animal could have provided if sold for meat or for breeding when mature. Even when compensated monetarily, some farmers may perceive they are not receiving fair compensation for the trauma, time or hardships they face protecting their assets, or the emotional loss of losing their livestock (Kruuk 2002; Thirgood *et al.*, Chapter 2). The programme in Switzerland described above addressed this by calculating the full potential market value of livestock lost to lynx and wolves, even if the animal killed was not yet mature. The State of Wyoming, USA, now pays more than market value for losses of livestock verified to be caused by grizzly bears to make up for the losses that cannot be verified or found. Programmes in France and Spain have adopted similar policies for bear-related compensation. In Norway, damage above average non-predation mortality is eligible for compensation if other predator-caused mortality has occurred in the area, predators reside permanently in the area, or the area has a history of chronic loss from predators (Linnell and Brøseth 2003).

Less successful programmes typically paid much less than market value (Nyhus *et al.* 2003). In Botswana, one programme reportedly paid out only about one-quarter the value of the livestock lost. In India, one state-run scheme reportedly set compensation at an even smaller fraction of the market value of each domestic animal. Expectations among local people about what the government could do increased, but human–wildlife conflict was not alleviated.

Human–wildlife conflict also leads to the loss of human life. Zhang and Wang (2003) report that while elephants (*Elephas maximus*) rarely kill people in the Simao region of China, the potential for such tragedy scares people and intensifies negative attitudes toward wildlife. Putting a value on a human life is both difficult and, according to some, immoral. Paying too little for a human death or injury may have no effect on reducing negative attitudes toward wildlife. Despite these obstacles, several nations compensate for the loss of human life. In Zimbabwe, families of 21 victims killed in 2001 while protecting their fields from elephants were each paid Z$15 000

(~$273) (The Sunday Mail 2002), and in Malaysia, the government recently established a fund to pay families RM5000 (~$1300) for injuries and RM10 000 for death resulting from wildlife (*The Star* 2003).

Ultimately, a compensation programme must ensure that local people are part of the overall management of the problem and participate in determining what constitutes appropriate compensation. A conservation programme manager in Botswana noted that attempting to 'buy' community compliance with conservation will not work unless decisions related to wildlife and money are made by the communities themselves. Of course, participatory approaches are easier said than done. Rural residents have an incentive to inflate their claims and, unlike participants in a competitive commodity market, conservationists cannot simply walk away and find a more reasonable seller.

Payments and sustainability

Wildlife damage is unlikely to disappear with time and thus a compensation programme must have a sustained source of sufficient funds. The annual cost of maintaining a compensation programme – particularly if damage is frequent and extensive – can be substantial and can range from a few thousand dollars for small programmes to more than a million dollars for regional multi-species programmes. In Pakistan, one non-governmental organization (NGO) spends approximately $2000 annually on a snow leopard project and the Italian government has spent $2 million on compensation for wolf, feral dog and bear damage.

The cost of compensation programmes is not limited to the payment for wildlife damage. Management of the programme and verification can involve major costs also. In Asturias, Spain, rangers devote an estimated 1000 person–days per year assessing wolf damage for an area of 5000 km^2 with 200 wolves (Blanco 2003). As Karanth and Madhusudan (2002) note, the cost of administering schemes can be higher than the monetary value of the crops or livestock lost.

Funding for compensation schemes typically comes from public agencies (via taxes), non-governmental conservation organizations, private funds or fees paid by certain groups like hunters (de Klemm 1996). In some cases, compensation is in the form of live animals. For example, in Turkmenistan, the Worldwide Fund for Nature (WWF) purchased 196 sheep to provide a long-term source of compensation (Lukarevsky 2003).

As in any conservation initiative, stable long-term funding is essential to achieving success. Wagner *et al.* (1997) reported that six states or provinces in North America had to cancel programmes because of budget cutbacks or changing priorities. Programmes established when wildlife populations are low and conflict is relatively rare may face problems if wildlife populations, and thus conflict, increase. Uneven distribution of wildlife damage incidents, as occurs with geese in the UK, can confound estimates of the amount of funds needed to sustain a programme, but such estimates are an important means of securing adequate funding (Hanley *et al.* 2003; Cope *et al.*, Chapter 11).

Incentives

Even in situations where rapid and accurate verification and adequate payments are possible, a successful compensation programme has to wrestle with three thorny issues related to incentives.

First, if compensation provides full insurance to rural residents against wildlife damage, but rural residents are able to influence the probability of loss (e.g. by using guard dogs for livestock), then a problem that economists call 'moral hazard' arises. Knowing that they will be compensated for wildlife damage, some individuals may be less likely to adopt new – or improve existing – management practices that would discourage conflict in the first place. Not only are residents less likely to improve their current methods, but they may also reduce their current level of investment in management practices (e.g. they might graze animals closer to predator habitat). In the absence of incentives for rural residents to protect their assets, a permanent state of conflict is assured.

In some regions of Spain, farmers are compensated regardless of whether they leave their livestock unattended or they watch their livestock closely (Blanco 2003). A rational farmer might not bother to invest in costly methods of protecting his livestock. Participants who already use good management techniques may then criticize programmes when they are not compensated for the time and money they invest in good management practices, while others who take no action are fully compensated.

One solution to the moral hazard problem is to force farmers to act to reduce the risk of losses. This can be achieved by requiring participants to adopt observable risk-reducing investments before they are eligible for compensation. For example, payment may be denied to those with poor practices (e.g. not putting livestock in enclosures at night) or to those who do not follow guidelines for damage prevention after their first reported incident. The key aspects of these requirements are that they must be observable and effective at reducing risk. One problem with requiring specific investments, however, is that farmers may know better than outsiders how best to reduce wildlife damage. In the absence of consultation, external pressure may do little to alleviate the costs of moral hazard.

Another way to reduce moral hazard is to force rural residents to bear some of the risk by establishing a deductible (i.e. the resident pays part of any loss). Deductibles encourage farmers or ranchers to protect their crops and livestock. In effect, many compensation programmes already do this by not paying the full value of damages. However, risk-sharing may not be acceptable to farmers who resent incurring substantial costs on behalf of others who value the wildlife. Even without such beliefs, residents who incur a cost from wildlife damage may still have incentives to kill wild animals. Furthermore, mandatory changes in livestock husbandry that are not perceived to be worth the cost incurred will simply not be implemented (de Klemm 1996). In a study of the willingness to pay of Namibian households for an electric fence to guard against wildlife damage, Sutton *et al.* (2002) concluded that most households would never choose to invest in electric fences unless they were heavily subsidized. Of course, for compensation to be an option of choice, logical and sustainable subsidies should not be considered anathema. Ultimately, any costs from moral hazard should be weighed against the possible conservation benefits of a compensation programme.

A second incentive problem arises from compensation's role as an agricultural subsidy. Compensation increases the net returns from agricultural production and thus may provide incentives to convert natural habitat to agriculture. In a theoretical simulation, Rondeau and Bulte (2003) found that under certain assumptions (e.g. free trade, open-access land, price effects from compensation), compensation had a negative effect on wildlife. Furthermore, as a subsidy, compensation can delay the exit from farming by individuals who ought to exit (because the activity is not profitable) and encourage entry by those who, in the absence of compensation, would not farm. In less economically developed countries, compensation programmes might attract outsiders hoping to benefit – a 'magnet effect' – and ultimately increase the number of people exposed to wildlife damage. Ensuring that compensation is not available to unsanctioned new settlements and enforcing land-use agreements (e.g. keeping protected areas from being turned into agriculture) might be ways to address this problem.

Finally, cash itself can have undesirable effects. An influx of funds can generate conflict within communities or among communities and conservation practitioners. Such conflict is likely if programmes are not transparent or if perceived distribution of funds is not equitable.

Private insurance

To manage compensation programmes more efficiently and to sustain them over time, conservationists have looked at the insurance industry as a model. Several countries in Europe have used such an insurance-based approach (de Klemm 1996; Fourli 1999; Blanco 2003).

In most situations, however, an unsubsidized, purely private system of insurance is not viable. The difficulties associated with verifying damage and farmer investments in damage avoidance are substantial and thus lead to problems of fraudulent claims, moral hazard and adverse selection. Adverse selection occurs when insurance agents cannot differentiate high-risk from low-risk clients and thus cannot provide insurance to low-risk clients at a reasonable price. For example, a study of the Sariska Tiger Reserve in Rajasthan, India (Sekhar 1998), found that villagers experienced on average a loss from wildlife depredation of Rs.370 while the annual premium to insure livestock would amount to approximately Rs.1168. This was in part because a small number of households experienced considerable damage while a larger percentage experienced slight or no damage.

Even if private insurance could be made available, rural residents may view it as too expensive given their perceptions of the likelihood of damage, their limited budgets and their access to alternative means of 'insurance' (e.g. poisoning animals). In a contingent valuation survey of households living near wildlife in Zimbabwe (Muchapondwa n.d.), 131 households had a zero willingness to pay for insurance and the mean willingness to pay for insurance for the remaining 439 households was Z$215 (~$4) per year – too low for a viable private insurance market. Furthermore, even if insurance was affordable, many would be unwilling to finance compensation for damages caused by wildlife perceived to be the state's responsibility. Thus, successful implementation of private insurance may require subsidization, either by a government agency, an NGO or members of a community not affected by wildlife damage (e.g. see Miquelle *et al.*, Chapter 19).

Project Snow Leopard in Pakistan (Hussain 2003) claims to have overcome some of these problems by relying on community participation and an innovative financial design. Farmers pay premium contributions to a community-managed fund per head of livestock. A second fund, operated jointly by the community and Project Snow Leopard, generates income from ecotourism. The committee verifies kills and makes compensation recommendations. Claims that exceed the claimant's accumulated premium amount are paid from the second fund. Biological surveys suggest snow leopard (*Uncia uncia*) populations are stable or increasing in the area, but without a control area it is difficult to know whether this is a direct result of the compensation programme.

MEASURING SUCCESS

Documenting and defining the success of any conservation initiative can be a challenge. The goal of many compensation programmes is to increase tolerance for wildlife, but they may not actually eliminate conflict from occurring in the first place (Klenzendorf 1997; Wagner *et al.* 1997). Even when compensated, farmers may continue to kill wildlife illegally (Kruuk 2002; Naughton-Treves *et al.* 2004). There is also a risk that people will be more frustrated at the failure of an inadequate compensation programme or cessation of a successful one than if none had been in place at all (Wagner *et al.* 1997).

Thus, despite many attempts to implement compensation schemes of different kinds, little empirical evidence of their success or failure is available. Many programmes cannot objectively quantify the impact they are having on people's attitudes or the impact on the wildlife populations of conservation interest. Ultimately, more rigorous methods to evaluate the success of compensation schemes are needed, and managers should be encouraged to build into their programmes appropriate evaluation measures that are ideally linked to the conservation status of the 'problem' animal. Comparative assessments of local attitudes, as well as of the health and size of target wildlife population(s) before and after a programme has started, are sorely needed. Assessments of the number of animals of conservation interest killed in retribution for attacks or via illegal harvesting might be additional indicators, but it is notoriously difficult to monitor such retaliatory, illegal killing because it is often carried out clandestinely. Such assessments are further complicated by the many other factors that influence wildlife numbers, such as weather or disease.

ALTERNATIVES TO COMPENSATION

The logic of compensating victims of wildlife damage is appealing, but conservationists, wildlife managers and even some recipients of compensation question whether it is the most effective and efficient use of scarce conservation funds. Alternatives, such as tourism or trophy-hunting, may potentially provide better investments in time, money and results (see Walpole and Thouless, Chapter 8, and Leader Williams and Hutton, Chapter 9, for discussion of these issues). In some circumstances, alternative methods such as building wildlife barriers or providing additional habitat may be more cost-effective than depending on individual farmers to adequately protect their farms. In a theoretical analysis of wildlife damage in high-income nations, Dyar and Wagner (2003) show that purchasing land to increase the amount of habitat between wildlife and crops/livestock can be more efficient than compensation.

Karanth and Madhusudan (2002) argue that just modifying human or animal behaviour ultimately does not address the root cause of the problem.

They suggest that in some situations, such as where high densities of people and potentially lethal animals like tigers (*Panthera tigris*) coexist, and where voluntary resettlement of communities from enclaves within protected areas is possible, preventive spatial separation of people and wildlife may be the most cost-effective and practical solution.

Compensation, if available, is typically just one part of a larger human–wildlife conflict management strategy. In the western USA, lethal control remains a central part of management plans for wolves and bears (e.g. see Bangs *et al.*, Chapter 21). Compensation may result in a landowner giving a large carnivore additional chances or encourage discussions of how to prevent conflict, but lethal control may ultimately still be used. Although the popularity of compensation programmes has been driven in part by the perceived ineffectiveness of education and moral persuasion, well-targeted participatory information campaigns may still be an effective use of scarce funds.

Performance payments

Rather than compensate rural residents for wildlife damage, practitioners may want to consider making payments that are conditional on wildlife abundance (Ferraro and Kiss 2002). In this approach, payments are tied to wildlife populations (e.g. for every elephant that exists in the district, communities are paid $\$x$; for every wolf den on a landowner's property, the landowner is paid $\$y$). A good example of this approach comes from Sweden, where Sami reindeer herders are paid for successful wolverine (*Gulo gulo*) dens within their areas (Swenson and Andrén, Chapter 20). Similarly, farmers in Scotland are paid to maintain protected populations of geese on their land (Cope *et al.*, Chapter 11).

Although this approach has been most frequently used for habitat protection, it may be amenable to wildlife protection in situations in which wildlife populations can be monitored (e.g. Musters *et al.* 2001). Unlike the damage compensation approach, the payment does not decrease a farmer's incentive to reduce the probability of damage. Instead, rural residents would have an incentive to adopt whatever mechanisms (including doing nothing) are most cost-effective to reduce the amount of damage that wildlife imposes on their assets. Furthermore, performance payments do not increase the profitability of agriculture and thus, unlike damage compensation, do not increase the incentives to expand agriculture (they may in fact provide disincentives).

Payments tied to wildlife abundance could also improve the welfare of rural residents by diversifying the household portfolio and reducing household exposure to risk. If agricultural production is relatively more variable than the wildlife populations to which performance payments are tied, then risk-averse farmers would be willing to accept lower relative returns from the wildlife payments in return for the payments' ability to reduce risk.

The performance payment approach, however, suffers from many of the same constraints faced by more traditional compensation schemes. Timely and accurate verification is still necessary, but instead of verifying damage, practitioners have to verify population size and changes over space and time. Other challenges include devising the rules for payment (e.g. is payment tied to a population estimate from a given method? do payments change linearly or non-linearly from a baseline estimate?), determining the appropriate payment levels, and deciding whom to pay. A performance payment approach must determine how to differentially pay residents according to their (unobservable) costs from living with wildlife. In the case of wide-ranging wildlife, to whom should go the rewards for the protection of a particular animal? In some cases, such as paying for animal nesting sites, payments can be targeted at the household level; in other cases, such as paying conditionally on the population abundance of a species in a given area, payments might best be channelled to community-level benefits (e.g. reduced taxes or funds for schools and health clinics). A community-level payment is not necessarily undesirable because such payments may induce strong peer pressure effects on individuals: if one person kills an animal, this person has reduced the benefits that everyone in the community receives. Local people often have better information for monitoring their own members, but frequently have little incentive to do so. Practitioners can draw from research on designing incentive systems in situations in which one cannot monitor every individual's actions and thus must resort to rewarding and penalizing individuals as a group (e.g. Segerson 1988).

Despite the potential advantages of a performance payment approach in mitigating human–wildlife conflict, thus far there have been few economic analyses like the wolverine example cited above, or field experimentation to guide practitioners in implementing such an approach. Our intention in highlighting performance payments is simply to encourage researchers and practitioners to think more deeply about the potential of such an approach as an alternative to damage compensation, and to experiment with it in the field.

CONCLUSION

Human–wildlife conflict is an important threat to the survival of many species and requires innovative, practical and cost-effective solutions. Compensating farmers for damage caused by wildlife has been tried as one such solution with mixed success around the world. By shifting the economic burden away from local people, at least in part to the broader public that wants to see these species conserved, compensation supporters believe it encourages constructive engagement by the people most closely tied to the future of the world's large, dangerous and endangered species.

Unfortunately, the issues surrounding such conflict are typically a complex mix of behaviour (human and wildlife), ecology, socio-economics, politics and geography – making the resolution of these conflicts extremely difficult. While there appear to be general guidelines that can aid wildlife managers in implementing effective compensation schemes, it is important to be sensitive to and to incorporate site-, species- and culture-specific issues.

Key determinants of success for compensation schemes typically include the accurate and rapid verification of damage, prompt and fair payment embedded in a transparent process, a long-term source of funding capable of responding to variations in damage over time, clear rules and guidelines that link payment to sound management practices, an appreciation of the cultural and socio-economic context, and an ability to actively monitor the wildlife population of interest.

In many areas of the world, the constraints to implementing successful compensation schemes are formidable. Corruption inhibits transparency. Insufficient funding precludes fair payment. Inadequate staff and training impede accurate verification. Awareness of these pitfalls may help practitioners to find workable solutions. Without such solutions, rural residents living near wildlife will continue to have considerable incentives and motivation to kill problem animals and to convert wildlife habitat.

Ultimately, compensation is only one tool in the human–wildlife conflict mitigation toolbox. To work most effectively, it needs to be part of a comprehensive approach that includes proactive measures to prevent conflict in the first place, options to control offending animals and incentives to change land-use practices.

Several alternative approaches may provide opportunities for further testing. Lessons from the private insurance industry may help practitioners to more effectively estimate demand for and total costs of a programme over time. A performance payments approach, more frequently used for habitat protection, may sidestep some of the constraints, like moral hazard, faced by standard compensation programmes. Linking incentives to wildlife abundance may have positive effects on the welfare of rural residents as well by diversifying their sources of income. Regardless of which approach, or combination of approaches, is used, it is clear that much remains to be learned about the effectiveness of using economic incentives to reduce the negative impact of human–wildlife conflict. The ultimate measure of a compensation scheme's effectiveness, as for any conservation tool designed to reduce human–wildlife conflict, is whether it keeps fewer tigers, elephants or gorillas (*Gorilla gorilla*) from being killed. But in most situations today we do not know if this is happening, and we have no way to compare whether the approach used was the best or most cost-effective strategy.

There appears to be a clear need for carefully designed, empirical research on the use of compensation payments to achieve conservation goals. In particular, studies with adequate controls are sorely needed. For example, it would be desirable to compare villages in similar areas that do or do not have access to compensation, to compare communities where compensation is paid at below- or above-market rates, or to explicitly compare traditional compensation schemes to an incentive-based performance payment approach. Practitioners would benefit from well-planned research on the joint effectiveness of several strategies aimed at reducing human–wildlife conflict. A publicly accessible database that would permit a transparent comparison of true costs and true benefits of various approaches could be a powerful step in this process.

Compensation programmes, under certain circumstances, may provide one component of an effective response to human–wildlife conflict, particularly when endangered species are the main 'offender' and lethal control is not a desirable option. However, if carried out with inadequate attention to certain key factors, these schemes can be a waste of resources destined to do more harm than good.

ACKNOWLEDGEMENTS

This chapter was based in part on an article that appeared in *Conservation in Practice* (2003) 4 (2) 37–40. We gratefully acknowledge the assistance we have received from Alistair Bath, Joshua Ginsberg, Ginette Hemley, Richard Hoare, Rodney Jackson, Sybille Klenzendorf, Caroline Mitten, Lisa Naughton-Treves, Judy Mills, Paul Paquet, Christine Rastas, John Seidensticker, Mingma Sherpa, Barry Spergel, Gail Carlson and all of those survey respondents from around the world who were willing to share their experiences, good and bad, regarding compensation schemes with us.

8

Increasing the value of wildlife through non-consumptive use? Deconstructing the myths of ecotourism and community-based tourism in the tropics

MATTHEW J. WALPOLE AND CHRIS R. THOULESS

INTRODUCTION

It has long been recognized that the costs of living with wildlife, particularly for poor, rural communities in the developing world, are rarely offset by any material benefits (Ghimire and Pimbert 1997). Whilst the economic benefits of wildlife and biodiversity are diffuse and accrue to society in general, and financial benefits generally accrue to governments and external entrepreneurs, many of the costs are acute and borne locally (Dixon and Sherman 1990; Wells 1992; Balmford and Whitten 2003). This does nothing to improve tolerance among local communities and, as a result, wildlife continues to decline through persecution and habitat destruction.

In recognition of this problem, conservationists have embraced local economic development as a strategy for wildlife conservation and conflict resolution (IUCN/UNEP/WWF 1980). The new era of 'conservation with a human face' (Bell 1987) eschewed earlier protectionist policies and adopted the mantra that wildlife should be conserved for the benefit of local communities who bear the costs of coexistence (Anderson and Grove 1987; Western and Pearl 1989). This paradigm shift has gained ground in successive global conferences and strategies (McNeely and Miller 1984; World Commission on the Environment and Development 1987; IUCN/UNEP/WWF 1991; McNeely 1993). Sustainable utilization and equitable benefit sharing are now enshrined within the Convention on Biological Diversity (2005) that was launched at the 1992 Earth Summit in Rio.

The concept of integrated conservation and development assumes that utilization will provide a sustainable economic incentive for communities to tolerate coexistence with wildlife, thereby addressing both poverty and conservation (Wells and Brandon 1992). This apparent win–win scenario has gained widespread support in developing countries with high biodiversity

People and Wildlife: Conflict or Coexistence? eds. Rosie Woodroffe, Simon Thirgood and Alan Rabinowitz. Published by Cambridge University Press.

and economic underdevelopment (Western and Wright 1994). Some initiatives promote consumptive use of wildlife and other natural resources (Leader-Williams and Hutton, Chapter 9). Most, however, rely on non-consumptive use of wildlife, principally tourism (Zube and Busch 1990).

The promise of tourism for wildlife and people

Tourism has provided an economic rationale for government-sponsored landscape and wildlife conservation ever since modern protected areas were first established in North America and Africa (Runte 1987; MacKenzie 1988). Over the past 50 years, tourism has been one of the most consistent global industries, growing from 25 million to 693 million international arrivals between 1950 and 2001. This has been accompanied by a geographical shift towards developing countries. Over the past two decades, Africa, East Asia and the Pacific, and the Middle East have all increased their market share at the expense of traditional Western destinations (World Tourism Organization 2002).

Equally, nature-based tourism, 'with the specific motive of enjoying wildlife or undeveloped natural areas' (World Travel and Tourism Environment Research Centre 1993), is an expanding sector. This is due to increasing accessibility and promotion of large, well-stocked national parks in Africa, together with the increasing number and popularity of wildlife television documentaries (Ceballos-Lascurain 1993).

Concurrent with this growth has been a change in emphasis within the industry towards more socially and environmentally responsible tourism, that has become collectively known as 'ecotourism' (Giannecchini 1993). In theory, ecotourism provides tangible economic benefits from wildlife to offset the costs of protection *and* coexistence, 'providing revenue to the local community sufficient for local people to value, and therefore protect, their wildlife heritage as a source of income' (Goodwin 1996: 288).

The reality of tourism for wildlife and people

Most wildlife tourism takes place within protected areas. Recently, however, there has been increasing tourism development outside protected areas. Some of this has taken place on private land, chiefly fuelled by the desire for a more 'exclusive' experience (e.g. Londolozi Reserve in South Africa, Save Valley Conservancy in Zimbabwe and Lewa Wildlife Conservancy in Kenya). Many of these areas were formerly cattle ranches, and it seems clear that investing in wildlife tourism has improved wildlife tolerance among these relatively wealthy landowners (Frank *et al.*, Chapter 18).

However, the real battlegrounds between people and wildlife are outside parks and private reserves where poor rural people coexist with wildlife.

Communities adjacent to parks and private reserves may benefit from employment and business opportunities in tourism within such areas, and are sometimes offered a share of revenues in return for tolerance of wildlife. Alternatively, communities may engage in 'community-based' tourism development beyond protected area boundaries. These initiatives, based largely around the provision of overnight accommodation and wildlife viewing, range from pure commercial concessions, through joint venture partnerships, to wholly community owned and managed operations. They generally emphasize community participation and empowerment, thereby promising a greater range of financial and non-financial benefits than parks-based tourism (Table 8.1). As a result, there has been an enormous investment in community-based tourism by donors.

However, the assumption that tourism-induced poverty alleviation will generate conservation benefits by reducing or offsetting the costs of human–wildlife conflict is almost entirely untested. There is remarkably little empirical evidence of the success of these indirect approaches in improving community tolerance of wildlife, and little understanding of the necessary conditions for success.

The ability of tourism to deliver its promises will depend upon three critical conditions common to most, if not all, indirect incentive schemes that attempt to integrate economic development and conservation. First, enterprises must generate net benefits for communities. Second, the distribution of benefits amongst individuals should take into account variation between individuals in the cost of living with wildlife. Third, there should be a clear understanding of the linkages between the receipt of benefits from wildlife and the need to conserve it.

This chapter reviews the evidence and assesses the likelihood that each of these conditions can be met in practice. We consider the contribution that parks-based tourism makes to neighbouring communities, as well as community-based enterprises beyond park boundaries. Evidence is amassed from tropical developing country examples, primarily but not exclusively from Africa and Asia. This reflects a weighting in the literature and the authors' own experiences towards these continents. However, there are common themes in wildlife tourism across the globe, and the lessons of this review are just as applicable elsewhere.

GENERATING A NET BENEFIT FROM TOURISM FOR COMMUNITIES

If tourism is to act as an economic incentive for tolerance of wildlife then it must generate profits that offset the direct and indirect costs of living with

Table 8.1. *Some potential benefits from various types of wildlife-tourism venture for rural communities coexisting with wildlife*

		Community-based		
Tourism benefits	Parks-based[a]	Commercial concession	Public–private partnership	Community-owned
Individual financial				
Employment (guards, guides, porters, waiters, etc.)	•	•	•	•
Peripheral business opportunities (handicrafts, transport, etc.)	•	•	•	•
Communal financial				
Revenue-sharing (as household dividends or community projects)	•			
Lease/concession fees		•	•	
Bed night fees		•	•	
Entrance fees		•	•	•
Commercial profits			•	•
Non-financial				
Development inputs (roads, water, etc.)	•	•	•	•
Livelihood diversification	•	•	•	•
Capacity-building			•	•
Institutional strengthening			•	•
Security			•	•
Local pride			•	•

[a] Including tourism within state-protected areas and private ranches and reserves.

wildlife. This will depend on the balance of benefits and costs, and on the way that they are locally distributed among individuals. This section considers the first of these issues.

Commercial viability

For tourism enterprises to succeed they must generally be commercially viable – the income must exceed the operating costs plus the initial financing costs, although the latter may be provided by donor aid. The gross financial benefits of tourism are a function of demand, i.e. the number of visitors who are prepared to pay the price asked for the product. This in turn is dependent upon a number of factors.

Access

The vast majority of wildlife tourists are international visitors from Europe and North America (Goodwin *et al.* 1998). Therefore, demand is likely to be higher for a destination that is cheaply and easily accessible both internationally and within the destination country. Access is usually easier to protected areas that are already established destinations. Communal land some distance from protected areas is likely to have less well-developed access. For example, a community lodge in Papua New Guinea, developed with non-governmental organization (NGO) backing, failed to attract visitors because access to the remote site was expensive and operators were unwilling to incorporate the venture into their itineraries (Pearl 1994).

Security

Tourists will not visit a destination unless it is politically stable, with few risks to health and security. These factors can change dramatically, and national or international events can have significant effects on the local economy (Kayanja and Douglas-Hamilton 1984; Goodwin *et al.* 1998). Recent examples include the foot-and-mouth disease outbreak in the UK, and terrorist threats in Kenya and Indonesia. Community initiatives may suffer more than protected areas because the latter usually have well-established security networks and may be considered more secure.

Quality

The presence of easily visible wildlife, and in particular flagship species such as elephants or tigers, is a priority for most wildlife tourists (Goodwin and Leader-Williams 2000). Since such flagships are often those most in conflict with people, tourism may be well placed to directly offset the costs of living with them (Walpole and Leader-Williams 2002). However, outside protected areas where wildlife occurs at low density and is often persecuted, it is less easy to view. This may constrain community-based tourism (WWF 2000).

Uniqueness

Sites with unique wildlife, such as the Galapagos Islands, generate higher demand than those with relatively ubiquitous species for which substitute destinations exist. Community-based ventures can rarely compete with established protected areas in terms of wildlife viewing opportunities, and must develop their own niche that complements existing products. One of the potential problems with the current explosion of community-based tourism in East Africa is a focus almost exclusively on relatively high-end 'eco-lodges' (Watkin 2002). This market may already be saturated, and benefits will decline as increasing numbers of initiatives carve up a limited market.

Existing markets

Regardless of competition, the presence of an existing tourism circuit within a country or region is valuable since it provides a local market to tap (Ashley *et al.* 2001). This is in itself dependent upon the issues of access and security described above. For example, since protected areas in the Congo Basin are unable to reap significant tourism revenues (Wilkie and Carpenter 1999), it is unlikely that community-based ventures could do so either.

Marketing

Tourists will only visit if they know about an attraction and like the sound of it. High-profile examples of community-based tourism, such as Il N'gwesi in northern Kenya, have benefited from significant publicity and international awards. However, as similar initiatives develop, they will be less able to rely on the novelty value that pioneering examples have exploited. High-quality marketing is expensive for communities to sustain, and generally requires an umbrella organization, such as NACOBTA (Namibian Community-Based Tourism Association) in Namibia, to promote a collection of initiatives (Nicanor 2001).

Price

Price has a high impact upon demand and therefore revenue. Protected areas have traditionally been seen as a public amenity, and pricing policies that promote affordable universal access are one reason why protected areas have historically failed to generate significant revenues for conservation or communities (Walpole *et al.* 2001). Equally, overpricing may deter visitors and limit revenues. As protected areas in Africa increasingly charge premium fees for visitor entry, communities are being encouraged to do the same for access to their lands outside parks. Unless quality is maintained to an equivalent level in community areas, high pricing may be unsustainable.

Economic and social viability

Economic benefits and costs can be both financial and non-financial. Even if tourism is not financially profitable, it may still be viewed positively if it brings other, non-financial benefits. These may include capacity-building both individually and as a community, development inputs such as roads, power, etc., improved security, livelihood diversification, increased natural capital, reinforced socio-cultural values of wildlife and a sense of local pride. However, to be an incentive for conservation they must still be perceived to offset the costs of living with wildlife.

Wildlife conservation costs

Where community-based ventures are isolated from protected areas, they will have to bear the costs of ensuring that there is sufficient wildlife for visitors to view, which can severely reduce their viability. Even within protected areas, tourism rarely offsets the full costs of wildlife management (Walpole *et al.* 2001; Wells 1993). In general such costs, just like tourism development costs, tend to be bankrolled by governments and donors.

Wildlife conflict costs

Conservationists perceive tourism as an alternative to less wildlife-compatible livelihoods, thereby reducing conflict. However, tourism is often developed as a livelihood diversification strategy alongside traditional activities. If tourism acts as an incentive for conservation such that wildlife populations increase and become less wary of people then, paradoxically, it may exacerbate conflict, as has happened with lions in India (Saberwal *et al.* 1994). This is almost never taken into account when tourism is promoted.

Opportunity costs of forgone activities

Usually, however, some incompatible activities must be forgone in order to permit wildlife tourism. The loss of access to grazing and water resources set aside for wildlife, and the loss of culturally significant activities such as hunting, can represent a significant cost for communities (Gibson and Marks 1995; Berger 1996). Such opportunity costs will be higher in areas of high agricultural potential, where the relative economic value of wildlife will be lower even though biodiversity value may be high (Norton-Griffiths and Southey 1995).

Unmet expectations

Tourism is only likely to be acceptable where the costs described above are minimal, i.e. in low productivity areas with low human and livestock densities, and where local livelihoods are compatible and wildlife causes little conflict. With the widespread promotion of tourism as a conservation and development panacea, a culture of expectation and dependency has evolved among communities that appears to disregard both the risks and impact that the costs of tourism and conservation can have on net benefits. When these expectations are not met, disenchantment can ensue (Western 1994; Oates 1999). By displacing other development options and then underperforming, tourism brings with it the danger of reduced economic flexibility and increased poverty (Hackel 1999).

Limitations and trade-offs for community-based ventures

Community-owned operations generally offer more non-financial benefits than those where the community merely receives revenue and employment from a commercial operation on their land or from tourism in adjacent protected areas (Table 8.1). This is one of the reasons why donors tend to encourage communities to develop their own ventures, or to engage in partnerships leading towards community ownership and management of enterprises. However, community-based ventures are often at a competitive disadvantage compared with commercial operations even where conditions are favourable to generate demand. Some of the reasons for this are given below.

Lack of access to capital

Whilst direct financial benefits of tourism are often small, costs are often significant. The initial, direct costs of infrastructure development alone are usually beyond the resources of local communities, and require significant external investment that is unlikely to be recouped in the short term. Moreover, the recurrent costs of running the operation, maintenance and marketing all eat into profits. Even for an established venture, these costs will consume 70% or more of income (Ashley and Roe 1998). Unless projects have received donor support for development costs, profits may not be generated for several years. During difficult times, commercial operations will not declare dividends and may be forced to take out loans to survive. These options are not readily available to community operations.

Non-commercial priorities

The choice between using profits for commercial activities such as marketing, reinvestment or upgrading facilities, and provision of social benefits, is a difficult one for communities. In an extreme case, given an alternative between using a vehicle to take a sick person to hospital or using it for a tourist game drive, a community would be more likely to take the decision to help the sick person from their own community than would a commercial operator. The need to provide social benefits may lead to cuts in the quality of service.

Lack of understanding of tourism

Where communities have attempted to undertake their own tourism enterprises they have often failed due to a lack of professionalism, a lack of attention to product quality, a lack of understanding of what tourists come to see and a lack of appreciation of market realities (Roe *et al.* 2001).

Equally, where communities enter into partnership with commercial operators to avoid these pitfalls, they may expose themselves to commercial risks that they do not understand and have no control over. Profit-sharing partnerships that provide the greatest community 'ownership' are likely to yield little net income at first, even if gross income appears considerable (Ashley and Roe 1998).

Conversely, land concessions that involve little local participation do provide regular and uncomplicated lease fees and bed night levies (Reeve 2000). This exposes something of a paradox for the empowerment and participation lobby. Communities may be better off forgoing a certain level of ownership, control and participation in return for more stable or guaranteed incomes.

Ultimately, however, the overzealous promotion of community-based tourism reflects an unrealistic estimate of the size of the market for such products. Community-based tourism remains a niche market, inhabited by those clients who have a socially responsible ethic or a cultural interest, and who are prepared to substitute a community-based product for an alternative part of their itinerary. This sector of the industry may well be growing, but the majority of clients are still going to opt for a safari in a game park and a trip to the beach rather than squeezing in a few days at a community lodge.

IMPROVING WILDLIFE TOLERANCE AMONG INDIVIDUALS

Even where tourism is financially, economically and socially viable, it will only improve tolerance towards wildlife where those benefits reach the people bearing the costs, and where people understand and act upon the linkages between tourism benefits and wildlife conservation. This section considers these two critical conditions in turn, and the limitations of current approaches in achieving them.

Targeted benefit distribution

Business and employment

Despite the rhetoric, tourism rarely offsets the costs of living with wildlife for individuals. Tourism development in and around protected areas has traditionally been dominated by external commercial interests, with the result that much of the economic benefits are removed from the local economy (Bookbinder *et al.* 1998) (Box 8.1). Even where business and employment opportunities are open to local people, those that benefit are not necessarily those who suffer the greatest costs of living with wildlife. The former tend to be more wealthy and educated, whereas the latter tend to be poorer and more dependent upon resource exploitation and subsistence livelihoods (Mehta and Kellert, 1998).

Box 8.1. Parks-based tourism: dragon-viewing in Komodo National Park, Indonesia

The Komodo dragon (*Varanus komodoensis*), the world's largest lizard, is confined to the islands of Komodo National Park (KNP) and nearby Flores, in eastern Indonesia. With a population of approximately 3000, it is a threatened species, and a flagship for conservation in the region. However, it is also a large predator that occasionally attacks the people and livestock living within the park.

On the surface, it appears that tourism offers a means to offset the costs of living with dragons. Between 1984 and 1997 mean annual tourism growth was 26%. The 30 000 visitors in 1996 contributed over $1 million to the local economy around the park, partially supporting over 600 jobs and 30% of the local population. Local residents recognized the link between the presence of the dragons and the existence of tourism, and appeared to have a positive attitude towards KNP (Walpole and Leader-Williams 2002).

However, closer inspection revealed a significant mismatch in local benefits and costs. Whilst the island residents within the park suffered the loss of access to terrestrial resources within the park, and some conflict with dragons, 99% of the locally captured tourism benefits (ignoring the 80% or more of total tourism benefits that bypassed the local community altogether) flowed to the residents of the gateway towns outside the park through which tourists passed (Walpole and Goodwin 2000). Moreover, whilst receipt of tourism benefits improved stated attitudes towards tourism, it did not affect attitudes towards conservation (Walpole and Goodwin 2001). Finally, a management decision to halt the feeding of dragons during tourist viewing removed one of the only forms of revenue open to villagers, namely the sale of sacrificial goats (Walpole 2001).

Thus it would seem that, despite huge growth and economic potential, tourism offered few real incentives to local villagers to tolerate the dragons and

Box 8.1. *(cont.)*

the park, or to prevent access by external poachers. This illustrates the general failure of *laissez-faire* tourism development to engage the rural poor due to their lack of skills, capital and access to the market. Equally, as an export-oriented industry, tourism exposes both parks and people to the vulnerability of changing market demand. In this case, arrivals to KNP dropped by 25% in 1999 after political and economic instability in Indonesia. Demand is likely to be further dampened after the October 2002 terrorist bombing in Bali, given the importance of Bali as a regional access hub for KNP.

Communal benefits

Equally, communal benefits do not offset individual costs from living with wildlife. Such benefits are generally social rather than financial. Thus, under most circumstances, and usually with encouragement from NGOs and governments, benefits from community-based tourism operations will go towards communal projects such as schools, bursaries and clinics rather than being distributed as cash. Kenyan subsistence farmers whose children remain at home to protect crops from wildlife do not benefit from community schools built with tourism revenues (Walpole *et al.* 2003). Even where household dividends are paid, as in Uganda, they are likely to be too thinly spread to offset individual costs (Archabald and Naughton-Treves 2001).

Deciding how to distribute the benefits of tourism is complicated by the difficulty of identifying the appropriate beneficiaries. Defining the boundaries and membership of a community, and quantifying the costs and who bears the disproportionate majority is notoriously difficult. As a result, equitable distribution of benefits is rare, and those who pay the greatest costs of adopting non-consumptive utilization of wildlife through tourism are unlikely to reap a net benefit.

Understanding the linkages between tourism benefits and wildlife conservation

Changing perceptions

There is some evidence that tourism benefits improve people's stated attitudes towards wildlife and conservation (Infield 1988; Archabald and Naughton-Treves 2001). However, it is equally clear that costs associated with wildlife decrease local support for wildlife (Newmark *et al.* 1993; de Boer and Baquete 1998). Thus, where wildlife conflicts are high, people may support tourism development whilst concurrently clamouring for lethal control or consumptive use (Mehta and Kellert 1998; Archabald and Naughton-Treves 2001). The latter may construct more direct linkages than non-consumptive tourism by providing a greater sense of ownership over resources whilst directly offsetting wildlife conflict by removing problem animals (Leader-Williams *et al.* 2001).

Changing actions

The extent to which a change in attitude results in greater tolerance and less persecution of wildlife is unclear. In many cases, benefits are simply insufficient or too inequitably distributed to prevent wildlife persecution and unsustainable utilization. Moreover, the commercialization of wildlife through tourism, coupled with unmet financial expectations, risks destabilizing traditional cultural and spiritual relationships that may have helped to maintain wildlife in the past (Boonzaier 1996; King and Stewart 1996).

A further problem is that the relationship between wildlife numbers and tourism benefits is not a simple one, and there may be a significant lag between changes in the former and changes in the latter. Evidence from Komodo National Park suggests that a declining visibility of dragons may threaten future tourism revenues, and the park management recognize and are addressing this issue (Walpole 2001). However, it may be less clear to a pastoralist living among high predator densities that killing cattle-rustling lions will decrease his income from tourism. Indeed, given that his welfare will be maximized in the short term by taking the tourism revenue *and* killing the lion, he is likely to do both, as examples from Zambia and Kenya suggest (Lewis and Phiri 1988) (Box 8.2). By the time revenues begin to decline, it may be too late. Community-based tourism that relies on alternative cultural or environmental attractions may further undermine any perceived link between wildlife and tourism benefits.

Box 8.2. People and parks: revenue-sharing and buffer zone development around Masai Mara National Reserve, Kenya

The Masai Mara National Reserve (MMNR) is world famous for its huge density and diversity of wildlife, and in particular the wildebeest migration from the Serengeti and the large carnivores that depend upon it. In return for tolerance of wildlife outside the park, and the loss of access to MMNR itself, neighbouring Maasai agro-pastoral communities are entitled to a 19% share of local government income from tourism in MMNR, as well as revenue from tourism development outside the park.

MMNR generates enormous tourism demand, and millions of dollars of revenue annually. However, very little of that reaches local communities. Corruption and mismanagement of the park and its tourism reduced revenue flows to a trickle of their potential. Furthermore, opportunities for private tourism enterprise within the group ranches are controlled by powerful local elites who expropriate the best land for themselves at the expense of the less powerful and well connected (Thompson and Homewood 2002).

Box 8.2. *(cont.)*

Tourism is currently not offering local people an incentive to tolerate wildlife. As a result of continued livestock predation and crop damage, wildlife is being killed. Moreover, the opportunity costs of wildlife and tourism outweigh the benefits, particularly on the peripheries of the ecosystem. As a result, land is being transformed for intensive cultivation, and wildlife populations are in decline (Homewood *et al.* 2001).

Ironically, tourism itself may be contributing to the problem. The expectation of tourism benefits has fuelled immigration to the borders of MMNR, particularly around the entrance gates. This has increased local cattle densities and their incursions into the park, resulting in further wildlife disturbance and increased retaliatory killing of wildlife (Walpole *et al.* 2003).

Despite massive economic potential, tourism is failing to deliver its promises for people or wildlife in the Mara. In this example the cause is the failure of local government ownership and common property resource management systems to combat corruption, address individual needs and aspirations, and implement adequate controls and conservation measures. The future for the Mara ecosystem rests in a fledgling public–private partnership, The Mara Conservancy, which is implementing transparent revenue collection and dispersal, and reinvesting resources in security and management (Walpole and Leader-Williams 2001).

Preventing subversion

Even where tourism benefits act as an incentive for conservation for the majority, they are unlikely to yield conservation benefits without mechanisms to reduce the effects of subversion by the few, whether they be powerful individuals within a community or outsiders who gain more from overexploitation or persecution of wildlife than from conservation (Leader-Williams *et al.* 1990a). There is a clear need for strong institutions to prevent wildlife overexploitation and ensure equitable benefit distribution. However, wildlife departments often lack the resources or will-power to enforce regulations either within or beyond protected area boundaries, whilst communities themselves are often too weak to resist either internal or external overexploitation (Barrett *et al.* 2001). Under these circumstances conservation benefits will not be realized.

The limitations of communal approaches to conservation and development

The neo-liberal development policies of the past few decades have generally failed to engage poor rural communities in the free market economy. This is as true for wildlife tourism as it is for agriculture. In the search for a solution, many rural communities are being encouraged to adopt communal approaches based largely on socialist economic principles.

In most parts of the world, socialism – which has been defined as 'an economic theory or system in which the means of production, distribution and exchange are owned by the community collectively' (McLeod 1982: 1098) – has been discredited as a path to economic advancement, and abandoned in favour of more individual-based systems. Community-based conservation in Africa is one of the few areas where entire communities are encouraged to run businesses for their mutual self-interest, and to achieve equitable division of the proceeds from these ventures.

Expecting individuals to work for the common good without a corresponding reward has been identified as a key weakness of socialist systems. They are almost always open to individual exploitation and corruption (Box 8.2). This so-called 'tragedy of the commons' takes place because the benefits accrue to the individual whilst the costs are shared amongst the community. Thus the individual seeking to maximize personal welfare will gain more in the short term by overexploitation (Hardin 1968). This can be overcome if communities have a strong sense of unity and strong institutions for the enforcement of regulations (Ostrom *et al.* 1999; Barrett *et al.* 2001). Some small-scale traditional societies may have the former, and industrialized societies are more likely to have the latter. However, many rural communities in the developing world are in a state of transition and experiencing weakened traditional values and institutions, population pressures, increased social diversification and heterogeneity, and increased access to destructive technology. Under these circumstances, the rural poor have little choice but to discount the future heavily and focus on immediate personal benefit rather than long-term communal or societal welfare, whilst the powerful minority are able to exploit the system to their own advantage (Western and Wright 1994; Murphree 1995).

Many commentators advocate the decentralization of land tenure and wildlife proprietorship to the community, with the justification that giving communities ownership engenders responsibility (Lynch and Alcorn 1994). Nevertheless, without effective management and policing such devolution carries an equal if not greater risk of exploitation and wildlife decline.

CONCLUSIONS AND RECOMMENDATIONS

Wildlife tourism promises benefits to offset the costs of living with wildlife, thereby providing incentives to tolerate and conserve wildlife. However, for most rural communities living with wildlife in the tropics, the reality is very different. Benefits are few, and do not offset individual costs. Moreover, the benefits that do materialize are often too indirect to act as an incentive for conservation, and are easily subverted.

These problems span the spectrum of wildlife tourism models, from traditional parks-based tourism, through revenue-sharing with communities

Box 8.3. Community-based tourism: Torra Conservancy, northern Namibia

In 1996, the Namibian government amended its legislation to give rural communities the right to utilize wildlife. Prior to that, wildlife had remained the property of the state, and illegal killing of wildlife for subsistence and commercial benefit, and in retribution for conflict, had severely reduced wildlife populations. The legal ruling paved the way for the formation of community 'conservancies' upon which various forms of legal wildlife utilization could be practised alongside traditional livelihood options. To date, 13 conservancies have registered and over 20 more are under development. A major component of their wildlife utilization plans is non-consumptive tourism, either through community campsites or, increasingly, partnerships with the private sector (Roe *et al.* 2001).

Economic forecasts suggest that tourism has a high potential to be profitable both financially and economically (Barnes *et al.* 2002). For example, in Torra Conservancy in Kunene region, a partnership has been agreed with a commercial tourism operator. Benefits are flowing to individual community members through employment and business opportunities associated with a safari camp. Additionally, a nominal lease fee and 10% of the camp turnover is also given to the conservancy. Wildlife is recovering and the conservancy is almost financially independent (Humphrey and Humphrey 2003).

This and other examples in northern Namibia have been widely publicized. However, the detailed evidence suggests that non-consumptive tourism alone is unlikely to bring about widespread conservation and conflict resolution. The wildlife recovery actually preceded the tourism development as a result of donor-funded community game guard schemes initiated in the 1980s. Equally, donor investment has been critical in establishing the conservancy, its infrastructure and its early running costs. Tourism potential in this site is relatively high due to the recovering wildlife and relatively low human population density, and the low opportunity costs of the marginal, arid environment. Furthermore, trophy-hunting and other forms of consumptive use contribute a significant proportion of revenues (Arnold 2001; Humphrey and Humphrey 2003).

It may be too soon to properly judge initiatives such as Torra Conservancy. Although disagreements over revenue distribution have been overcome, and household dividends paid, the recovering wildlife may be causing increased conflict. A widespread review of community-based tourism in Namibia suggests that such initiatives are more likely to succeed as part of a multiple-use wildlife strategy involving appropriate, flexible and well-understood private partnerships where costs are equitably compensated (Roe *et al.* 2001).

around parks, to fully community-based initiatives. Some examples of the latter have been relatively successful (Box 8.3). However, these have required considerable long-term external support, and may not be replicable. Moreover, the most successful initiatives have comprised both non-consumptive and consumptive utilization in order to be profitable.

There is a growing shift in donor and NGO policy away from direct conservation intervention in favour of development interventions, on the assumption that poverty alleviation alone will solve environmental problems. In turn, many community-based tourism initiatives focus on generating benefits for people without developing the necessary linkages with conservation. As a result, ensuring that tourism acts as a mechanism for development overshadows consideration of whether it acts as an incentive for conservation, and some development advocates admit that it may not (Ashley and Roe 1998).

Relying on poverty alleviation to achieve conservation objectives is likely to be dangerous whether or not it is achieved. Small benefits generated at the outset of a tourism venture, together with the promise of greater benefits to come, may create a climate in which communities support conservation and tolerate wildlife. However, it is common that these expectations are not met. Equally, as benefits rise, so do aspirations, making it even more unlikely that tourism can satisfy whole communities indefinitely.

Without clear and unambiguous incentives and sufficient regulation, improving rural livelihoods may be too indirect a mechanism to generate conservation benefits. On this basis, some commentators are now advocating a direct-payments approach that more clearly rewards conservation performance (Ferraro and Kiss 2002). However, better use could be made of tourism benefits to provide clearer links. Development inputs such as schools and roads should be the responsibility of the state, not the wildlife or tourism sectors. If tourism benefits were used to mitigate conflict directly then they might be of more use to conservation.

The three critical conditions (providing net benefits that are appropriately distributed and clearly linked to positive conservation outcomes) are most likely where:

- tourism potential is high
- wildlife conflict is low or easily mitigated
- opportunity costs are low or reduced
- the 'community' is a well-defined and cohesive constituency of beneficiaries
- external investment and expertise are harnessed in public–private partnerships in a way that is comprehensible and minimizes risk
- local entrepreneurial activities are encouraged through training and micro-investment schemes
- linkages with wildlife conservation are clear and well understood
- benefits are performance-related, so as to act as direct conservation incentives

- benefits are used to mitigate conflict and ensure livelihood and food security
- leadership is strong, representative and non-corrupt
- cultural values support, rather than contradict, non-consumptive use
- regulatory and enforcement mechanisms are strong.

These conditions can be applied to the full spectrum of tourism models, from parks-based to wholly community-based enterprises. However, they are unlikely to be met in all, or even the majority, of cases. Whilst tourism is a valuable use of wildlife, it is unlikely to be a justification for conservation on its own (McNeely 1989; Brown 1998).

Box 8.4. Tourism as part of a broader conflict mitigation strategy

Tourism as an indirect development-based incentive for conservation suffers many limitations. However, it may achieve greater success as part of a multifaceted strategy for conflict resolution, in the following ways.

Funding mitigation

Direct efforts to prevent livestock predation and crop damage, such as guarding, fencing and other deterrents, represent a substantial cost for subsistence communities. Tourism offers the potential to provide resources to fund such direct mitigation activities, although it is currently rarely applied in this way.

Targeted compensation

Compensation rarely solves conflict, and may exacerbate it by removing any incentive for crop and livestock protection. However, compensation might be turned into an incentive for conservation through spatial zoning, with no payment for losses within a buffer zone adjacent to parks or tourist camps, but payment for verified incidents further afield. This might act to separate crops and livestock from areas of highest wildlife value, and would be a valuable use of tourist income. Such schemes are being considered on a small scale to mitigate carnivore conflicts in Kenya and India (C. Cottar and U. Karanth pers. comm.).

Insurance and direct incentives

Insurance schemes jointly supported by local communities and tourism ventures may be more sustainable than compensation. Such a scheme has been established to mitigate snow leopard conflict in Baltistan (Hussain 2000). A similar project has gone further and established conservation incentives programmes with local communities in Mongolia. In areas with significant carnivore–livestock conflict, tourism is providing income through handicraft sales. Cash bonuses are paid in return for verifiable tolerance of carnivores and adoption of ungulate-friendly pastoral practices (T. McCarthy and P. Allen pers. comm.).

Consumptive use

Hunting or cropping may provide a more direct perceived link between wildlife and benefits than non-consumptive nature-based tourism. Moreover, it provides a means of reducing conflict and improving tolerance through problem animal control (Leader-Williams *et al.* 2001). A zoned system of non-consumptive tourism in parks and buffer zones, with limited hunting in outlying areas, would diversify wildlife utilization, providing greater and more secure benefits over the longer term.

This does not mean that tourism has no role to play. Community-based tourism enterprises are intended to break even, provide social benefits and support the costs of wildlife conservation. This places them at a disadvantage compared to commercial operations, which only have to make money. In most cases it will not be possible for a wildlife enterprise to succeed in all three objectives. This apparent lack of financial sustainability does not mean that an initiative is unsuccessful – it may still be providing conservation benefits more cheaply than any other way. Moreover, if communities are acting as the custodians of wildlife for an international community which values its continued existence, then the amenity values captured through tourism represent only a part of the total economic value of wildlife. Thus, there is an argument that tourism alone should not be expected to foot the entire bill for wildlife conservation. If so, the international community must decide whether the biodiversity being conserved is worth a long-term financial input.

In the final analysis, tourism is capable of generating a range of benefits for people coexisting with wildlife. However, it is only one tool for improving wildlife tolerance. Its conservation legacy is likely to be greatest where those benefits are deployed as part of a battery of conflict mitigation methods that includes direct incentives, targeted compensation, consumptive use and conflict-avoidance strategies (Box 8.4).

9

Does extractive use provide opportunities to offset conflicts between people and wildlife?

NIGEL LEADER-WILLIAMS AND JON M. HUTTON

INTRODUCTION

The use of wildlife remains in something of a cleft stick as a possible solution to contemporary problems in conservation (Hutton and Leader-Williams, 2003), such as offsetting the costs of conflict between people and wildlife. On the one hand, the list of abuses suffered by many species of wildlife when used commercially seems endless (Milner-Gulland and Mace 1998; Bennett and Robinson 2000). At the same time, the Convention on Biological Diversity promotes the role of sustainable use in providing people with the necessary incentives to conserve biodiversity on land, which ultimately requires decisions about the opportunity costs of different forms of land use (McNeely 1988, Swanson 1994; Hutton and Leader-Williams 2003; Convention on Biological Diversity 2005).

The Convention on Biological Diversity has, nevertheless, based its aspirations on situations where wise use has led to positive incentives for conservation. For example, the catastrophic losses of native species after the colonization of North America led sportsmen to protect their interests in the early and mid nineteenth century, by seeking to reduce the numbers of game animals killed and establish preserves (Gray 1993). Sportsmen who fished and hunted for pleasure, rather than for commerce or necessity, became one spearhead for formal policies to conserve wildlife and its habitats (Reiger 1986; Jackson 1996). Likewise, following the colonization of Africa, formal conservation policies in many countries in the late nineteenth and early twentieth century sought to regulate hunting and to establish game reserves (MacKenzie 1988; Leader-Williams 2000), and subsequently in southern Africa to re-establish species on private land to create further hunting opportunities (Bothma 2002; Lewis and Jackson, Chapter 15). These approaches built, in a more inclusive way, on those

People and Wildlife: Conflict or Coexistence? eds. Rosie Woodroffe, Simon Thirgood and Alan Rabinowitz. Published by Cambridge University Press.

policies followed over many centuries by royalty and aristocrats in Europe and Asia, of establishing royal forests for hunting (Gray 1993).

An alternative incentive for protecting wildlife and wilderness areas emerged in North America in the 1870s, with the establishment of national parks as 'pleasuring grounds' for the people (Runte 1987; Leader-Williams *et al.*, 1990a). Worldwide, non-consumptive tourism has increasingly provided an economic and social incentive for government-sponsored landscape and wildlife conservation, given that wildlife tourism is now one of the fastest-expanding sectors of the world's largest formal industry (Honey 1999; Walpole and Thouless, Chapter 8). In turn, this has led to a clash of ideals among conservationists over whether wildlife should or should not be killed to promote conservation (Hoyt 1994). Unfortunately, this pits opposing positions against each other (Conover 2002), even though most conservationists are fully engaged in the common objective of finding incentives to conserve wildlife and its habitats.

In this chapter, we will not probe the relatively fixed positions of individuals within this clash of ideals (Box 9.1). Instead, we aim only to examine how different forms of extractive use might contribute to the common commitment of seeking to offset the costs of conflict for those living with wildlife, and to serve as an incentive to conserve wildlife habitats (Gray 1993; Hutton and Leader-Williams 2003). Throughout the chapter, we favour the generic term *extractive use*, which equally refers to the removal of individual animals or plants, alive or dead, as well as the extraction of parts and derivatives thereof. This avoids any confusion over the term *consumptive use*, which is often only equated with the killing of individuals (e.g. Gray 1993). However, the biological effect of removing individuals entirely from the wild population is the same whether this involves lethal extraction through hunting or fishing, or non-lethal extraction through collection or live capture (see Hutton and Leader-Williams 2003). Furthermore, throughout the chapter, we stress the critical importance of monitoring decisions about land-use changes, including changing patterns of illegal offtake, to indicate the success of incentives to promote conservation (see Hutton and Leader-Williams 2003), including offsetting the costs of conflict between people and wildlife.

HOW CAN EXTRACTIVE USE OFFSET CONFLICTS?

People have long benefited from direct extractive use of wild living resources, and such use still remains an imperative for many people (Hutton and Leader-Williams 2003). Where extractive use is an imperative for local societies, the benefits of living with wildlife are presumably

Box 9.1. Hunting with hounds as an incentive to conserve lowland habitats in Britain

Hunting with hounds has followed its current traditions in Britain since the seventeenth century, and is heavily embedded in British folklore. Most of the 200 registered packs of hounds hunt foxes (*Vulpes vulpes*) across lowland Britain, while a small number of packs still hunt deer (*Cervus elaphus*), mink (*Mustela vison*) and hares (*Lepus europaeus*). The two polarized sides to recent debates on whether to ban hunting have focussed, on the one hand, on claims of cruelty inflicted on the quarry species and, on the other hand, on claims of its utility in controlling pests, as well as generating opportunities in the countryside. Until recently, much evidence on hunting with hounds was either speculative or contradictory (Burns *et al.* 2000), particularly where research funds were provided by special interest groups on opposing sides of the debate. Here, we concentrate on conflict and incentives to conserve, rather than on issues of animal welfare, which permeate so many debates.

A temporary halt to fox-hunting during the foot-and-mouth disease outbreak of 2001 suggested that a permanent ban would not result in a dramatic increase in fox numbers (Baker *et al.* 2002). Hunting accounts for only 6.3% of the 400 000 foxes that are killed annually in Britain, and over five times as many foxes are killed by shooting and snaring than by hunting with hounds (Burns *et al.* 2000). Therefore, it is not surprising that fox-hunting is an ineffective form of pest control. Instead, fox-hunting harvests a sustainable offtake and might represent a form of traditional, community-based conservation (Leader-Williams *et al.* 2002). Genuine community-based conservation projects should improve local tolerance towards wildlife and provide an incentive to maintain biodiversity without the need for statutory regulation and recurrent public funding.

The defence of hunting with hounds on conservation grounds relies on two main predictions in the event of a ban. First, that landowners' tolerance of quarry species, particularly those claimed to be pests, would decline, thereby increasing their persecution by other potentially less humane methods and, counter-intuitively, reducing quarry numbers. Second, that landowners would be less likely to voluntarily maintain biodiversity-rich habitats used by quarry species, such as woodlands and hedgerows (Leader-Williams *et al.* 2002). However, is there any evidence to support these lines of argument?

An example of the first prediction is the herd of nearly 3000 wild red deer that live on Exmoor, in southwest England (Thomas 2002). Much of the 700 km^2 over which the herd roams comprises over 250 small farms through which the deer move freely for grazing and shelter. These farms suffer crop damage, particularly to maize grown for silage, to young trees and to hedge banks. The economic costs of this damage have been offset by community interest in the traditional management of the deer by hunting with hounds (National Trust 1993; Thomas 2002). Farmers claim that the hunt disperses deer from their land. Furthermore, landowners and farmers who support the hunt are keen to preserve the deer and ward off poachers from their land. A public inquiry has claimed that a permanent ban would lead to increased shooting of the deer by farmers, and to less protection from poachers (National Trust 1993) and, by implication, greater cruelty from wounding

(Urquart and McKendrick 2003). The temporary ban on hunting over that part of Exmoor owned by the National Trust, as a resulted of subsequently disputed claims of cruelty (Bateson and Bradshaw 1997), offers the chance to test these claims in an opportunistic experiment.

A recent study has provided published evidence supporting the second prediction (Oldfield *et al.* 2003). Much of Britain's biodiversity exists on privately owned land, where its conservation relies on appropriate habitat management by landowners, who do indeed voluntarily conserve biodiversity-rich habitats through the incentive of participating in field sports, such as fox-hunting and game-bird shooting, which require natural habitat used by quarry species (Tapper 1999). In three sites across central England, landowners participating in field sports maintained the most established woodland and planted more new woodland and hedgerows than those who did not, despite the equal availability of subsidies. Unsurprisingly, the effect of hunting with hounds, which is an important social activity for landowners, was less than that for game-bird shooting, which is also an important source of revenue, but the effect of hunting with hounds was nonetheless significant (Oldfield *et al.* 2003). This suggests that taking sustainable offtakes by hunting with hounds provided, an incentive for habitat conservation and increased tolerance of quarry species. Full implementation of the recent ban made on animal welfare grounds will allow these predictions to be tested experimentally.

accepted alongside the costs. However, commercial forms of extractive use have been developed more recently as part of the choice in modern conservation paradigms (Gray 1993; Hutton and Leader-Williams 2003; Lewis and Jackson, Chapter 15). Most of these forms of extractive use have not been purposefully designed to solve conflicts between people and wildlife. Nevertheless, some could incidentally serve to offset conflict directly, or to improve tolerance towards wildlife. However, if extractive use is to serve as a financial or social incentive to increase tolerance to wildlife, then it must generate benefits that offset the direct and indirect costs of those people living with wildlife (Emerton 2001; Walpole and Thouless, Chapter 8). This in turn will depend on the balance of benefits and costs, and on the way that they are locally distributed among individuals (Emerton 2001). In this section we examine some of the general issues associated with sanctioned forms of extractive use in contemporary conservation, as a possible tool to offset conflict, and how they may differ in this regard from wildlife tourism.

Extractive use and conflicts in contemporary conservation

Modern conservation paradigms often promote rules and regulations that outlaw extractive use in high category protected areas (IUCN 2003). Hence, the real battlegrounds between rural people and wildlife are found outside parks where rural people coexist with wildlife (see also Walpole and

Thouless, Chapter 8). While neighbours may not wish to see their local protected area abolished, the negative attitudes they hold towards such protected areas often arise from a complex of reasons (Conover 2002). In Africa and Asia, these reasons include lost access to resources, the lack of these or other local benefits from protected areas, the actions of protected area staff towards neighbours, and the loss of life and livelihoods caused by wildlife originating from protected areas (Infield 1988; Nepal and Weber 1993; Newmark *et al.* 1993; Kothari 1996; de Boer and Baquete 1998; Gillingham and Lee 1999). In theory, the first three effects could be separated from the fourth, if it were possible to compare attitudes outside pairs of protected areas that were similar in almost every respect but for being fenced (e.g. Infield 1988) and unfenced (e.g. Gillingham and Lee 1999). Unfortunately, the latter two studies are from two countries, apartheid South Africa (Infield 1988) and independent Tanzania (Gillingham and Lee 1999), with such different policy backgrounds that meaningful separation of these effects is not possible. Therefore, it appears best at this stage to consider how extractive use can help offset the complex of negative attitudes towards protected areas, including conflicts caused by wildlife, if indeed this is a separate effect.

In the case of conflicts between wildlife and human interests on private land, the situation is much less complex, because – in rural Africa and Asia at least – private landowners do not usually face the same problems with protected areas as do local communities. However, conflicts with wildlife may well occur on private land (Conover 2002), for example by herbivores that compete with livestock or damage crops (Conover 2002) (Box 9.1) or top charismatic predators that prey on livestock (Rabinowitz, Chapter 17) (Box 9.2). In situations where extractive use is sanctioned for private landowners, it is easier to disentangle their land-use decisions than it is for communities around protected areas.

Could extractive use help to offset conflicts?

Direct extractive use of species causing conflict

Extractive use of problem species, or of those individuals causing conflict, has the potential to reduce conflict directly through lethal control, as well as providing compensatory benefits such as meat. Thus hunting of problem species may be a routine part of a farmer's life, for example for the Maya in Mexico (Jorgenson 2000) or for local people in North Sulawesi (Lee 2000). In both these situations, where the meat from hunted animals is of dietary importance, the levels of offtake may be high enough to reduce agricultural damage. In contrast, the offtake of problem species hunted for sport, such as red deer in Britain (Box 9.1) or elephants (*Loxodonta africana*) in Zimbabwe (Box 9.3),

Box 9.2. Sport hunting to help reduce control of problem cheetahs on Namibian farmland

From a global population of perhaps 15 000 cheetahs (*Acinonyx jubatus*), some 20% occur in Namibia (Nowell and Jackson 1996; Marker *et al.* 2003c), of which some 95% range freely on commercial livestock and game farms (Marker-Kraus 1994; Marker *et al.* 2003b). In the 1980s, 43% of Namibian commercial farmers viewed cheetahs as a serious pest (Marker-Kraus *et al.* 1996), although the threat might be exaggerated (Marker *et al.* 2003a), and farmers removed as many as 7000 cheetahs between 1980 and 1991 (Marker *et al.* 2003b). With awareness building by the Cheetah Conservation Fund, only 300 cheetahs were removed annually from 1986 to 1995 (Marker *et al.* 2003a). This level of offtake was probably sustainable from a population of 3000 cheetahs, but it would have required very little reduction in population growth rate or increase in offtake for problem animal control to become unsustainable (Marker *et al.* 2003b). As importantly, it continued to be a wasteful use of a valuable asset and fostered an approach that provided no financial incentive farmers to limit the killing of problem cheetahs. Consequently, after Namibia acceded to CITES in 1992, it presented a strong case for Parties to allow sport hunting of cheetahs, and a quota of 150 animals was agreed in 1994. Only 25% of farmers thought the cheetah was a serious pest by 1993 (Marker-Kraus *et al.* 1996), but this does not imply that the new incentive provided by the CITES quota was solely responsible (Maryland 1997), as awareness building had also played a part in the change of perceptions (Marker *et al.* 2003c).

The offtake by sport hunters has risen from 20 to 101 cheetahs between 1994 and 2003, so the agreed annual quota of 150 cheetahs has never been fulfilled (Marker *et al.* 2003c). In order to link the trophy hunting of cheetah to a conservation programme that would enhance its status in Namibia, a contract was developed between the Namibian Professional Hunters Association and the US-based Conservation Force. This Enhancement Agreement makes efforts to link sport hunters with commercial farmers who have problem cheetahs, and adds US$1000 for each animal to the current trophy fee of US$2000. As a result, some US$356 000 has accrued for cheetah conservation activities since 2000. Meanwhile, Namibia's population has increased to some 4500 ± 1250 cheetahs, although 150–200 problem cheetahs continue to be shot annually.

The sport hunting of cheetahs has not fulfilled its potential in terms of fully using the quota nor of reducing problem animal control since the mid-1990s. However, the majority of the world's sport hunters live in North America, and legal offtakes and earnings could be increased if cheetah trophies were allowed into the USA. In 1995, the Government of Namibia petitioned the US Fish and Wildlife Service to downlist the status of Namibian cheetah from 'endangered' to 'threatened', which would have achieved this aim. Their case was well supported by a study from the University of Maryland (Maryland 1997). The petition was considered and finally rejected in 2000 (US Federal Register 2000). This outcome appears negative for cheetah conservation, because the costs of conflict, as evidenced by numbers shot on problem animal control, exceed the opportunity costs of unfulfilled benefits, as evidenced by an unfilled quota that is well within the limits of sustainability of an increasing cheetah population.

Box 9.3. CAMPFIRE helps to reduce conflict with wildlife and encourage conservation on communal land in Zimbabwe

The legislation for the Communal Areas Management Programme for Indigenous Resources (CAMPFIRE) had its origins in colonial Rhodesia. Because much wildlife occurred outside protected areas, and both private and communal land-holders had little or no incentive to manage it, wildlife habitat and numbers were declining seriously. Therefore, the Parks and Wildlife Act 1975 sought to confer custodianship of wildlife to owners or occupiers of alienated land, to encourage conservation through sustainable use. The 1975 Act initially conferred proprietorship (or 'Appropriate Authority') only to white farmers on private land. This led to a great increase in private land under wildlife management, through economic motivation and self-interest, without adding to formally protected areas and stretching state budgets for conservation further (Child, 1995).

At Independence, wildlife management authorities sought to extend similar rights to communal land residents, because significant wildlife populations and habitat remained on remote communal land near to protected areas. A 1982 amendment to the Act allowed the Minister to appoint a rural district council to be 'the Appropriate Authority for such area of Communal Land as may be specified'. This provided an enabling environment to extend economic benefits demonstrated for private land to communal land, albeit only at the level of the district council. Strictly, the CAMPFIRE programme applies only to areas under communal tenure, and seeks to promote rural development rather than conservation *per se*. Equally, if successful, conservation objectives will be met by ensuring that more communal land is managed under wildlife than other less appropriate forms of land use. Thus, CAMPFIRE was fully integrated to the national policy on wildlife, alongside achieving conservation objectives in state-run protected areas and on private land.

Two rural district councils were granted Appropriate Authority status in 1988, and CAMPFIRE has since encompassed some 30 rural district councils (Bond 2001), or some 36 000 km^2 of communal land outside protected areas. Before Appropriate Authority was extended to rural district councils, revenue from wildlife-based activities in communal lands was channelled to the central treasury. Local people only benefited from wildlife through non-sanctioned hunting and snaring. Consequently, communal-land farmers were primarily concerned with reducing conflicts that wildlife posed to human life, livestock and crops. From 1989 to 1996, district councils in CAMPFIRE earned a total of US$10 million (Bond 2001), primarily through sport hunting, which has provided over 90% of income through fees and through leases sold to commercial safari operators. Of this, 60% of income derives from the sale of elephants hunts (Bond 1994). The balance of income earned has come from the sale of tourism lease rights, and the sale of hides and ivory, and of products such as crocodile (*Crocodylus niloticus*) (Box 9.4) and ostrich (*Struthio camelus*) eggs (Bond 2001).

Once benefits started to flow in from CAMPFIRE to the first rural district councils, attitudes to wildlife among local communities began to change (Murphree 1993). Revenues were allocated for community projects, such as clinics and schools, and for household dividends, requiring searching decisions on which households to include and exclude in benefits from common property. As attitudes to wildlife began to improve, poaching was no longer seen as individual and entrepreneurial defiance of state regulations, but rather as theft from one's neighbour leading to reduction of household dividends. Therefore, economic incentives were central to a process of important institutional change under CAMPFIRE (Bond 2001).

In the early 1990s, the pattern of benefits accruing to the increasing numbers of rural district councils encompassed in CAMPFIRE changed (Bond 2001). While the total earnings flowing into CAMPFIRE rose between 1989 and 1996, the median benefit accruing to CAMPFIRE households fell overall in real terms, from US$19.40 in 1989 to US$4.49 in 1996 (Murombedzi 2001). Hence, the direct incentives at household level for managing wildlife and its habitat became increasingly marginal in all but a few communities when compared to other sources of income such as agricultural production (Murombedzi 2001). Two main constraints were identified (Bond 2001). First, wildlife-based income was inversely related to population density, and many communities were users of wildlife produced elsewhere. Second, Appropriate Authority was only devolved to rural district council rather than to producer level. While rural district councils are encouraged to devolve 50% of revenue earned from wildlife to ward level, they are not obliged to do so. Accordingly, much of the revenue earned from CAMPFIRE does not accrue directly to local communities or to households who suffer the costs of living with wildlife.

At the district level, however, income from CAMPFIRE has been an important source of locally earned revenue, accentuated further by structural and fiscal reforms that devolved greater independence to rural district councils. Consequently, many rural district councils have imposed decisions on communal-land farmers to manage land for the maintenance of wildlife and its habitats. In contrast, few units of production at sub-district level have had the incentives to make their own positive land-use planning decisions. The two most notable exceptions have been the high-income wards of Kanyuria and Mahenye, which have shown very positive institutional change due to the scale of their incentives (Murphree 2001).

Despite its success and reputation, therefore, some key structural issues have dogged more widespread progress in CAMPFIRE. A key principle of CAMPFIRE, that appropriate proprietorial units were to be communities of collective interest, has never been legally enshrined, and these communities still cannot be granted Appropriate Authority. Instead, only rural district councils, which are administrative units rather than production units, can enjoy this status. Thus a legal discrimination still exists between communities and the former private-land farmers, who were both proprietorial and production units. In turn, this has given rural district councils, and increasingly central government, the opportunity to appropriate revenue, thereby replicating the extractive practices of colonial and post-Independence governments. Furthermore, it is unclear what the current political and economic impact will be on wildlife in the communal lands. *Inter alia*, there has been a country-wide breakdown in land tenure, and this will inevitably have an impact on emerging common-pool resource management units in communal land.

may not directly reduce agricultural damage. Nevertheless, hunting for sport may serve to increase tolerance, and thereby reduce the intensity of other forms of retribution or problem animal control (Conover 2002; see also Treves and Naughton-Treves, Chapter 6). Conversely, promoting use of problem species in order to improve their status can have the unexpected consequence of later increasing conflict by better conserving wild populations (Box 9.4).

Box 9.4. Ranching of crocodiles to increase sustainable offtakes while offsetting conflicts

The crocodile family comprises 23 extant species of tropical and subtropical aquatic reptiles, many of which can kill people. Women who collect water and wash clothes in crocodile habitats are particularly vulnerable (Ross 1998). In turn, crocodilians are particularly prone to over-hunting by people, because adult crocodilians are generally long-lived, slow to reach sexual maturity and experience low natural mortality (Hutton 1992). Therefore, over-hunting of crocodiles for their hides, coupled with more recent problems of habitat loss, had resulted in most of the 23 species of crocodile becoming endangered, depleted or declining in numbers by the early 1970s, and some species were on the brink of extinction. Excessive exploitation was rampant, regulated harvests were almost non-existent, and illegal trade in crocodile products was the norm (Hutton and Webb 2003).

Efforts to conserve crocodilians began with *in situ* protection in certain range states in the 1960s. Most species (17 out of 23) of crocodilian were included on Appendix I of CITES in 1975, and the remainder were included on Appendix II, to regulate international trade in adult skins. However, the IUCN Crocodile Specialist Group increasingly believed that long-term conservation of crocodiles was only possible through sustainable-use programmes that showed conservation benefits from wild populations, rather than by protection of wild populations alone (Ross 1998). Therefore, ranching programmes were developed as a positive measure under CITES for those crocodilian species originally listed on Appendix I (Hutton 1992). In terms of biological sustainability, ranching is a technique that ideally suits the life history the crocodile family. Because eggs and juveniles of this family experience very high mortality, removing up to 92% of eggs and juveniles from the wild can have less impact upon the wild population than removing breeding adults. The effects of harvesting can be further offset by returning some larger ranched animals to the wild (Hutton 1992). In terms of their economic and social benefits, these programmes comprised the collection of eggs, and sometimes hatchlings from the wild, and rearing of the resulting young on a ranch produces a crop of skins and meat. Local communities remain the custodians of eggs and young outside strictly protected areas, and can control the supply of eggs to the ranchers who pay them for eggs. In terms of conservation benefit, local people, now deriving benefit from the resource, had less incentive to kill adults. As a result of these efforts, illegal trade in crocodilian skins vanished, and several populations of the Nile crocodile (*Crocodylus niloticus*) and the saltwater crocodile (*C. porosus*), and one species, the American alligator (*Alligator mississippiensis*), have been subsequently downlisted to Appendix II. As importantly, the status of many species has improved in the wild, and those crocodilians that are the subject of sustainable-use programmes are considered the least threatened (Ross 1998).

While generally a positive situation, problems are emerging. Not all crocodile range states have successfully invested in a ranching programme (Hutton and Webb 2003), while the demand for ranched skins now appears fully met by those range states which first engaged in ranching. Furthermore, the custodians of the eggs derive little benefit from the final price of the crocodile skin product, particularly if the ranches are far removed from the sites of egg production. Finally, the numbers of recorded fatal conflicts have increased as crocodile populations have recovered and more people use crocodile habitats. For all these reasons, more range states are requesting CITES for adult quotas, to shift the balance more towards greater local benefits and reduced local costs.

Extractive use of species other than those causing conflicts

Use of wild species or natural resources other than the problem species cannot reduce the conflict directly *per se*, but may help to provide alternative benefits to offset the costs of conflict. Thus, sanctioned collection of natural products from the forest may help offset the conflict caused by species that cannot be used extractively (Box 9.5). Furthermore, elephants are usually not the most important cause of crop damage in Africa (Naughton-Treves and Treves, Chapter 16), but their use, either through problem animal control (Box 9.6) or through sport hunting (Box 9.3), at levels above the net damage they cause in relation to other species that crop-raid, may help offset the conflict more commonly caused by other species (Gillingham and Lee 1999).

Extractive use as part of a suite of measures to offset conflicts

Even though extractive use may provide benefits, these may not alone be sufficient to increase tolerance of conflict. Indeed, extractive use is often implemented as part of a suite of measures to provide benefits to communities. Hence, other conservation tools such as education, tourism and law enforcement may be deployed alongside extractive use, and may also play a role in helping to offset conflict (Conover 2002). For example, benefits from gorilla (*Gorilla gorilla*) tourism and from extractive use of natural products may help to offset the costs of crop-raiding by gorillas (Box 9.5), while education of private landowners (Marker *et al.*, 2003a) and sport hunting of cheetahs may help to offset the costs of livestock loss to cheetahs (Box 9.2).

Box 9.5. Harvest zones as part of an integrated approach to reduce conflict with mountain gorillas around the Impenetrable Forest

Bwindi Impenetrable National Park is renowned for its population of some 300 mountain gorillas that are the prime tourist attraction of Uganda, as well as the flagship species for efforts to conserve the forest (McNeilage *et al.* 2001). However, local communities complain of frequent crop raiding to the southwest of Bwindi, and of their fear of gorillas that come to feed on banana plants in cultivation patches. Furthermore, elephants also frequently raid crops in the rainy season to the east of Bwindi, while baboons (*Papio* sp.) and bush pigs (*Potamochoerus* sp.) are the most common and serious crop-raiders around Bwindi (Biryahwaho 2002; Siriri 2002). Conflicts arising over crop-raiding may pose a direct threat to gorillas, as well as increasing contact, and the risk of transmitting diseases, between people and gorillas (Kalema-Zikusoka *et al.* 2002; Guerrera *et al.* 2003).

Various conservation measures adopted at Bwindi have also resulted in conflict between local people and the national park. Local people were formerly employed in sanctioned pit-sawing and gold-mining activities, and also freely collected major forest products, such as firewood and building poles, and minor non-wooden forest products, such as honey, basket-making materials and bamboo, from the forest (Cunningham 1996). After periods as a Crown Forest, and then as a Forest Reserve, the Bwindi Impenetrable National Park was declared in 1991. This fuelled anger among local people, once local human activity was completely prohibited in the forest under national park status, with the subsequent loss of employment and legal access to forest resources. Villagers refused food and community services to national park staff, who were also often attacked by angry mobs of pit-sawyers and miners. Resentment by villagers also resulted in direct threats against the gorillas, and the deliberate setting of forest fires (Wild and Mutebi 1996).

Sanctioned collection of minor forest products from Bwindi was reinstituted in 1994 as a conflict resolution strategy, primarily because local communities depend heavily on forest resources for most of their subsistence needs. Under collaborative management arrangements between the national park and certain parishes, harvest zones have been established in some 20% or 66 km^2 of Bwindi. Each participating parish has a memorandum of understanding with the national park that defines the roles and responsibilities of the resource users, particularly regarding the records of harvested resources and the reporting of illegal activity. Sanctioned uses comprise bee-keeping, and the collection of herbal medicines and basket-making materials (Bensted-Smith *et al.* 1995; Wild and Mutebi 1996).

Later programmes have sought to further improve local livelihoods and social conditions. A programme to share revenue from gorilla tourism began in 1995, and had contributed funds to 19 community projects by 1999. In 1996, a donor lodged funds in a conservation trust for Uganda's two mountain gorilla national parks, and this has also helped to fund a variety of community projects around Bwindi. Meanwhile, strict law-enforcement measures and development of gorilla tourism have continued inside the national park (Muruthi *et al.* 2000).

Bwindi has been a national park for over 10 years, and conservation of its flagship gorilla population remains a primary goal. Bwindi has been hailed a success in protected area management through its integrated approach

(Borrini-Feyerabend 1997). The establishment of harvest zones was key in reducing conflicts around Bwindi, and in improving the attitudes of local communities towards the national park, and to crop-raiding (Bensted-Smith *et al.* 1995; Wild and Mutebi 1996; Blomley 2003; Makombo 2003). Their success in improving attitudes probably arose because they restored benefits directly to those who suffered the costs of denied access and of crop-raiding. However, the impacts of integrated programmes on illegal activities, and on the subsequent distribution of flagship species, are still not known. Consequently, further questions still remain about the effectiveness of the integrated policy for conserving Bwindi and its gorillas, and research needs to determine whether improved attitudes among the local community have translated into improved conservation and increased tolerance of crop-raiding on the ground.

Box 9.6. Trophy and meat hunting in and around Selous Game Reserve in Tanzania

The Selous Conservation Programme was established in 1988 in response to the unsustainable commercial poaching of black rhinos (*Diceros bicornis*) and African elephants living in one of Africa's largest protected areas during the 1970s and 1980s (Krishke *et al.* 1996). Comprising some 50 000 km^2 of predominantly miombo woodland in the southeast of Tanzania, the Selous' national designation allows sustainable use of its resources. The Selous is also internationally designated as a World Heritage Site. As a large and inaccessible area infested with tsetse flies and underlain by poor soils, yet formerly holding one of Africa's largest big game populations, the Selous has developed a considerable reputation as a safari hunting destination. The Selous was divided into 47 hunting blocks in the 1960s. Tourist hunting was temporarily banned in Tanzania from 1973 to 1978, but has since remained open and the industry has grown steadily in size (Leader-Williams *et al.* 1996a). More recently two of Selous' former hunting blocks have been given over to leases for non-consumptive tourism, which is also growing in the Selous (Leader-Williams *et al.* 1996b).

The primary aim of the safari hunting industry in Tanzania is to generate some form of sustainable income from the use of wildlife. The government can charge a range of right-to-use fees, comprising game fees for each animal shot, as well as observer fees, conservation fees, permit fees and trophy-handling fees. In addition, the hunter has to pay a daily rate to the professional hunting company that runs his safari. The key species for a safari hunter coming to Tanzania are lion (*Panthera leo*), leopard (*P. pardus*) and buffalo (*Syncerus caffer*), which each made up 12% of the total licence fees, as well as elephants where the population can support a quota. However, to be allowed to hunt such a combination of species, a hunter must book a classic 21-day safari (Leader-Williams *et al.* 1996a), which may cost in the range of US$100 000 per client. Since 1992, regulations have stipulated that around 9% of the game fees must be paid to local district councils (Leader-Williams *et al.* 1996a), as a form of protected area outreach (Barrow and Murphree 2001). In 1992, Tanzania earned over US$5.3 million in fees, which should have resulted in payments of

Box 9.6. *(cont.)*

some US$300 000 to rural district councils throughout Tanzania. However, the likelihood of such money reaching individual households who suffer the costs of living with wildlife outside the game reserves, or the poachers who use the resources inside the game reserve, was small.

The Selous Conservation Programme sought, therefore, to engage with local communities sharing a border with, or living in close proximity to Selous, in order to safeguard the ecological integrity of the game reserve, and to promote good relations between the reserve and neighbouring communities. The project aimed to encourage community participation in wildlife management and local access to benefits through providing a quota of animals shot for meat (Krishke *et al.* 1996). There were perceptions of widespread crop damage, but because the villagers did not have any property rights for the wildlife on village lands, they were dependent on district game scouts for the control of problem animals. In the past, large numbers of problem elephants were shot around the Selous, around 1000 per year up to the mid-1970s (Rodgers and Lobo 1982). However, by the mid-1990s, crop-raiding was mainly by medium-bodied species rather than large mammals, and was not nearly so extensive on the ground as reported by local people (Gillingham and Lee 2003). Furthermore, although access to game meat had a positive influence on perceptions about the benefits of wildlife, it had no significant effect on local perceptions of the game reserve or of the activities of wildlife authority staff. These negative attitudes arose from the inequitable distribution of benefits from the Selous Conservation Project, and the limited nature of community participation in wildlife management (Gillingham and Lee 1999). Indeed, most communities did not appear interested in wildlife management, or in conserving wildlife on their land under the conditions of state ownership prevalent at the time (Songorwa, 1999). Tanzania has since adopted a policy of establishing Wildlife Management Areas (Leader-Williams *et al.* 1996b), in which local people should gain use rights over wildlife and other natural resources on their land. If fully implemented, this may allow more equitable sharing of benefits from wildlife managed on village land.

Political difficulty of extractive use

Despite its possible advantages, it may be politically difficult, or even impossible, to sanction or implement extractive use of some groups of species in some national situations, because of the strong feelings it attracts for and against (Hoyt 1994; Conover 2002). Some countries, including India and Kenya, have banned hunting and other forms of use because of heavy losses of wildlife, as well as irresponsible behaviour on the part of, and lack of control over, the user community (Price Waterhouse 1996; see also Western and Waithaka, Chapter 22, Karanth and Gopal, Chapter 23). In India, the Wildlife (Protection) Act of 1972 bans all hunting of wild animals, and regulates trade in species listed in an attached schedule. In 1999, India banned all exports of CITES listed species, regardless of their status in the CITES Appendices. Nevertheless, the use of many species, including high-profile endangered species causing conflict like tigers (*Panthera tigris*), continues illegally (Misra 2003).

Does extractive use differ from wildlife tourism as a tool to offset conflicts?

Focus and sustainability

In contrast with most commercial forms of wildlife tourism (Walpole and Thouless, Chapter 8), extractive use can more often directly target those mega-vertebrates that cause most conflict, because these species can produce the charismatic hunting opportunity and trophy that sport hunters seek, or the meat that pays back the local farmer for the loss of his or her crops (Conover 2002). However, the underlying patterns of sustainability of wildlife tourism and extractive use may differ. There are many examples where extractive use has led to overuse of the harvested species (Milner-Gulland and Mace 1998; Bennett and Robinson 2000), but no clear examples of where this has happened as a result of wildlife tourism. Equally, wildlife tourism, as a growing sector of the multinational tourism industry, can have direct impacts on the species that attract the tourists, as well as many other intrusive impacts on natural areas. These include indirect impacts through infrastructure, facilities, waste disposal and litter, all necessary to service, or arising from, large numbers of visitors (Roe *et al.* 1997).

Remoteness, security and infrastructure

In contrast with commercial forms of wildlife tourism (Walpole and Thouless, Chapter 8), extractive use usually needs little infrastructure and capital investment. Furthermore, wildlife tourism may only be economically viable in areas where good access, large mammals and good visibility combine. However, extractive use can occur in less accessible areas where other forms of land use or wildlife tourism would not be viable. Even in countries such as Tanzania, that are currently visited by many wildlife tourists, hunting for sport greatly diversifies options to pursue an alternative high-earning form of wildlife use by many fewer tourists in remoter, more wooded and more tsetse-fly-infested areas that game-viewing tourists would not visit (Leader-Williams *et al.* 1996a). Furthermore, sport hunters tend to be less fickle in visiting wildlife areas in times of trouble and unrest. For example, hardly any foreign wildlife tourists would currently visit the Democratic Republic of Congo for reasons of security, access or visibility, so sport hunting provides perhaps the only option to earn foreign exchange from visitors taking part in a commercial form of wildlife use (Wilkie and Carpenter 1999). Because of its minimal requirement for infrastructure, very few forms of extractive use attract multinational players, in contrast with the many multinationals involved in providing facilities for wildlife tourism. Therefore, there may be less leakage of revenue (Gossling 1999) from extractive use, allowing more benefit to remain with national, local and wildlife economies.

CAN EXTRACTIVE USE GENERATE BENEFITS TO OFFSET CONFLICTS?

Many different forms of extractive use are now practised as choices in modern conservation (Hutton and Leader-Williams 2003). Most of these uses are carried out legally and bring in accountable revenue. However, the balance of benefits and costs, and the way that they are locally distributed among individuals, differ between forms of extractive use (Table 9.1), and between different systems of conservation. Under centrally controlled and legislated wildlife management systems, many of these uses are generally of little direct benefit to local communities (Table 9.1), as with much wildlife tourism (Gossling 1999; Walpole and Thouless, Chapter 8). Consequently, local communities may seek direct benefits by illegally killing or collecting considerable numbers of animals for meat (Bennett and Robinson 2000) or trophies, providing they escape detection and punishment (Milner-Gulland and Leader-Williams 1992). In contrast, problem animal control is often carried out legally but brings in no accountable revenue, yet can be of direct benefit to local communities (Table 9.1). If the other sanctioned forms of extractive use are to be effectively deployed, their benefits need to accrue more fully to people bearing the costs of conservation (Table 9.1). In turn, such people's attitudes to wildlife should become more positive, encouraging positive land-use decisions (Conover 2002), such as reducing illegal offtake, or conserving wildlife habitats rather than converting them to other less compatible forms of use (Box 9.1). This section reviews the benefits arising from, and the costs associated with, some different forms of extractive use under these different situations.

Less commercial forms of extractive use

Problem animal control

Local communities and private landowners view loss of life and livelihoods as a major cost of living with wildlife. However, national policies on problem species of wildlife, which are often among the charismatic flagship species, vary widely and may, *inter alia*, depend on the existence values attached to particular species by dominant groups, and the stage of national development (Leader-Williams and Dublin 2000; Conover 2002). Some countries allow recourse to killing individuals of most problem species, and allow certain products such as meat to be used by rural communities as a direct,

Table 9.1. *The legality and economic status, and potential benefits to local communities, of some forms of extractive use under centrally controlled and legislated wildlife management systems*

	Legal and economic status			Benefit to local communities		
Form of use	Legal with revenue	Legal without revenue	Illegal	Full	Partial	Employment
Meat poaching			+	+		
Trophy poaching			+		+	
Problem animal control		+		+		
Natural product collection		+		+		
Ranching	+				+	(+)
Live capture	+				+	(+)
Cropping	+				+	(+)
Resident hunting	+					(+)
Tourist hunting	+					(+)
Tourist game viewing[a]	+					(+)

[a] See Walpole and Thouless, Chapter 8.
Source: Based on Leader-Williams (2000).

legal benefit (Table 9.1). In some African countries, large numbers of animals, especially elephants, have been (Box 9.6), and are still, killed under this pretext (Leader-Williams *et al.* 2001), due to the lack of other legal benefits that rural communities currently derive in centrally controlled protected area systems (Table 9.1). However, local communities through community-based natural resource management (CBNRM) programmes (Box 9.3), and private landowners (Box 9.1), can in some instances directly benefit from producing problem species of wildlife on their land. In such situations, they have the option of realizing the inherent conflict that exists between the objective of maximizing earnings and reduced productivity that may result from excessive control of problem animals (Murphree 1993; Leader-Williams *et al.* 2001).

Some countries do not allow the killing or hunting of species that are nationally protected because of their conservation status, or of species that attract high existence values (for example top carnivores or great apes), while others enact rules that allow the killing of only a few problem species (Mishra 1984; Saberwal *et al.* 1994; Misra 2003). In such situations, richer nations or international non-governmental organizations (NGOs) promoting conservation or existence values may choose to offset some or all of the costs of living with problem species through schemes that directly compensate particular types of loss (Conover 2002; Nyhus *et al.*, Chapter 7). Where developing countries cannot afford to implement compensation schemes nor allow the killing of pests, general conflicts between protected areas and people could be partially offset by sharing benefits from wildlife tourism (Walpole and Thouless, Chapter 8), or by sanctioning certain forms of extractive use. These may include natural product collection (Box 9.5) or ranching (Box 9.4) that seek, respectively, to indirectly or to directly offset the costs of living with problem species.

Natural product collection

Allowing access to key resources can provide direct benefits for local communities (Table 9.1). The collection of natural products such as thatch, poles, medicines or honey has been allowed under situations of protected area outreach or co-management (Barrow and Murphree 2001) in some countries for several years (Nepal and Weber 1993; Kothari 1996) (Box 9.5). Such approaches recognize the imperatives of those who depend on extractive use of resources (Nepal and Weber 1993), and help offset the general conflict felt between neighbours and protected areas, particularly where direct benefits cannot be gained from controlling problem species (Box 9.5). However, the extent to which positive attitudes to sanctioned resource use promotes positive decisions towards land use, including the reduction of illegal use, requires wider testing. The approach of providing or restoring direct benefits from species or natural resources other than those causing conflict may (Box 9.5) or may not (Mehta and Kellert 1998) serve to improve general attitudes to protected areas. Nevertheless, published evidence of increased tolerance to species causing conflict, or to positive changes in land use, often seems to be lacking (Box 9.5).

More commercial forms of extractive use

Ranching

Ranching is the rearing in a controlled environment of specimens, usually of eggs or young, taken from the wild with the intention of engaging in wildlife trade. Ranching is best suited to species which are the source of products of high commercial value and which also exhibit high juvenile mortality in the wild (Hutton 1992). Ranching may involve species that directly cause conflict, or species that indirectly offset the costs of conflicts caused for other reasons (New 1994). Although ranches are usually sited on land that is privately or communally owned or leased, eggs or young can be regularly harvested from areas over which rural communities have control. For example, crocodilian eggs are regularly harvested from community areas to supply ranches in Zimbabwe (Box 9.4) and make a minor contribution to CAMPFIRE earnings (Box 9.3), possibly directly offsetting some of the costs of living with a problem species. Likewise, endemic butterflies are ranched to offset indirectly the opportunity costs of not clearing a biologically very important area of coastal forest at Arabuko-Sokoke in Kenya, as well as to offset indirectly the costs of crop raiding by elephants and baboons resident in the forest (Gede National Museum 2004). In both cases, published evidence of improved attitudes to problem species, as well as to more general conflicts with the neighbouring protected area, or of improved land use decisions, appears to be lacking.

Live capture

Live capture of animals has two fairly distinct components, and this is one reason why we use the generic term *extractive* rather than *consumptive* use (Hutton and Leader-Williams 2003). Live animals caught for trade, even though not intentionally killed as in hunting, are still lost to the wild population, with the same biological consequences as hunting. Some countries allow live capture from areas outside protected areas, primarily for export to the pet trade. Such trade is sometimes partially justified on the basis, say of crop damage caused by grain-eating birds (Thomsen *et al.* 1992; Leader-Williams and Tibanyenda 1996). Supplying the pet trade commercially usually results in a high-volume, low-priced trade, with a large illegal component. Few financial benefits derive to local harvesters who are closest to those that suffer the costs of living with any pest species captured for sale, while most benefits derive to local exporters, and particularly to pet retailers in developed countries (Thomsen *et al.* 1992; Leader-Williams and Tibanyenda 1996). The conservation benefits of such use remain questionable in centrally controlled systems (Table 9.1). If greater benefits were to accrue to local harvesters under devolved systems of ownership, wiser land-use decisions might result.

In contrast, some countries promote live capture of animals from protected areas as a tool for restocking other protected areas or private land with wildlife. In such cases, live sales can generate considerable revenue, while keeping source populations well below ecological carrying capacity. In southern Africa, in particular, positive land-use decisions have resulted in tremendous conservation gains through establishing hunting opportunities on private land from live captures made from protected areas (Child 1995; Lewis and Jackson, Chapter 15). Where ownership of wildlife is fully devolved to private or to communal landowners, they too can make informed decisions about whether to sell live animals at market prices or to opt for other forms of extractive use, such as selling hunting opportunities (Bothma 2002).

Cropping, meat hunting and resident hunting

Animals may be harvested for their products including meat and a range of by-products, such as skins and horn, whether for subsistence or market purposes. Meat hunting is often indiscriminate with regard to sex and age class, and is often unsustainable (Bennett and Robinson 2000). When legally sanctioned, it usually takes place outside protected areas, on private or communal land. When managed under central (as opposed to devolved) control, it may require the payment of right-to-use fees. In developing countries, this creates opportunities for the resident hunter who is part of an urban elite that can afford fees, while further disenfranchising local hunters who cannot afford such fees to gain legal access to wild meat (Leader-Williams 2000). In developed countries in North America or Scandinavia, by contrast, resident hunters may pay right-to-use fees that help offset the costs of sustainable management (Edwards and Allen 1992; Conover 2002).

In tropical countries, experience indicates that large-scale cropping exercises, involving a high offtake, sophisticated meat-processing and distribution systems and an overpriced product, are uneconomic (Eltringham, 1994) and provide few local benefits (Table 9.1) except employment. On the other hand, small-scale cropping exercises undertaken under the direct control of rural communities or local hunters can be quite sustainable (Eltringham, 1994) and be of direct benefit to local communities. However, as with problem animal control, if more commercially beneficial opportunities arise, local communities may realize the inherent conflict between the objective of maximizing earnings and reduced productivity that may result from using larger numbers of animals for meat and by-products (Murphree 1993; Leader-Williams *et al.* 2001).

Hunting for sport

Although wild animals have been hunted by people for hundreds of years, recreational hunting has only recently become a commercial pursuit (Gray 1993). Before North America was colonized, hunting for sport was the prerogative of the ruling classes in Europe and Asia. Such class privileges lent an air of exclusivity to hunting and also provided protection to the forests used by royalty. The deferential attitudes of commoners towards game animals began to change as colonists found abundant wildlife in North America and Africa where, in the absence of legal or moral constraints, game animals were killed without restraint (MacKenzie 1988; Gray 1993). Sportsmen have since moved to protect their interests, not only in North America and Africa, but in many other countries where sport hunting remains sanctioned.

Where legally sanctioned, sport hunting can occur in lower categories of protected areas established for sustainable use (IUCN 2003). It can also occur on communal land or on private land (Leader-Williams *et al.* 1996b; Lewis and Jackson, Chapter 15), thereby offering a range of opportunities to offset the costs of conflicts between people and wildlife (Conover 2002). Sport hunting is open to domestic and international hunters, for example North Americans hunting their native grizzly bears (*Ursus arctos*) in North America or hunting the 'Big Five' on an African safari (Edwards and Allen, 1992; Leader-Williams *et al.* 1996a). Sport hunting is promoted as a low-volume and high-earning form of extractive use, that is usually aimed at producing trophy animals. People may also sport hunt for its social benefits (Box 9.1). When sold commercially, sport hunting can produce considerable benefits to local communities, through programmes such as CAMPFIRE and ADMADE (Lewis and Jackson, Chapter 15) (Box 9.3). Nevertheless, the means by which benefits are shared (Baker 1997), and how they are used, is critical. In many instances, revenue from sport hunting is not shared with the unit of production (Murphree 1993), but with another centralized bureaucracy that extracts management fees and its own rent, without passing on the full benefit to those suffering the costs of living with wildlife (Gibson and Marks 1995) (Box 9.3). Furthermore, evenly addressing the welfare needs of whole communities, such as the building of schools and clinics, does not address individual household needs. Indeed, even if household dividends are forthcoming, dividing them evenly across households does not address the uneven spatial distribution of conflicts across the producer community (Sitati *et al.* 2003).

Nevertheless, many community-based projects involving sport hunting claim success in terms of improved attitudes to wildlife and protected areas (Murphree 1993, 2001), of reductions in poaching through employment of village scouts (Lewis and Alpert 1997), of improvements in community welfare through socially based projects such as building schools and clinics, of greater understanding of management issues by community leaders, and of stronger private-sector commitment to the resource. While such claims may be true and probably represent real progress for community-based approaches, the question of their success can only be answered by showing that individual households are adopting appropriate land-use practices, including reducing illegal offtakes. Published analyses are generally lacking (Lewis and Phiri 1998) (Box 9.3). Nevertheless, results from this type of analysis, whether for extractive use or wildlife tourism, could help guide community-based programmes to adopt policies and legal structures that will advance more compatible land-use practices on lands being managed for wildlife benefit.

Such analyses have proved somewhat easier, though not very common, for different forms of hunting on private land (Conover 2002) (Boxes 9.1 and 9.2). The situation on private land is less complex, because the often interlinked causes of negative attitudes of neighbours towards protected areas are simplified to conflicts caused by wildlife on private land. When hunted for sport, the status of a species that causes conflicts by preying on livestock has been enhanced through a positive measure promoting use, even though its success remains limited because trophies cannot be exported to the country where most sport hunters originate (Box 9.2). Similarly, those who hunt for sport conserve more wildlife habitat within an agricultural setting than those who do not participate in such sports (Box 9.1).

CONCLUSIONS

Extractive use of wildlife and other natural resources still remains an imperative for many people. Equally, newer forms of extractive use have been introduced as choices within the modern conservation paradigm, although none except problem animal control has been specifically designed to directly offset the costs of conflict with wildlife. In many instances, extractive use is politically difficult to implement, whether at a national level because of opposition to extractive use, or for particular species, such as large apes or top predators, whose sanctioned killing would not be acceptable to the majority. Nevertheless, some forms of extractive use may offer advantages in providing benefits over wildlife tourism, because extractive use requires little infrastructure, and can occur in remote areas lacking security.

At the same time, extractive use generally attracts less multinational involvement, and may be less prone to leakage of revenues than is wildlife tourism. However, revenue from commercial forms of extractive use on communal land often pays for social projects, rather than providing household dividends for those that most suffer the costs of living with wildlife (although the same is true of wildlife tourism). While local attitudes might improve, we find there is little published evidence of positive land-use decisions, including reductions in illegal offtake, from sanctioned extractive use on communal land (and again the same is true of wildlife tourism). Equally, it is difficult to show an effect because neighbours hold negative attitudes to their local protected areas for a complex of reasons, besides conflicts with wildlife. The situation on private land is less complicated, and there is some, although perhaps surprisingly little, published evidence of positive land-use decisions as a result of extractive use that targets species causing conflict for private landowners.

Two key recommendations arise from this review. First, the success or otherwise of promoting use of wildlife, whether through extractive use or wildlife tourism, to offset conflicts must rest with showing positive land-use decisions, including showing a reduction in illegal use. There is, however, little such published evidence for the key battlegrounds between protected areas and local communities, and we hope that future research will address this issue. Second, the benefits from the more commercial forms of extractive use seem generally to be targeted towards social projects, rather than towards those who disproportionately suffer the costs of conflict. Until this is redressed, it is unlikely that extractive use (or indeed wildlife tourism) will produce the positive land-use changes for which there is some evidence on private land, where the unit of benefit is closer to the unit of production.

ACKNOWLEDGEMENTS

We would like to thank Julia Baker, Ivan Bond, Christian del Valle, Peter Erb, Benjamin Mancroft, Rowan Martin, Bill Morrill, Guy Parker and Hugh Thomas for help and advice with case study material, and an anonymous referee for constructive comments.

10

Zoning as a means of mitigating conflicts with large carnivores: principles and reality

JOHN D. C. LINNELL, ERLEND BIRKELAND NILSEN, UNNI STØBET LANDE, IVAR HERFINDAL, JOHN ODDEN, KETIL SKOGEN, REIDAR ANDERSEN AND URS BREITENMOSER

ZONING: A CONCEPTUAL INTRODUCTION

Conflicts in land use are an inevitable consequence of the presently high human population densities living on a planet of finite size. Within this finite space, land use planners struggle to integrate as many potentially conflicting elements as possible using two approaches: the multi-use concept where compatible land uses can occur in the same area, and zoning. Zoning is any form of geographically differentiated land management where different forms of potentially conflicting land use are given priority in different areas. For example, in modern town planning some areas are zoned as residential, others as commercial, industrial, agricultural or recreational. Zoning has been widely used in biodiversity conservation in the creation of national parks, nature reserves and other protected areas. The focus of this chapter is to examine how zoning can be applied to the conservation of large carnivores. This requires balancing the twin goals of conserving viable populations of large carnivores, and minimizing conflicts with humans, which is proving to be an exceptional challenge in our crowded world.

LARGE CARNIVORES AND HUMAN ACTIVITY: CONFLICTS, COMPATIBILITY AND CONTEXT

Conflict

Zoning is only an issue because large carnivores cause conflicts with some human activities and interests throughout the world. These conflicts have been described in detail elsewhere (Woodroffe *et al.*, Chapter 1, Thirgood *et al.*, Chapter 2) but here we shall list the most important conflicts relevant for the discussion on zoning. An important aspect of these conflicts is the

People and Wildlife: Conflict or Coexistence? eds. Rosie Woodroffe, Simon Thirgood and Alan Rabinowitz. Published by Cambridge University Press.

extent to which they can be mitigated given the input of extra resources and new techniques.

(1) *Livestock depredation.* This issue potentially affects all livestock from cattle, sheep and goats to bees, depending on the predators present (Jonker *et al.* 1998). Technical approaches are available to mitigate much of the depredation, providing alternative wild prey are present, although in many cases it may be very expensive to adapt husbandry (Linnell *et al.* 1996; Svensson *et al.* 1998; Breitenmoser *et al.*, Chapter 4). Impacts on some forms of husbandry, such as herding of semi-domestic reindeer in Fennoscandia, may be difficult to mitigate (Pedersen *et al.* 1999; Nybakk *et al.* 2002).

(2) *Competition for game animals.* In many areas, hunters and carnivores have a real or perceived competition for harvestable game (Mech and Nelson 2000; Solberg *et al.* 2003; Thirgood and Redpath, Chapter 12). This is a conflict that is virtually impossible to mitigate, as carnivores cannot be conserved if they cannot access their prey. However, the conflict may be reduced if large carnivore populations are regulated by hunter harvest.

(3) *Predation on domestic dogs.* Many carnivores, especially wolves (*Canis lupus*), but also leopards (*Panthera pardus*) and cougars (*Puma concolor*), kill domestic dogs, both in hunting situations and those kept close to houses (Fritts and Paul 1989; Kojola and Kuittinen 2002). At present there is no effective way of mitigating these attacks.

(4) *Man-killing/mauling.* A wide range of carnivore species (leopards, tigers (*Panthera tigris*), lions (*P. leo*), wolves, cougars and bears) can attack and sometimes kill humans (Linnell *et al.* 2002; Quigley and Herrero, Chapter 3). Some of this can no doubt be mitigated, but a part of it will be inevitable where humans and these species overlap. For some species like wolves, there is also the added risk of attack by rabid individuals which can result in the transfer of a life-threatening infection or result in the death of the victim outright (Linnell *et al.* 2002).

(5) *Fear, dislike and wider social conflicts.* Many people fear, and simply dislike, large carnivores (Linnell and Bjerke 2002), and kill them (sometimes illegally) for this reason. Any overlap with large carnivores' distributions will directly reduce the quality of life of these people, even in the absence of any material or economic conflict. In addition to these directly tangible conflicts, there are many wider social conflicts associated with large carnivore management other than the large carnivores *per se.* These include conflicts over knowledge systems,

the feeling of rural communities being overruled by urban societies and a general resentment of outside influence in local affairs (Skogen and Haaland 2001; Brainerd 2003; Skogen *et al.* 2003). While some of these conflicts may be reduced through education and experience, it is unlikely that they will ever be eliminated.

Compatibility

With all the focus on conflicts, we tend to forget that there are many human activities with which large carnivores need not necessarily come into conflict. Despite public misperception, wilderness is not always necessary for large carnivore conservation. Industries based on harvesting natural resources such as fishing, logging and harvest of wild ungulates need not conflict with large carnivores if they are carefully regulated (Fritts *et al.* 1994). Even extraction industries need not have direct negative effects on carnivores (McLellan 1990). Some forms of agriculture, such as crops and vegetables, intensive horticulture or poultry production may also be compatible with some large carnivore species, depending on the degree of habitat modification. Even cattle are immune to depredation from many of the smaller carnivores, and are rarely killed to the same extent as sheep and goats. Recreation and tourism are also generally compatible with large carnivores as long as some special disturbance situations are avoided (White *et al.* 1999), and the presence of large carnivores may increase the ecotourism potential of an area (Goodwin *et al.* 2000). Finally, and not without controversy (Rabinowitz 1995), recreational or trophy harvest of the large carnivores themselves may be compatible with their conservation in many regions (Creel and Creel 1997; Hofer 2002; Large Carnivore Initiative for Europe 2002).

Not all carnivores are equal – in conflict

For each of the conflict areas mentioned above there are clear species differences in the extent to which they cause conflicts. For example, people are more afraid of bears and wolves than of wolverines (*Gulo gulo*) and lynx (*Lynx lynx*) in Norway (Røskaft *et al.* 2003). Tigers, leopards and bears are probably associated with most cases of man-killing (Quigley and Herrero, Chapter 3). Bears and honey badgers (*Mellivora capensis*) are the main species that regularly damage beehives (Jonkel *et al.* 1998; Cape Nature Conservation 2001). In temperate areas, wolves can cause the greatest problems for sheep-farmers or reindeer-herders, are among the most widespread dog killers, and evoke the strongest social conflicts. Different types of livestock have very different vulnerabilities to different carnivore species. It is therefore necessary to consider that each species will have different degrees of compatibility with human activities. The picture can be further complicated by regional differences.

Not all carnivores are equal – in conservation

It is also important to consider that not all large carnivore species are endangered with imminent extinction. While many species such as the giant panda (*Ailuropoda melanoleuca*) or Iberian lynx (*Lynx pardinus*) are approaching or have reached critical status (interestingly neither of these species causes direct conflict with human interests), many other species (e.g. brown bears (*Ursus arctos*), black bears (*U. americanus*), wolves, leopards and cougars) are widespread and their numbers occur in the tens of thousands (Nowell and Jackson 1996; Servheen *et al.* 1999). The status of a species can also vary within its range, it being very common in some areas and endangered in others. Accordingly, the appropriate management scenario for a particular species can vary from reintroduction and strict protection through to sustainable harvest and even population reduction. It is therefore vital to consider each species in each region as a specific case.

HOW IS ZONING MEANT TO MINIMIZE CONFLICTS AND CONSERVE CARNIVORES?

The basic concept of zoning is to reduce the spatial overlap between large carnivores and unmitigated sources of conflicts by differentiating the use of management tools directed at both the carnivores and their conflict in a complementary manner. In the present context there are three sets of actions that can be utilized to achieve this separation:

(1) manipulate large carnivore density (this may include everything from reintroduction to large-scale lethal control)
(2) adjust the way conflicting activities are conducted (e.g. mitigation of conflict)
(3) remove the potentially conflicting activity from carnivore range.

In a zoning scenario, these three groups of actions would be implemented in different ways in different places, such that carnivores are given preference in some areas, whereas human interests are given preference in others. In a simple two-zone scenario, carnivores would be given preference in one area. In the 'carnivore zone', this would involve protection (or conservative harvest/selective control: Treves and Naughton-Treves, Chapter 6) of the carnivore(s), at the same time as reduction of the conflict potential in the area. For conflicts that can be mitigated, like livestock depredation, this could entail providing advice and economic assistance to improve husbandry practices, or providing bee-keepers with electric fences. Other sources of conflict can be removed from zones where carnivore densities are intended to be high. On one level, this could involve exchanging grazing areas, or using a change in subsidies to encourage a switch from sheep to cattle. On another level, it has involved the movements of thousands of villagers to make way for Project Tiger reserves in India (Panwar 1987; Karanth and Gopal, Chapter 23). The advantage of zoning in this scenario is that it allows the concentration of expensive conservation and conflict-reduction/mitigation measures into a limited area.

In some areas, sources of conflict may exist that either cannot be mitigated, are too expensive to mitigate on the required scale, cannot be moved, or are given political priority over large carnivores. Here, the option would be to create zones where carnivores are not allowed to reach high population levels, even though they otherwise might have been able to occur in significant numbers. In some areas, carnivores' presence may not be tolerated at all. Despite the public perception of large carnivores as being symbols of the 'wilderness', many species are able to adapt to very heavily modified landscapes (e.g. Mech 1995). The combination of this adaptability, and carnivores' capacity for long-distance dispersal (see below) implies that individual carnivores are going to be turning up far from the areas where they are to be given preference. How to deal with these individuals is going to be increasingly difficult in the future, as the 'protectionist' movement gathers strength in developed countries (Mech 1995). One non-lethal option is translocation; however, this is rarely a suitable strategy, and, with the exception of some highly endangered species or special situations, is more a public-relations exercise than an effective management tool (Linnell *et al.* 1997). Sterilization and immuno-contraception are both expensive and impractical on a large scale (Haight and Mech 1997). The implication is that lethal control will remain the main management tool for reducing carnivore density (Mech 1995, Treves and Naughton-Treves, Chapter 6).

These scenarios with 'carnivore areas' and 'no-carnivore areas' are extreme. In reality, any zoning system will probably involve a graded series of zones in many shades of grey, rather than a simple black-and-white, two-zone system. In effect, most discussions will focus on the number of different types of zone, and the degree to which their management contrasts, rather than on if zoning will be used or not. In effect, zoning can be viewed as different management practices in different areas to suit different local conditions.

LARGE CARNIVORES AND SCALE

Home range size

It is clearly vital to scale management zones to the size of the biological process that they are designed to manage (Thiollay 1989; Schwartz 1999). Being on top of the food chain, it is not surprising that large carnivores should live at relatively low densities, and utilize relatively large home ranges that can vary in size from 10 to many thousand square kilometres (Hefner and Geffen 1999; Walton *et al.* 2001). Most adult large carnivores occupy stable home ranges that do not vary dramatically in location from season to season or from year to year . This spatial stability is further enhanced by the fact that most species of large carnivore show some form of intra-sexual territoriality (Sandell 1989). However, there are a number of exceptions with some species following migratory prey (Ballard *et al.* 1997; Cook *et al.* 1999; Pierce *et al.* 1999; Walton *et al.* 2001) and others making frequent extraterritorial excursions (Messier 1985; Hofer and East 1993; Boyd *et al.* 1995; Landa *et al.* 1998) that can result in movements over distances from 10 to 100 km. The implication is that individual large carnivores need lots of space, especially in habitats with low productivity. When we are thinking about carnivore zones, we must think in terms of bigger land areas than we are used to thinking about for any other terrestrial species' group.

Dispersal

While most adult large carnivores occupy stable home ranges, juveniles of most species show long natal-dispersal distances that are best measured in the tens to hundreds of kilometres in length (Boyd *et al.* 1995; Sutherland *et al.* 2000). Dispersal can be a great advantage for a zoning-based carnivore conservation system as it allows for connectivity between non-contiguous populations and zones (Van Vuren 1998) which contributes positively to both genetic and demographic viability of populations (Durant 2000). However, dispersal can also cause problems as it limits our ability to confine large carnivore presence to their intended zones. The result of long dispersal distances is that there will be a wide area around any carnivore zone where dispersing individuals will occur from time to time. In the absence of predator-proof fencing, this will make it hard to maintain steep gradients of carnivore density, and sharp borders between zones (Saberwal *et al.* 1994).

The utility of protected areas

Throughout the world there is a wide network of existing protected and wilderness areas. It is apparent that reserves have been instrumental in saving several species or populations from extinction during the conservation crises of the last 50 years (Saberwal *et al.* 1994; Seidensticker *et al.* 1999). Protected and wilderness areas, especially when networked (Soulé and Terborgh 1999), may well provide an insurance of some base population level and, where they exist, should be managed in a manner that favours large carnivores (Fritts and Carbyn 1995; Miquelle *et al.* 1999). However, they will never be able to achieve the task of carnivore conservation alone, and in some regions existing protected areas have very little relevance to large carnivore conservation (Fritts and Carbyn 1995; Linnell *et al.* 2001a). This is largely due to the relative size of protected areas compared with the area requirements of carnivores. A further problem that exists is that comparatively few protected areas or 'wildernesses' are free of conflict within their borders. In many regions, people live, farm, graze livestock, gather forest resources and hunt game within national parks and nature reserves (Khan 1995; Warren 1998). Even those carnivores that are found within protected areas may not be immune from what happens beyond the area's borders, as poachers can intrude, diseases can be transferred from neighbouring domestic animals, and extinction can occur through random events (Newmark 1995, 1996; Woodroffe and Ginsberg 1998, 2000; Brashares *et al.* 2001).

Therefore, most populations and species of large carnivores will need to be conserved, at least partially, in multi-use landscapes where the potential for conflict with human activities is higher. Creating differences in large carnivore management between reserves and unprotected areas outside is a relatively established management system; that of creating different zones within multi-use landscapes is far more controversial. It is in this context of the multi-use landscape that we need to evaluate the potential utility of zoning.

ZONING OF LARGE CARNIVORES IN PRACTICE

Dingoes in Australia

Australia has one of the most dramatic zoning systems for any large carnivore. Dingoes (*Canis lupus dingo*) are a major predator on sheep, to such an extent that it is widely regarded that dingoes and free-ranging sheep are incompatible. Dingo depredation on cattle is far less serious and can generally be tolerated by producers (Corbett 1995; Allen and Sparkes 2001). In order to allow free-ranging sheep-farming, there has been a continent-wide policy of zoning dingoes out of the major sheep-farming areas in western and southeastern Australia. To limit constant immigration, a 5600-km-long fence (formerly 8600 km) has been built from the south Australian coast almost all the way to the eastern coast of Queensland.

This encloses the majority of the major sheep farming areas in the country. Within this fence, dingoes are so rigorously controlled (by poisoning, trapping and shooting) that they are basically absent. In western Australia and a parts of Queensland that are outside the fence, poison baits are distributed in a 20-km-wide buffer area surrounding sheep-breeding areas, to prevent dingo occupation and immigration into the sheep-farming area. Outside the fence and poison-buffers dingoes are allowed to persist. The methods seem to work in that it allows sheep farming to continue and large dingo populations to persist (Thompson 1986; Corbett 1995; Allen and Sparkes 2001).

Wolf recovery zones in North America

Following the 1973 listing of the wolf under the Endangered Species Act in the lower 48 states of the USA, a series of federally mandated recovery plans have guided wolf population restoration. This has led to the active reintroduction of wolves into Yellowstone, central Idaho and Arizona/New Mexico, and the natural expansion of wolves in Montana, Minnesota, Wisconsin and Michigan. The recovery has been so successful that, less than 30 years after recovery was initiated, wolves in the northern Rocky Mountains and the Great Lakes area have been downlisted from 'endangered' to 'threatened' status, and plans are under discussion to delegate wolf management to the individual states in these areas. During this recovery process, zoning has been used in all areas. In the west, three recovery areas were originally designated, in northern Montana, central Idaho and the Greater Yellowstone Ecosystem. In the east, Minnesota was divided into five zones. The main difference between the zones was in the degree of protection given to wolves inside, versus outside. In order for the reintroductions to be acceptable it was necessary to create a special 'experimental, non-essential' population designation for wolves in the reintroduction areas. This gave managers much greater flexibility to control wolves that caused problems in these areas, in contrast with areas where wolves were recolonizing naturally and were therefore afforded the maximum protection of the Endangered Species Act (Fritts *et al.* 1995, 2001; Brown and Parsons 2001). In the final reclassification rule, the lower 48 states were zoned into three 'distinct population units' – with wolves being reclassified as threatened in two, but remaining endangered in a third (US Fish and Wildlife Service 2003c). The 'experimental, non-essential' status has remained in place for sub-regions surrounding the reintroduction sites in two of these regions.

As a part of the proposed reclassification process, each state with wolf presence in the western and eastern distinct population segments was required to develop wolf management plans. In their initial wolf management plans, different states proposed to tackle the issue of zoning in different ways. Montana and Idaho chose to remove the zoning structure and actively manage wolves wherever they colonized within the state (Idaho Legislative Wolf Oversight Committee 2002; Montana Fish, Wildlife and Parks 2003). By contrast, Wyoming proposed to manage wolves as trophy/ game animals in protected and wilderness areas while treating them as 'predators' in the rest of the state, implying very different management regimes and degrees of protection (Wyoming Game and Fish Department 2003). Wyoming's draft plan was, however, rejected by the federal Fish and Wildlife Service. Wisconsin opted for four zones and Minnesota has simplified its four-zone system into a two-zone system, with different degrees of protection (Minnesota Department of Natural Resources 1991; Wisconsin Department of Natural Resources 1999). In several of these situations, *de facto* 'no-wolf' areas will be maintained

Brown bear core protection area in Slovenia

In 1966, an area of approximately 3000 km^2 in southern Slovenia, bordering the area of bear distribution in neighbouring Croatia, was designated as a core protection area. At the time there were approximately 160 bears in Slovenia, 95% of which were in the core area. Within this zone, bears were harvested in a manner intended to allow population increase; outside the zone, bears were given less protection. Within the zone a system of bait sites was established (at least one per 60 km^2) where carcasses were provided for the bears. By 2000, the population of bears in Slovenia had increased to over 500, of which only 75% were still found within the core protection area (Huber and Adamic 1999; Anonymous 2002a; M. Adamic pers. comm.). Clearly, the extra protection of this core area allowed a small population to recover in a small country that has a relatively highly modified landscape. However, the expansion of bears into areas outside the traditional core area has lead to greatly increased conflicts with humans and calls for increased hunting quotas (Adamic 1997; Anonymous 2002a). This has not only caused increased tensions within Slovenia, but also with neighbouring Italy and Austria who have hoped that Slovenia could allow bear population expansion in a northwestern corridor to provide linkage with their small bear populations (Adamic 1994).

Cougar management in North America

For large carnivore species that are so abundant that they can be managed as game animals, zoning is widely used in the form of harvest management units. Since cougars were upgraded to game animals in New Mexico, USA, in 1971 the harvest management system has evolved into a system with 15 regions, each with a specific quota and hunting season. These quotas are set in an attempt to balance three different objectives, apart from the overall goal of conserving cougar populations. First, the system aims to provide hunting opportunities for recreational and trophy hunters. Second, there is an aim to limit depredation on livestock. Third, because cougar predation on bighorn sheep (*Ovis canadensis*) appears to be threatening several small and isolated populations with extinction, there is a desire to reduce predation levels in these bighorn sheep ranges (New Mexico Department of Game and Fish 2005).

A similar system of zone-specific quotas is in use in most of North America where cougars are hunted (e.g. 11 units in Alberta, 53 units in Utah: Ross *et al.* 1996). Washington State has nine different cougar management units, where goals vary from maintaining the present population density (in five units), reducing population density because of human-safety issues (in three units) and having no real goal of cougar presence in an area considered to be unsuitable habitat (one unit) (Washington Department of Fish and Wildlife 2003).

The rise and fall of a wolverine conservation zone

An isolated wolverine population is found on the Snøhetta plateau in southern Norway (Landa *et al.* 1998, 1999, 2000c). Recognizing the vulnerability of this isolated population, and the increasing conflicts with livestock, the government established a 13 505-km^2 core conservation area in the region in 1991. By 2000, the population had increased to the extent that eight reproductive events were being documented each year. During the same time the conflicts with sheep had increased even further. By 2000, losses of a total of 8772 sheep were being compensated annually following wolverine depredation. In 1998, licensed hunting, intended to limit wolverine numbers and thereby depredation, had begun in the areas outside the core area. By 2001, the depredation levels were so high that the core area was removed under political pressure from the agriculture lobby, thus exposing all wolverines to potentially unsustainable licensed hunting.

During its 10-year life, the core area had effectively safeguarded the isolated population, allowing it to increase in numbers and distribution, yet it had failed to reduce conflicts. This was simply because there were no serious attempts to change sheep husbandry practices, or to decrease sheep numbers inside the core area. In fact, the numbers of sheep being released each summer into the core area increased constantly (by a factor of 4 from colonization in 1979 until core area establishment, and by a further 30% following core area establishment). The message is clear – if you are going to create a zone for carnivores you must address the conflict potential within the zone.

PUBLIC ACCEPTANCE OF ZONING

Zoning may in many cases appear to be a logical way to minimize the impact of large carnivores on livestock, and reduce the costs associated with introducing mitigation measures. However, the overall problem of large carnivore conflicts with people concerns far more than *material conflicts*. Conflicts over large carnivores are frequently interpreted as conflicts between animals and people; however, this is only half the story. Experience with large carnivore recovery in Western Europe and North America has shown that the social conflict dimensions can be even more important, and even harder to deal with, than the material conflicts (Kellert *et al.* 1996; Bjerke and Kaltenborn 1998; Kaltenborn *et al.* 1998, 1999; Bath 2001; Bath and Majic 2001; Skogen 2001). The carnivore controversies are obviously also *social conflicts*; conflicts between different groups of people. It is quite easy to see that they involve people with strongly diverging perspectives on land use, and perhaps on nature itself. Basically, some want the predators removed because they threaten economic interests or even lifestyles, while others see them as vital parts of a wilderness that is shrinking all too rapidly. Studies have indicated that people who are opposed to current carnivore protection frequently direct more aggression toward its human proponents that toward the animals themselves. Other aspects of this social conflict are related to friction within processes of economic and cultural modernization, urban–rural tensions and conflicts between hegemonic and subordinate forms of knowledge (Wilson 1997; Skogen 2001). Zoning systems for carnivores are generally set up to prevent or minimize material damage and economic losses, primarily concerning livestock. But whether or not they are successful in this respect, they may have limited impact on the general level of conflict throughout the range of the species in question – inside and outside the zones. Indeed, as we shall now argue, they may very well increase the level of conflict.

Studies in Norway

Norwegian studies have shown that, concerning the wolf at least, the level of social conflict may be just as high in areas with no significant livestock losses (indeed without grazing livestock) as in areas where such losses have been substantial (Skogen and Haaland 2001). This is partly due to social mechanisms hinted at above. But even in such areas (which are precisely the ones that are likely to be within carnivore zones), the conflicts retain a material core: wolves, for example, attack dogs, eat huntable game and scare quite a few people. And the groups that are most strongly affected by these problems, that is, hunters, dog-owners and people who are afraid to go out, are often not subject to compensation or damage-prevention programmes in the same way that livestock herders are.

The Norwegian studies leave little doubt that one of the measures that potentially could have the greatest conflict-reducing effect is carnivore hunting in a form that is open to local hunters. Hunting would contribute to controlling carnivore population growth, and possibly increase shyness. But hunting would also give local people a sense of active involvement in the curbing of a problematic situation. The feeling of powerlessness on one's own turf has come across as the most disturbing part of local experiences in our studies, and a more active role for local hunters could most likely alleviate this situation. This would affect not only the hunters themselves, but also considerable segments of rural communities where hunting is tightly interwoven with local outdoor practices and local culture in general (Skogen and Haaland 2001; Krange and Skogen 2001; Skogen 2003).

Any extreme form of zoning is likely to greatly limit the possibilities for carnivore hunting within the 'protection' zones. This type of zoning will therefore remove the possibility for introducing a social-conflict-reducing measure, carnivore hunting, and increase the feeling of social injustice for those that have a zone 'forced' onto the area where they live. However, a less extreme form of zoning, where carnivores are zoned out of areas with the absolute highest conflict potential, but are actively managed over larger areas (e.g. some form of recreational harvest, limited control allowed) may be far more acceptable from the social-conflict point of view, at least in some cultures. The downside is that this would require mitigation measures focussed on livestock production to be spread over larger areas, greatly increasing costs. Even the use of the word 'zoning' generates negative reactions from many people, leading us to rephrase things as either 'carnivore management (or conservation) units' or 'geographically differentiated management' in place of zones or zoning.

Different cultures may well react very differently to zoning of carnivores. However, there are clear parallels with the Norwegian situation in central Europe (Breitenmoser 1998; Bath 2001; Bath and Majic 2001) and the western USA (Wilson 1997), with the possible difference that conditions are even more polarized in these situations due to the existence of more extreme pro-conservation and often anti-hunting interest groups. Unfortunately, there are relatively few comparative human-dimensions data from developing countries – study in these areas would also be complicated by the fact that 'carnivore zones' in these regions usually correspond to protected areas – such that it would be impossible to evaluate attitudes towards the zoning of carnivores separately from other issues such as resource access.

CONCLUSIONS

Zoning can represent a valuable tool for conserving large carnivores while minimizing the conflicts that they inevitably cause. The advantages include:

(1) Zoning allows the resources for costly conflict reduction and intensive conservation measures to be concentrated into limited areas. Mitigation can be very expensive. In this context, zoning is most useful to mitigate livestock depredation.
(2) Zoning provides for simplified management procedures where responses depend on where conflicts occur, and can be initiated without time-consuming investigation.
(3) Zoning provides predictability for people, so that they can make long-term plans and economic investments knowing to what extent large carnivores will be a part of their future. This may also allow people to become accustomed to the presence of large carnivores, and thereby reduce fear levels. This concerns both those inside a carnivore zone (who can adapt to carnivores' presence) and those outside (who know that they will not have to adjust to carnivore presence in the future).
(4) Zoning allows regional adjustment of large carnivore density to regional conflicts.

However, there are a number of clear disadvantages to zoning. These include:

(1) As local people will rarely welcome being included inside a zone where carnivores are given preference, it will create an atmosphere of an urban majority overruling a rural minority. This will lead to the creation of an 'us against them' mentality that may result in increased poaching and increased social conflicts. Furthermore, the concentration of carnivores into an area may actually increase fear and other social conflicts.
(2) Concentration of large carnivores into an area, especially if multiple species are zoned sympatrically, may increase the competition between game hunters and large carnivores. This is also likely to be the case for game ranches and semi-domestic reindeer herding.

There are also a number of technical challenges associated with zoning. The large spatial scale at which large carnivores use the landscape presents many challenges for their conservation. If we are to consider zoning as a conservation strategy, we need to think about big zones, and because of carnivores' dispersal potential we should consider that it will be impossible to produce zones with sharp borders (but see the dingo example above). The result is that conservation and management units need to be much larger than those normally used for other species. This will require cooperation between different administrations and agencies, and in many cases different countries.

On the whole, we find it inevitable that some form of zoning will be needed in most places, as it seems impossible to imagine a future where carnivores are not subject to some form of management, and where this management can be applied equally in all areas. The real issues are to what extent it will it be a black-and-white system, or a system with many shades of grey, and the relative sizes and placements of the zones. The most controversial issues will be with the extremes, especially if carnivores are to be zoned out of an area. Experience with carnivore recovery in Europe and North America in recent decades has revealed many surprises as to where carnivores are able to survive, and where conflicts do or do not occur. Many apparently unsuitable areas may well turn out to be suitable, and vice versa. Caution should therefore be used, and flexibility maintained, when planning zone priorities before carnivores have returned.

Finally, while the adoption of zoning may lead to increased protection of carnivores in some areas it may also lead to an an increased need to control individual large carnivores in other areas, both in response to conflicts, and potentially in advance of conflicts (Treves and Naughton-Treves, Chapter 6). While some people will view this control as unethical, others may regard it as the provision of hunting opportunities. Either way, the diversity of public opinion will ensure that there will be controversy. In this context, there is no universally applicable formula. It will depend on the ecology of the species in question in the given areas, the extent to which sources of conflict exist, and on the cultural and socio-economic status of the human population. There are no magic formulas or perfect solutions in large carnivore conservation, just a lot of more or less acceptable, and often controversial, compromises.

11

From conflict to coexistence: a case study of geese and agriculture in Scotland

DAVID COPE, JULIET VICKERY AND MARCUS ROWCLIFFE

INTRODUCTION: THE GOOSE-AGRICULTURE CONFLICT IN SCOTLAND, UK

Scotland hosts six species of geese for all or part of the year which comprise nine distinct breeding populations (Table 11.1). Six of these populations breed in arctic regions (Greenland, Iceland, Svalbard or Russia) and winter in Scotland, while the Scottish and naturalized populations of greylag geese (*Anser anser*) and Canada geese (*Branta canadensis*) are resident. Many populations are protected under the 1979 European Union Birds Directive (Directive 79/409/EEC) and the 1981 Wildlife and Countryside Act in the UK. Greenland white-fronted geese (*Anser albifrons flavirostris*) and both populations of barnacle geese (*Branta leucopsis*) are afforded the highest level of protection under Annex I of the EU Directive, while bean geese (*Anser fabalis*), greylag geese, Canada geese, pink-footed geese (*Anser brachyrhynchus*) and white-fronted geese are listed under Annex II/2 and may be legally hunted in the UK.

Most populations are increasing, largely as a result of hunting controls (Owen 1990). However, their ranges in Scotland remain very restricted in time and space (Fig. 11.1, Table 11.1). The entire Svalbard barnacle goose population, for example, winters in a 50 × 5 km coastal strip on the Solway Firth, in southwest Scotland (Fig. 11.1). Similarly, even highly mobile species like the pink-footed goose, with a winter range extending from northeast Scotland to southeast England, remains concentrated around a few important roost sites.

Agricultural damage caused by geese is of two broad types: (i) on grass, standing crop may be reduced and sward structure damaged, and (ii) on arable crops such as wheat and oilseed rape, yield may be reduced. Grazing by barnacle geese can result in a reduction of up to 83% in standing crop of

People and Wildlife: Conflict or Coexistence? eds. Rosie Woodroffe, Simon Thirgood and Alan Rabinowitz. Published by Cambridge University Press.

Table 11.1. *The size and trend of goose populations in Scotland*

English name	Scientific name	Population size (1999–2001)	Trend
Bean goose[a]	*Anser fabalis*	300[b,c]	Stable
Pink-footed goose	*Anser brachyrhynchus*	245 000[d]	Increasing
Greenland white-fronted goose	*Anser albifrons flavirostris*	20 660[b]	Increasing
Greylag goose (Icelandic population)	*Anser anser*	80 000[d]	Stable
Greylag goose (Scottish population)	*Anser anser*	6 500[b]	Increasing
Greylag goose (naturalized population)	*Anser anser*	21 000[b,e]	Increasing
Canada goose[a]	*Branta canadensis*	50 000[b,e]	Increasing
Barnacle goose (Greenland population)	*Branta leucopsis*	37 800[b]	Increasing
Barnacle goose (Svalbard population)	*Branta leucopsis*	21 000[f,g]	Increasing

[a] These species do not cause significant goose–agriculture conflict in Scotland, and are not considered in the text.
[b] The Wetland Bird Survey 1999–2000 (Musgrove *et al.* 2001).
[c] Scottish population is around 150 individuals (Musgrove *et al.* 2001).
[d] The 2000 census of pink-footed geese and Icelandic greylag geese (Hearn 2002). About 80% of pink-footed geese and over 90% of greylag geese were found in Scotland.
[e] UK totals; most naturalized greylag geese and Canada geese are found outside Scotland.
[f] WWT Svalbard barnacle goose project report 2000–2001 (Griffin and Coath 2001).
[g] The wintering range of Svalbard barnacle geese covers both England and Scotland, with more than half using the Scottish range (Griffin and Coath 2001).

grass in early spring and subsequent losses of silage of up to 38% (Percival and Houston 1992). Intensive grazing by pink-footed and greylag geese on barley and wheat have resulted in mean yield losses of 7% and 15% of grain and 14% and 20% of straw for barley and wheat respectively (Patterson *et al.* 1989). The financial costs of goose damage include late turn-out of stock, loss of silage, hay or grain production, and extra costs of fertilizer applications and reseeding. Estimated costs range from £6000 (about US$10 800) per farm per year at Strathbeg to £12 000 (about US$21 600) per farm per year on Islay (Daw and Daw 2001) (Fig. 11.1). Furthermore, many of the 'conflict hotspots' are in areas where agricultural productivity is low and farming is already marginal. Thus, the conflict may be localized but is often intense, with a small number of farmers incurring a large proportion of the

Figure 11.1. The main locations of wintering wild goose populations in Scotland.

damage. However, yield losses within a site are highly variable between years, even when goose density remains constant, as losses vary with weather conditions and the timing of grazing. For example, Percival and Houston (1992) found that average loss of silage yield due to barnacle geese varied from 13% to 38% between sites and years, and that there was no consistent relationship between local grazing intensity and yield loss. Such unpredictability has typically been found in studies of this kind (e.g. Patton and Frame 1981; Patterson *et al.* 1989; Owen 1990; Patterson 1991).

MANAGING THE GOOSE–AGRICULTURE CONFLICT: POSSIBLE SOLUTIONS

A number of solutions to the goose–agriculture conflict have been suggested in recent years. These include culling and scaring geese to reduce numbers on agricultural land, compensating landowners for the losses incurred and managing land as reserves or sacrificial crops.

Culling

By culling, we here refer to systematic efforts to reduce population sizes (see also Treves and Naughton-Treves, Chapter 6). To be effective in resolving the conflict, this approach must reduce goose populations to levels at which little or no damage is incurred. Geese are relatively long-lived birds, and changes in survival rates therefore have a strong effect on population growth (Rowcliffe *et al.* 2000), as demonstrated by the heavy impact of hunting on goose populations in the past (Ebbinge 1991). Culling by encouraging existing wildfowling and lethal crop protection is being used to reduce 'overabundant' white goose (*Chen* spp.) populations in North America, through greatly liberalized legislation on bag limits, timing and methods of hunting (US Fish and Wildlife Service 2001a). Although the success of this approach is yet to be assessed, there would be many practical difficulties with this approach in Britain. Owen (1990) estimated that an effective cull of pink-footed and greylag geese, for example, would require 90 000 birds to be shot in each of the first five years and a maintenance cull of 40 000 per year thereafter. This would require a large number of highly active hunters, probably far more than are currently active in the UK where around 32 000 grey geese (*Anser* spp.) are shot annually.

The main arguments against culling are political. Most goose populations are considered 'at risk' under the EU Birds Directive. Even though member states can derogate from the protection rules to prevent serious damage to crops, shooting under the derogation is licensed only where it has been demonstrated that there is no other satisfactory solution. In effect this means that numbers shot by farmers are kept to a minimum required for localized crop protection. Initiating a systematic cull would require first demonstrating that populations are excessively high, well beyond anything that could be considered at risk in the long term, and second, that levels of damage are also unacceptably high. This is arguably the case for North American white geese, which number in the millions and are causing widespread irreversible habitat degradation on their Arctic breeding grounds, to the detriment of coexisting species (US Fish and Wildlife Service 2001a). No such arguments can be made in the case of any European goose population.

The current presumption against culling does not rule out shooting as part of an integrated management solution. Wildfowling can generate considerable income for local economies in areas with wintering geese (Royal Society for the Protection of Birds and British Association for Shorting and Conservation 1998) but more research is required into the practicalities and implications of harvesting of legal quarry species. The current and potential roles of sustainable wildfowling and licensed shooting in the management of, respectively, quarry and fully protected populations are discussed in more detail below.

Scaring

A range of methods have been deployed for scaring geese off agricultural land, from the use of scarecrows or gas guns, barriers in or above fields (streamers on sticks or kites), to the employment of a full-time human goose-scarer (Vickery and Summers 1992). However, geese become habituated to most scaring techniques relatively quickly (Vickery and Summers 1992) and, since scaring is rarely coordinated between farms it usually serves simply to move the problem elsewhere. Where it has been trialled, coordinated scaring has proved extremely costly (Percival *et al.* 1997) and only effective in reducing damage levels when it is deployed alongside dedicated refuge areas for geese (Vickery and Summers 1992; Vickery and Gill 1999).

Compensation schemes

Direct financial compensation for goose damage is a third option, but estimating damage is costly, complex and time-consuming (Owen 1990; Patterson 1991; Nyhus *et al.*, Chapter 7). Due to the high variability of yield losses discussed earlier, compensation is likely to be subject to a high degree of uncertainty, regardless of the quality of the sampling regime. Goose density or dropping densities have been used as surrogates for actual yield losses, for example on Islay, but agreement on the timing and frequency of counts, and hence payment levels, were difficult to reach (National Goose Forum 1998a). Overall, direct compensation for losses has not proved effective in reducing goose–agriculture conflict, either in Britain (National Goose Forum 2000) or abroad (Van Eerden 1990).

Management of reserves for geese

Alternative feeding areas or refuges can be established and managed to attract geese off agricultural land and reduce damage levels (Owen 1990; Percival 1993; Vickery *et al.* 1994). Such refuges have existed in Scotland for some time. Caerlaverock National Nature Reserve in southwest Scotland, for example, was designated in 1957 and extended by the Wildfowl and Wetlands Trust (WWT) in 1970. Its establishment has been largely responsible for the increase in barnacle geese at these sites from around 300 to over 20 000 individuals since 1948/9 (Owen and Norderhaug 1977; Griffin and Coath 2001).

A great deal is known about how to optimise the design and management of grassland refuges (for a review see Vickery and Gill 1999). Most species require fields of at least 5 ha, near traditional roost sites with minimal disturbance. Their attractiveness can be greatly enhanced by managing the sward through cutting or grazing and fertilizer application. However, even well-managed reserves are unlikely to attract and support the entire target population all winter so agricultural damage is still likely to occur on land adjacent to refuges. Furthermore, on a local scale the areas required are large and although they can be reduced by effective management (Vickery and Gill 1999), such management can be costly.

MANAGING THE GOOSE–AGRICULTURE CONFLICT: A COORDINATED APPROACH

From the preceding sections it is evident that: (i) on their own neither culling, scaring, compensation nor reserves is an adequate solution; (ii) the conflict is likely to escalate as goose populations increase; (iii) the problem is highly localized and very variable with respect to damage levels so solutions must be site specific; (iv) at the national level, there must also be a degree of commonality and equity across sites and (v) because many of the species are migratory and protected by international laws, solutions must meet international conservation obligations.

Other countries in Europe suffer from goose–agriculture conflict, and the most successful resolution of the conflict appears to be closely linked to the provision of a coordinated approach. A mixture of techniques are used to reduce conflict, from uncoordinated shooting or non-lethal scaring (e.g.

Poland, France, Romania), through regional coordination of compensation and management (e.g. Germany), to nationally coordinated schemes that include scaring, compensation and land management (e.g. the Netherlands, Sweden, Norway). Under the Dutch system, the legislation that regulates the hunting of geese also provides a public body that advises farmers on how to avoid damage. This public body also pays compensation to those farmers adversely affected by geese, which is partly funded by hunting licence fees. However, compensation payments increased rapidly from 1977 to 1986, so other techniques were developed, such as providing reserves and changing crops away from those favoured by geese (Van Eerden 1990). A national policy was implemented in 1991 that encouraged the planting of less sensitive crops, the use of non-lethal scaring, compensation and dedicated waterfowl reserves (National Goose Forum 1998b). In this way, geese could be managed in reserve areas by both attracting them into the reserve through positive land management, and moving them away from commercial farmland by scaring and planting crops unlikely to be damaged by geese. This coordinated approach inside and outside reserves, involving payments, appears to have helped considerably in reducing goose–farmer conflict in the Netherlands.

The approach in Scotland has been to develop a national framework with locally devolved schemes, ensuring a high degree of input from stakeholders at all levels. Local goose management schemes were launched in Scotland in the early 1990s. These local schemes used a combination of scaring, shooting and refuge management alongside payments to farmers, but initially were independent of one another and uncoordinated at a national level, with no forum for sharing experience. In 1997 the National Goose Forum was formed to develop a National Policy Framework for the management of geese and agriculture in Scotland (National Goose Forum 1997a). The aims of this policy were to: (i) meet the UK's nature conservation obligations, (ii) minimize the economic losses to farmers and (iii) maximize the value for money of public expenditure (Scottish Office Agriculture, Environment and Fisheries Department 1996). The development of this national policy was underpinned by five 'guiding' principles (Box 11.1). As well as advising ministers on this national policy, the National Goose Forum (i) provided advice to local goose management groups; (ii) oversaw population monitoring of geese in Scotland; and (iii) coordinated research on geese and agriculture (National Goose Forum 1997b). Members included those with farming, conservation, wildfowling and government interests.

The National Goose Forum delivered the National Policy Framework in 2000 (National Goose Forum 2000), at which point government ministers approved the continuation of local goose management schemes. The

Box 11.1. Key principles

The development and implementation of the National Policy Framework for goose populations in Scotland

(1) The strategy should be subject to regular review.
(2) Geese should be actively managed.
(3) Management schemes should be introduced, if appropriate, for specific populations and should aim to reduce agricultural damage and conserve goose populations that require protection.
(4) There should be a clear basis for the calculation of any payments made under management schemes.
(5) The strategy must avoid any detrimental effect on the benefit to local communities from geese whilst minimizing loss to farmers.

The approval of local goose management schemes by the National Goose Monitoring and Review Group

(1) The conservation status of the goose population.
(2) The number of geese in the proposed scheme area.
(3) The density and duration of geese on farmland.
(4) The presence of a reserve managed for geese in the area.
(5) The damage caused by geese on farm businesses.
(6) The level of interest in participation in a scheme.

National Goose Monitoring Review Group replaced the National Goose Forum as the national coordinating organization, with the remit of assessing, on an annual basis, the proposals from local goose management groups for conformity to the guidelines set out in the National Policy Framework. These local groups develop proposals tailored to local needs, and are given the primary responsibility for administering schemes on the ground (Scottish Office Agriculture, Environment and Fisheries Department 1996). Thus the general parameters for goose management schemes are set at the national level, determined by constraints, while local groups ensure the details match local needs.

The approval of new schemes is guided by a set of agreed principles, based on the conservation status of the goose populations involved, levels of damage prevalent, and the willingness of stakeholders to participate (Box 11.1). Information on the latter two points is provided by local groups when applying for the approval of a scheme, while conservation status is determined by a dedicated population viability analysis process. This is used to assess current extinction risk, allowing populations to be prioritized in terms of their need for protection (Scottish Office Agriculture, Environment and Fisheries Department 1996; National Goose Forum 1998c, d). This process has led to a general presumption against management schemes for populations at low risk (such as pink-footed geese), although it is accepted that schemes for these populations may still be appropriate near large, established roosts and reserves where high levels of damage can be demonstrated (National Goose Forum 1998e).

Schemes are reviewed annually at a national level, as are specific management agreements and any payments for individual farmers at a local level. However, schemes are assured for longer periods to provide security for farmers and goose populations, and there is a guaranteed one-year withdrawal notice if the scheme is no longer considered to be required or appropriate. The National Policy Framework itself is to be reviewed at five-yearly intervals to ensure that it meets the challenges of any changes to the goose–agriculture conflict in Scotland. To assess accurately the changes in the level of conflict over time, monitoring of Scottish goose populations is overseen by the National Goose Monitoring Review Group (National Goose Forum 2000). This is particularly important in the case of wild geese for several reasons: (i) changes in shooting practice may alter conservation status, (ii) migratory populations could be affected by events outside Scotland and (iii) increasing population sizes of geese may also alter conservation status. Changes in the status and distribution of geese will impact on local schemes so there must be flexibility in the system.

MANAGEMENT SCHEMES IN PRACTICE

Five pilot local goose management schemes were already in operation when the National Goose Forum was established in 1997: Islay (Greenland barnacle and white-fronted geese), Solway Firth (Svalbard barnacle geese), South Walls in Orkney (Greenland barnacle geese), Loch of Strathbeg (pink-footed geese), and The Uists (Scottish greylag geese). In 2000, three of these five pilot schemes (Islay, Solway Firth and South Walls) and one new scheme (Kintyre) were approved by the National Goose Monitoring Review Group under the new administrative system (see Fig. 11.1 for locations of the schemes).

Since 2000, the fundamental framework in each of the schemes has been the establishment of scaring zones and undisturbed feeding refuges (National Goose Forum 1998a, d) (Boxes 11.2 and 11.3). The former may involve shooting (under licence where appropriate) to reinforce scaring, and modifications of land management to reduce its attractiveness to geese (e.g. reduce fertilizer inputs or plant unfavourable crops). Conversely, in the refuge zones, land is managed to enhance its attractiveness to geese and minimize disturbance. In all schemes, payments are made on a per-hectare basis for positive management rather than as compensation for lost yield,

Box 11.2. Solway Firth Barnacle goose management scheme

The Solway Firth is a large, relatively undeveloped estuary on the border of England and Scotland encompassing extensive saltmarsh (merse) that is the traditional winter feeding area for the entire population of Svalbard barnacle geese (Fig. 11.1). In the winter of 1948/9, only 300 individuals were found on the Solway Firth (Owen and Norderhaug 1977). Since then conservation measures throughout their range have resulted in an increase to around 20 000 (Griffin and Coath 2001). The population increase has been accompanied by a habitat shift, and geese now also graze improved pasture bringing them into conflict with the local farming community as they compete with livestock for spring grass (Owen *et al.* 1987). The Caerlaverock National Nature Reserve and other protected areas provided alternative feeding areas, thus alleviating the conflict (Owen 1977, 1980) but the reserves can no longer accommodate the entire population and conflict has re-emerged (Owen *et al.* 1987).

A pilot management scheme was implemented in 1994 and focussed on two existing non-governmental organization (NGO)-run reserves with adjacent feeding zones separated from scaring zones by buffer zones. The aim was to protect the recovering goose population, whilst managing their distribution to reduce conflict with farming interests. Farmers placing fields in feeding and buffer zones received area payments from Scottish Natural Heritage (the government conservation agency) to cover the additional costs of fertilizer and overwinter feed incurred as a result of delayed turn-out onto fields. Local Scottish Natural Heritage staff loaned scaring equipment and advised on its use in the scaring and buffer zones.

Thus scaring is used to redistribute geese alongside payment for the positive management of alternative feeding areas. This affords protection to the geese (which also serve as a tourist attraction) and rewards farmers for their contribution to this protection. Payment rates for managing feeding zones are set to cover the costs of reseeding pastures more frequently, adding extra fertilizer in autumn and spring to promote extra grass growth for geese, and of delayed turn-out of livestock to pasture in spring – a rate of £195 (about US$350) per hecture. Buffer zones receive £50 (US$90) per hecture (increasing to £100 (US$180) where traditional use by geese is high, and non-audible scaring is ineffective). There are no payments for land in scaring zones, but Scottish Natural Heritage provides assistance with scaring and habitat management. The major cost is for management options that benefit barnacle geese and help to concentrate them in core areas, with total costs of around £120 000 (US$216 000) per year.

The population size and distribution is monitored through each winter and the results provide the basis for zonation in the following year. The scheme has succeeded in increasing goose numbers on land entered into the feeding and buffer zones whilst maintaining goose numbers on the reserve itself (Cope *et al.* 2003). This scheme has been highlighted as an example of best practice in managing goose–agriculture conflict (National Goose Forum 2000; National Goose Monitoring Review Group 2000).

Box 11.3. Loch of Strathbeg goose management scheme

The Loch of Strathbeg, northeast Scotland, is a freshwater roost site for pink-footed and greylag geese (Fig. 11.1). The main focus of conflict is the pink-footed goose population which reaches 50 000 and 20 000 individuals in autumn and spring respectively. The loch is protected as a Special Protection Area, and 90 ha of the surrounding land is owned and managed by the Royal Society for the Protection of Birds (RSPB). A scheme was initiated in 1994 to reduce the conflict in spring, when grass yield loss was most costly to farmers wanting to turn livestock out onto fields. Pink-footed geese are highly mobile on a day-to-day basis, and undisturbed feeding zones, centred on farms with the highest goose densities, were created as several dispersed refuges rather than being concentrated around one location. To enhance the attractiveness of these feeding zone pastures for geese, fertilizer (80 kg N per hectare) was applied in late February (one to two months earlier than under normal management), and they were left ungrazed in spring (Patterson and Fuchs 2001). Legal scaring methods were encouraged outside the feeding zone including professional bird-scarers (Raynor *et al.* 1996). Unlike barnacle geese, pink-footed geese are a quarry species and shooting of geese in the close season (1 February to 31 August in inland areas) was used to improve the efficacy of scaring through a streamlined hunting licence procedure. Initially upfront payments of £75 (about US$135) per hectare for the feeding zone did not provide enough incentive for farmers to maximize the numbers of geese on these fields. Payments were therefore increased to £80 (US$144) per hectare for fields with the highest goose densities, with reductions to £50 (US$90) per hectare for fields with lower densities, paid in arrears (Raynor *et al.* 1996). Goose density was estimated by dropping counts, conducted under an independent contract.

Between 1994 and 1996 around 250–300 ha of land was managed in feeding zones, supporting 11–13% of geese in March and April compared with 5–6% on the smaller RSPB reserve (Raynor *et al.* 1996). Under the current scheme these figures rose to around 20% and 7.5% respectively in 1997 (Patterson and Cosgrove 1997). The current scheme comprises around 500 ha of feeding zone and 500 ha of buffer zone, supplemented by adjacent scaring zones. Whilst this scheme was controversial, as it was aimed at the management of a population not of high conservation priority, the scale of agricultural losses justified the continuation of this scheme (Box 11.1). The need for this scheme is reflected in its increasing popularity amongst local farmers, shown in the increase in area under management agreements.

with fields entered into feeding zones receiving the highest payments and no area payments in the scaring zones. Within the scaring zone, the farmer is primarily responsible for the scaring regime, but logistical support or capital grants for scaring equipment may also be provided, and payments to assist with extensification can be considered (National Goose Forum 1998d). Schemes are open to all farmers and crofters in the scheme area on a voluntary basis. This framework has shifted the role of the farmer from

Table 11.2. *The location, scale and cost of goose management schemes in the year to March 2002 (SNH 2002)*

Scheme area	Total cost (£)[a]	Total number of geese	Area covered by scheme payments (ha)	Cost per goose (£)[a]	Cost per ha (£)[a]
Islay	629 699	43 935[c]	6 199	14.33	101.58
Solway Firth	117 887	21 000[d]	921	5.61	128.00
South Walls	15 380	1 150[e]	610	13.37	25.21
Loch of Strathbeg	74 000[b]	20 000[f]	703	3.7	105.26
Kintyre	53 154	3 427[g]	582	15.51	91.33

[a]£1 is equivalent to approximately US$1.8 at current exchange rates.
[b]Projected cost from Loch of Strathbeg Goose Group.
[c]Islay Goose Management Group; comprises 32 815 barnacle geese and 11 120 white-fronted geese.
[d]Griffin and Coath (2001).
[e]South Walls Local Goose Management Group.
[f]Loch of Strathbeg Goose Group, peak spring count.
[g]Kintyre Goose Management Group; comprises 2912 white-fronted geese and 515 Icelandic greylag geese.
Source: Scottish Natural Heritage (2002).

a victim of conflict to an agent of the conflict resolution. Farmers are, in effect, receiving rewards for managing their land partially for the benefit of goose conservation, providing farmers with a real benefit from the presence of geese.

The costs of the schemes differ greatly according to the intensity of management, the area they cover and the number of geese involved (Table 11.2). In 1997, the combined cost of all five schemes, met by Scottish Natural Heritage, was about £350 000 (US$630 000) per year (National Goose Forum 1997a). Since this time, the cost of some schemes has increased in line with the number of geese and the level of interest amongst farmers to a total payment for the year to March 2002 of £890 000 (about US$1.6 million) (Scottish Natural Heritage 2002). It seems possible that these costs will continue to rise since there is little evidence that overall numbers of geese in Scotland are approaching a natural equilibrium. Indeed, in the past, populations believed to have been nearing carrying capacity (e.g. Svalbard barnacle geese: Owen and Black 1991) have entered further phases of rapid population growth (Pettifor *et al.* 1998). Density dependence observed at the scale of the breeding colony (e.g. Larsson and Forslund 1992) cannot be translated directly to the population level because of uncertainties about the availability of breeding habitat.

With the rapid increase in the total cost over the period 1997 to 2002, the long-term financial viability of these goose management schemes is uncertain. Within the current National Policy Framework, the payment mechanism is through national conservation budgets (NGF 1999). However, with the many needs of conservation competing for limited resources, it may be argued that the narrow focus on reducing goose–agriculture conflict that has developed could have a negative impact on the resolution of other issues. By adopting a wider perspective when dealing with the goose–agriculture conflict, other funding sources could become available, as well as giving the opportunity to address other issues at the same time.

The intensification of agriculture in recent years has been at least in part responsible for exacerbating the conflict by attracting geese from semi-natural feeding areas onto farmland (Owen 1980). An alternative long-term vision for Scotland's agriculture is the encouragement of wider extensification through agri-environment schemes. There is a growing need for rural policy within the European Union to be integrated, and geese represent a major opportunity for this, encouraging investment in rural areas in green tourism, sport and environmental initiatives on and off farms. Promoting the positive aspects of geese to farmers will also assist in adjusting to the more 'environmental' focus of recent reforms in the Common Agricultural Policy whereby payments are likely to be increasingly linked to the provision of environmental benefits. This would reduce goose–agriculture conflict and provide wider biodiversity benefits. In addition, the use of the Agri–Environment Regulation to pay for goose management would allow 50% of funding to come from the European Union.

This broader approach would focus more strongly on increasing the benefits of geese, alongside strategies to minimize the costs to agriculture, since there are direct and indirect economic benefits to be derived from geese. Under the current situation, farmers receive direct benefits from the presence of geese by farming for geese. These benefits could be extended by using the presence of geese to generate further income for local economies from tourism and wildfowling (RSPB and BASC 1998). Tourism in Scotland is a major and growing industry and recent estimates suggest that total tourism expenditure related to Scotland's natural heritage is around £105 million (about US$189 million) per year (RSPB and BASC 1998). Around 25 000 staying visitors and 19 000 day visitors visit goose reserves in Scotland each winter, spending around £3 million (about US$5.4 million) in local economies, half of which can be attributed directly to the presence of geese (RSPB and BASC 1998). Sport shooting in Scotland accounted for 7993 geese and around 108 000 ducks shot in 1988/9, with a total associated expenditure of £30.8 million (about US$55 million) (RSPB and BASC 1998). On Islay, the reduced pasture yield caused by geese equates to a loss of between £336 973 (US$606 000) and £787 733 (US$1.4 million) per year. Expenditure by birdwatchers on Islay is estimated to be between £269 000 (US$484 000) and £346 000 (US$623 000) per year (RSPB and BASC 1998). Although this suggests there is a net cost of geese, there is clearly scope for tourist and wildfowler expenditure to be directed, at least in part, towards landowners who suffer the costs of geese, potentially improving their willingness to participate in management schemes with lower financial support from central government.

There are also indirect or non-market (existence) values quantified in terms of public 'willingness to pay', in this case for the protection of geese. Surveys using contingent valuation suggest countryside management is of relatively low priority compared with health and education (Hanley *et al.* 2003), and protecting wildlife was considered less important than providing employment and food when just considering countryside management. Considering wildlife protection on its own, protecting geese was considered less important than fish stocks and birds of prey. However, despite this low ranking, geese are still afforded an existence value. Hanley *et al.* (2003) found that the general public would be willing to pay £11 (US$20) mean (£4 (US$7) median) per household per year to prevent a 10% fall in numbers and £16 (US$29) mean (£5 (US$9) median) per household per year to generate a 10% increase in numbers of endangered goose species. The existence value attached to wild geese in Scotland differed between residents in areas of conflict and the general public. Most groups favoured policies that were focussed on endangered species rather than all geese, but whereas the general public placed high value upon management strategies without shooting, residents accepted shooting as part of management. Residents tended to be against large (50%) increases in geese and in favour of maintaining current levels, while visitors were in favour of a moderate (25%) increase in goose population sizes. Thus the existence value of geese does, perhaps, provide additional justification for the use of public funds in supporting measures to protect goose populations in Scotland.

CONCLUSIONS

The conflict between goose conservation and agricultural interests in Scotland (and more generally in northwest Europe) is in many ways typical of several such issues involving common herbivorous species feeding on agricultural land. However, the goose–agriculture conflict is unusual in that the wildlife populations involved are given a relatively high degree of conservation importance, and the areas of conflict are very localized. It is these facts that make the relatively elaborate management schemes described above both necessary and possible. Culling is not a viable solution practically or politically, and options such as scaring, compensation or the establishment of reserves are inadequate if implemented in isolation. The exact nature of the problem and the level of damage vary widely between sites and years and with goose species, habitat and weather. Thus solutions must be site-specific, flexible and tailored to local needs. However, since several conflict hotspots exist in Scotland and many species are migratory and protected under national and international law, coordination at a national level is required.

The key strength of the policy adopted for goose management in Scotland is its combination of top–down and bottom–up management whereby government ministers control the general parameters and local groups ensure the details are tailored to local needs. This approach should be widely applicable to other cases of wildlife conflict. The clear definition of agricultural, conservation and budgetary goals at the national level has led to a marked reduction in perceived conflict without resorting to a strategy of extermination. At the same time, the fact that local stakeholders in conflict areas have strong input to the form of local schemes has been crucial to their widespread acceptance. This is perhaps best demonstrated by the fact that levels of interest in goose management schemes have increased, as has the geographic spread of proposed schemes (a new pilot scheme has recently been established on the Hebridean islands of Coll and Tiree, for example). The Scottish goose management policy is constantly evolving, through regular reviews of policies and practices in the light of changing attitudes and goose populations. This review process has undoubtedly also contributed to the success of the approach.

The biggest potential challenge is the long-term financial viability of the schemes. The marked reduction in perceived conflict has been achieved at a considerable financial cost. This has been lessened, to some extent, by the move away from direct damage-based compensation, as used in the pilot Islay scheme, towards area-based payments (see Nyhus *et al.* (Chapter 7) for further comparison of payment methods). Even so, the costs of the approach are still substantial, and rising as goose populations increase and interest grows. While it can be argued that these costs are at least partially justified by the indirect (existence) value of geese to the general public, it is questionable whether the conservation budget that currently funds the schemes will be able to justify indefinitely increasing costs, given the many other urgent issues that require support. The long-term stability of the approach to conflict resolution in Scotland will depend on the sourcing of alternative and sustainable funding. The current solution has provided farmers with rewards for managing their land positively for geese. Local economies derive direct benefits from the presence of geese through tourism and wildfowling, and mechanisms should be established whereby more of those benefits are experienced by those land managers who have put the effort into making their land attractive to geese. Encouraging wider extensification through agri-environment schemes may also play a future role in reducing the conflict by reducing the local density of geese, and also providing wider biodiversity benefits. The total cost of management schemes, at around £890 000 (US$1.6 million), is only 3% of the total Agri-Environment budget for Scotland (Scottish Executive Environment and Rural Affairs Department 2002), and if the Agri-Environment Regulation is used in this way, it will allow 50% of funding for goose management schemes to come from the European Union.

ACKNOWLEDGEMENTS

We are grateful to many people who supplied information: G. Banks, B. Bremner, G. Churchill, G. Dalby, J. Doherty, L. Griffin, R. Hearn, E. Laurie, J. Love, M. MacKay, I. Patterson, E. Rees, A. Robertson and F. Younger. N. Read, S. Gough and R. Hooper helped in the production of the manuscript. JAV's involvement was part funded by the Joint Nature Conservation Council.

12

Hen harriers and red grouse: the ecology of a conflict

SIMON THIRGOOD AND STEVE REDPATH

INTRODUCTION

One of the most contentious conservation issues in the UK concerns the conflict between the conservation of hen harriers (*Circus cyaneus*) and commercial hunting of red grouse (*Lagopus lagopus*) (Thirgood and Redpath 1997; Thirgood *et al.* 2000a). Hen harriers are rare raptors in the UK and are of high conservation importance. Despite their legal protection, harriers and other raptors are killed on grouse moors because they are thought to have a negative impact on grouse-shooting. Grouse-shooting is an important land use in the uplands and benefits conservation because it retains heather moorland, a habitat of conservation significance in Europe. Shooting also generates income for landowners and is important in some rural economies. The conflict between raptor conservation and grouse management thus brings into opposition two powerful groups of stakeholders: 'hunters' who own much of the uplands; and 'conservationists' who attach value to birds, particularly raptors. The conflict is highly politicized and often the entry point to debates about land use and landownership. In this chapter, we review 10 years' work that has quantified the problem of raptor predation on grouse and tested management solutions. We begin by placing the harrier–grouse conflict into a wider context by assessing the conservation significance of heather moorland, grouse management and lethal control of raptors. We then examine the impact of raptor predation on grouse populations and grouse-shooting and go on to investigate potential solutions to the conflict. Finally, we address the interaction between science and politics in the management of human–wildlife conflict and assess why this particular conflict has not been resolved.

People and Wildlife: Conflict or Coexistence? eds. Rosie Woodroffe, Simon Thirgood and Alan Rabinowitz. Published by Cambridge University Press.

CONSERVATION BACKGROUND

Heather moorland

The conflict between hen harriers and red grouse is played out on the heather moorlands of the UK. Heather moorland is one of Europe's most distinctive habitats and has considerable conservation, economic and aesthetic value (Thompson *et al.* 1995). It is dominated by the dwarf shrub heather, the global distribution of which is largely restricted to the UK, and much of which has been lost to forestry and grazing. The conservation significance of heather moorland is largely due to the birds that it supports. The birds are a mixture of arctic, boreal and temperate species and include internationally important populations of golden eagles (*Aquila chrysaetos*) and peregrine falcons (*Falco peregrinus*) and high densities of waders such as golden plovers (*Pluvialis apricaria*) and curlews (*Numeinius arquata*).

Grouse management

Heather moorland comprises about 25% of the uplands (15 000 km^2) and 50% of this (7500 km^2) is managed by private landowners for grouse-shooting (Hudson 1992). 'Walked-up' shooting is conducted when grouse density is low and involves hunters walking the moors shooting grouse as they are encountered. 'Driven' shooting involves hunters remaining in hides shooting grouse that are driven over them by beaters; it requires densities of grouse in excess of 60/km^2 and can generate high revenues. The production of grouse populations at high density requires professional gamekeepers whose main tasks are to manage heather and to control predators. Heather management involves burning to produce a mosaic of different-aged stands providing food and cover for grouse. Predator control involves the legal killing of red fox (*Vulpes vulpes*), stoats (*Mustela erminea*) and crows (*Corvus corone*). Traditionally predator control was extended to raptors and, despite legal protection in 1954, this practice continues on many moors.

Central to any debate about raptors on grouse moors is the more fundamental question of whether or not grouse management benefits biodiversity conservation (Thirgood *et al.* 2000a). There is little evidence of any direct effects of grouse management on bird species other than red grouse. Comparisons of bird densities, vegetation and management variables on 320 moorland sites revealed higher densities of golden plover, lapwing (*Vanellus vanellus*), curlew and red grouse on managed moors but higher densities of meadow pipit (*Anthus pratensis*), skylark (*Alauda arvensis*), whinchat (*Saxicola rubetra*) and crow on unmanaged moors (Tharme *et al.* 2001). Furthermore, correlations of bird abundance and grouse management may result from the effects of habitat management or predator control, or indeed from other characteristics of the areas studied. Two conclusions about grouse management and biodiversity have, however, emerged from the literature. First, if many moorland estates had not been managed for grouse they would either have been planted with exotic conifers or farmed for sheep, resulting in loss of heather moorland (Robertson *et al.* 2001). Second, raptor persecution on grouse moors has resulted in a marked decline in the distribution and abundance of harriers, peregrines and eagles (Thirgood *et al.* 2000a).

Lethal control of raptors

Killing raptors was commonplace through the late nineteenth and early twentieth centuries, reducing the distribution and abundance of many species and, despite legal protection, it continues today (Newton 1979). We focus here on hen harriers, the cause of most controversy, but peregrines and golden eagles are also killed (Scottish Raptor Study Groups 1997). Harriers were formerly widespread in the UK but persecution during the nineteenth century restricted the breeding population to the Scottish islands (Watson 1977). Harriers recolonized the mainland in the 1940s, reached their current distribution in the 1970s, and surveys during 1988 and 1998 estimated their population at 670 females with 600 of these in Scotland (Bibby and Etheridge 1993; Sim *et al.* 2001). They are rare breeding birds in parts of the uplands, particularly on the grouse moors of northern England and eastern Scotland. Breeding success of harriers has been shown to be three times lower on grouse moors than on other moorland, with persecution accounting for 30% of breeding failures (Etheridge *et al.* 1997). In the same study, annual survival of female harriers on grouse moors was half that on other moorland, each year between 55 and 74 females were killed, and estimates suggested that the harrier population would increase by 13% per year if persecution ceased.

DEFINING THE CONFLICT

Harriers are killed on grouse moors because hunters believe that harrier predation reduces grouse harvests. But is this belief justified? What is the impact of predation on grouse populations and shooting bags? Although research indicated that harriers could remove 24% of grouse chicks from low-density populations (Redpath 1991), the prevailing view was that predation had no impact on grouse harvests (Jenkins *et al.* 1964; Picozzi 1978). It was clear that the relative abundance of raptors and grouse was important in influencing the impact of predation. No information was available on the

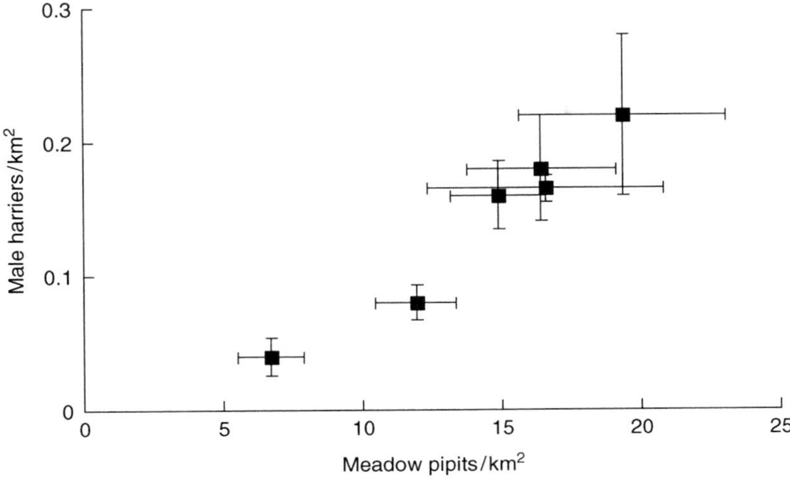

Figure 12.1. Mean (± SE) density of breeding male hen harriers in relation to mean (± SE) abundance of meadow pipits on six moors. Only data collected after four years of harrier protection from persecution were included. (From Redpath and Thirgood 1999.)

mechanisms involved or on the factors influencing raptor numbers and diet on moorland. We conducted a study during the period 1992–7 with these aims in mind, funded by a unique consortium of conservationists and hunters (Redpath and Thirgood 1997).

What influences the density and diet of hen harriers on moorland?

The breeding density of hen harriers varies from 0–5 females per 10 km^2 on moorland and this variation is important in determining their impact on grouse populations. We compared the breeding densities of harriers to the abundance of their prey – grouse chicks, meadow pipits and field voles (*Microtus agrestis*) – over five years on six moors. Harrier numbers did not increase with the abundance of grouse, but breeding densities were highest where meadow pipits were most abundant (Redpath and Thirgood 1999) (Fig. 12.1). Combining this result with the finding that meadow pipits were more common on moors where there was a mosaic of heather and grass suggested that harriers would also breed at higher density on these grassy moors (A. A. Smith *et al.* 2001). There was also considerable annual variation within moors in the numbers of breeding harriers that was related to fluctuations in field vole abundance (Redpath *et al.* 2002) (Fig. 12.2).

Harriers are generalist predators whose main summer prey are grouse chicks, meadow pipits and field voles (Redpath and Thirgood 1997). We

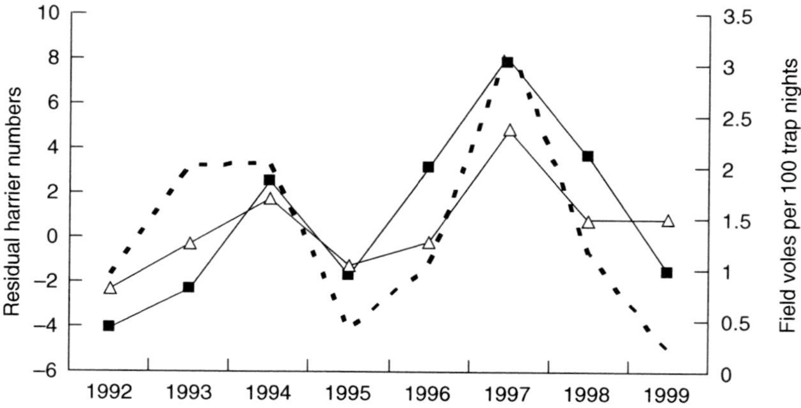

Figure 12.2. Residual numbers of harriers (Δ males, ■ females) attempting to nest in relation to the number of field voles trapped per 100 trap nights from 1992 to 1999 at Langholm. (From Redpath *et al.* 2002.)

compared the provisioning rates of grouse chicks to harrier nests with the abundance of grouse chicks on eight moors during 1992–6. Grouse chicks increased in the diet of harriers as grouse density itself increased and this had important effects on the impact of predation on grouse populations (Redpath and Thirgood 1999). The functional response of harriers to grouse chicks was sigmoidal with provisioning increasing at 50 chicks/km^2 and stabilizing above 100 chicks/km^2 (Fig. 12.3). The proportion of grouse chicks taken by individual harriers was greatest at densities of 67 chicks/km^2 – equivalent to about 12 pairs of grouse/km^2 (Fig. 12.4). The implication is that the effect of harrier predation would be greatest on grouse populations at low density.

In summary, the breeding density of hen harriers was related to the abundance of alternative prey and thus the impact of harrier predation on grouse varied spatially and temporally. On moors with no lethal control of raptors and high prey densities, breeding densities of harriers were high. If these grouse populations fall during cyclic troughs to densities of less than 12 pairs/km^2, then harrier predation can limit these grouse populations at low density (Redpath and Thirgood 1999).

Does raptor predation limit red grouse populations and reduce shooting bags?

The numerical and functional responses described above predict that predation by harriers can limit grouse populations at low density. But does this actually happen? We assessed the impact of raptor predation on grouse populations at our intensive study site at Langholm in southern Scotland

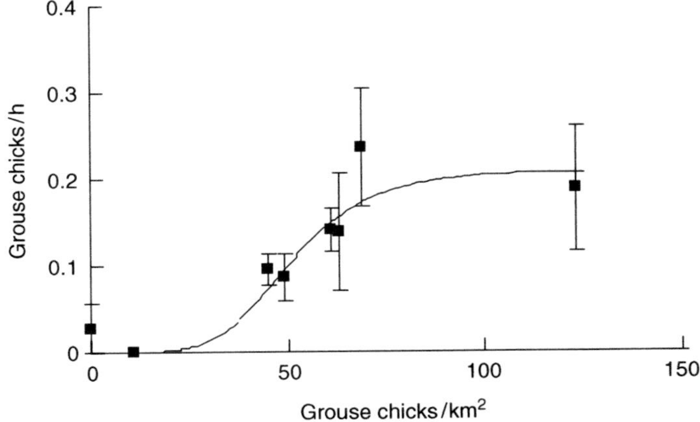

Figure 12.3. The rate at which grouse chicks were provisioned to harrier nests in relation to the density of grouse chicks. Each point represents a mean (± SE) of a number of nests observed on one moor in one year. The Type III functional response curve was fitted with the equation: $y = 0.21x^{5.1} / (51^{5.1}+x^{5.1})$. (From Redpath and Thirgood 1999.)

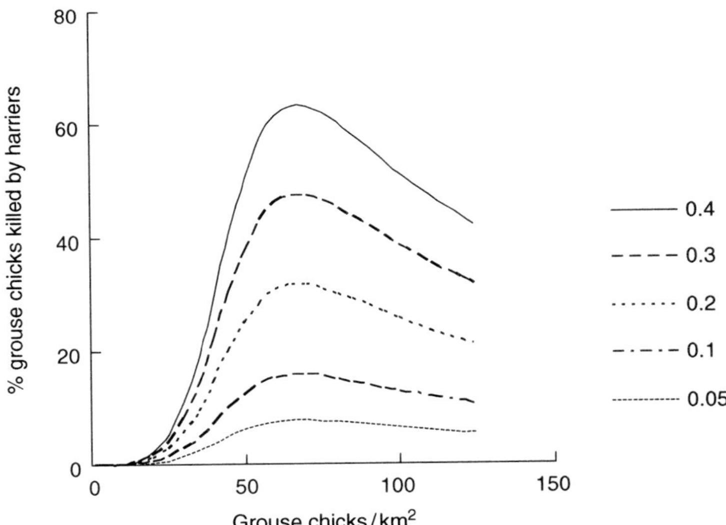

Figure 12.4. Estimates of the percentage of grouse chicks removed by harriers over the nestling period in relation to the density of grouse chicks. Lines are derived from the functional response equation for five different densities of harriers (per km²) which reflect the observed variation in harrier density. (From Redpath and Thirgood 1999.)

during 1992–8 (Thirgood *et al.* 2000b). Numbers of breeding female harriers increased with protection from two in 1992 to 20 in 1997, and peregrines also increased from three to six pairs during this time. Grouse density in April, July and October declined during 1992–8. Summer losses of adult grouse averaged 30% between April and July and were density-dependent. Raptors were the cause of more than 90% of the summer mortality of adult grouse, although we could not distinguish between harrier and peregrine predation. Summer losses of grouse chicks averaged 45% between May and July and were not density-dependent. Harriers killed 37% of grouse chicks by late August. Summer raptor predation appeared to be additive to other causes of mortality and we estimated that it reduced autumn grouse density by 50%. Harriers and peregrines may overwinter on moorland and grouse are important in their diet. Winter losses of grouse between April and October at Langholm averaged 33% and were density-dependent. Raptors were the cause of 70% of winter mortality and they killed 30% of the grouse present in October. We were unable to distinguish between winter predation by harriers and peregrines or whether winter mortality to raptors was additive to other losses. We developed a model that combined the estimated reduction in autumn grouse density caused by raptor predation with the observed density-dependence in winter loss of grouse at Langholm from 1992 to 1996. The model predicted that, in the absence of raptor predation from April to October, grouse density in spring would be 1.9 times greater, and grouse density in autumn would be 3.9 times greater, than in the presence of raptors (Fig. 12.5). The model

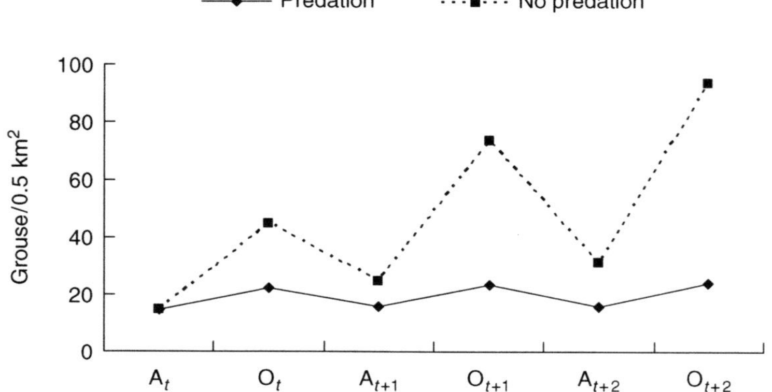

Figure 12.5. Predicted effects on grouse density in April (A) and October (O) in year one (*t*+1) and year two (*t*+2) of removing summer predation by harriers and peregrines at Langholm. All figures refer to grouse/0.5 km^2. (From Thirgood *et al.* 2000b.)

suggested that raptor predation prevented the grouse population from increasing and was thus a limiting factor (Thirgood *et al.* 2000b).

These different strands of evidence suggest that raptor predation had a major effect on grouse populations. But what impact did raptor predation have on grouse-shooting? We were fortunate in having annual records of grouse-shooting for Langholm and two nearby moors that differed in their treatment of raptors. First, however, we had to tease apart the effect of raptor predation from other factors effecting grouse-shooting (Thirgood *et al.* 2000c). The numbers of grouse shot on Langholm declined during the period 1913–90 and this was not due to harriers and peregrines as they were controlled throughout this period. The long-term decline in grouse bags coincided with the conversion of half of the heather moorland to grass from 1950 to 1990. Given the association between grouse and heather, a link between heather loss and declining grouse bags was suggested. Harrier and peregrine density at Langholm increased to high levels following protection in 1990 whilst grouse bags declined each year until shooting was abandoned in 1998. The prediction of a peak in bags at Langholm in 1996, based on the cyclical patterns of bags during the period 1950–90, was supported by the observed peaks in 1997 on two nearby moors with few raptors which formerly cycled in synchrony with Langholm (Fig. 12.6). This large-scale manipulation of harrier density provides convincing evidence that whilst long-term declines in grouse bags were probably due to habitat loss, the high levels of raptor predation subsequently limited the grouse population and suppressed a cycle (Thirgood *et al.* 2000c).

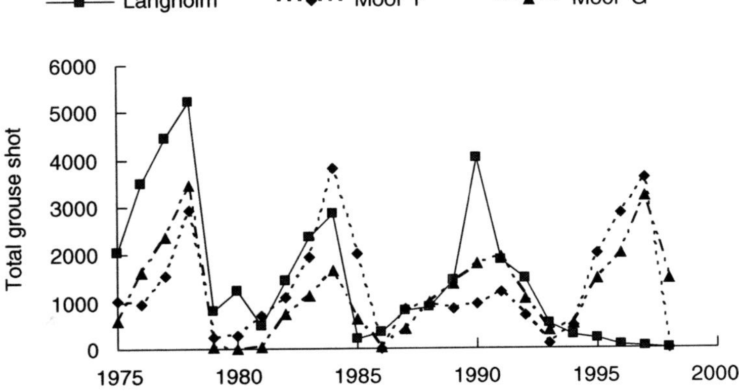

Figure 12.6. The numbers of grouse shot at Langholm from 1975 to 2000 in comparison to the numbers shot on two nearby moors 'F' and 'G' during the same period. Raptors were protected at Langholm during the period 1990–2000. (From Thirgood *et al* 2000c.)

In summary, detailed analysis of demographic data combined with analysis of shooting records and a large-scale manipulation of raptor density provided evidence that predation by raptors, particularly hen harriers, limited red grouse populations and reduced shooting bags (Thirgood *et al* 2000b, 2000c). These results, in conjunction with the predictions of the numerical and functional responses (Redpath and Thirgood 1999), clearly indicated that genuine conflicts existed between grouse management and raptor conservation (Thirgood *et al.* 2000a).

TESTING THE SOLUTIONS

Our research demonstrated that predation by hen harriers can reduce grouse harvests. But can this conflict be resolved or at least reduced? There are several potential solutions to the conflict, some of which are constrained by the legal protection afforded to hen harriers by UK and European Union legislation. Here we focus on the ecological aspects of the mitigation techniques and turn later to the legal and political issues.

Doing nothing

One approach to the conflict is to do nothing. This is essentially what is currently happening but a lack of pro-active management has led to two negative outcomes. First, illegal killing of harriers on grouse moors is common and severely threatens their status in the UK. Alternatively, if harrier control were to cease without remedial measures in place, some grouse moors would become unviable because of increased predation with consequences for the conservation of heather moorland.

Paying to reduce conflicts

Financial inducements to compensate for loss or increase tolerance are widely used to mitigate conflicts between people and wildlife (Nyhus *et al.*, Chapter 7). Could compensation help to reduce conflicts between harriers and grouse? In theory at least, compensation for lost revenue caused by predation might increase tolerance for harriers on grouse moors. There are, however, several obstacles to overcome. First, for some hunters, money could never compensate for the loss of shooting. Second, even if hunters are willing to accept compensation, the amounts of money involved are very large. For example, the value of grouse-shooting at Langholm was estimated at US$150 000 per year (Redpath and Thirgood 1997). This includes only the income generated annually from shooting and not the capital value of the moor. Third, it is unlikely that it would be politically acceptable for landowners to receive public money to prevent them from killing legally protected raptors. Grouse-shooting is perceived as a minority sport pursued by the wealthy and as such does not receive widespread public support. Finally, in the unlikely event that it was acceptable, it is unclear who would pay the compensation.

Habitat management to reduce susceptibility to predation

Red grouse live in a heterogeneous environment of heather and grass. Could this habitat mosaic be manipulated to reduce the susceptibility of grouse to predation? Management of nesting habitat is widely used to improve the breeding success of waterfowl (Greenwood *et al.* 1995) and current work is testing whether manipulating farmland can reduce hen harrier predation on grey partridge (*Perdix perdix*: E. Bro pers. comm.). In theory, habitat could influence predation on nests, chicks and adult grouse and be managed to reduce losses. We assessed the influence of habitat on nest site selection and nest success of red grouse (Campbell *et al.* 2002). Grouse nested in vegetation that was taller, denser and with greater canopy cover than was available at random. However, nest site characteristics had no effect on nesting success, which at 80% was high for wild galliforms. It is possible that nesting habitat might be more important in the absence of mammalian and avian predator control on grouse moors.

We also investigated whether habitat influenced predation on grouse chicks and adult grouse (Thirgood *et al.* 2002). Harriers were more likely to encounter grouse chicks in a mixture of heather and grass than expected from the distribution of grouse chicks on moorland. However, having encountered a chick, there was no effect of habitat on subsequent strike success. There was no evidence of an effect of heather cover or vegetation height and density on the likelihood of individual grouse surviving the winter, or of habitat directly influencing grouse mortality rates at the population level. However, grouse density was higher and winter losses were lower on areas with more heather, probably due to dispersal. Thus habitat characteristics may have an important effect on local compensation of winter mortality, although the effects at the population scale are unknown. Overall the effects of habitat on grouse susceptibility to predation appear limited, and habitat manipulation does not, therefore, appear to be a very promising approach to reduce conflicts.

Diversionary feeding to modify harrier diet

One method with potential to reduce predation on grouse is to provide alternative food to harriers. Diversionary feeding has been used to reduce wolf predation on moose (Gasaway *et al.* 1992). We tested the effects of providing harriers with carrion prior to incubation and during the nestling stage during 1998–9 to assess whether:

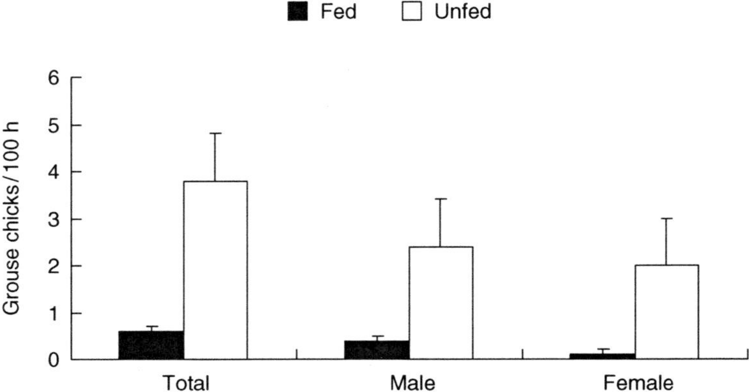

Figure 12.7. Mean (±SE) rate at which harriers provisioned grouse chicks to their nests. Data are shown separately for males and females with and without diversionary food. (From Redpath *et al.* 2001.)

(1) feeding harriers reduced their predation on adult grouse and grouse chicks, and
(2) feeding harriers increased their breeding density and breeding success.

Providing harriers with food in spring had no effect on mortality rates of adult grouse. Raptors were the main cause of mortality for adult grouse during April and May, killing 18% of the grouse overall and 78% of those grouse which died. There were no differences in adult mortality rates between areas where harriers were fed and where they were not fed. Providing harriers with food in summer reduced provisioning rates of grouse chicks to harrier nests sevenfold (Fig. 12.7). Over the two years of the experiment, fed harriers delivered on average 0.5 grouse chicks/100 h, compared to 3.7 grouse chicks delivered to nests without supplementary food. Despite the decrease in predation by harriers, grouse chick survival remained poor and grouse density continued to decline during the two years (Redpath *et al.* 2001).

Feeding harriers had no significant effect on the breeding densities of either male or female harriers, although feeding was associated with increased density in one area in one year. Fed harriers had larger clutches than unfed harriers but there was no evidence that feeding led to more harrier chicks returning to breed in subsequent years (Redpath *et al.* 2001). Despite these promising results, hunters are reluctant to adopt diversionary feeding as a management tool. There is concern that feeding will increase harrier numbers, which is exacerbated by the refusal of conservationists to contemplate limiting harrier densities, and worries that feeding will encourage other predators that prey on grouse.

Using predators to limit harrier density

Intraguild predation is recognized as an important factor influencing carnivore communities (Palomares and Caro 1999). Conversely, the removal of top predators may cascade through ecosystems with mesopredators increasing in density and depleting prey populations (Palomares *et al.* 1995; Crooks and Soulé 1999). Studies on intraguild predation in raptor communities are rare (Hakkarainen and Korpimaki 1996) and we have little understanding of their effects in the harrier–grouse system (Thirgood *et al.* 2003). Hen harriers interact competitively with other raptors and are occasionally killed by golden eagles (Watson 1997). There is also evidence that harriers may avoid nesting in areas used by eagles (Fielding *et al.* 2003). If eagles suppress harrier densities, could the presence of eagles on grouse moors reduce harrier predation on grouse? There may be a trade-off involving increased loss of grouse to eagles, but an eagle is unlikely to kill as many grouse as a hen harrier and the home range of an eagle pair could hold many pairs of harriers (Thirgood *et al.* 2000a). At present this is a speculative hypothesis, but it would be quite feasible to test with a trained eagle.

Similar uncertainty surrounds the impact of mammalian predators on hen harrier population dynamics. Potts (1998) reported several cases where harrier populations increased following the control of red fox and suggested that predator control on grouse moors benefited harriers. This idea was not supported by Green and Etheridge (1999) who showed that the nest success of harriers on unmanaged moorland was no higher in regions in which nests were close to grouse moors than in those in which they were far from grouse moors. This issue is best resolved with experiments in which harrier nest success is monitored at sites before and after changes in predator control.

Habitat management to reduce harrier density

There is widespread agreement that improved habitat management is essential to retain viable grouse moors in the long-term (Scottish Natural Heritage 1998). Reduction in heather moorland caused by afforestation and overgrazing have almost certainly contributed to declines in grouse bags (Thirgood *et al.* 2000c). More controversial is the suggestion that habitat management could reduce densities of harriers and thus predation on grouse. Harriers breed at high densities on grouse moors where meadow pipits and field voles are abundant (Redpath and Thirgood 1999) and these species are common on moors with a mosaic of heather and grass

(A. A. Smith *et al.* 2001). The implication is that habitat management – such as reducing grazing pressure – to increase heather will reduce prey for harriers and thus reduce harrier density and predation on grouse. This approach has been promoted by conservationists (British Trust for Ornithology *et al.* 2001), but a number of important questions remain (Thirgood *et al.* 2000a). First, how much and how fast will reducing grazing increase heather? Second, to what extent will habitat management influence harrier densities? Third, what are the biodiversity implications of increasing heather on moorland? Finally, will increasing heather do more harm than good? There is considerable variation between individual harriers in predation on grouse chicks and recent analysis suggests a positive relationship between predation rates and the proportion of heather around nests (Amar *et al.* 2004). This analysis predicts that increasing heather by 50% would increase grouse chick predation by 174%. Modelling is needed to investigate the trade-off between the numerical and functional responses of harriers to changes in prey densities in relation to habitat management.

Direct human intervention to limit harrier density

Having decided that it is desirable to limit harrier density as a component of an integrated conflict-resolution strategy, a pragmatic approach is for direct intervention rather than reliance on intraguild predation or habitat management. This is, essentially, what hunters have been doing illegally for decades. Potts (1998) suggested the compromise of a 'harrier quota' in which all grouse moors would support a given density of harriers and breeding attempts above this would be prevented. Potts (1998) suggested that a density of 0.04 female harriers/km^2 (one nest/25 km^2) would meet harrier conservation and grouse management purposes, and applied across all suitable habitats in the UK, would result in some 1600 female harriers, double the current population (Potts 1998). A more quantitative attempt to answer the question of what harrier density would be compatible with grouse management indicates that a wider range of harrier densities could coexist with grouse-shooting (Redpath and Thirgood 2003). A deterministic model suggests that breeding densities of harriers of 0.3/km^2 (one nest/3.3 km^2) could coexist with viable shooting (Fig. 12.8). This assumes that all harrier nests are equivalent in terms of grouse predation, which may not be the case, and it may be beneficial to target intervention on those nests with the greatest proportion of adjacent heather (Amar *et al.* 2004).

How would harrier densities be limited? Potts (1998) suggested that harriers could be translocated from grouse moors where they were abundant to other moorland where they were rare. Whilst appealing, a translocation programme seems unlikely to play a major role in reducing conflicts on

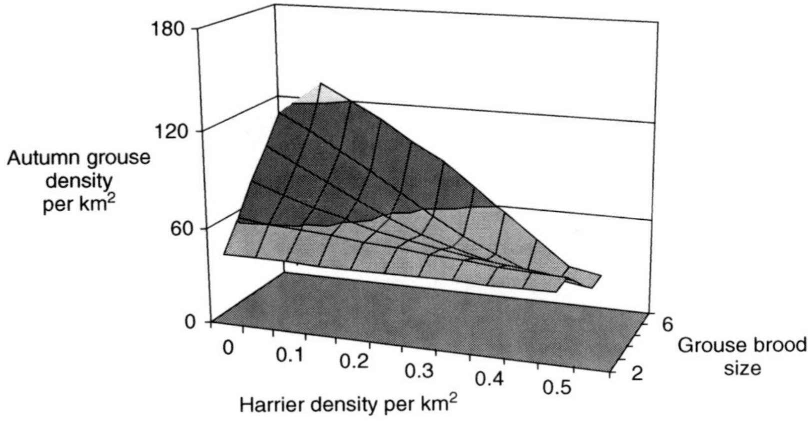

Figure 12.8. Outputs from model giving autumn density of grouse over a range of harrier densities and grouse brood sizes. (From Redpath and Thirgood 2003.)

grouse moors because it would be an imprecise method of reducing harrier numbers and would have to continue indefinitely despite there being a finite number of recipient sites (Watson and Thirgood 2001). Although politically unattractive, it appears that lethal control may ultimately be necessary to limit harrier densities to levels compatible with grouse-shooting. Lethal control could in theory be based on the removal of eggs from nests and not involve the destruction of adult harriers. Not surprisingly, stakeholders are polarized on the desirability of harrier quotas and lethal control.

Management systems and zoning

The management solutions described above should not be considered as alternatives. Reducing harrier densities 'naturally' through intraguild predation or habitat management is compatible with direct intervention to limit harriers through quotas. Put simply, if increasing heather or introducing eagles were successful in reducing harrier densities then quotas would not be exceeded and lethal control would be unnecessary. In the same way, providing diversionary food for harriers is a good management strategy regardless of the density of harriers breeding on a grouse moor. The reluctance of grouse managers to adopt diversionary feeding is primarily due to concerns about increasing harrier numbers. These concerns would be reduced if diversionary feeding were combined with a system to limit harrier density.

The harrier–grouse conflict takes place on private land outside protected areas. Debate has focussed on solutions that can be applied across the entire harrier range in the UK. Would it be more appropriate to think in terms of areas managed primarily for harriers and areas managed primarily for grouse? Zoning is widely used in the management of species that come into conflict with humans (Linnell *et al.*, Chapter 10). European directives obliging member states to designate Special Protection Areas for birds of conservation concern have already started this process. Special Protection Areas have already been designated for hen harriers in areas chosen theoretically as ideal harrier habitat that can be managed to maximize harrier density. Perhaps different conservation targets could be set in upland areas that are primarily managed for grouse?

Modelling the uncertainty

Although the harrier–grouse system has been thoroughly studied in the field, only preliminary attempts have been made to model the outcomes of different management scenarios (Thirgood *et al.* 2000b; Redpath and Thirgood 2003). Computer-intensive techniques are now available which can incorporate all aspects of uncertainty that result from measurement and estimation error, model mis-specification and lack of knowledge (Harwood 2000) and are currently being applied to the harrier–grouse system (K. Graham and S. Smout unpubl. data).

Uncertainty also surrounds the acceptability of different management solutions to the various stakeholders and there seems to be little agreement between hunters and conservationists about possible solutions. Multiple-criteria decision models that quantify stakeholder viewpoints showed, not surprisingly, that the hunters preferred direct intervention to limit harrier densities whilst conservationists preferred to allow harriers to attain natural densities (Redpath *et al.* 2004). However, both groups agreed that diversionary feeding could be a valuable technique to reduce predation, with the caveat that uncertainties over harrier numbers had to be addressed. The decision-modelling workshops highlighted the value of objective techniques in identifying the options with the highest likelihood of agreement and developing dialogue and trust between stakeholders.

CONSERVATION SCIENCE MEETS CONSERVATION ACTION

Has 10 years' effort and the expenditure of US$1 500 000 on research done anything to reduce the conflict between harrier conservation and grouse management? The answer to that question is a qualified yes. The research described in this chapter has certainly increased our understanding of harrier–grouse dynamics and conflicts between raptors and game birds generally. We have moved on from the 1990s debate about whether or not harrier predation reduced grouse harvests to the current debate about how the impact of harriers can be reduced. The reality on grouse moors, however, is that harrier persecution is at least as bad now as it was 10 years ago. Indeed it is possible that illegal killing has intensified because hunters are aware of the research findings and are adopting the precautionary principle.

Whilst illegal killing by hunters is the proximate cause of the poor conservation status of hen harriers in the UK, it can be argued that the inflexibility of both hunters and conservationists is the ultimate cause. Our research has demonstrated that genuine conflicts exist between hen harrier conservation and grouse management and potential solutions to the conflict are available but have yet to be implemented. The results of the diversionary-feeding experiments were encouraging, but further tests of the applicability of the technique are required to convince stakeholders. Hunters are reluctant to undertake such trials because they are concerned that harrier densities will increase with feeding and they will be unable to reduce them legally. The current situation is effectively a stalemate, with further feeding trials acknowledged as desirable by all, but conservationists unable or unwilling to offer the compromises over harrier numbers that are demanded by hunters before the feeding trials go ahead on private land.

What should now be done to break the deadlock and move the conflict forward? It is essential to initiate further feeding trials on a range of sites at different harrier and grouse densities. These trials could, in theory, be accompanied by derogation from legislation to allow harrier densities to be reduced on experimental sites should feeding prove to be insufficient to maintain the viability of grouse moors. If such derogations are not forthcoming, as seems likely, it will be necessary for hunters to accept the financial losses resulting from increased predation, or for the entire conservation community to look for ways of financially compensating these losses. Regardless of such derogations, the efficacy of diversionary feeding needs to be tested before intervention to limit harrier numbers will be considered by conservationists.

Having tested the efficacy and applicability of diversionary feeding, the next stage would be to test methods of locally reducing harrier numbers on moorland. There is extreme reluctance on the part of conservationists to contemplate direct intervention to limit harrier densities. Although understandable in the context of historical persecution and mistrust between conservationists and hunters, this reluctance seems to us rather misplaced on conservation or ethical grounds, given the enthusiasm shown by the same conservationists for indirect approaches to limiting harrier densities through intraguild predation and habitat management. The attitude of the British conservationists appears out of step with modern approaches to dealing with human–wildlife conflict, where lethal control is typically retained within a toolbox of non-lethal techniques, to be applied when non-lethal approaches are ineffective (Conover 2002; Treves and Naughton-Treves, Chapter 6, Woodroffe *et al.*, Chapter 24).

ACKNOWLEDGEMENTS

We thank the Game Conservancy Trust, Centre for Ecology and Hydrology, Scottish Natural Heritage, Royal Society for the Protection of Birds, Joint Nature Conservation Committee, Department of the Environment Transport and Regions, Natural Environment Research Council and the European Union for funding. The Buccleuch Estates generously provided their grouse moor and considerable logistic help. Peter Hudson, Ian Newton and Dick Potts offered guidance and support at critical moments. Numerous assistants helped with fieldwork. ST was based at the Centre for Conservation Science at Stirling University whilst writing this chapter and at the Frankfurt Zoological Society Africa Regional Office in the Serengeti whilst revising it. Comments from Arjun Amar, Bea Arroyo, Juliet Vickery and Rosie Woodroffe greatly improved this chapter.

13

Understanding and resolving the black-tailed prairie dog conservation challenge

RICHARD P. READING, LAUREN MCCAIN, TIM W. CLARK AND BRIAN J. MILLER

INTRODUCTION

In 1999 the US Fish and Wildlife Service recommended designating the black-tailed prairie dog (*Cynomys ludovicianus*) as warranted for listing as threatened under the US Endangered Species Act, but precluded from such listing by other, higher priority species (US Fish and Wildlife Service 1999). This 'warranted, but precluded' finding flamed a management controversy that had been brewing for years and instigated a flurry of activity by agricultural interests, government agencies, conservation non-governmental organizations (NGOs), scientists and others. Stakeholders became polarized between those who want to conserve prairie dogs and those who want to limit them. Although ecologists have noted for decades the huge decline of black-tailed prairie dogs, as recently as 10–15 years ago the status of prairie dogs and their management was largely neglected and therefore not controversial. How did this issue move to the forefront of conservation controversies in this country? In this chapter we use a policy sciences approach (Lasswell and McDougal 1992) to describe and analyse the controversy surrounding prairie dog conservation and management by examining the context of the issue, the key stakeholders, and the processes being used to understand and address the problem. We end with recommendations to improve prospects for black-tailed prairie dog recovery and conservation.

THE CONTEXT OF PRAIRIE DOG CONSERVATION: ORIENTING TO THE 'PROBLEM'

Fully understanding the challenge of black-tailed prairie dog recovery requires a comprehensive assessment of the context. Our analysis contains more than a simple biological and ecological review, we also incorporate

People and Wildlife: Conflict or Coexistence? eds. Rosie Woodroffe, Simon Thirgood and Alan Rabinowitz. Published by Cambridge University Press.

historical, social, economic, political, organizational and process assessments. Some people perceive prairie dogs as a threat to their livelihoods and lifestyles, whereas others view them as ecologically important members of the prairie ecosystem. Differences in perception stem from different ideologies, value systems and situations of the people involved. However, before delving into the complex human context, we summarize the underlying prairie dog ecology.

Focus of the conflict: prairie dogs

Black-tailed prairie dogs are 1-kg ground squirrels that live in colonies of strongly defended family groups, known as coteries (Hoogland 1995). Black-tailed prairie dogs are the only species of five that inhabits the Great Plains of North America, ranging over most of the short and mid-grass prairies from southern Canada to northern Mexico (Hoogland 1995; Miller *et al.* 1996). Black-tailed prairie dogs are the most widespread and most communal of the species. They eat grasses and clip down other vegetation to detect predators.

Ranchers look at prairie dog colonies, see less grass and assume that prairie dogs decrease livestock carrying capacity. Early eminent biologists lent support to these understandable perceptions, but provided no data (Merriam 1902; Bell 1921; Bonnie *et al.* 2001). Yet, O'Meilia *et al.* (1982) found no statistical difference in weight gain in steers raised on pastures with and without prairie dogs. Other studies found a 1–8% impact on livestock production from prairie dog activities depending on circumstances and the species of prairie dog (Crocker-Bedford 1976; Hansen and Gold 1977; Uresk and Paulson 1988). Although prairie dogs and livestock eat many of the same forage plants, prairie dogs avoid many plants preferred by livestock and vice versa (Coppock *et al.* 1983a, b; Uresk 1984). More importantly, much of the lost biomass on prairie dog colonies is apparently compensated by improved forage quality in the form of better digestibility, growth rates and nutrient content (Bonham and Lerwick 1976; Coppock *et al.* 1983b; Detling 1998). Improvements may result from increased plant diversity, soil productivity and water capacity on prairie dog colonies relative to surrounding prairie (Munn 1993; Outwater 1996; Weltzin *et al.* 1997b; Detling 1998). Even if prairie dog–livestock competition is greater than suggested by this body of research, an economic analysis found that prairie dog control is not cost-effective (Collins *et al.* 1984).

Traditional economic analyses do not consider the benefits of prairie dog activities to livestock producers or the contributions of prairie dogs to local ecosystems (Miller *et al.* 1990). For example, prairie dogs promote plants more tolerant of grazing (Bonham and Lerwick 1976) and stem the spread of plants considered undesirable to livestock producers, such as mesquite (*Prosopis* spp.), sagebrush (*Artemesia* spp.), and other woody shrubs (Coppock *et al.* 1983a; Weltzin *et al.* 1997a). Prairie dog activities significantly increase species diversity across landscapes inhabited by prairie dogs – grasslands consisting of a mix of colonized and uncolonized habitat (i.e., increased beta diversity) (Kotliar *et al.* in press).

Prairie dog decline and human perspectives

When Europeans first arrived in the Great Plains, they encountered prairies teeming with life, including millions of bison (*Bison bison*) and billions of prairie dogs (Hoogland 1996). Biologists estimate that prairie dogs inhabited 410 000 km^2 in the USA in 1900, but by 1960 that area had declined to about 6000 km^2 – a reduction of more than 98% (Marsh 1984). Yet the decline continued; by 1998 prairie dogs covered only 2800–3200 km^2 (Biodiversity Legal Foundation *et al.* 1998; National Wildlife Federation 1998). Though accepted by most experts, others dispute these data. Some US state agencies have argued that the area is substantially higher than these estimates, suggesting that over 15 000 km^2 remains (Luce 2003).

Prairie dog decline began with the introduction of livestock to the Great Plains in the late nineteenth century. Earlier visitors admired the industrious habits and social structure of prairie dogs (Jones 1999). Livestock producers, however, viewed prairie dogs as pests and initiated extermination programmes. Farmers also converted prairie dog habitat into cropland (Miller and Reading 2002). Control efforts quickly expanded into massive government-sponsored programmes that covered millions of hectares (Dunlap 1988; Miller *et al.* 1996). Poisoning remains the largest factor responsible for prairie dog declines (US Fish and Wildlife Service 1999). Government eradication programmes continue to this day, albeit at reduced levels.

Around 1900, plague (*Yersinia pestis*) was introduced to the USA from Asia via rat (*Rattus* spp.) vectors and expanded its range until about 1950 (Cully and Williams 2001). Prairie dogs have exhibited no resistance to this introduced pathogen (Barnes 1993). Plague devastated prairie dog populations as it spread, often killing over 99% of individuals in affected colonies (Cully and Williams 2001). Large colonies appear particularly susceptible to outbreaks (Cully and Williams 2001). Many conservationists believe that effective management of plague represents the greatest biological challenge to maintaining large populations of prairie dogs today.

Other locally important causes of prairie dog decline include urban development and recreational shooting. Although human populations are dropping in much of the prairie dog range, urban sprawl affects some areas, especially in Colorado (US Fish and Wildlife Service 1999). Finally, prairie dog shooting has become a popular pastime in a small but growing sector of society. Preliminary research suggests that shooting, especially if heavy, can negatively affect prairie dog social structure, behaviour and population dynamics (Knowles 1988; Vosburgh and Irby 1998). Prairie dogs cannot adapt to continued habitat destruction, plague epizootics and human persecution (Miller and Reading 2002).

Prairie dog Management: from control to conservation

Together with state wildlife agencies (Luce 2003), we believe that successful prairie dog conservation requires establishing and maintaining viable populations of prairie dogs distributed throughout their historic range in a manner than enjoys broad and enduring public support. This is perhaps a working approximation of the common interest. With the exception of plague, the biological aspects of this challenge are relatively easy to address. Yet the socio-political aspects of the challenge are formidable.

For over 100 years, prairie dog management in the USA meant lethal control. Ranchers and farmers along with state and federal agencies cooperated to rid the Great Plains of this perceived agricultural pest. The growing conservation, environmental and animal welfare and rights movements over the course of the twentieth century brought new participants to the prairie dog policy arena. The new players call for conserving this keystone species as a means of protecting grassland ecosystems (Miller *et al.* 1996; Dolan 1999), or what some call the Prairie Dog Ecosystem (Clark *et al.* 1989; Miller *et al.* 1994).

The first substantial conservation activities for prairie dogs arose from efforts to recover the critically endangered black-footed ferret (*Mustela nigripes*) (Miller *et al.* 1996; Clark 1997). Ferrets are obligate associates of prairie dogs, feeding almost exclusively on them and using their vacated burrows as dens and refugia (Forrest *et al.* 1988). Ferret decline resulted from prairie dog decline. By 1987 ferrets were probably extinct in the wild (Miller *et al.* 1996; Clark 1997). As recovery efforts proceeded, conservationists realized that few large prairie dog complexes (groups of colonies in close proximity) remained to support ferrets. Ferret recovery therefore depends on prairie dog restoration. The US Fish and Wildlife Service ferret recovery plan calls for establishing 10 separate populations of 150 or more adult ferrets by 2010 (Forrest 1988). By 2002 only four sites existed with enough prairie dogs to support even half that goal (Predator Conservation Alliance 2001). As a result, many people originally involved in ferret recovery shifted their attention to prairie dog conservation.

Prairie dog destruction angered animal rights and wildlife advocates (Prairie Dog Coalition 2002), who eventually resorted to litigation to list the species under the Endangered Species Act (Biodiversity Legal

Foundation *et al.* 1998; National Wildlife Federation 1998). The federal government designated black-tailed prairie dogs as 'warranted, but precluded' for listing under the Endangered Species Act in 2000 (US Fish and Wildlife Service 2000). This mobilized state wildlife agencies to try to prevent listing under the Act and thereby avoid losing local control over prairie dog management (Van Pelt 1999; Luce 2001, 2003; Miller and Cully 2001). They created an 11-state Black-tailed Prairie Dog Conservation Team. The US Fish and Wildlife Service supported this initiative and delegated prairie dog management to the states, with a directive that 'doing nothing' was not an option. This action moved the states to centre stage. The federal US Fish and Wildlife Service retains oversight and reviews the status of prairie dogs each year (US Fish and Wildlife Service 2001b, 2002).

The states produced a conservation plan, the 'Black-Tailed Prairie Dog Conservation Assessment and Strategy', with subsequent addenda over the past four years (Van Pelt 1999; Luce 2001, 2003). The goal of preventing listing of the species under the Endangered Species Act (and hence losing control over management) arguably takes precedence over recovery of the species – a classic case of goal substitution driven by competition for power (Van Pelt 1999; Luce 2001, 2003; Miller and Cully 2001). The interstate plan notes a lack of biological information to determine area occupied and distribution goals to conserve prairie dogs over the long term (Luce 2003). However, the plan goes on to list a number of biological objectives without providing the basis for how those numbers were determined, other than stating that some were constrained by socio-political considerations (Luce 2003). Instead, it suggests that the use of adaptive management will permit the states to refine their numbers as they acquire more information (Luce 2003). However, the plan never clarifies how success or failure in adaptive management will be assessed, nor how the plan will be adapted or terminated. The plan also calls for conducting additional research and monitoring, identifying focal areas that contain high densities of prairie dogs, creating financial incentives for cooperating landowners who conserve prairie dogs, and increasing regulation of and oversight over prairie dog shooting and poisoning (Luce 2001, 2003). However, the plan also permits unrestricted shooting and calls for providing money to cooperating landowners for poisoning, even if a state remains below its target objectives for prairie dog acreage.

During the interstate planning process, some state and federal agencies began taking conservation action (Luce 2003) and some progress has been realized. For example, a few states removed 'pest' species designations from prairie dogs and others are working toward that end (Predator Conservation Alliance 2001). Some state and federal agencies also started regulating prairie dog poisoning and shooting, which was formerly unlimited (Luce

2003). Arizona (where black-tailed prairie dogs are extirpated), Colorado and South Dakota banned some shooting, primarily on public land (Luce 2003). The US Forest Service declared a temporary moratorium on poisoning prairie dogs within National Grasslands. The US Bureau of Land Management also ceased poisoning prairie dogs on land it manages, and both agencies began more active prairie dog conservation. Agencies and advocacy groups are now considering prairie dog conservation incentives for landowners (Bonnie *et al.* 2001; Luce 2001), advocated as early as 1990 (Miller *et al.* 1990). For example, in 2002 Colorado started a $600 000 pilot programme using lottery money to provide financial incentives to landowners who conserve prairie dogs (Davis 2002).

Despite these achievements, current policy and various state and federal management actions are insufficient to establish and maintain viable populations of black-tailed prairie dogs in a manner that enjoys public support. Indeed, even the US National Park Service continues to permit poisoning as a 'good neighbour' policy. Prairie dog conservation also continues to face hostile state legislatures (e.g. in 1999 Colorado passed a law restricting translocation as an alternative to lethal control), agricultural communities (e.g. few landowners joined the Colorado landowner incentive programme and county governments are considering legislation to discourage involvement; also see Frost (n.d.)), and wildlife commissions (e.g. in 2002 the Nebraska Game and Fish Commission barred the state from engaging in any conservation activities or planning for black-tailed prairie dogs). So, despite limited progress, major problems remain.

Each of the threats noted above remains important, and will persist into the future if not addressed. However we argue that the most important barriers to prairie dog conservation include a narrow focus on biological issues to the exclusion of socio-political factors: government agency inertia, powerful political forces lobbying against conservation, prevailing negative values and attitudes toward prairie dogs, and the polarized conflict that prevents cooperation among stakeholders.

ANALYSIS OF PRAIRIE DOG CONSERVATION AND CONTROVERSY

Our analysis of the prairie dog conservation challenge includes an appraisal of the social and decision processes used by the people involved.

Stakeholders: social process analysis

Understanding the social process in prairie dog conservation requires understanding the participants to help us understand barriers and devise improvements. Those who are most active and have the greatest stake in the outcome of an issue are usually referred to as key stakeholders. In analysing social process, we want to know as much as possible as about these stakeholders with respect to the problem at hand (see Lasswell and McDougal 1992; Clark and Wallace 1998, 2002). How do they see the issue? What are they expecting to get out of the process? What are their values, attitudes and perceptions? With whom do they identify? How are they organized? What are their assets and liabilities? What strategies do they use to try to get what they want? How have decisions affected them and the institutions in which they interact?

Participants in prairie dog management policy include the agricultural community (especially ranchers), conservationists, animal rights activists, wildlife and land managers, recreational shooters, urban developers, the general public and the prairie dogs themselves. Native Americans hold diverse values and attitudes and therefore fit better as subgroups within the other categories. Different stakeholders are contesting the way the prairie dog problem is being defined. Although each group has diverse memberships, we can draw some generalizations.

Ranchers

The agricultural, and especially livestock, industry began expanding into the range of the prairie dog in the late nineteenth century and viewed these animals as pests that required eradication or at least control (Jones 1999). This viewpoint remains prevalent (Reading *et al.* 1999). The perception is that prairie dogs severely restrict forage for livestock, despite scientific evidence to the contrary (see above; Reading *et al.* 1999). Ranchers also fear loss of traditional lifestyles and control over public grazing lands (Reading and Kellert 1993; Reading *et al.* 1999). Both are occurring, and increasing numbers of ranchers are unable to make a living in the western Great Plains (Licht 1997). The extent to which prairie dogs are responsible for these trends is debatable, but probably not high.

Conservationists

In almost direct opposition to ranchers, conservationists view prairie dogs as a native, keystone species of the prairies that deserves protection (Reading and Kellert 1993; Reading *et al.* 1999). These values and attitudes evolved primarily from the sport hunting movement of the nineteenth century, as modified by game management principles and doctrines of the first half of last century and the environmental movement of the 1960s and 1970s. Conservationists arrived on the Great Plains more recently than ranchers, but are growing in numbers. They demand increased involvement in public land management and often resort to lawsuits, media publicity and appeals to popular pressure, including ballot initiatives to achieve their goals.

Animal rights activists

Animal rights activists are primarily rectitude driven and base their attitudes and perceptions of prairie dogs on strong ethical and ecological values toward wildlife. They want to end human-inflicted pain and suffering to animals. They support extending legal rights to animals (Wise 2000), including prairie dogs, now reserved for humans. The ideology of animal rights activists can be traced to the animal welfare movement, and developed into a more influential organized interest in the last few decades (Rudacille 1998). This group is growing, and animal rights activists often work with conservationists, using similar strategies.

Agency personnel

Federal, state, local and tribal agency wildlife and land managers hold personal views that vary greatly, but can be quite similar within a single agency or profession. These managers face a volatile, polemic issue in managing prairie dogs. Despite multiple-use mandates, most agencies are strongly influenced by special interests. This influence is manifest in policies that often clash with the common interest (Meier 1993). Historically, many agencies were recruited into large-scale prairie dog poisoning campaigns (Jones 1999). An anti-prairie dog attitude remains strong among many managers (Reading 1993), but this is changing. Recent research suggesting that prairie dogs compete little with livestock and function as keystone species has prompted some agency personnel, particularly wildlife biologists, to begin viewing these animals differently. Yet even wildlife biologists are sensitive to ranchers' attitudes toward prairie dogs, because ranchers control access to land used by hunters (their main constituents). Agencies also struggle for power among themselves (Fischer 2000). For example, state, local and tribal agencies in the West compete for power with the federal government.

Recreational shooters

Recreational shooters form a small but vocal stakeholder group that values prairie dogs as abundant live targets for target-shooting (Vosburgh and Irby 1998). Shooters identify with the agricultural community and view themselves as highly skilled hunters and agents of control for agricultural pests (Hawes-Davis 1998). They mostly embrace a world-view that suggests many species are 'varmints' that require active control. Although limited in number, shooters tout their skill and the money they spend to influence management. They exert influence because they actively promote their interests, enlisting the support of ranchers, gun rights activists and local businesses.

Developers

Developers play a restricted role geographically in the prairie dog management challenge, but they are key stakeholders along Colorado's Front Range. Developers view prairie dogs as pests that interfere with their ability to expand urban developments and maximize profits on both private and public lands.

General public

The American public is diverse, and most citizens are unaware of the prairie dog conservation problem. However, public support for conserving wildlife is strong. For example, a survey by Czech and Krausman (1999) found 84% of the public support the current Endangered Species Act or would like it strengthened. Small sectors of the public, such as homeowners living near urban prairie dog colonies, are a part of the development debate. Several other subgroups exist among the public. For example, many Native Americans with traditional beliefs consider themselves intimately interconnected with prairie dogs, and believe that the tribes are obligated to protect and restore native wildlife communities.

Prairie dogs

Black-tailed prairie dogs are participants in this issue too, as are a myriad of other associated species. Their interests, to the extent they can be known, appear to be for continuation of the species and individual well-being. Right or wrong, humans will decide prairie dogs' fate. Nevertheless, we suggest that a conservation approach in the common interest should include considerations of other life forms, including prairie dogs and their many associated species, not just humans.

Relations among stakeholders

Relationships among the main stakeholder groups in prairie dog management have become increasingly conflict laden. Conflict intensified after black-footed ferret listing and recovery programme implementation. Agencies overseeing prairie dog control suddenly found themselves responsible for conserving enough prairie dogs to restore ferrets. State and federal agencies engaged in a battle for control of the ferret recovery programme (Clark 1997). Issues of tribal sovereignty emerged, especially with respect to prairie dog control and proposed ferret reintroductions on tribal lands.

Arguments over the importance of prairie dogs (e.g. pest vs. keystone species) between or even within agencies resulted in conflicting actions (e.g. both poisoning and conserving prairie dogs) and agency personnel giving out conflicting information (e.g. prairie dogs do or do not compete with livestock) (Miller *et al.* 1994, 1996).

Conservationists advocated protection and restoration of prairie dogs and ferrets, and performed 'watchdog' activities on government actions (Reading 1993). Animal rights activists fought to protect both ferrets and prairie dogs, with some staging protests on related issues. Conservationists and animal rights groups sued government agencies to list prairie dogs and associated species under the Endangered Species Act. Ranchers elicited the support of their elected representatives to pass laws and regulations preventing prairie dog conservation activities, fearing the restrictive nature of the Endangered Species Act and its implications for their ranching practices and lifestyles (Reading and Kellert 1993). Recreational shooters feared losing access to public land in the name of ferret conservation and fought efforts to restrict shooting (R. Reading unpubl. data). On Indian reservations, those calling for prairie dog and ferret restoration as a part of tribal heritage fought with tribal ranching interests.

Participants in the ferret recovery programme tried to improve relations among stakeholders and reduce conflict, and achieved limited progress. Yet each of these issues resurfaced as other species associated with prairie dogs were petitioned (e.g. swift fox, *Vulpes velox*) or proposed for listing (e.g. mountain plover, *Charadrius montanus*). During this time, black-tailed prairie dog populations continued to decline. The unresolved conflict among stakeholders intensified following the 1998 petitions to list black-tailed prairie dogs as threatened.

Decision-making in prairie dog conservation: decision process analysis

Any conservation programme involves decision-making (Clark and Brunner 1997). Prairie dog conservation is about 'how the prairie dog ecosystem will be managed' and 'who gets to decide.' In other words, it is partly a political problem that centres on power and authority relationships. Each participant approaches the situation with a unique perspective, interacts in various situations (e.g. in media, courtroom, bureaucracy), uses assets (e.g. money, knowledge, skill), employs strategies (e.g. science, education, lobbying, litigating) and seeks outcomes (e.g. more or fewer prairie dogs, more power, respect) that differ from those of other participants (Lasswell and McDougal 1992). Improving the decision process can help improve any programme by helping to save valuable resources, managing conflict and addressing problems in a timely fashion.

Organizational arrangements

The decision-making process in prairie dog conservation occurs within organizational arrangements, both formal and informal, that dictate how information is or is not shared, how resources are distributed, who is involved, how participants interact, and ultimately how decisions are made and implemented. Understanding these variables helps identify the best opportunities for improving decision-making.

Bureaucratic structures and decision-making increasingly dominate prairie dog conservation and management. This is not surprising given the increasingly prominent role that government agencies are playing (and subsequent challenges to their activities through litigation by NGOs). Historically, however, informal decision-making was common, particularly among private landowners (Jones 1999). Private landowner control over prairie dog management remains important locally, but agencies are becoming increasingly involved (Luce 2001). Indeed, much of the conflict over prairie dogs results from landowner concerns that they are losing control over their ability to manage prairie dogs as they see fit on lands they own or utilize for grazing (Reading *et al.* 1999).

Following the 1998 petitions to list the black-tailed prairie dog as threatened, states within the former range of the species organized the Interstate Black-Tailed Prairie Dog Conservation Team and hired a coordinator (Interstate Team: Van Pelt 1999). This team's structure was based on a similar effort by several states to avoid an Endangered Species Act listing for the swift fox (US Fish and Wildlife Service 2001b). The Interstate Team largely continues past patterns of interaction, old power arrangements, bureaucratic designs and modes of participation (Van Pelt 1999; Luce 2001). Ultimately, the effectiveness of the interstate effort will only be known after the next few years, but there are signs that this approach will prove inadequate. The organizational structure leaves power and authority fragmented among the many states, severely constraining the ability of the coordinator to make progress, despite great effort on his part. As a result, the current plan is constrained by a traditional state approach to conservation that is neither innovative, comprehensive, nor inclusive. In addition, some states face hostile state legislatures and commissions. For example, by 2001 Nebraska, Wyoming, North Dakota and Colorado had, in one way of another, withdrawn from the official interstate effort and Montana, North Dakota and Oklahoma rejected the interstate plan's objectives. These actions call into question the new organization's ability to coordinate prairie dog conservation effectively. Despite these problems, federal agencies (especially the US Fish and Wildlife Service) and some NGOs promote the interstate approach.

The interstate plan is a clear attempt by the states to avoid greater federal involvement in prairie dog management. By contrast, most conservation and animal activists are calling for just the opposite. Moreover, ranchers, recreational shooters and tribal governments desire greater local control by local landowners or counties. Tribal governments, which are independent of most federal and state legislation and strongly influenced by ranching interests on reservations, hope to ensure that tribal organizations predominate on reservations. As a parallel to the interstate initiative, eight tribes formed an Inter-Tribal Prairie Ecosystem Restoration Consortium to coordinate prairie dog conservation and management on tribal lands (Predator Conservation Alliance 2001).

Aside from a day-long meeting sponsored by an NGO and held soon after the US Fish and Wildlife Service issued their 'warranted, but precluded' designation, there has been little attempt to organize all stakeholders into the conservation management process. The Interstate Team has chosen to leave this issue, and many others, to the individual states to address (Luce 2003). As a result, prairie dog conservation has been dominated by state and federal wildlife and land management agencies and, to a lesser extent, academics and large, national conservation organizations. As a result, current organizational arrangements exclude several key stakeholders from the decision process, especially agricultural agencies and interest groups (e.g. stock-growers associations), powerful local politicians, recreational shooters and a large number of smaller animal rights and conservation organizations. An assessment of the decision-making process to date suggests that current organizational arrangements limit involvement and have begun to strain relations among stakeholders, increase unproductive conflict, and hinder prairie dog conservation success. (Not all conflict is bad. Well-managed conflict can result in productive discourse and lead to greater innovation and creativity.)

Overall, current prairie dog conservation efforts offer little that is new, creative, or helpful to maximize cooperation among stakeholders. Instead, the programme chosen appears to be as conservative and close to the status quo as possible. The interstate effort lacks socio-political goals, despite their crucial importance in this issue (Luce 2003). Ironically the states use socio-political and economic constraints to help define biological goals, which they admit lack a solid biological basis (Luce 2003). As a result, the interstate effort is failing to ensure the survival and viability of prairie dog ecosystems in ways that benefit from broad public support.

Alternatives to current agency efforts have received less attention, but may offer opportunities for improvement. The Predator Conservation Alliance (2001) put out a plan for prairie dog restoration that calls for an interconnected system of prairie dog focal areas, prairie dog reintroduction, and measures to reduce prairie dog poisoning, prairie dog shooting and the effects of plague. Other non-profit organizations are purchasing land and working to restore prairie dog populations on private land, and the nine tribes of the northern Great Plains include prairie dog management as a major component of their Intertribal Prairie Ecosystem Restoration Consortium. Still, most of the alternatives being considered or implemented focus on the biological aspects of the challenge and do little to address values and attitudes, organizational arrangements, and power and authority structures that underlie the more immediate biological threats.

IMPROVING PRAIRIE DOG CONSERVATION: APPLYING TOOLS FROM THE SOCIAL SCIENCES

Addressing conflict in conservation is difficult, as competing interests are usually firmly entrenched. There is no 'silver bullet.' Instead, we suggest focussing on improving processes that manage conflict and move toward conservation in the common interest. The prairie dog conservation challenge is complex and contentious, and it probably will not yield to conventional problem-solving. Yet the issue is being treated as though it was simpler, better structured, and amenable to traditional approaches (cf. Van Pelt 1999; Luce 2001, 2003). The problem now is to reduce and better manage conflict. We stress that this will not be easy nor quick, but depends on building new cooperative relationships and expanding on successful practices to date. We suggest altering organizational structures and employing a 'practice-based' approach to continually improve prairie dog conservation management.

Improving organization and management

Prairie dog conservation probably would benefit from changes to the current style of organization and management. We recommend broadening the formal and informal structures to include all stakeholders who wish to participate. Adequate communication is vital: face-to-face communication is best. At the national level we recommend creating a national working group to serve as a forum for conveying values, opinions and demands, for exchanging ideas and information, for expressing legal and other obligations, for managing conflict more productively, and for identifying opportunities to improve conservation prospects. For practical reasons, this will likely be an outgrowth of the Interstate Team; however, we caution that it must function much more broadly than it currently does.

Additional organizational structures are necessary at local levels. Most states have, or are in the process of creating, working groups or similar structures for prairie dog conservation. We envision, and the interstate plan calls for, more direct action by state organizations (Luce 2003). Broad participation by diverse interests is even more crucial at this level. Two types of working groups might be required. A formal, decision-making body would include representatives of all organizations legally required to be involved. Such a group is necessary to increase coordination and cooperation, and also because agencies generally cannot delegate their legal authority to others. A second, less formal advisory group would include representatives of all interested stakeholder groups. Although it might be necessary to limit membership to keep the size manageable, the advisory group should represent the full range of interests and values. Group members, however, must be willing to work cooperatively with diverse participants, some of whom may embrace very different values. Obviously, managing such a group competently is crucial. This group should function as an interactive, brainstorming body that could identify and manage conflict before it becomes unproductive. Direct input to the decision-making body from the advisory group would increase programme effectiveness. In addition, we recommend forming special, high-performance teams to address specific tasks.

Practice-based conservation (adaptive management)

The interstate plan calls for adaptive management (Luce 2003), and practice-based conservation is adaptive management at its best. A practice-based strategy emphasizes procedures that can be adapted to changing circumstances (Simon 1983). It involves finding and taking advantage of opportunities that exist or can be created to address problems. Practice-based conservation involves three steps, each of which requires ongoing evaluation (Kleiman *et al.* 2000). First, participants identify the 'best practices' employed to address the prairie dog conservation challenge. Second, these practices are adapted and applied to similar circumstances elsewhere. Finally, the most effective practices are diffused as widely as possible, where professionals continue to adapt, refine and upgrade them relying on their own experience. Thus, the prudent way for prairie dog conservation to proceed is to avoid fixed strategies, but instead continually upgrade performance by developing, identifying and adjusting best practices (Clark and Brunner 1997).

Developing pubic support, managing opposition

Effectively managing and reducing conflict represents perhaps the greatest challenge facing prairie dog conservation. Prairie dogs elicit strong emotional responses from several stakeholder groups, reflecting the strength of the underlying values and issues that prairie dogs symbolize. If society hopes to conserve prairie dogs, an effective public-relations programme is necessary. The overall strategy probably requires a careful mix of what Cutlip and Center (1964) refer to as pressure, purchase and persuasion (Reading *et al.* 1999). By pressure we mean the judicious use of laws, regulations and control over resources (e.g. public lands) to promote prairie dog recovery without alienating people (Chin and Benne 1976). Such sanctions should be invoked firmly, but sparingly and only after other tactics fail. Purchase might entail using financial incentives to stakeholders that maintain prairie dogs on land they own or control (see Miller *et al.* 1990; Bonnie *et al.* 2001). However, such financial incentives generally do little to alter underlying values and attitudes. Finally, persuasion requires an innovative public-relations/education component that differs substantially from most past programmes.

Often, biologists and others argue that working to influence other people's values and attitudes is ethically or morally wrong. Such a position supports the status quo, permitting other people and institutions to influence values and attitudes (Lamb *et al.* in press). We argue that while we should respect the values and attitudes of others, it is important to recognize that people are constantly trying to influence each other. In the USA, society has chosen to conserve its native wildlife, including black-tailed prairie dogs, using a series of broadly popular laws, such as the Endangered Species Act. Thus, working to develop a conservation programme that all stakeholders can accept, based on the methods we are promoting, is warranted, and in the public's interest.

These three recommendations – improving organization and management, using practice-based approaches and developing public support – are well-recognized tools to improve poorly performing programmes like the current prairie dog management effort. Few, if any, new resources are required to implement our recommendations. It will, however, require reallocating existing resources more effectively.

ACKNOWLEDGEMENTS

Support for this work was provided by the Denver Zoological Foundation, the Northern Rockies Conservation Cooperative, and the Southern Plains Land Trust. Denise Casey and Rosie Woodroffe provided comments that greatly improved the manuscript.

People and elephants in the Shimba Hills, Kenya

TIMOTHY J. KNICKERBOCKER AND JOHN WAITHAKA

INTRODUCTION

The 253-km^2 Shimba Hills National Reserve in the Coast Province of Kenya serves as a powerful example of both the value of protected areas for conserving Africa's elephants and the complexities of human–animal conflicts inherent in such an approach. Boasting one of the richest forests in terms of biodiversity in Kenya, containing endemic, threatened and endangered flora and fauna, and serving as the primary water source in the area, Shimba Hills is one of the most important representatives of the remnant humid tropical forests in the East African coastal region. The biodiversity of the Shimba Hills ecosystem, however, is threatened with impoverishment, and the 600-plus elephants currently confined to the reserve contribute to its deteriorating condition. The vast majority of people that live near the reserve teeter in the economic balance, relying on small-scale agriculture for their livelihood. Destruction of their crops by animals catapults them into economic deprivation, and sometimes elephants directly threaten their lives.

Soaring human populations in the area combined with cultural changes in resource use and the politicization of land-tenure issues in the Kenyan state have led to more land settlement and cultivation, thus restricting elephant migrations and increasing the frequency of human–elephant conflicts. The completion of an electric fence in 1999 that virtually surrounds all of Shimba Hills National Reserve, and the creation and fenced annexation of the community-owned Mwaluganje Elephant Sanctuary in the same year, have reduced the extent of human–elephant conflict. It is clear that the fence is buying much-needed time for conserving biodiversity and saving and improving human lives in the Shimba Hills region; however, it is crucial to see the fence as merely part of larger natural resource

People and Wildlife: Conflict or Coexistence? eds. Rosie Woodroffe, Simon Thirgood and Alan Rabinowitz. Published by Cambridge University Press.

management processes. As with other elements of natural resource management, the fence has limitations and consequences on biodiversity conservation that vary according to the environmental and socio-cultural context into which it is placed.

This case study explores issues related to fencing as a strategy for reducing human–wildlife (primarily elephant) conflict in the Shimba Hills region in two ways. The first presents the relationship between the fence, conservation strategies and biodiversity. The second looks at the ways in which the fence project affects, and is affected by, the culture of indigenous peoples (the Digo) living near the reserve. In terms of biodiversity conservation, the fence creates an island habitat, highlighting the need for more aggressive management of elephant populations that are about three times the optimal number for this protected area. There is also evidence that the fence has increased (or at least heightened the awareness of) the number of human–primate (primarily baboon) conflicts in the region as they reverse the intent of the fence by using it as a protection barrier against humans.

Analysis of the cultural context is paramount to predicting the long-term effectiveness of the fence. Changing expectations of the Digo and other local peoples, ideas they have about Kenya Wildlife Service and the Kenya Forest Department and their visions for 'development', challenge the framework for community involvement in maintaining the fence, and perhaps more importantly, increase resource exploitation strategies among individual Digo. Other natural resource management solutions must include incentives that most Digo perceive are viable options given their perceptions of previous trends.

BRIEF DESCRIPTION OF SHIMBA HILLS

The Shimba Hills Ecosystem is located in Kwale District within the Coastal Province of Kenya (Fig. 14.1). The Shimba Hills National Reserve proper covers about 192 km^2. It runs parallel to the Indian Ocean coastline at an average distance of 15 km from the shores. The 25-km^2 Mkongani North and West Forest Reserves are annexed to it, and there is a corridor that links Shimba Hills National Reserve to the community-owned (yet managed by the Kenya Wildlife Service) Mwaluganje Elephant Sanctuary (which together have an area of 36 km^2). Hence, the entire Shimba Hills conservation area is 253 km^2 and is jointly managed by the Kenya Wildlife Service, the Kenya Forest Department and the Mwaluganje Elephant Sanctuary Committee. (As a matter of convenience, for the remainder of this chapter the total protected area in the Shimba Hills ecosystem will be referred to as the reserve or simply Shimba Hills.)

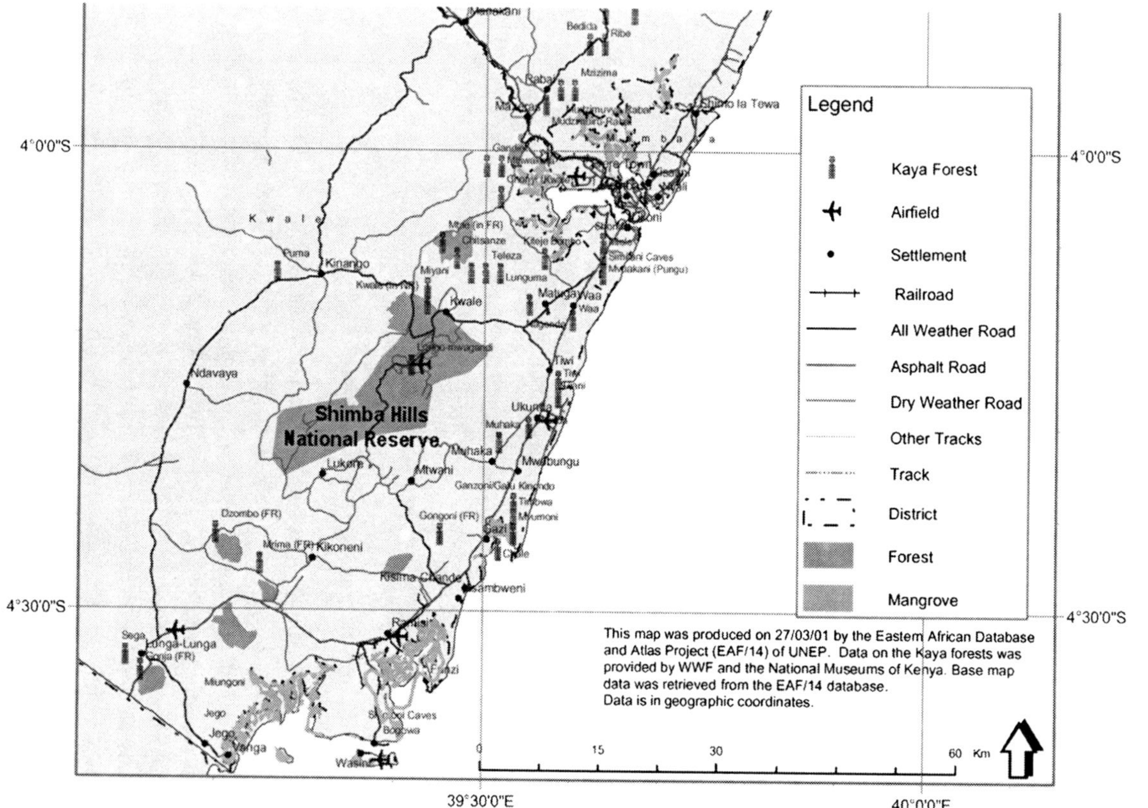

Figure 14.1. The Shimba Hills National Reserve and Mwaluganje Elephant Sanctuary area with Digo and Duruma Kayas (sacred forest groves) and other forest patches. (After United Nations Environmental Programme 2001.)

History

Shimba Hills is one of the richest forests in terms of biodiversity in Kenya. Its unique importance was recognized by British administrators who demarcated the forests for protection in 1903, 17 years before Kenya was declared a British colony. When Kenya attained political independence in 1963 a proposal was put together for enhancing the conservation status of the Shimba Hills forests and the wildlife within it. In support of the government's decision, Risley (1966), the then Kwale District Commissioner, wrote: 'Even if there were no single four-footed animals in Shimba Hills, the area would assuredly merit National Park status in any country of the world because of its beauty, its scenery and its ornithological and botanical interests.' The area was finally gazetted as a national reserve in 1968. Top of the ten management problems identified during the creation of the park was the need to monitor the Shimba Hills habitat (Glover 1968).

The year Shimba Hills was made into a national reserve, visionary scientists raised the concern that the area was too small to adequately perform its ecological functions. For example, Makin (1968) noted that the area was relatively small and proposed that the government maintain adequate land for elephant dispersal to prevent habitat deterioration. He foresaw an increase in human activities in the adjacent areas and doubted whether big game could survive in the Shimba Hills after the surrounding areas had been developed for agriculture. He proposed the establishment of game corridors along the important elephant migratory routes and buffer zones in areas that were unsuitable for agriculture.

The status of conservation outside the reserve changed so drastically that, commenting on Makin's idea later in the year, Glover (1968) lamented that the situation in the Shimba Hills was a 'salvage' operation, noting that it was probably getting too late to establish corridors and buffer zones. He urged the government to step in immediately to save the situation. Thirty years have passed and still very little action has been taken though Makin's predictions have come true. Many activities with strong negative conservation impacts have increased exponentially inside and outside the reserve.

Indigenous views and use

The Digo and the Duruma are each one of nine groups that make up a loosely structured ethnic federation of indigenous coastal peoples referred to as the Mijikenda. Even though the colonial administration formally evicted people from the Shimba Hills when the national reserve was created at the turn of the century, most Digo (and other Mijikenda groups such as the Duruma) view Shimba Hills as part of their traditional territory to which they have certain rights. For example, there are at least five specific forest patches in Shimba Hills that the Mijikenda call kayas (two of which are of primary importance to the Digo cultural heritage). Managed by a council of elders, the kayas are ritual centres where Mijikenda ancestors putatively first lived when they migrated to the region centuries ago. Traditionally, they are small clearings and space under the canopy of forest patches. Some Mijikenda elders use kayas to communicate (via prayer, sacrifices, offerings, etc.) with spirits or spiritual forces on behalf of the community.

Each of the kayas in Shimba Hills (and hence within the fenced area) has experienced severe impact from elephants. The Duruma Kaya Mtae in Mwaluganje Elephant Sanctuary, in particular, is nearly completely deforested. The first, and arguably the most important, Digo kaya (Kaya Kwale) is in the Shimba Hills National Reserve. It too is deforested and overrun by elephants, causing Digo elders to use it less than they were accustomed to prior to the growing elephant population.

The Forest Act permits 'traditional' or limited use of forest resources by surrounding communities; however, there is some ambiguity within the interpretation of the Act. Typically, the rights include: collection of dead wood for firewood, thatching materials, climbers and lianas for construction; select and limited logging for timber and building poles; gathering of wild fruits, roots, herbs, honey and mushrooms for consumption and medicine; etc. According to Digo informants, permits have been more difficult to attain over the last two years; however, most informants have continued to use forest products informally.

Digo ideas about the Shimba Hills National Reserve and the animals and resources within it are connected to their cosmological and socio-political understandings. A detailed discussion of Digo world-views is beyond the scope of this chapter; however, to facilitate discussions about land issues it is convenient to divide general ideas the Digo have about their land into four categories of space: *social space, sacred space, vacant space* and *foreign space.*

Social space is where the mundane affairs that constitute the bulk of Digo life take place. It includes homesteads, fields, gardens, paths, shops, markets, administrative centres and schools. Widely accepted Digo values and norms dictate patterns of interaction and resource use in social space. For example, there is an underlying ideal of resource equality that is connected to a complex array of mechanisms (e.g. general reciprocity, conspicuous giving, inconspicuous resource possession, envy, sorcery, etc.) that mediate between individual ownership and community obligations of resources management in social spaces.

Sacred space is the areas where Digo perceive or establish a portal to supernatural realities that undergird their landscape. Traditionally, sacred space for the Digo was connected to physical places of significance. This includes large trees, caves, rock formations, springs and hilltops, in addition to the previously mentioned kaya forests. Traditionally, there were prohibitions on unchecked resource use for these spaces. Spiritual specialists interpreted and enforced the prohibitions, by way of spiritual threats for individuals and the community.

Vacant space refers to areas that are not cultivated or used by people, yet it is within what most Digo consider as their traditional territory. There were (and are in varying degrees) rules about claiming vacant land but there were few restrictions on using the resources within these areas. *Foreign space*, on the other hand, is space that is not considered part of Digo territory. Typically, this is land that was purchased by individuals or companies and the rules of resource use vary according to the owners and the community's relationship to them.

The establishment of the Shimba Hills National Reserve has created a foreign space that for the Mijikenda lies somewhere between sacred and vacant. *Waganga* (traditional spiritual specialists) have overseen activities that provide individuals and the community benefits from the sacred spaces. Hunters, resource gatherers and settlers have benefited from the vacant space, and they have shared these benefits with other community members. Social space, of course, is by definition beneficial, albeit conflictual, since it provides shelter and land for farming. For the most part, the Digo do not see the foreign space of the Shimba Hills National Reserve as beneficial since it has been poorly, if at all, integrated into their cultural beliefs and practices, but in as much as they can see it as sacred or vacant they have clearer ideas about its benefits.

IMPACT OF ELEPHANTS ON PEOPLE

Between 1980 and 1994 (before the fence was constructed) some 2171 cases involving crop and property damage, human deaths and injury were recorded in the area around the Shimba Hills National Reserve (Waithaka and Mwathe 1995). Out of these, 57.3% involved elephants while the rest involved other animals. In the same period elephants from the reserve reportedly killed 18 people and injured at least 22 people (Waithaka and Mwathe 1995). All other animals killed three people during this time. To add to this, many cases of crop and property damage go unreported, as do many injuries caused by elephants to people who were inside the reserve (for fear of legal reprisal).

Mnene (1992) estimates that farmers living next to the reserve during this time lost US$90 000 annually from crops damaged by elephants and other wildlife. This estimate may be conservative. Crop damage is severe and widespread, with incidents reported as far as 15 km from the reserve boundary. Though elephants have been the primary animal causing crop damage, other problem animals include the buffalo, baboon, bush pig, sable antelope, bushbuck, monkeys (e.g. Colobus, Sykes) and bush babies. Crops damaged include coconut palms, cashew trees, citrus fruits, bananas, cassava, maize, beans and other vegetables – in other words, every crop the farmers rely on is susceptible to losses due to wildlife in the reserve. Since 1989, compensation from the Kenya Wildlife Service is only considered for incidents involving loss of human life (which is about US$375) or limbs (which averages about US$177) due to wildlife.

The dramatic increase in human–elephant conflicts around the Shimba Hills National Reserve is positively correlated with growing elephant populations confined to the reserve, and the ensuing habitat degradation. The number of elephants living in the reserve is estimated to lie between 650 and 750 based on a 48% forest formation (Litoroh 1997). Other studies over the last few decades that tried to establish elephant numbers using direct methods were consistently lower, ranging from 200 (Glover 1968) to about 300 (Kiiru 1995) individuals. Indirect dung count studies estimated between 500 and 600 elephants with the mean around 450. An aerial count in 1997 counted 464; however, the pilot and a Kenya Wildlife Service representative participating in the count felt that a considerable number were missed due to thick cover. Most experts estimate that there are definitely more than 600 elephants in the reserve, which is three times the optimal number of about 200 individuals (Litoroh 1997; O. K. Macharia pers. comm.).

IMPACT OF PEOPLE ON ELEPHANTS

The overpopulation of elephants in the reserve is directly correlated to increased human activities in lands directly adjacent to the reserve, and in lands that have traditionally offered elephants corridors into other habitats such as the Abrabuko-Sokoke Forest and the Tsavo West Ecosystem (Kiiru 1995). The large-scale external pressures that have pushed elephants into fragmented habitats throughout sub-Saharan Africa have been well documented elsewhere (Hanks 1981; Parker and Graham 1989a, b; Waithaka 1993; Barnes *et al.* 1998; Hoare and Du Toit 1999). Shimba Hills is no exception. Even prior to the completion of the national reserve fence in 1999, cultivation in the surrounding areas kept elephant populations hedged into fragmented habitats, but now the fence clearly delimits their habitat range. In short, this area is an environmental hotspot that remains vulnerable to the combination of desperately poor economic conditions, fast-growing population, and perhaps the breakdown of values that may have helped to curb wanton exploitation.

In addition, the mostly Digo and Duruma human populations around the Shimba Hills have been growing rapidly over the last few decades. The current population of the 8322-km^2 Kwale District is about 520 000 people (over half of whom are Digo), which averages out to 60 people/km^2 (Wakajummah 2000). This relatively low average is deceiving. Most locations in the immediate vicinity of Shimba Hills have about 120–150 people/km^2 because, among other things, of the ample rainfall and relatively inexpensive land. This higher density is present despite the fact that in most regions near the Shimba Hills people tend to live in homesteads surrounded by cultivation fields, rather than in nucleated villages. Between 1969 and 1999, the population growth rate has consistently been over 3% per year in Kwale District (Foeken and Owuor 2000). (With a population of 3500, Kwale has the largest concentration of people living near the fence. It is also the District Administrative centre, and the Kenya Wildlife Service has offices there.) Presently, those under 20 years of age make up a larger proportion of the population than other groups, suggesting that the numbers will continue to rise. In addition, an influx of up-country peoples (groups from the interior) contributes to the growth figures (Wakajummah 2000), resulting in increases in the ethnic diversity of what was up to 50 years ago a relatively homogenous cultural area. This increase in population has a dual negative effect. It increases the strain on natural resources and it breaks down social structural values, norms and rules that once were more widely shared amongst individuals – values and norms which conceivably could be tapped into to help meet conservation goals.

ELECTRIC FENCING: THE SOLUTION, A PARTIAL SOLUTION, OR A PART OF THE PROBLEM?

By the late 1970s and early 1980s, it became apparent that something drastic needed to be done about the growing elephant problem in the Shimba Hills. The number of elephants outside the protected area within the district progressively declined from 2000 in 1973, to 1420 in 1977, to 182 in 1987, and finally to zero by 1989. Many elephants sought refuge in the Shimba Hills National Reserve while at the same time more and more people were growing crops and living at the periphery of it. It was a recipe for disaster. Within the reserve itself increasing deforestation and the attraction of coconuts, cassava and other elephant favourites encouraged elephants to wander into neighbouring fields (many of which were cultivated) for available food. The Kenya Wildlife Service and others promoted and tried a number of methods to minimize conflict between people and elephants, including game moats, high tensile steel fences (which were erected in particularly problem-ridden areas but were by no means an attempt to strictly confine elephants to the reserve) and traditional control measures such as shooting, scaring, etc. Local people relied primarily on scaring elephants away with bows and arrows. All of these were insufficient to contain the escalating problems. A decision to fence the elephants within the Shimba Hills ecosystem was reached by 1995 as part of a wider conflict-resolution strategy. This exacerbated management issues within the reserve itself, and it marks the entrenchment of the major problem between elephants and the threats to the Shimba Hills forests.

Another key component of this wider conflict resolution strategy was to work more closely with the indigenous communities around the reserve, which was a strategy consistent with the Kenya Wildlife Service's policy trends in the mid-1990s. The community focus prompted donors of the fence project (such as the European Union) to insist that the community take 'ownership' of the fence by, among other things, promising to maintain it. The Kenya Wildlife Service conducted workshops for community leaders, facilitated the organization of fence committees and held informative meetings with community members at large to educate them about the benefits of the reserve. Although there was some opposition from segments of the community regarding the commitment to maintain the fence, the eagerness to have the fence built (and elephants kept out of the crops and settlements) was virtually unanimous. This intensified relations between the community, the Kenya Wildlife Service, and other conservation agencies, and it marks the growing awareness of the socio-cultural divide in natural resource management strategies between the Digo and conservation agencies in Kenya.

The fence as a solution to human–wildlife conflicts

Once the multiple sections of fence (built at varying times from 1994 to 1999) were connected in 1999, completely enclosing Shimba Hills, the electric fence was a resounding success in drastically reducing the number of human–elephant conflicts in the Shimba Hills National Reserve region. Although the fence was designed to completely surround the reserve, in fact, to this date there remains a section, less than a kilometre long, that is not fenced due to the combination of a controversy involving private claims of land ownership and the particularly challenging terrain in this area. There have been isolated incidents of elephants using this gap to move to and from Mwaluganje Elephant Sanctuary. This does provide another corridor from the Shimba Hills reserve to the elephant sanctuary, but one that takes elephants across private landholdings. In March of 2000 one elephant that used this corridor killed a woman who unfortunately crossed its path.

Both the crops destroyed and the humans injured or killed by elephants have drastically been reduced. The numbers went from over 18 people killed by elephants between 1980 and 1994 to two people killed by elephants in the reserve from 1999 to present. The community perceptions are also positive. Over 200 people living adjacent to the reserve were asked whether or not they were happy that the fence was there and without exception the responses were positive. People feel safer. Most of those who had abandoned their homes and fields near the border of the reserve returned to work the land with a renewed optimism. Squatters have also moved into the area in waves. For example, of the 39 homesteads that border one 10-km section of the fence, 14 were established or re-established after the fence was completed in 1999, and of these nine were established by squatters.

The fence as a partial solution to human–wildlife conflicts

The number of elephants within the reserve has not significantly changed since the fence was erected, meaning that deforestation within the reserve and the potential for future conflicts are real threats. In addition, there have been numerous incidents of elephants either breaking through the fence when the electrical current is absent or finding other means of making incursions into crops adjacent to the reserve (e.g. passing under the fence at a ravine or moving a log onto the fence, allowing elephants to pass over it). In this sense the fence is only a partial solution – it must be combined with an elephant management programme that can reduce the number of individuals to an optimal level, and increase maintenance efforts at problem areas. Of the three most suitable methods for reducing elephant numbers in Shimba Hills (culling, fertility control and translocation), the Kenya Wildlife Service has preferred to use translocation. There is some experimentation with fertility control methods ongoing but the prospects look dim, especially for the short term. Culling, although the least expensive approach and the method recommended by many leading Kenyan conservationists, remains unpopular to policy-makers in the Wildlife Service and a large constituency of international conservation organizations that are directly and indirectly influential in Kenya. Kenya, and the Wildlife Service in particular, relies on benefits from the tourist industry so there is a hypersensitivity to policies that may adversely affect it. Further, the translocation projects are largely funded by international or bilateral donors so the greater expense that the Kenya Wildlife Service might have to incur to use this method is mitigated.

Maintenance of the fence at problem areas also requires additional outside funding. The Kenya Wildlife Service budgets KSH 50 000 (US$675) per kilometre annually for maintenance of the fence over and above the maintenance provided by the local communities, which is done at little or no cost to the Kenya Wildlife Service and its donors. Further, there was only one technician for the whole Shimba Hills fence with the expertise necessary to repair the electrical fence when it was damaged by elephants or natural phenomena. Between the spring of 2000 and the summer of 2001 this technician could recall only a few weeks when he was aware of no problem areas where elephants could get out – the remaining time he had knowledge of multiple areas that elephants could, and in numerous incidents did, cross the reserve boundary into privately held land used for farming. It is clear that the expertise, training and funding for proper maintenance of the fence are inadequate to the task.

The fence as a part of the problem

There are several ways in which the fence can be seen as a part of the problem regarding human–wildlife conflicts in general. The first point is related to the previous discussion. If optimal numbers of elephants are not reached quickly in Shimba Hills then it is likely that food resources for the elephants will become scarce to the point that they would aggressively return to browsing outside the reserve for their food. Depending on the severity of the shortages of habitat and food (e.g. in cases of famine, etc.) this re-emergence of the elephant conflicts could pose an even greater threat to human crops and lives – since the number of humans settling nearer to the reserve has risen since the completion of the fence.

Another problematic aspect of the fence is that baboons have learned new strategies for using the fence as a barrier against humans. Baboons can easily get through the fence by climbing the wooden posts, by jumping from nearby tree limbs, or by finding inevitable gaps in the fence in the diverse terrain of Shimba Hills. Observations and conversations with farmers living near the reserve reveal that baboons have learned to stake out humans from the other side of the fence until they see a safe opportunity to cross it to raid crops. When detected, the baboons simply scurry over the fence and stare back at their pursuers from the safety of their protected area. In nearly one-third of the 40 households questioned informants stated that they have seen more baboons and bush pigs since the fence was completed. Even if this is based on their perceptions rather than statistically significant figures, the implications are the same – the fence is associated with an increase in human–wildlife conflict for some people who live near its borders. This inversion of the purposes of the fence – keeping people on the outside rather than containing animals inside – does little to promote principles of community ownership of the fence.

Finally, the cultural ramifications of the fence can be construed as negative in as much as the fence is seen as the primary connection point to the community – the way in which the Digo communities will cooperate with Kenya Wildlife Service officials in natural resource management efforts. The main factors that lead to such reasoning by the community are the lack of benefits from the protected area and the lack of effectiveness of the fence in stopping crop-raiders.

The fact is, many Digo see the reserve *fence* as foreign – after all, Digo are traditionally opposed to fences and hard boundaries between 'nature' and 'society'. Such a claim may appear unfounded since the overwhelming majority of Digo living in communities near the reserve desperately wanted the fence erected. In fact, they had been pleading over a decade for a fence. Further, the donors (e.g. Eden Wildlife Trust and European Union) and managers (the Kenya Wildlife Service) of the project thoughtfully designed it on the condition that the community must take ownership of the fence (primarily to ensure that it would be maintained). Despite the good intention and the intelligent design, just as the protected area it surrounds, the fence fails to be sufficiently integrated into Digo cultural institutions. Instead, the fence project has unintentionally led to a rise in unrealistic expectations of the community, demands on community group formation without attractive incentives or the establishment of necessary facilitating mechanisms, and further alienation of Digo from resource management of the protected area itself.

In addition, due to environmental and historical circumstances, Digo relations with the outsiders, who are primarily in charge of managing the Shimba Hills reserve and the fence, are tainted with mistrust and discord. The resentment between Digo and up-country Kenyans and white settlers, the pattern of unstable short-term relations with Western donors, development workers and conservationists, and the growing ambivalence toward Swahili and Arabs who hold power on the coast, leave the Digo with few allies (e.g. other Muslim Mijikenda) with whom they can identify. Much of this stems from the fact that most Digo occupy a relatively low socio-economic position compared with these other groups. A further complication of relations that Digo have with outsiders who are managing the fence is that their marginalization (real and perceived) from political representation in the Kenyan state has woven patterns of non-cooperation into their relations with government officials (e.g. Kenya Wildlife Service Staff) regardless of, or in addition to, sentiments about their ethnic affiliation.

CONCLUSIONS AND RECOMMENDATIONS

This case study of the Shimba Hills National Reserve fence project brings together issues of survival for three very important yet different types of resources: (1) elephants; (2) crucial and threatened habitat with all its endemic and diverse flora and fauna; and (3) an indigenous culture and indigenous people's right for self-determination. The fence is buying time for the survival of elephants but it does so at the cost of the habitat. The fence project, and greater attention to the local area, is part of the indigenous Digo people's processes of cultural change but it runs the danger of further alienating the Digo because it involves them in management of the boundary of the protected area rather than in direct natural resource management issues that are at the core of the reserve's very existence. To resound a familiar symbol in development studies, the only way for these three resources to survive in the reserve is for the local community members to perceive benefits from the resources. The fence is not designed for this, and it certainly is not enough to ensure that the local community values the reserve. Hence, a restructuring of resource use and management strategies involving the local community is a necessity. The following recommendations may be a start toward these goals:

(1) *Translating the legitimacy of the existence of the Shimba Hills National Reserve into Digo cultural beliefs and values.* One way to do this would require reconceptualizing space according to Digo ideas and then develop frameworks that bridge the gap between Digo conceptions of sacredness and Western ideas about the 'sacredness' of nature. Of the three spatial categories of the Digo, only sacred space has definitive boundaries (even that is questionable in many instances.) Sacred space has traditionally existed for the community and those who enter it (with very few exceptions) must follow specific protocol to avoid negative sanctions for themselves or their community. It would seem a translation of natural space and sacred space could find a common ground on which Westerners and Digo (including those who do not hold beliefs in the sacred space as supported by Digo traditional religion) could operate. Educational programmes, such as those given to fence committee members before the fence launching ceremony, that inform community members of why the Kenya Wildlife Service and conservationists believe that the reserve is beneficial (e.g. it is a water catchment area, trees are part of the process that brings rain, biodiversity is important to future generations, etc.) are a good starting-point but they must go to the next level, and involve listening to what

benefits the Digo believe the reserve can bring (spiritual or tangible). Whatever strategy is used to translate the legitimacy of the Shimba Hills reserve to the Digo it is imperative the Digo cultural beliefs and practices are taken seriously – rather than as vestiges of a superstitious past or similar views that have marred relations between Digo and persons implementing government or outside projects in Digo communities.

(2) *Finding ways to allow the Digo to benefit directly from the resources in the Shimba Hills National Reserve or maintenance of the fence.* Even if the Digo are convinced of the legitimacy of the reserve's existence, it does not necessarily lead to action that is consistent with the Kenya Wildlife Service's policies. Many Digo living near the reserve are desperate. The only way for the community to monitor individuals who may be illegally using resources in the reserve is if they perceive that they are benefiting in a tangible way from the reserve. There are a plethora of models for how resources can be shared with the communities around parks (e.g. sharing gate proceeds, developing sustainable-use programmes, hiring members of the community as guards, etc.). The idea that the community was to take ownership of the fence without hiring any of them or paying for community projects is a mistake in an environment where there is resentment of outsiders and where few economic opportunities are available. Obviously, this will raise the cost of fences (or any conservation project that attempts to engage the community seriously) but it is a necessary component if the fence model has any chance of moving beyond its current short-term solution status.

(3) *Fostering relations between the Digo and the Kenya Wildlife Service, Kenya Forest Department and NGOs in terms that are not dictated from the latter groups.* This is a reiteration of the first point but it goes beyond it in an important way. The Digo are socially, politically and economically marginalized from the main actors in running the Kenyan state. Marginalized communities find it difficult to get a voice in policies and decision-making processes that involve management of resources, institutions or programmes that have a direct effect on them. Good governance puts the onus on those who hold power (e.g. policy-makers) to find ways for those who are underrepresented (e.g. the marginalized) to be heard. This does not have to require more time and energy on the part of policy-makers in the long run, but it does require more in the early stages of decision-making processes or programme implementations. What actually needs to take place is that seeds of civil society need to be sown into the community so that

> grassroots groups can develop as they are coming to terms with planned action in their community from outside forces (e.g. government agencies). In other words, true empowerment (i.e. making it possible for them to get what they want) of the Digo must be initiated by those who hold the power (in this case, the Kenya Wildlife Service, Forest Department and NGOs) as the starting-point for fostering better relations between themselves and the Digo communities.

Ideally, these three recommended foci would have been weaved into the fence project from its inception, and it is recommended new fence projects in any location consider these same three areas; however, it is not too late for both sides of the Shimba Hills National Reserve fence project to reconceptualize their ideas about space, resource distribution and community relations with government, conservation and development agencies.

15

Safari hunting and conservation on communal land in southern Africa

DALE LEWIS AND JOHN JACKSON

INTRODUCTION

Safari hunting, sometimes called tourist hunting, is a form of land use that is capable of generating exceedingly high market value for wildlife resources. Well-established trade practices exist worldwide for marketing hunts to maximize the economic value of wildlife, and to satisfy tourist demand for hunting adventures and trophy quality animals. Revenues from these sales account for over US$500 million annually worldwide, and in South Africa alone, annual revenues exceed US$50 million (Van der Walt 2002). The direct revenue to the government of Tanzania from safari hunting exceeds US$10 million annually (E. Severre pers. com.).

On private land, tourist hunting has become a major factor for wildlife recovery and reintroduction in many parts of the world. Examples of these successes are in the tens of thousands and have unequivocally proven to be an important economic engine for conservation. In South Africa, for example, over 9000 private game ranches occupy 10 364 km^2 (Bothma 2002), and have contributed to the recovery of blesbok (*Damaliscus albifrons*), white rhinoceros (*Ceratotherium simum*), black wildebeest (*Connochaetes gnou*) and the Cape mountain zebra (*Equus zebra zebra*) (Flack 2002). Today, South Africa has more wildlife than it has had in the past 100 years (Bothma 2002). It is estimated that game ranches in South Africa contain 1.7 million large mammals on 13.3% of the land formerly designated for agriculture, but subsequently converted to private game ranches (Eloff 2002). Gross annual income from these ranches exceeds US$84 million with major income sources derived from tourist hunting, biltong and live-animal sales (Bothma 2002).

In contrast, the economics of sustaining wildlife production from tourist hunting on community or customary land, especially in less developed parts

People and Wildlife: Conflict or Coexistence? eds. Rosie Woodroffe, Simon Thirgood and Alan Rabinowitz. Published by Cambridge University Press.

of the world, face a number of constraints, some of which can greatly influence the marketability of a hunting area and the conservation benefits of tourist hunting. For example, low-income communities who share their land with wildlife often face livelihood needs that increase the costs of living with wildlife (see Thirgood *et al.*, Chapter 2), and lower household support for wildlife production. Habitat destruction and wildlife poaching are common manifestations of this problem. Households also contend with the opportunity costs of pursuing alternative uses of the land. Such decisions are weighed against not only the possible incentives for producing wildlife, but also the tenure and resource rights that ensure these incentives are guaranteed (Freese 1996; Tietenberg 1996). If these conditions are not met, households are apt to pursue agricultural or livestock-based land uses through more reliable local markets, often at irreversible costs to wildlife habitats. This is where private game ranches have a distinct advantage, because laws generally support the rights of private landowners to produce and market wildlife without external interference (Freese 1998). Under private ownership, conversion of private farmland to game ranching adds approximately 5000 km^2 per year to wildlife production in South Africa (Flack 2002), primarily because, on average, wildlife revenues exceed livestock revenues by over 50%, due largely to a wider diversity of marketable products (Bothma 2002).

In many countries, such as Zambia, Tanzania and Botswana, wildlife is state-owned and tourist hunting on customary land is controlled by a government management authority. Under these arrangements, the hunting industry typically depends on a number of stakeholders who compete for the same pot of safari revenues. This tends to increase the complexity for sharing and co-managing hunting profits to reduce local threats to wildlife (Gujadhur 2001), and increases the risks of unfair trade deals that marginalize the community as a primary stakeholder and wildlife producer for the industry. As a result, the community becomes less of a resource owner and more of a landlord able to receive only fixed ‘rent’ fees from commercial parties, who exploit, administer and largely control wildlife markets. Such ambiguity of ownership and incentives for producing wildlife hinder the capacity for tourist hunting to reach its full potential as a tool for conservation and rural development (Honey 1999). Finally, rural communities on whose land tourist hunting is practised usually have the least knowledge about tourist hunting among the stakeholders. Such disparity lowers community interest in becoming more actively engaged with the private sector as a cooperating partner, and also limits their ability to seek more equitable terms of commercial partnerships (Freese 1998).

In recent years, tourist hunting advocates and wildlife conservationists have sought increased collaboration with government authorities to make tourist hunting more responsive to these constraints. Much of this joint effort was pioneered through community-based programmes that enhanced community involvement in wildlife conservation through economic incentives and joint-enterprise development from tourist hunting. Key examples of these programmes were CAMPFIRE (Communal Areas Management Programme for Indigenous Resources) in Zimbabwe, ADMADE in Zambia and the Chobe Enclave Conservation Trust in Botswana.

All these experiences reveal the need to make safari hunting accountable and more responsive to rural development needs. This chapter highlights key lessons from these experiences in southern Africa, and offers them as a framework for developing the needed policies, management criteria and partnerships for sustaining benefits from tourist hunting.

PRIMARY REQUIREMENTS OF TOURIST HUNTING POLICIES THAT SUPPORT CONSERVATION

Modalities for conducting safari hunts are reflected in such policy instruments as concession agreements, procedures for awarding concessions, stakeholders' role in allocating and marketing hunting quotas, licensing controls and fee structures. All are fundamental to ensuring the industry is competitive and profitable. Without the technical capacity and the political will by government to develop, monitor and enforce these instruments through legally binding agreements with the stakeholders, attempts to achieve conservation results through safari hunting will probably fail. Experience shows, however, that this condition is not easily met and can expose local communities to increased uncertainties and risks by committing their land to wildlife production.

Zambia, for example, suffered from a two-year ban on safari hunting due to a range of problems related to ambiguous procedures for renewing concession agreements as well as weak controls for regulating hunting quotas and issuing animal licenses. Despite an annual injection of about US$800 000 into rural development needs from safari revenues, local authorities were not consulted when the ban was imposed. Without alternative revenues to sustain community participation in wildlife conservation, wildlife numbers suffered serious losses (Zambia Wildlife Authority, unpubl. aerial surveys) and consumer confidence in Zambia as a safari hunting destination declined sharply (Causey 2001). In Tanzania, approved procedures for awarding highly favoured hunting concessions were thwarted by senior government officials to favour Arab hunting interests. As a result of severe over-hunting in these concessions, wildlife stocks for a number of species became depleted (Honey 1999). In Botswana, a key economic species, the lion (*Panthera leo*), was removed from the safari hunting quota for reasons attributed mostly to non-hunting sentiments rather than technical merits (Jackson 2002).

Economic incentives can provide a highly effective tool for controlling or even eliminating threats that limit wildlife production. For incentives to work, accountable local institutions need to target the very people who contribute to these conflicts with incentives designed to minimize the occurrence of conflicts. This adds another important level of complexity to policy development, because both institutional skills, and an understanding of the relationships between conservation and local livelihoods, are needed to drive this approach.

A growing body of experience suggests that lack of household livelihood security is a primary cause of illegal wildlife use and wildlife habitat loss (Lewis and Tembo 1999; Jagt *et al.* 2000; Lewis 2001). Addressing both household livelihoods and wildlife conservation requires formal agreements with local leaders to use safari revenues effectively to support household livelihoods as a basis for reducing threats to wildlife. Murphree (1993) argues against passively earned benefits as opposed to linking benefits with community effort to produce wildlife. Policies in the USA, for example, work to increase wildlife producers' compliance with management conditions that promote wildlife production and, in exchange, landowners are allowed to market their animals for fee hunts (Rasker and Freese 1995).

Strategic policy interventions that can promote an approach linking revenue incentives with wildlife production on community land include:

(1) By-laws or constitutions that mandate community leaders on procedures for using safari hunting revenues to mitigate conflicts that arise from livelihood needs, and which pose threats to wildlife production.
(2) Transparent procedures for remitting revenue shares to community accounts in ways that enhance local knowledge that wildlife production is a profitable land use.
(3) External controls that prevent financial mismanagement by community leaders or non-compliance by the private sector to agreements with the community.

Despite legislative support for communities to receive revenue incentives from safari hunting throughout much of southern Africa, reliable and transparent procedures for returning these revenues to local communities are often weak. Botswana and Namibia have addressed this problem by ensuring transactions of wildlife sales take place within the community, with direct payments to community institutions through their representative leaders. In such cases, communities are empowered to sell hunting quotas and witness the full economic value of wildlife through public auctions or tenders (Gujadhur 2001).

Ensuring that wildlife revenues reach target groups in ways that promote wildlife production is commonly recognized as a problem for most Community Based Natural Resource Management (CBNRM) programmes in southern Africa. In Botswana, for example, dominance by community power elites to invest funds properly pose a threat to household-level benefits (Jagt *et al.* 2000), and in Zimbabwe the local council authority may restrict the flow of revenue shares to village communities (Freese 1998).

Potential for corruption

Where the above-mentioned policy requirements are lacking, or their enforcement is inadequate, interference by vested interests who seek a preferential share of tourist hunting profits can become a serious constraint to the industry's contribution to conservation. This is especially true in countries with weak management institutions unable to control such manipulation or corruption of the industry. In such cases, the community often becomes marginalized and reluctant to control land uses that conflict with wildlife production. The tourist hunter, who may think his hunting fees are contributing to conservation, instead becomes an unwitting contributor to a potentially corrupt industry that siphons an unfair share of these revenues into private accounts, leaving the local community unrewarded for its efforts to produce wildlife.

Protecting the community from such abuse, and promoting their rights as wildlife producers, requires legally recognized community institutions for managing and benefiting from wildlife. Institutions that are democratically elected and fully representative encourage local participation in decision-making, increasing the legitimacy of such institutions as civil society groups. If such institutions are given the full legal rights to challenge government or private-sector decisions that communities feel are unfair, they can bring about a much-desired balance among wildlife stakeholders for promoting community incentives to produce wildlife.

Zambia has instituted policies to help achieve this balance by creating Community Resource Boards as local wildlife management authorities. Though they have been in existence for only a few years, these Boards have already exercised their powers by seeking court interventions on several issues, including payment of unpaid safari revenue shares, and rejection of selection procedures for safari concession holders.

Without such a balance, a more centralized control of the industry by government can lead to private-sector manipulation of lease agreements, as well as the leaseholder selection process itself, and can contribute to serious consequences for wildlife conservation, including corruption costs that reduce private-sector investment in wildlife production, reduced criteria standards for selecting leaseholders, and lowered investment in wildlife production caused by subcontracting to more experienced operators.

Defining government's role

Providing the right policy environment for safari hunting to support conservation requires a clear role for government's own involvement in the industry. A direct role in owning and operating hunting concessions by government authorities has produced a dismal track record in southern Africa. Self-serving interests tend to manipulate and undermine management controls of hunting quotas, pricing structures and private-sector regulations (Honey 1999). The consequences of these problems often lead to increased wildlife management costs, which governments must shoulder, while disenfranchising local communities from taking a more active management role themselves. Local solutions to conserving wildlife are usually less expensive and more cost-effective (Lewis 1999).

Interventions by government should therefore be carefully designed to motivate community stakeholders to produce wildlife, and increase partnerships that can produce competitive tourist hunting products which ultimately support conservation. In the USA, for example, the Federal Duck Stamp Program generates US$22 million annually from duck-hunters for the purchase and protection of wetlands (Burnett 2002). Similar programmes financed by hunters support mountain sheep (*Ovis canadensis*) and elk (*Cervus elaphus*) conservation.

Ensuring interventions and policy guidelines are followed has proven difficult in remote rural areas of southern Africa where tourist hunting is allowed. Though the level of supervision by government may be logistically difficult, failure to provide such supervision increases the risks of distrust among cooperating partners (Jagt *et al.* 2000), and lowers the level of stakeholder commitment to key obligations for enhancing partnerships. The ADMADE programme in Zambia relied on community audits to assess local leadership commitment to wildlife threat reduction and to assess management performance in terms of law enforcement effort, compliance with land-use plans and data management. The safari operators' successes in meeting basic targets of conservation commitments were also subjected to audit (Lewis 1999).

Different user-right models for benefiting from tourist hunting

Numerous cases exist where communal authorities operate their own safari hunting companies with partners they choose themselves. Most of these cases are in North America, where ownership of wildlife is invested in Native American societies living on reservations. White Mountain Apache Reservation in Arizona, USA and the Inuit Hunting Councils in Nunavut, Northwest Territories, Canada are examples of indigenous communities having total ownership of their tourist hunting enterprise. Both offer some of the most expensive tourist hunts on the market, due largely to a highly successful programme of resource protection and management. A single elk hunt on the White Mountain Apache Reservation, guided by Native Americans from the reservation, can sell for as much as US$16 000. A polar bear (*Ursus maritimus*) hunt in Nunavut generates US$24 000 for the Inuit Council.

The community ownership model described above requires full wildlife ownership rights of the resource. For most customary landowners in southern Africa, wildlife is state-owned. This remains a major constraint upon wildlife conservation on communal lands in Africa. Despite this limitation, there is a trend throughout much of the region for promoting user-rights through various joint-venture models. Such models take advantage of the marketing skills of the private sector, but give key powers to community authorities through stipulated user-rights of the resource. In Botswana, for example, the community has the right to sell their quotas both to safari clients and to local citizens, according to a required minimum allocation level for both categories. Any balance of animals may be allocated and sold in the way the community chooses.

Joint-venture agreements are viable policy instruments for building community–private-sector relationships for increasing local commitment to wildlife production. Much of this potential is in the details of the agreement itself. Facilitating such agreements often requires a neutral facilitator to ensure that any disparity of knowledge or understanding is not exploited during the negotiation process. Another approach is to link the duration of a hunting concession to a process that initially limits the duration but provides a graduated extension of the concession based on measured improvements in the joint-venture agreement and its compliance.

Policy compliance: the role of international agreements and marketing conventions

With the exception of trade laws protecting selected species (such as CITES, the US Endangered Species Act and European Union Regulations), and conventions to enhance government commitment to sustainable use of biodiversity (Convention on Biological Diversity 1994), international agreements that set standards for tourist hunting to support conservation are non-existent. Given the potential for abuse of the system on communal land, there is clear justification for hunting and conservation advocates to seek such standards, especially since rural communities generally lack the capital or know-how to run a safari hunting concession or to effectively secure their needs from the industry. A number of individual hunting organizations prescribe ethical approaches to hunting by their own members, but in general these pertain to aspects of fair hunting or standards of services required by hunting operators.

Because of their strategic importance to the success and characteristics of global tourist hunting products, safari hunting marketing conventions could apply economic pressure for international compliance to conservation standards by requiring verifiable proof before accepting marketing agents from a country under review. Such pressure could operate at the national level or at the individual operator level. Alternatively, a certification based on different grades of compliance could be the basis for discerning safari clients to select operators or country destinations that support conservation. Although certification proposals based on a 'green marketing' concept have been proposed (Lewis and Alpert 1997), the tourist hunting industry to date has not undertaken any adoption of such ideas.

Policies that promote local participation and knowledge in the industry

Throughout much of southern Africa, there is a strong cultural tradition for hunting. Safari operators often view local hunters – who are usually respected and knowledgeable people in their own communities – as poachers in conflict with the interests of running a safari hunting business. This perception is reinforced by wildlife laws which often require local hunters to buy licences that they cannot afford; hence they often do become poachers. Such alienation denies the industry a potentially constructive and valued asset to the safari operator. One basic starting-point for remedying this problem is for the safari operator to identify and employ the top local hunters as trackers, skinners and assistants to the professional hunting guide. Wise safari operators have long done this, but not systematically or programmatically. A more long-term approach is to ensure local residents can achieve more senior positions in a safari operation, including as hunting guides.

This cultural divide can be widened still further by conflicting policies that undermine joint-venture relationships with the community. A good example comes from policies that allow non-safari hunts at low, subsidized values on the same land and key species as are allocated to tourist hunting. As a result, the same animals marketed to international tourist hunters at full market value are sold at greatly reduced prices to the local market. Where such overlap limits quotas for selected species important to the industry, or contributes to hunting disturbances that lower the hunting area's marketability, industry profits suffer and revenue incentives for the community producer declines. Not only are these local hunts difficult to control and regulate, but they also discourage private-sector commitments to community partnerships (Lewis 1999).

In Botswana, local authorities enforce local guidelines for controlling the adverse effects of low-cost hunts on their land. Regulations include the type and number of vehicles allowed per licensed hunter, the number of passengers, the timing and duration of the hunt, where the hunting may take place, conditions for monitoring the hunt, and so forth. As a result, there is a high degree of private-sector commitment in Botswana to help train community members in skills needed to occupy senior positions in the hunting enterprise (Jagt *et al.* 2000).

Reducing threats to key economic species of conservation importance

Use of wire snares to kill wild animals is a common way for farmers in many rural areas of Africa to compensate for failed grain crops, by exchanging game meat for cereals such as maize or sorghum. Snaring is an especially destructive form of poaching, because it is indiscriminate and wasteful. For example, when a potential prey animal is snared, predators such as the large cats or hyaenas (*Crocuta crocuta*) are attracted and are often snared themselves since multiple snares are normally set in the same area (Lewis and Tembo 2000). A pack of wild dogs (*Lycaon pictus*), for example, was snared in Upper Lupande Game Management Area, Zambia in 1991 and this species has not been sighted there since (Chilikati, scout patrols). Records also exist of entire family groups of kudu (*Tragelaphus strepsiceros*) killed by snares (Phiri, scout patrols). Without some form of economic compensation and incentives awarded to residents for keeping wild animals safe from snares and outside poaching pressures, these threats can lead to local wildlife extinctions.

Under the right policies, tourist hunting could pay communities special producer fees for every animal legally hunted and regarded as a serious livelihood threat, such as lion, crocodile (*Crocodylus niloticus*), elephant (*Loxodonta africana*) or hippo (*Hippopotamus amphibius*). The community, in turn, would have the responsibility to use the funds to support compensation costs, and related livelihood support, to households most affected by these problem animals. Compliance could be measured by the absence of sighted predators scarred by snares, fresh poached carcasses of elephants found poached in the hunting area, and so forth. Such policies promote stakeholder cooperation to reduce mortality threats to wildlife, by minimizing human costs of living with problem animals. In Zimbabwe under the

CAMPFIRE programme, 50% of the licence fee for hunting an elephant is returned to the local community as an incentive to reduce illegal hunting and as compensation for possible livelihood costs (Maveneke 1996).

INVESTMENT STRATEGIES FOR PRODUCING WILDLIFE ON COMMUNITY LAND: CASE STUDIES IN ZAMBIA

From 1997 to 1999, safari hunting revenues generated on average US$71 536 for resident communities in seven different hunting concessions in the Luangwa Valley, Zambia, supporting a total population of about 15 000 households. Less than 10% of this money supported direct needs of individual households. The balance was allocated to law enforcement and to community capital projects such as schools, clinics and vehicles. During the same period, approximately 70 safari clients responded to a questionnaire to assess human disturbances that affected their hunting success, and 62% complained of encountering snares. Livelihood studies among these same communities showed that 40–60% of all farmers were unable to grow enough food to last them from one harvest to the next, and that a substantial number of these households used wildlife snaring as a coping strategy to meet their food needs (Lewis and Tembo 2000). This suggested that safari revenues reinvested back into the community were having little positive impact on food production as a key factor influencing the use of snares. We examined alternative ways that investment of safari hunting revenues could address such livelihood needs as a basis for reducing threats to wildlife. The following sections provide an overview of this ongoing work on different investment strategies that result in significant savings in wildlife numbers.

Food security

Based on the results described above, we reasoned that farmers who produced sufficient food to feed their families would be more likely to cooperate with efforts to produce wildlife than those who experienced food shortages. To test this notion, we identified approximately 4000 households as food-insecure residents, from five different hunting concessions in the Luangwa Valley in 2001. Selected households were assisted with improved farming skills, farming inputs and supplementary food support on the condition that they adopt these skills and reduce their dependence on wildlife. By 2003, 67% of these farmers had become self-reliant in food production and demonstrated their commitment to wildlife production by voluntarily surrendering over 12 000 snares and 76 firearms, which represented a potential annual saving of over 1500 wild animals.

Transforming poachers

Within the five safari concessions mentioned above, approximately 75 residents were identified by local community leaders as hunters who actively poached, on average killing 35 animals annually. Lack of access to legal markets forced them to sell their meat within the local community at an extremely low value of about US$5 per animal. Most of these hunters operated with relative impunity, because local informers assisted them with information to avoid arrest. When they were arrested and punished by a court of law, such deterrence was often not effective and they resumed poaching when returning home.

As an alternative to law enforcement, 17 of these local poachers underwent a formal two-month training in alternative livelihood skills and were promised inputs to help start their new livelihoods if they gave up poaching for at least eight months. From these initial recruits, 15 remained compliant for three years, and are now fully engaged in alternative livelihood strategies. The programme has now trained 97 local poachers and is expanding to most parts of the country.

The success in transforming the initial 17 poachers represented a saving of about 525 animals with a total market value of well over US$100 000. In comparison with the US$4250 required to train all 17, plus an additional US$1700 to support their livelihood input support for the first year, the potential profit returned from this investment offers the private sector a clear rationale for investing in local hunters and developing them as an asset for the safari hunting industry.

Poultry versus game meat

Among the communities residing in the five hunting concessions mentioned above, poultry is an important marketable commodity and also represents a major source of animal protein throughout the year. Newcastle's disease, which is extremely infectious and usually fatal to chickens, is an endemic disease throughout these concession areas. It normally kills over 80% of a household's chickens and can have a devastating effect on a community. Households are left without a reliable source of animal protein and potential income is lost for at least a full year. Without alternative protein or ways of generating income, the loss of poultry often translates into increased rates of wildlife snaring.

The vaccine for Newcastle's disease costs only US$3 per vial, which vaccinates up to 1000 birds for six months. Its use in rural areas, however, is restricted because refrigerated storage is required and its supply is from distant urban markets. Safari operators could easily solve these problems for their community partners at a negligible cost, and, by doing so, communities would increase poultry yields and thus reduce their dependence on game meat.

CONCLUSIONS

Because of its enormous economic value, safari hunting cannot afford to isolate itself from rural development, nor is it sufficient to simply remit revenues to a community if the revenues themselves do not address household livelihood needs. In this chapter, we recognize various ways for improving synergies and levels of cooperation to enhance the conservation value and economic benefits of safari hunting. Collectively, these methods represent a more holistic approach to reducing wildlife mortality threats attributed to land-use practices around wildlife protected areas. As management interventions, they require the targeting of selected households who represent the greatest potential threat to wildlife production. Such interventions also require greater commitments of time and expertise than most safari operators may have to offer. Such constraints may necessitate improved networking and collaboration by the safari industry with other partners, including community-based organizations, to address the livelihood needs of those households representing the greatest threat.

Perhaps the most serious underlying threat facing safari hunting is the industry itself. The industry's low operating costs relative to the extremely high market value of safari hunts make it vulnerable to intense pressure, by competing private-sector parties, to use whatever influence or means available to secure a concession. This pressure creates ideal conditions for corruption and corrupted management systems, and usually leads to lowered rates of wildlife production and ultimately lowered profits. The problem is not uncommon in southern Africa, where political interference and deal-making often override long-term conservation goals, and erode community support for safari hunting.

While the industry should be in the forefront, advocating policies that promote ownership rights and working partnerships with local communities, government has the ultimate responsibility to make such policies happen. Otherwise, safari hunting will give way to more rewarding ways for households to use the land. If such a fate befalls safari hunting, inadequate policies that fail to protect the industry from greed and lack of understanding of what drives rural people to degrade their resources will be the reason. Perhaps more than photo-tourism, which requires high overheads and aesthetic attractions, safari hunting can be an indispensable way of securing large land areas around national parks for wildlife management.

International marketing conventions drive a major share of safari hunts sold worldwide. These conventions could influence policy development in support of conservation by requiring minimum compliance to policy standards for nation–states sending trade delegations to market their hunts. Verifying compliance will require closer cooperation between conservation groups and hunting organizations in the monitoring of safari hunting policies and practices.

ACKNOWLEDGEMENTS

This chapter is a contribution to wildlife conservation policies in southern Africa supported by Wildlife Conservation Society and USAID. We recognize the significant contributions to this paper by the Zambia Wildlife Authority through its close partnership and support of the ADMADE programme and efforts to help reform community-based wildlife management policies.

16

Socio-ecological factors shaping local support for wildlife: crop-raiding by elephants and other wildlife in Africa

LISA NAUGHTON-TREVES AND ADRIAN TREVES

INTRODUCTION

Human–wildlife conflict is often viewed as a local problem involving the misbehaviour of people or animals (e.g. elephants transgress park boundaries to raid neighbouring crops, or farmers plant crops in wildlife habitat). Framing the issue this way tends to promote technical solutions like fencing and buffer crops; useful but often inadequate measures for promoting the long-term coexistence of people and wildlife (Breitenmoser *et al.*, Chapter 4; Osborn and Hill, Chapter 5). Geographers, anthropologists and other social scientists can illuminate the deeper causes of conflict and help guide long-term management solutions in several ways. First, social scientists can reveal the driving forces of land use change that impel people to plant crops or raise livestock in high-risk areas. Additionally, they can also assess the severity of the conflict by documenting the spatial and social distribution of wildlife damage, and the varying capacity of individuals to cope with such losses. Finally and more broadly, they can illuminate the social factors that intensify human–wildlife conflict or favour coexistence (Knight 2001).

In this chapter, we analyse the socio-ecological factors that shape rural African citizens' tolerance of crop loss to wildlife, particularly elephants (*Loxodonta africana*). Elephants are the focus of much human–wildlife conflict research in Africa. They deserve special consideration as an Appendix I CITES species and a tourist, 'flagship' species. We first survey 26 reports from 15 African countries to identify factors that intensify human–wildlife conflict, and to compare losses between elephants and other 'pests' at different scales. We also draw from the general literature on pests and risk in African peasant agriculture to better understand why some communities may be unable or unwilling to tolerate crop losses to wildlife. We then test the predicted patterns of vulnerability in the area

People and Wildlife: Conflict or Coexistence? eds. Rosie Woodroffe, Simon Thirgood and Alan Rabinowitz. Published by Cambridge University Press.

around Kibale National Park, Uganda, where farmers risk crop loss to a variety of wildlife, including primates, bush pigs (*Potamochoerus* spp.) and elephants. Our case study and review indicate that elephants and other large mammals generally cause far less damage to regional agricultural production than do rodents and invertebrate pests. However, aggregate measures of damage may be misleading. People's perception of risk is as important as actually losses, and their perceptions more often focus on rare, extreme damage events (e.g. a catastrophic raid by elephants) than persistent, small losses that cumulatively may be greater. Moreover, large ungulates and large carnivores are often viewed as highly charged symbols of state intervention and coercion; thus the damage they cause is especially resented (Newmark *et al.* 1994; Naughton-Treves 1997; De Boer and Baquete 1998; Nchanji and Lawson 1998; Gillingham and Lee 1999). The Kibale case study also reveals that when risk is absorbed at the individual household level, material wealth, and in particularly landholding size, determines who is able to cope with major losses to wildlife.

COPING WITH WILDLIFE 'PESTS' IN RURAL AFRICA

Contemporary factors intensifying human–wildlife conflict in Africa

Human–wildlife conflict is not a new problem. During the pre-colonial period, in some areas of Africa, crop-raiding by elephants and other large animals caused food shortages, displaced settlements or prevented agriculture altogether (Game Department of Uganda 1924; Osmaston 1959; Naughton-Treves 1999). By contrast, relatively few African farmers today regularly confront large wildlife on their land unless they live near protected areas or in remote regions. Ultimately, habitat loss and the extirpation of large species have reduced the overall area of conflict. (There are important local exceptions to this general trend. In areas where wildlife conservation rules are enforced, the zone of conflict may expand as wildlife populations recover. Such is the case in several southern African regions where elephant numbers have grown significantly over the past decade (see Osborn and Hill, Chapter 5).) Why, then, do leading conservationists now identify human–wildlife conflict as a primary threat to conservation in Africa (Hoare 1995; Kangwana 1995; Tchamba 1995; Barnes 1996; Western 1997)? Because where conflict persists today, its consequences are amplified for both wildlife and people. For example, laws designed to protect rare or endangered species (e.g. hunting prohibitions) often compromise people's ability to defend their crops or livestock. Meanwhile, wildlife survival may be threatened by lethal control or fencing campaigns (Woodroffe and Ginsberg 1998). In some cases local citizens' protests over wildlife damage can undermine regional or national conservation programmes (Anonymous 1994; Tchamba 1995).

Research on the underlying causes of human–wildlife conflict in Africa reveals the variable and complex interactions between rural populations and wildlife (Table 16.1). No single factor or condition explains conflict across the continent. Moreover, despite growing attention to human–wildlife conflict, uncertainty persists about the actual magnitude of the problem. Some experts claim that farmers consistently exaggerate crop damage to wildlife (Wakeley and Mitchell 1981; Bell 1984a; Roper *et al.* 1995; Siex and Struhsaker 1999). Others suggest that elephants and other large ungulates are unjustly blamed for damage, and that smaller animals, such as rodents or primates, cause greater losses over time (Mascarenhas 1971; Gesicho 1991; Hawkes 1991; Gillingham and Lee 1999). Unfortunately, the database on crop damage amounts and patterns is poor and burdened by ill-defined methods that limit comparisons between species and between sites. Too often, researchers exaggerate impacts by extrapolating results from crop-raiding 'hotspots' to entire regions, and rarely do they compare farmers' reports with systematic field measurements. To understand rural citizens' complaints, we must examine the spatial distribution and extent of crop loss, as well as the socio-ecological factors that shape local coping strategies and perceptions of risk.

Crop loss to wildlife versus other pests in the tropics

The term 'pest' is typically defined as any animal that consumes crops during any stage of the agricultural cycle, from planting to post-harvest storage (Porter and Sheppard 1998). Definitive comparisons of the economic impact of wildlife in comparison with other pests in tropical bush-fallow or shifting agricultural systems are difficult due to scarce data and extreme variability in crop yields and losses across farms, communities and regions. However, the literature on 'pests' provides rough estimates for the magnitude of non-wildlife losses, and reveals important factors shaping local coping capacity and tolerance of pests.

Farmers in tropical environments are exposed to a greater variety of pests than are temperate farmers, although the density of any given pest species is usually lower (Porter and Sheppard 1998). Tropical farmers also tend to be exposed to elevated and chronic levels of loss, in contrast with the periodic outbreaks of single pests in temperate agro-ecosystems (Oerke *et al.* 1995; Yudelman *et al.* 1998). For example, 60% of Tanzanian farmers ($n = 916$) rated pests as their primary economic problem, above low crop prices, lack of transport, failed rains and poor soils (Porter 1976). In Zimbabwe, local farmers ranked pests (including wildlife) first among

Table 16.1. *Explanations for intensifying human–wildlife conflict in Africa*

Changes in land use	Changes in wildlife behaviour and ecology	Changes in human socio-political systems
Agriculture expands into wildlife habitat, driven by population growth, voluntary or state-sponsored settlements, or shift to farming by pastoralists (Barnes 1990; Campbell *et al.* 1999; Gachago *et al.* 1995; Graham 1973; Hill 1997; Kiiru 1995; Tchamba 1996; Thouless 1994; Western 1997)	Wildlife populations 'packed' into parks and reserves by heavy hunting outside boundaries face food shortages (Gachago *et al.* 1995; Barnes *et al.* 1995; Mwathe 1992; Chiyo *et al.* (unpubl. data); Naughton-Treves *et al.* 1998)	Centralized, state ownership of wildlife lowers local tolerance of wildlife (Western 1997; Naughton-Treves 1997)
Agriculture intensification reduces availability of wild foods (Campbell *et al.* 1999; Naughton-Treves 1998)	Wildlife subject to intense hunting or culling form large groups and cause greater damage to local crops and vegetation (Southwood 1977)	Privatized land ownership erodes traditional collective coping strategies for wildlife pests (Agrawal 1997; Bell 1984b; Lahm 1996; Mubalama 1996)
Agricultural decline yields extensive second growth favourable to crop-raiding wildlife; remaining farms are isolated amidst 'bush', and more vulnerable to raiding (Lahm 1996; Mascarenhas 1971)	Protected wildlife species lose fear of humans and forage among settlements and farms (Gachago *et al.* 1995; Tchamba 1996; Naughton-Treves 1998; Kangwana 1995)	Urban employment opportunities draw men away from guarding fields (Lahm 1996)
Logging yields abundant second growth favourable to crop-raiding wildlife (Lahm 1996; Barnes *et al.* 1991)	–	Better access to schools releases children from their traditional role as guards and sentinels against raiding wildlife (Goldman 1996)
Artificially maintained water sources attract wildlife to human settlements during droughts (Thouless 1994)	–	Politicians now pay closer attention to local citizens who complain loudly against elephants, and this raises public awareness of the conflict (Kangwana 1995; Anonymous 1994; Hoare 1995; Barnes 1996)
Construction of canals, power installations and cattle fences cuts off migration routes and leads to 'aggressive' wildlife behaviour (Kangwana 1995; Kothari 1996; Lahm 1994)	–	War may displace wildlife from forests into agricultural areas (Tchamba 1995)

30 obstacles to improved quality of life (Wunder 1997). While there is general consensus that pests reduce agricultural productivity significantly in the tropics, losses are rarely measured precisely, particularly in peasant agricultural systems. Estimates range from 10% to 50% of total crop production, with an average estimate of 30% loss (Porter and Sheppard 1998; Yudelman *et al.* 1998). Another comprehensive survey estimated even higher losses for African farmers; roughly 51% of production was lost due to insects (15%), pathogens (13%), weeds (13%) and other pests, including rodents (10%) (Oerke *et al.* 1995). These data lack precision, but they suggest the general order of magnitude of losses.

Crop yields and losses in peasant agriculture are difficult to measure and compare because farmers typically plant complex polycultures in fields of ill-defined area. Planting densities vary greatly within and between fields. Pest infestations happen sporadically and often coincide with changes in climatic conditions. Given the spatial and temporal complexity of peasant agricultural systems, calculating average pest losses is not only difficult, it may be misleading. One farmer may easily tolerate a 15% loss in maize, while her neighbour cannot (Goldman 1996). A 28% loss during a drought may cause hunger, but not during a good planting season (Scott 1976). In sum, explaining local tolerance to wildlife via average percentage crop losses is inadequate because it masks the vulnerability of certain individuals and the more fundamental factors shaping public perception of risk. One must also address the socio-economic factors that influence local capacity and willingness to cope with crop damage to elephants or other animals.

Collective versus individual strategies for coping with risk

The social significance of crop loss to wildlife may best be understood in terms of *vulnerability*, a concept used in environmental hazards research to encompass risk of exposure and capacity to cope. Cutter (1996: 532) defines vulnerability as 'the interaction of the hazards of place ... with the social profile of communities'. In other words, vulnerability is shaped by both biophysical and social conditions (Liverman 1990; Carter 1997). For example, a farmer might face high levels of risk because he plants crops in an area frequented by hippopotami (*Hippopotamus amphibius*), but he is not necessarily vulnerable if he has other substantive sources of income or food. A highly vulnerable farmer is someone who plants crops in risky places and has limited capacity to cope. Carter (1997) goes on to describe risk as a 'mechanism of differentiation', meaning that communities are internally differentiated by individual exposure to risk and individual capacity to cope with risk, *and* that risk in turn can further differentiate members of communities. Results of research on drought hazards in Africa highlight two key factors determining individual vulnerability: *insurance* and *wealth*. These factors are directly relevant to human–wildlife conflict.

The vulnerability of smallholder farmers to elephant crop raiding can be mitigated by two insurance strategies: (1) *individualist self-insurance* (e.g. field scattering, crop diversification, employment of guards on individual property), and (2) *social reciprocity between households* (e.g. voluntarily sharing public spaces and labour, and aiding less fortunate neighbours) (Scott 1976; Carter 1997). Individualist self-insurance strategies depend heavily on individual access to land, labour, capital, etc. By contrast, social reciprocity insurance depends on traditions of sharing, close community relations and communal land management. Of course there is overlap between individual and social insurance strategies, and farmers may participate in both. However, given the shift toward private landholding and markets, and the decline of sharing and communal property regimes, the tendency in rural Africa today is toward greater reliance on individualist self-insurance (Carter 1997). This suggests a trend toward individualization of risk. In Malawi, Bell (1984) observed that large extended families on traditional farms neighbouring a park suffered 80% less crop damage to wildlife than did families on individual plots in neighbouring government settlements. Thus a group of farmers may be able to collectively cope with crop losses to elephants, while individual households cannot.

The capacity of individuals or households to absorb risk depends largely on wealth (social and physical endowments) and political influence. In peasant agriculture, farm size is an index of wealth and may be the most important endowment for coping with risk. A case study from southern Africa showed that only 10% of individuals in the upper quartile of landholding size suffered food scarcity during drought, while 85% of the bottom quartile suffered food scarcity (Carter 1997). Land availability is also an important predictor of farmers' capacity to cope with crop losses in Kenya (Goldman 1996). As long as farmers had sufficient access to land, they continued to tolerate 15% losses of their maize yields to invertebrate pests. As land became scarce, individuals bought pesticides or changed to another crop (Goldman 1996). Wealth can also be measured in access to capital or labour. Capital permits smallholder farmers to hire guards or build barriers. In contrast, the poorest households face *compounding vulnerability* (Carter 1997; Naughton-Treves 1997). Without large landholdings they cannot buffer themselves from wildlife conflict, nor can they hire additional labour. For example, widows and invalids often suffer the greatest damage within communities and are least able to cope (Bell 1984a); L. Naughton-Treves unpubl. data). In short, subsistence farmers with minimal endowments (i.e. access to kinship or community labour and resources, or alternative incomes) are the most vulnerable (Scott 1976; Porter 1979).

Ranking wildlife pests

Another way to understand local tolerance to wildlife is to compare 'worst pest' rankings. In Table 16.2 we tabulate the results of 25 studies of wildlife pests in Africa. We selected only studies that explicitly ranked problem animals by species or group, and those from sites or regions where elephants are present. These 25 studies come from 13 countries and include both savanna and forest sites. They also include examples of each major type of human–wildlife interface (Hoare 1995): 'hard' edges of parks or reserves (e.g. Kenya and Ghana cases), mosaics of agriculture and natural habitat (e.g. Cameroon) and isolated agricultural settlements embedded in forest (e.g. Congo). Out of 38 types of animals ranked as problem animals, the five most frequently mentioned were: elephants (32 cases), monkeys (including baboons, *Papio* spp.) (30), rodents (19), bush pigs (18) and antelopes (11). The animals most frequently described as 'worst animal' were elephants (8), monkeys (including baboon) (8), bush pigs (5), cane rats (*Thryonomys swinderianus*) (2) and buffalo (*Syncerus caffer*) (2). Elephants' mean rank was 2.5 ± 1.5 ($n = 33$), and there was no significant difference between rankings at savanna versus forest sites ($n = 14$ savanna and 14 forest sites). There was also no significant difference in farmers' versus researchers' ranking of problem animals. The only apparent discrepancy was between the ranking of elephants at different scales of analysis. Elephants were not ranked 'worst pest' in any of the six nation-level assessments and in only two of the 15 provincial or district-level rankings. Bush pigs were the only large mammal to emerge in national-level rankings. By contrast, six of 16 studies at park borders ranked elephants worst. This suggests that elephants tend to be a significant pest at the local or possibly provincial level, but not at the national level.

Comparing 'worst pest' rankings between studies is problematic. For one, some studies focussed specifically on elephants, and may have biased results accordingly. Also, the scale of analysis varied from single villages to nations. Methods were often poorly defined. Many studies ranked animals only by interviewing local farmers. This is a valuable approach for learning about local attitudes, but respondents in such studies often hope for compensation and thus may inflate damage reports, particularly for large or highly symbolic species (Mascarenhas 1971; Gesicho 1991; De Boer and Baquete 1998). Other studies ranked animals by the relative amount or frequency of their damage. This approach may avoid the problems of inflated complaints, but it introduces other problems. For example, given the unpredictable pattern of raiding by wildlife, results from a single season or single year may not be representative. Thus the data in Table 16.2 are preliminary, and should be interpreted with caution.

Table 16.2. *Ranking problem wildlife in Africa*

Country	Site (habitat)[a]	Method[b]	Problem wildlife ranking (1=worst problem)					Elephant ranking	Sample size and unit	Reference
			1	2	3	4	5			
Cameroon	Banyang-Mbo Forest Reserve (f)	R	Cane rat	Buffalo	Porcupine	Bush pig	Bird	–	19 farms monitored, animals ranked by amount of damage	Naughton-Treves *et al.* 2000
	Banyang-Mbo Forest Reserve (f)	F	Elephant					1	Interviews and public meetings with 430 villagers around reserve	Nchanji and Lawson 1998
Congo	Parc National Nouabale-Ndoki (f)	R	Elephant					1	29 fields, elephants ranked first by damage amount	Madzou 1999
Democratic Republic of Congo	Okapi Faunal Reserve (f)	R	Primate	Elephant	Bush pig	Buffalo		2	40 farmers in 29 villages	Mubalama and Hart 1995
	Okapi Faunal Reserve (f)	R	Monkey	Bush pig	Elephant			3	40 farmers in 29 villages	Mubalama 1996

Table 16.2. *(cont.)*

Country	Site (habitat)[a]	Method[b]	Problem wildlife ranking (1=worst problem) 1	2	3	4	5	Elephant ranking	Sample size and unit	Reference
	Garamba National Park (s)	F	Elephant	Hippopotamus				1	48 interviews with field verification near park boundary	Hillman Smith *et al.* 1995
Gabon	Haut-Ogooué Province (s+f)	F	Cane rat	Porcupine	Elephant	Bay duiker	Bush pig	3	239 families in 15 villages	Lahm 1994
	Ngounie Province (s+f)	F	Cane rat	Elephant	Porcupine	Sitatunga	Chimpanzee	2	364 families in 20 villages	Lahm 1994
	Nyanga Province (s+f)	F	Cane rat	Elephant	Porcupine	Mandrill	Gorilla	2	333 families in 20 villages	Lahm 1994
	Estuaire Province (f)	F	Cane rat	Elephant	Porcupine	Sitatunga	Emin's rat	2	286 families in 19 villages	Lahm 1994
	Moyen Ogooué Province (f)	F	Cane rat	Elephant	Porcupine	Mangabey	Sitatunga	2	231 families in 19 villages	Lahm 1994
	Ogooué-Ivindo Province (f)	F	Cane rat	Elephant	Porcupine	Gorilla	Bay duiker	2	669 families in 38 villages	Lahm 1994
	Ogooué-Lolo Province (f)	F	Cane rat	Elephant	Porcupine	Gorilla	Suntailed monkey	2	298 families in 20 villages	Lahm 1994

	Ogooué-Maritime Province (f)	F	Elephant	Hippopotamus	Cane rat	Buffalo	Mangabey	1	210 families in 30 villages	Lahm 1994
	Woleu-Ntem Province (f)	F	Cane rat	Elephant	Gorilla	Chimpanzee	Squirrel	2	296 families in 37 villages	Lahm 1994
	nationwide (s+f)	F	Cane rat	Elephant	Porcupine			2	2926 families in 218 villages (sum of previous nine entries)	Lahm 1994
Ghana	Kakum/Assin Forest Reserve (f)	?	Bird	Monkey	Bush pig	Elephant		4	Regional appraisal (methods not specified)	Dudley *et al.* 1992
Kenya	Shimba Hills National Reserve (s+f)	F	Bush pig	Baboon	Elephant	Monkey	Buffalo	3	138 households on park boundary	Kiiru 1995
	SE Kajiado District (s)	F	Buffalo	Antelope	Elephant	Monkey	Wildebeest	3	137 households in 1977, frequency for those reporting conflict	Campbell *et al.* 1999
	SE Kajiado District (s)	F	Antelope	Elephant	Monkey	Zebra	Hyaena	2	223 households in 1996, frequency for those reporting conflict	Campbell *et al.* 1999

Table 16.2. *(cont.)*

Country	Site (habitat)[a]	Method[b]	Problem wildlife ranking (1=worst problem)					Elephant ranking	Sample size and unit	Reference
			1	2	3	4	5			
	Laikipia/ Samburu region (s)	F	Elephant					1	Methods unspecified	Thouless 1994
	nationwide (s+f)	F	Baboon	Monkey	Elephant			3	Interviews (sample size not specified)	KWS in Feral 1995
Malawi	Kasungu (s)	R	Elephant	Bush pig	Baboon	Vervet monkey	Eland	1	Damage amount in 80 farms adjacent to or within 1 farm of park	Bell 1984b
Mozambique	Maputo Elephant Reserve (s)	F	Bush pig	Hippopotamus	Elephant	Bushbuck		3	200 households in 4 settlements, ranked by frequency of complaint	DeBoer and Baquete 1998
Rwanda	Volcans National Park (f)	F	Buffalo	Cane rat	Jackal	Porcupine		–	181 farmers <2250 m of park boundary	Plumptre and Bizumuremyi 1996

Tanzania	Kilimanjaro National Park (f)	F	Monkey	Bush pig	Rodent	Bird	Baboon	–	1396 farmers residing < 12 km of reserve boundaries	Newmark *et al.* 1994
	Manyara National Park (s)	F	Elephant	Baboon	Buffalo	Warthog	Hippopotamus	1		Newmark *et al.* 1994
	Mikumi National Park (s)	F	Bush pig	Baboon	Monkey	Elephant	Hippopotamus	4		Newmark *et al.* 1994
	Selous Game Reserve (s)	F	Elephant	Hippopotamus	Buffalo	Bush pig	Monkey	1		Newmark *et al.* 1994
	Selous Game Reserve (s)	R	Vervet monkey	Bush pig	Bird	Rodent		–	Methods unspecified	Gillingham and Lee 1999
	nationwide (s+f)	F	Bush pig	Bird	Monkey	Insect	Rodent	8	916 enumeration areas, animals ranked worst in each	Agroclimato logical Survey of Tanzania 1972 in Porter 1998
Uganda	Kibale National Park (f)	R	Baboon	Bush pig	Red-tailed monkey	Elephant	Chimpanzee	4	93 farms within 500 m of park boundary	Naughton-Treves 1997
	Bwindi National Park (f)	F	Monkey	Baboon	Bird	Rodent	Elephant	5	316 farmers in 10 enumeration areas along park boundary	Bearsted-Smith *et al.* 1995

Table 16.2. *(cont.)*

Country	Site (habitat)[a]	Method[b]	Problem wildlife ranking (1=worst problem) 1	2	3	4	5	Elephant ranking	Sample size and unit	Reference
	Budongo Reserve (f)	F	Baboon	Bush pig	Vervet monkey	Bird	Bushbuck	–	245 farmers in villages adjacent to forest, forest fragments and plantations	Hill 1997
Zambia	Upper Lupande (s)	F	Bush pig	Monkey	Hippopo-tamus	Chimpanzee	Elephant	5	135 farmers within Upper Lupande Game Management Area	Balikrishan and Ndhlovu 1992
Zimbabwe	BuliliMamangwe (s)	F	Bird	Springhare	Elephant	Jackal	Warthog	3	966 farmers residing in up to the fourth village from	Hawkes 1991

Sebungwe, Zambezi Valley (s)	F	Insect	Bird	Elephant		Hippopotamus	3	unsettled wildlife area 2 villages	Wunder 1997
Nyami Nyami district (s)	F	Elephant	Buffalo				1	Frequency of formal animal complaints	Hoare 1995

[a] Official name of area (s, savanna + woodland; f, moist forest; s + f=mixture of both types).
[b] Method: how was the ranking generated? R, researcher measurements; F, farmer reports.

Table 16.3. *Factors shaping tolerance of pests*

	Higher tolerance	Lower tolerance
Socio-economic factors		
Land availability	Abundant land	Scarce land
Ownership of wildlife	God, self, community	Government or elite
Coping strategies	Varied, unregulated	Narrow, highly regulated
Social unit absorbing loss	Community, group	Individual or household
Labour availability	Abundant, inexpensive	Rare, expensive
Value of wildlife	High (game, tourism, etc.)	Low (pest, vermin)
Capital and labour investment in crop	Low	High
Type damage	Subsistence crop	Cash crop or livestock
Alternate income	Various	None
Ecological factors		
Wildlife body size	Small, non-threatening	Large, dangerous
Timing of raid relative to harvest	Early	Late
Wildlife group size	Solitary	Large
Damage pattern	Cryptic	Obvious
Crop preference of pest	Narrow, one crop	Any crop
Crop part damaged	Leaves only	Fruit, tuber, pith, grain
Circadian timing of raid	Diurnal	Nocturnal
Crop damage in each raid	Self-limited	Unlimited
Frequency of raiding	Rare	Chronic

Factors shaping local attitudes towards and capacity to cope with wildlife

To better understand farmers' attitudes to various wildlife species, and to explain their apparent intolerance of elephants, we reviewed studies that identified factors shaping tolerance of pests (Table 16.3). Some of these factors are obvious. For example, no animal taking human lives is tolerated. Livestock losses to wildlife are considered worse than crop losses. Tolerance is apparently shaped more by amounts of crop loss than by frequency of raids. Animals highly prized as game by the local population may be tolerated despite significant costs. For example, each year, white-tailed deer (*Odocoileus virginianus*) in Wisconsin cause > US$34 million in crop damage and US$92 million in damage to vehicles (38 000 deer–car collisions each year: WDNR 1994). Yet there is widespread support for maintaining a population of > 1.2 million deer due to the profitable and popular annual hunt (670 000 hunters participate and generate US$255m in sales). Other influential factors are less straightforward. For example, some studies conclude that farmers tolerate damage to high-value cash crops least (Blair 1979), while others suggest that raids on 'famine' crops like cassava cause greater resentment (Mascarenhas 1971).

Local intolerance for wildlife may also be amplified by institutional constraints on coping strategies. People are less tolerant of imposed risk than they are of risk they take on voluntarily. For example, Starr (1969) showed the public to be 1000 times more willing, on average, to accept voluntary risks (e.g. driving) than those imposed upon them (e.g. pollution). Farmers feel especially vulnerable to large, highly symbolic animals that are perceived to – and often do – belong to the government. For example, elephants are highly prized by tourists and wildlife agencies, but they inflict potentially catastrophic damage. The perceptions of farmers often reflect rare, extreme-damage events rather than persistent, small losses that cumulatively may be greater (Naughton-Treves 1997). The complex interplay of actual risk and the effectiveness of each farmer's coping strategies is filtered through a cultural and socio-economic perspective. When asked 'Which animal is worst?' or 'How severe are your losses to wildlife?', a farmer's answer is shaped not only by her previous experiences with wildlife pests, but also by her perceived status with respect to the park, conservation authorities and the researcher. The following case study from Kibale National Park explores the spatial and social distribution of crop damage to wildlife, and compares local risk perceptions and coping strategies. This case study illustrates many of the points identified in the broader literature regarding the distribution of damage and differentiated capacity of individual households to cope with risk.

LOCAL RESPONSE TO CROP DAMAGE BY WILDLIFE AROUND KIBALE NATIONAL PARK, UGANDA

Kibale National Park is a 760-km^2 forest remnant located in the Toro region of western Uganda (Fig. 16.1). Kibale is rich in primates and other species (Struhsaker 1997), including those notorious for crop-raiding, such as olive baboons (*Papio cynocephalus*), red-tailed monkeys (*Cercopithecus ascanius*), elephants and bush pig. Currently, 54% of the land <1 km of Kibale's boundary is used for smallholder agriculture (Mugisha 1994). Farmers in the area belong to two predominant ethnic groups, the long-present Batoro, and the immigrant Bakiga, who came to Kibale by the tens of thousands from southwestern Uganda during the 1950s and 1960s (Turyahikayo-Rugyema 1974). Chiefs of the Batoro people traditionally allocated land to immigrants on the outskirts of their settlements, in part to buffer Toro farmers from crop damage by wildlife (Aluma *et al.* 1989). Today, both groups plant more than 30 species of subsistence and cash crops: bananas, maize, beans, yams and cassava cover the greatest area. In both groups, women generally assume responsibility for food crops, whereas men tend cash crops, such as brewing bananas. Farm sizes are small – averaging

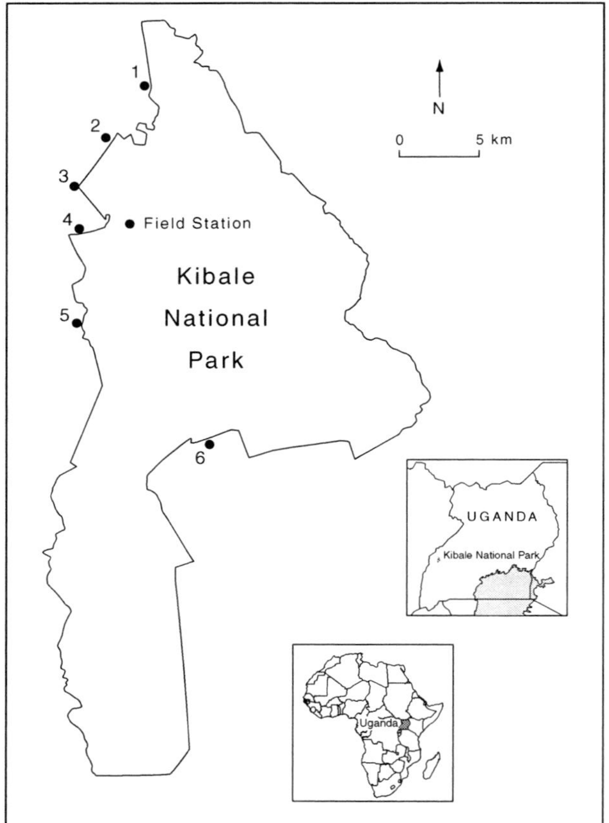

Figure 16.1. Map showing Kibale National Park and study sites.

1.4 ha – and population density is high: 94–272 individuals per km^2 around Kibale National Park (Mugisha 1994).

The social and physical landscape of Toro has profoundly changed this century (Naughton-Treves 1999). Where there were once isolated agricultural settlements amidst forest, today there are islands of forest embedded in agriculture. Natural habitat continues to shrink outside Kibale National Park. Edge species persist in agro-ecosystems (e.g. bush pigs, baboons and cane rats), but large or interior forest species are mainly confined to the park (Chapman and Onderdonk 1998). Despite regional declines in wildlife populations, farmers living within 1 km of Kibale complain bitterly about crop loss to animals. Anger about crop loss to wildlife is expressed most intensely during group discussions. People ask, 'Why should we starve so that baboons may eat?'

Research design and methods

This case study offers a synthesis of data collected during field research in 1992–4 and 1999 (Naughton-Treves 1997, 1998; Naughton-Treves *et al.* 2000). The basic aim of the research was to document systematically the amount and distribution of crop damage by wildlife in the communities neighbouring Kibale, and to then use multivariate analysis to predict vulnerability of loss at various scales (field, farm, village). We were equally concerned with understanding people's perception of risk and their varying capacity to cope with losses. The long-term nature of the study offered us an opportunity to assess response to damage over several years, and to test the hypothesis that a household's wealth powerfully shapes its coping capacity when risk is individualized (as per Carter 1997).

During 1992–4, crop damage to animals was monitored on 93 farms in six villages (Naughton-Treves 1998). Crop damage was measured each week by two assistants who walked transects through fields perpendicular to the boundary of Kibale (30 m wide extending 500 m from boundary). Along the transect, crop type and maturity were recorded. Every trace of crop damage by vertebrates (> 2 kg) was noted and its extent measured by pacing area or counting stalks. Raiding animals were rarely seen, so evidence from dung, tracks, bite marks and patterns of damage were used to infer the identity of the responsible species. Inter-observer reliability and damage measurement techniques are detailed in Naughton-Treves (1998). Also detailed there are techniques for identifying independent forays by animals. In brief, when adjacent transects crossed the same, large damaged area, only one event was noted (if the raiding species was the same). Similarly, if the same animal inflicted damage at multiple points along a transect, a single foray was recorded. These methods of determining independence do not inflate frequency estimates, particularly for animals that damage wide swathes of crops (e.g. elephants). We also conducted several public meetings and 145 interviews to appraise local attitudes to wildlife and coping strategies (Naughton-Treves 1997).

During 1999, the same team of field researchers resumed monitoring crop damage in three of the six original study villages (Fig. 16.1), this time on a monthly basis. In 1999 we also explored local farmers' long-term response to crop loss vs. other hardships by returning to all the original six villages to survey changes in land use and ownership. In essence, we traced the fate of 85 farms in relation to their history of crop-raiding. We assumed that farm abandonment was the most drastic response to crop-raiding, while field fallowing was a more moderate response. Note that in the local context, 'abandoning' a field means to leave it without crops for more than five years. 'Fallowing' a field refers to letting it rest for one to two years (short fallow), or three to five years (long fallow).

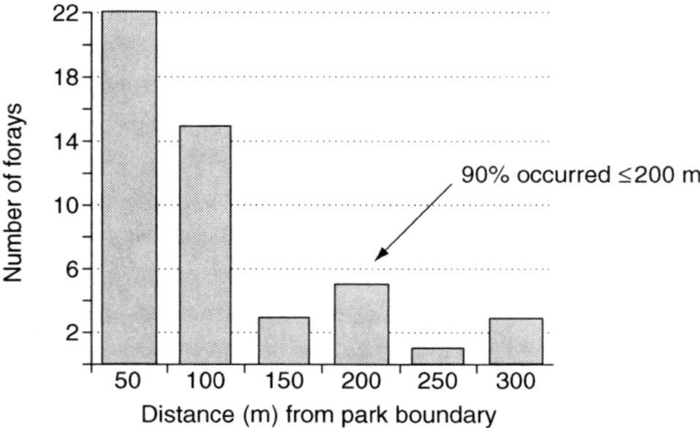

Figure 16.2. Frequency of elephant raids versus distance from the park.

Results

Amount and distribution of damage by wildlife

Across the two study periods, the strongest predictor of damage was proximity to the forest boundary. During 1992–4, 90% of damage events occurred < 160 m of the forest boundary, vs. 90% < 200 m during 1999. This pattern held true for elephants as well (Fig. 16.2). Households located within this 'high risk' zone lost 4–7% of crops per season on average in 1992–4, and 6–9.4% in 1999 (the average loss varies by crop type). In both sampling periods the distribution of damage was highly skewed such that some fields were on occasion completely destroyed, while many others were untouched.

The frequency and extent of crop damage varied markedly within and between villages, between species and between years. We recorded damage by 12 species, including livestock. Table 16.4 presents the results for the nine types of animals that caused damage more than once in 1999 (rodents are pooled). Goats damaged crops most frequently, but elephants did the most damage per foray (mean and maximum). Livestock caused almost two-thirds of the damage, while wildlife caused one-third. Among the wildlife, elephants accounted for the vast majority of area damaged (78%), but this was confined entirely to six farms at one village. Baboons were the most frequent raiders across villages. Figure 16.3 illustrates the variability between the two study periods.

Residents' coping strategies

Farmer households around Kibale generally manage their land individually. Collective planting, weeding or guarding is uncommon, although the immigrant social group (Bakiga) employ some collective land management strategies during certain seasons. Our previous analysis of individual

Table 16.4. *Crop damage by animals in farms neighbouring Kibale National Park, February to August 1999*

Animal	Frequency of crop damage			Area damaged m^2			Percentage of field lost per foray	
	Most frequently damaged crop	Percentage of farms damaged ($n = 51$)	Percentage of total forays ($n = 273$)	Total	Mean ± SE per foray	Largest single foray	Mean	Maximum
Goat	Banana, cassava	68.6%	62.3%	10 400	61 ± 7	668	6.0%	100.0%
Baboon	Sweet potato, maize	27.5%	7.7%	395	19 ± 10	211	5.2%	70.3%
Cane rat	Sweet potato, cassava	19.6%	4.4%	721	56 ± 20	200	16.0%	69.0%
Chicken	Bean, maize	15.7%	3.3%	298	33 ± 15	143	11.8%	16.2%
Red-tailed monkey	Banana, maize	13.7%	8.1%	212	10 ± 1	20	0.2%	0.6%
Wild birds	Groundnut, bean	11.8%	2.6%	112	16 ± 8	58	5.7%	15.1%
Elephant	Banana, cassava	11.8%	6.6%	5 207	289 ± 89	1475	5.7%	21.1%
Chimpanzee	Banana, maize	5.9%	1.5%	34	8 ± 3	13	0.6%	0.6%
Cattle	Banana, cassava	5.9%	3.7%	2 140	214 ± 79	790	33.0%	100.0%
Totals								
Amount of damage (m^2):								
Summed for all animals				19 519				
Summed for livestock (% of total)				12 838 (65.7%), of which 51% was caused by goats[a]				
Summed for wildlife (% of total)				6 681 (34.3%), of which 77.9% was caused by elephants				

[a] In 34 cases, we knew whose livestock did the damage and 19 (56%) of these were caused by neighbours' livestock.

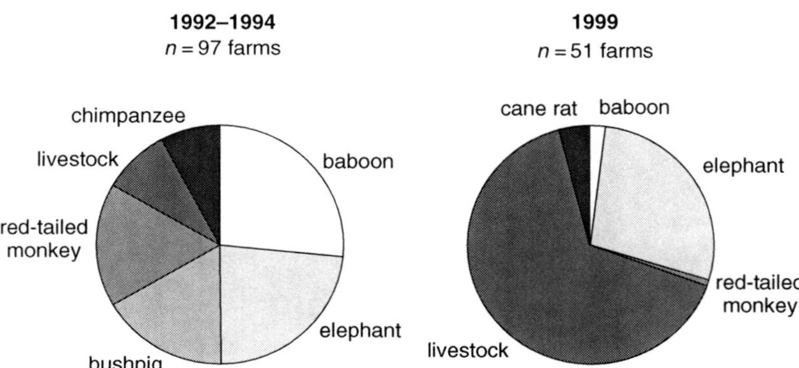

Figure 16.3. Pie charts comparing area damaged by different species. The 51 farms in the 1999 study were all part of the larger 1992–4 sample.

households' defensive strategies (e.g. hunting, strategic crop placement) showed that they could reduce damage by some species (e.g. bush pigs), but not others (e.g. elephants). In analysing people's actual and perceived risk of crop loss we learned that elephants inflict catastrophic damage to farms, but their forays are rare and highly localized. People's ranking of wildlife pests gave disproportionate weight to rare, calamitous raids by elephants (Naughton-Treves 1997). Another indication of the potential severity of elephant raids was that such events shaped people's attitude toward Kibale National Park. While the majority of farmers (83%, $n = 145$) believed that local people benefit from the park, those who suffered elephant damage were significantly less likely to identify benefits.

Differences between villages

Each village differed in the type and amount of pests they faced (Fig. 16.4). These data were analysed with a factorial design analysis of variance (ANOVA) incorporating village and proximity to forest as factors to predict the amount of damage in m^2. For all animals (wildlife + livestock), the villages differed significantly ($F_{2,982} = 12.4$, $p = 0.0001$). Villages still differed in the amount of crop damage when damage by wildlife and livestock were analysed separately (wildlife: $F_{2,971} = 7.4$, $p = 0.0007$; livestock: $F_{2,971} = 8.2$, $p = 0.0003$).

Direct and indirect costs of crop raiding

The direct, financial cost of crop-raiding can be estimated from the value of the crops per square metre multiplied by the area damaged (Fig. 16.5). Considering single forays, elephants inflicted the highest mean and maximum cost per farmer, but the overall cost of goat damage exceeded that of elephants and all other animals combined. Indeed, two-thirds of the

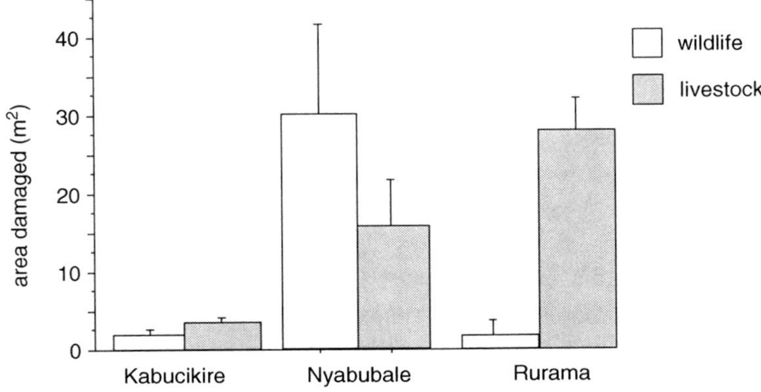

Figure 16.4. Area of crops destroyed by wildlife and livestock vs. village.

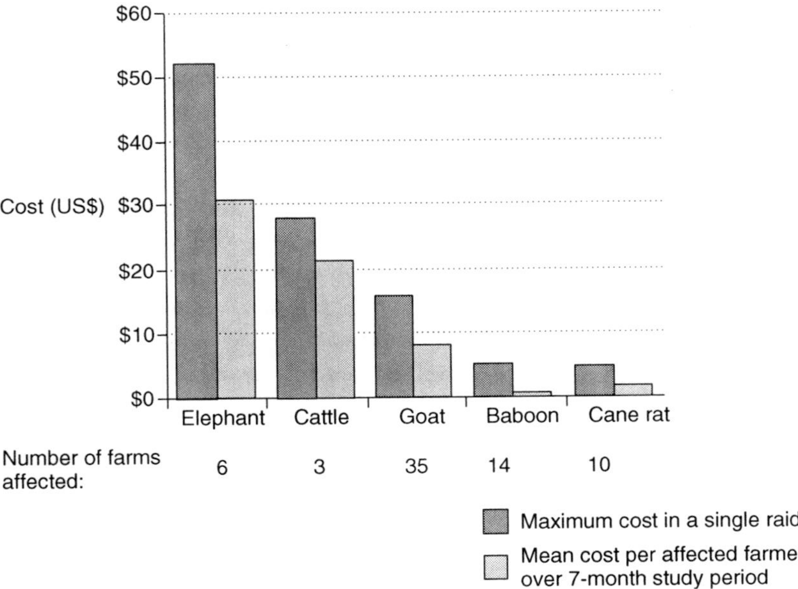

Figure 16.5. Direct costs (value per square metre multiplied by area damaged) of crop-raiding by various animals.

financial costs of crop damage were caused by livestock (goat, cattle, chicken and domestic pig combined).

In our assessment of general trends in land use in the six villages, we found that during the period 1994–9, farmers abandoned 32 fields ($n = 58$ farms, average $= 0.6 \pm 0.9$ fields per farm) and left 30 fallow ($n = 59$ farms, average$=0.5 \pm 0.8$ fields). By comparison, clearing of land led to the creation of 60 new fields or an average of 1.8 ± 0.8 fields per farm ($n = 84$, range 0–4). Hence, the clearing of new fields roughly equalled the combined abandonment and fallowing of old ones. There was a correlation between the number of fields cleared and the number fallowed (Spearman $r_s = 0.35$, $Z = 2.65$, $p = 0.008$), i.e. the same farmers who cleared new fields were the ones that fallowed older ones. However, there was no correlation between the number of fields cleared and the number abandoned ($r_s = 0.19$, $Z = 1.44$, $p = 0.15$).

There was good evidence that farmers abandoned fields because of wildlife crop-raiding. Farmers ($n = 67$ interviewed) stated that they abandoned fields because of baboons (36%), bush pigs (24%), banana weevils (15%), elephants (12%), poor soil (5%) or several rarer reasons: death, illness, chimpanzees (*Pan troglodytes*) and red-tailed monkeys (1.5% each). Abandoned fields averaged 52 m from the park boundary (SE = 10.1, $n = 43$).

Our measurements of damage support the idea that field abandonment followed crop-raiding. The three villages of the 1999 study differed significantly from one another in the number of abandoned fields (Kruskal-Wallis $H = 10.82$, $p = 0.0045$), and this corresponded to measured crop damage by wildlife (see ANOVA result above). Nyabubale suffered the most wildlife crop-raiding and had the most abandoned fields (mean of 2.7 abandoned fields, pair-wise comparisons $p < 0.003$ for each). The other two villages, Rurama and Kabucikire, did not differ significantly with means of 0.6 and 0.3 abandoned fields per farm respectively ($p = 0.31$). In 21 cases, entire farms were abandoned. The three villages differed significantly in the proportion of farms abandoned (5%–57%, df = 2, $\chi^2 = 16.5$, $p = 0.0024$). Again, Nyabubale contained more abandoned farms (45.5%) than either Rurama or Kabucikire (12% and 5% respectively). Forest regrowth on these abandoned farms is visible in Landsat images. Here the park edge appears to be expanding. Only 11 farmers could be interviewed about their reasons for abandoning their farms. Of these, six farmers blamed elephants and baboons together, one blamed elephants alone, three blamed a death in the family, and one simply blamed poverty.

We sought physical and social factors that might predict which farmers would abandon their farms (farm size, proximity to the park, ethnicity, and non-farm employment). Only the size of the farm predicted abandonment using univariate non-parametric tests. We had data on the size of 75 farms. Overall, abandoned farms were the same size as farms that were active (Mann-Whitney $U, Z = -0.52$, $p = 0.60$); however, this result is strongly biased by the significant differences in farm size between villages. To counter this bias, we compared the field size of abandoned and active farms within villages. The size of active farms was larger than the size of abandoned farms in every case (Wilcoxon signed-ranks test df=2, $Z = 2.02$, $p = 0.043$). The larger farms were significantly less likely to be abandoned, and this effect emerged most clearly beyond a size of 1.8 ha.

It appeared that farmers with large landholdings were less likely to abandon their land when faced by wildlife crop damage. This seems to reflect different land-use practices and greater flexibility in field management. Larger farms contained more abandoned fields ($r_s = 0.41$, $Z = 2.85$, $p = 0.0043$), slightly more fallowed fields ($r_s = 0.29$, $Z = 2.04$, $p = 0.041$) and many more newly cleared fields ($r_s = 0.50$, $Z = 4.22$, $p < 0.0001$). In effect, large farms were being maintained as small-scale systems of shifting agriculture.

DISCUSSION AND CONCLUSIONS

The Kibale case study highlights lessons from the literature. As around other African parks, the highly variable and localized nature of crop damage by wildlife around Kibale makes it difficult to quantify the economic impact of crop raiding for Kibale's neighbours. However, our field data from 1992–4 and 1999 and from Chiyo (2000) do clearly indicate that damage by elephants and other large species is tightly confined to < 200 m of the park boundary. This pattern concurs with observations at other 'hard edges' that having an active farm between you and the park is the best defence against crop raiding (Newmark 1996; Hill 1997). At Kibale, even within this high-risk zone, only a few farms (on the order of 10%) suffered elephant damage. At a regional level, Kibale's elephants have a negligible economic impact on agriculture relative to rodent and invertebrate pests.

From an international perspective, an annual loss of 4–9% of planted fields immediately along Kibale's boundary (equivalent to roughly US$6 per farmer, or US$100 per kilometre of border) appears a trivial price for maintaining elephants and other threatened wildlife. Moreover, most of Kibale's neighbours extract fuelwood and water from the park worth far more than US$6 per year (L. Naughton-Treves unpubl. data). But the farmers who live on Kibale's border poorly tolerate crop loss to wildlife, particularly because they cannot legally use the full range of traditional defensive strategies. They particularly resent damage from elephants who raid nocturnally and are potentially dangerous. Moreover, estimates of average losses mask the great variation in amounts lost by different farmers and villages. The farmers suffering crop loss to elephants absorbed an average cost of US$60 per year, a significant amount in an area where annual incomes average US$300 (National Environment Action Plan Secretariat 1995). A few individuals lost much more. To the farmer who has lost an entire year's production in a single night, average losses are meaningless. In some cases elephant damage caused families to abandon their land, particularly those who owned < 1.8 ha. Although elephant raids are relatively rare, their severe potential impact shapes attitudes among Kibale's neighbours. In Kenya, patterns of crop damage roughly similar to those observed around Kibale (localized but severe for certain communities) have led to a public outcry and the demand to build fences and/or reduce the size of national parks (Okwemba 2004). Such cases demonstrate that human–wildlife conflict can be a major obstacle to community support for conservation, and the hostility of a vocal minority can undermine regional conservation initiatives (Newmark *et al.* 1994; Naughton-Treves 1997; De Boer and Baquete 1998; Nchanji and Lawson 1998; Gillingham and Lee 1999).

In most of the communities neighbouring Kibale, people contend with wildlife damage on an individual household basis. As the literature on drought and vulnerability shows, when risk is individualized, an individual's landholding size becomes especially important. At Kibale, we found empirical evidence that larger landholders cope better with elephant crop damage (i.e. they are less likely to abandon their land after repeated elephant raids). This does not, however, mean that the larger landholders willingly tolerate elephants. Indeed, around Kibale, the wealthier, more powerful farmers were often the most vehement in their demands for compensation from the government. Hostility to elephants was intensified by general resentment of conservation authorities and the status of elephants as 'property' of the state. We also observed that the larger the assembled group of farmers, the louder the complaints and greater the estimates of elephant crop damages. This provides proof, once again, that measuring risk perceptions and tolerance of wildlife is a challenging endeavour.

Beyond building better fences at park boundaries or planting appropriate buffer crops, one of the most important strategies to ameliorating human–wildlife conflict is building local management institutions capable of balancing conservation objectives with the demands of local agriculturalists. Mitigating wildlife crop-raiding is inherently a communal endeavour, particularly for species like elephants. To minimize the incidence and impact of raids, farmers ideally would make collective land-use decisions (e.g. plant crops together in large blocks, and/or plant large buffer strips), or draw on traditional insurance systems based on social reciprocity (e.g. share not just the benefits of wildlife but the costs as well). More applied research is needed to test the viability of collective management of risk, and to identify political and institutional arrangements that foster community-level tolerance to elephant crop damage. Applied research is also needed to better predict the spatial pattern of elephant raids so that the costs and benefits of wildlife conservation are more equitably distributed. Unfortunately, the trend in much of rural Africa is toward individualized and private land management, making collective management difficult (Agrawal 1997). No doubt in situations where risk in entirely individualized among smallholder farmers, and wildlife is highly endangered, state agencies or conservation non-governmental organizations must compensate farmers for crop damage (see Nyhus *et al.*, Chapter 7). To avoid these situations, conservationists must lobby against land-use policies that create high-conflict situations, e.g., smallholder settlements placed on park boundaries. And whenever possible, they should promote community-level management of elephants for tourism or hunting, building on promising examples from eastern and southern Africa.

ACKNOWLEDGEMENTS

This study received formal clearance from the Uganda Council for Science and Research, as well as nine Local Councils bordering Kibale National Park. P. Baguma provided first-rate field assistance. The Wildlife Conservation Society, IUCN Elephant Specialist Group, Fulbright-Hayes Foundation, and the National Science Foundation (SBR #98–10144) provided financial support. Some of the text in this chapter is adopted from a report commissioned by the IUCN special task force on human–elephant conflict (Naughton-Treves *et al.* 2000). Comments from M. Walpole, S. Thirgood and R. Hoare greatly improved this chapter.

17

Jaguars and livestock: living with the world's third largest cat

ALAN RABINOWITZ

INTRODUCTION

The jaguar is the largest cat in the western hemisphere and one of the most powerful and recurring motifs of Central and South American cultures, past and present (Saunders 1989, 1991). The jaguar icon symbolized royalty and was thought to protect people against all other malevolent forces (Benson 1998; Saunders 1991). Still, the living jaguar, while considered a somewhat mystical being, was always feared and hunted as a dangerous predator.

However, unlike the other great cats, there have been no verified records of man-eating jaguars, and relatively few records of jaguars killing people. Hunters describe the jaguar as elusive and secretive, stating that, when stalked, it often tries to escape rather than fight (Singer 1916; Duguid 1932). Some hunters even captured jaguars without fear of injury by chasing and lassoing them (Fawcett 1954).

The real conflicts between humans and jaguars are relatively recent starting in the mid sixteenth century with the introduction of domestic livestock to the New World after European colonization (Arnold 1968). In the South American Pantanal, where the main economic activity is cattle production, the ranching industry and the cowboy culture, considered long-held and traditional, is in fact only 250 years old (Wilcox 1992; Dolabella 2000). As contact between jaguars and cattle increased, so did the problem of jaguars killing cattle. Now with nearly 4 million cattle in the Pantanal, the region has the largest ungulate biomass in the New World, and is considered one of the hotspots of jaguar–livestock conflict. The reaction to such conflict has always been to kill jaguars.

People and Wildlife: Conflict or Coexistence? eds. Rosie Woodroffe, Simon Thirgood and Alan Rabinowitz. Published by Cambridge University Press.

REALITY VERSUS PERCEPTION

The killing of cattle by jaguars is a well-documented phenomenon and has long been used as an excuse to hunt this cat for sport (Roosevelt 1926; Almeida 1990). But such hunting was never a solution to recurring jaguar–livestock conflicts, even when individual problem jaguars were eliminated. By the end of the 1970s, in reaction to large numbers of jaguars and other spotted cats being killed for the fur trade (Smith 1976), international trade conventions on spotted cat skins were adopted and jaguar hunting became illegal in most range countries (Nowell and Jackson 1996; Sunquist and Sunquist 2002). While this decreased the killing of jaguars, especially for profit, it exacerbated the jaguar–livestock conflict because people now felt that they had no legal recourse to fight against the incursions of this cat.

The idea that not all jaguars killed cattle and that a variety of factors affected jaguar behaviour was first put forth by Theodore Roosevelt (1926) who observed that jaguar depredation on livestock in Brazil was prevalent on ranches with a scarcity of wild prey but occurred infrequently in places where wild prey were abundant. However it was not until the 1980s that comprehensive research first started to give real insights on jaguar ecology and behaviour, and the relationship of jaguars to livestock (Schaller and Crawshaw 1980; Schaller 1983; Mondolfi and Hoogesteijn 1986; Rabinowitz 1986; Rabinowitz and Nottingham 1986; Crawshaw and Quigley 1991; Hoogesteijn and Mondolfi 1992). Additional studies over the next two decades continued to disprove many preconceived notions about jaguar depredation on livestock and provided a framework for mitigating at least some jaguar–livestock conflict.

Some of the earliest research showed that rudimentary management and poor husbandry practices by ranchers was perhaps the major factor in predisposing livestock to jaguar predation and in causing livestock loses that were unrelated to, but often blamed on, jaguars. Cattle left unattended to graze in forested sites became the dominant prey for jaguars in terms of available biomass (Schaller 1983; Schaller and Crawshaw 1980). This traditional practice of free-ranging also left the cattle exposed to many other causes of mortality. Research on ranches in Brazil (Schaller 1983) and Venezuela (Hoogesteijn *et al.* 1993) showed that losses of cattle to jaguar predation were, in fact, minor compared to other factors such as disease, malnutrition, abortion, snake, bat or vulture bites, accidents, and even puma depredation. On some ranches puma attacks on livestock were found to be considerably more common than attacks from jaguars (Fernandez 1995; Polisar *et al.* 2003), particularly when losses of calves were involved (Oliveira 1992).

Early research also uncovered that the hunting of jaguars and their prey actually contributed to livestock depredation. The practice of trying to eliminate all jaguars as a method of pest control often created problem animals when jaguars were wounded but did not die. In Belize and Venezuela, the majority of cattle-killing jaguar skeletal remains examined showed old shotgun wounds to their heads, while the same was not true with the remains of non-cattle-killing jaguars (Hoogesteijn *et al.* 1993; Rabinowitz 1986). Such injuries, mostly to the skull, probably made it difficult for jaguars to stalk and capture wild prey forcing them to turn to domestic livestock (Rabinowitz 1986). Hunting of other wildlife species that jaguars needed as food also predisposed jaguars towards livestock depredation, particularly when the abundance of key prey species was significantly reduced (Aranda 2002; Nunez *et al.* 2002; Polisar *et al.* 2003).

Despite efforts to study and protect jaguars, by the 1990s the future of jaguar conservation seemed bleak for a myriad of reasons:

(1) Existing protected areas were not, by themselves, saving jaguars throughout their range. Despite the success of some individual protected areas (Rabinowitz 2000), most protected sites were too small or too fragmented to maintain long-term viable jaguar populations, even those existing within extensive reserve frameworks (Ceballos *et al.* 2002). The possibility of designating many more protected areas in jaguar range was limited.

(2) Despite legal restrictions, jaguar hunting, and the hunting of jaguar prey, continued virtually unchecked, compromising the effectiveness even of protected areas for jaguar conservation. Government agencies prohibited the killing of jaguars without offering any assistance, explanation or alternatives for the conflicts people were facing.

(3) The practice of free-ranging cattle into jaguar habitat was still common. Husbandry and management practices that could have reduced jaguar–livestock conflicts had not been implemented throughout most of jaguar range.

(4) Ranchers and ranch managers had not changed their negative perceptions and feelings about jaguars, even when they were informed that most cattle mortality was caused by factors other than jaguars.

Thus, by the end of the twentieth century, jaguar populations were still threatened and little concerning jaguar–livestock conflict had changed (Weber and Rabinowitz 1996). Instead there were more outcries from ranchers, and increased lobbying efforts by trophy hunters claiming that continued conflict indicated that jaguars were increasing over most of their range and had to be controlled.

UNDERSTANDING JAGUAR–LIVESTOCK CONFLICTS

In 1999, the Wildlife Conservation Society called together many of the world's jaguar experts to examine the state of our knowledge regarding jaguar behaviour, status and distribution, and to set priorities for future research and conservation efforts (Medellin *et al.* 2002; Sanderson *et al.* 2002). The good news was that, despite the killing, loss of habitat and rapid decline of the prey base in many areas, jaguars still ranged from Mexico to Argentina throughout many areas of good habitat. A new jaguar population had been discovered and studied in the most northern part of their range (Brown and Gonzalez 2001), with individuals occasionally ranging into the southern USA (Rabinowitz 1999).

However, more jaguars were living outside rather than inside protected areas, a situation that was not likely to change soon. While the acquisition and protection of strategically located lands was still a high priority for jaguar conservation (Quigley and Crawshaw 1992), negative attitudes and perceptions by humans towards jaguars were clearly the greatest imminent threat to the species' survival (Woodroffe and Ginsberg 1998). It was imperative to initiate work with people on private lands and resolve conflicts at both the local and landscape level if viable jaguar populations were to survive.

By 2002 there were a number of studies spanning more than a decade relating to jaguar–livestock conflict issues. The most recent publications have examined the feasibility of more sophisticated mitigation techniques such as electric fencing (Saenz and Carillo 2002; Schiaffino *et al.* 2002; Scognamillo *et al.* 2002) or were written specifically to assist ranchers and other local people in addressing jaguar–livestock conflicts (Hoogesteijn 2002; Pitman *et al.* 2002). But the long-promulgated idea of mitigating jaguar–livestock conflict by simply changing or improving cattle husbandry and management practices, at little or no cost to the rancher, had still not resulted in any noticeable change throughout much of jaguar range. Clearly there was a serious flaw in government and privately sponsored outreach and jaguar conservation programmes that were trying to address this issue.

Discussions with ranchers indicated that the ranching community felt that they were being treated as part of the problem not part of the solution, and that there had been no real attempt to engage them or work with them cooperatively. Environmental non-governmental organizations (NGOs) and government agencies, on the other hand, felt that the ranchers, who were fully integrated into the modern world, knew enough about the problem and were wealthy enough to be doing 'the right thing' at their own expense. Consequently lines were drawn in the sand, and groups that should have been working together viewed each other as adversaries.

Only very recently have surveys in the Brazilian Pantanal closely examined the views of ranchers, or compared ranchers' views with those of other groups such as tourism operators and environmentalists (Zimmermann 2000; Marchini 2002). These surveys indicated significant differences of opinion regarding key issues affecting jaguars:

(1) Ranchers and managers wanted government assistance to help recover traditional cattle ranching, while tourism operators and environmentalists felt that there was no future for cattle ranching and that there were better uses for the land, such as reserves and ecotourism.
(2) Ranchers and managers believed that jaguar populations were doing fine, or even increasing. Tourism professionals and environmentalists, on the other hand, believed that jaguar numbers were declining and that the fault lay with the ranchers.
(3) Most ranchers, cowboys and managers considered jaguar to be the most detrimental species in the Pantanal; predators such as puma were rarely if ever mentioned. Environmentalists knew that puma and other factors often caused more livestock deaths than jaguars, and wrongly assumed that this fact mitigated some of the ranchers' negative feelings about jaguars.
(4) Ranchers, managers and cowboys valued biodiversity but assigned mainly intrinsic values to it, such as beauty, enjoyment and tradition. Tourism professionals and environmentalists constantly used scientific arguments to justify biodiversity conservation.
(5) Ranchers, managers and tourism professionals, in general, did not believe that environmental NGOs were beneficial. Some felt that the protectionist views of NGOs were detrimental to the environment and economy of the Pantanal.
(6) Ranchers believed that protected areas often have had a negative impact on terrestrial biodiversity while wildlife has benefited from infrastructure and management activities on ranch lands. The proof, ranchers asserted, was the fact that biodiversity already existed on ranch lands. Environmentalists and tourism operators thought that protected areas were, by definition, good for biodiversity, while ranches were bad.

Despite differences on important points, these surveys provided insight and opportunity for efforts to move forward with jaguar–livestock conflict issues. While ranchers felt that jaguar predation was a serious problem, they also valued their natural heritage. Most ranchers agreed that complete elimination of jaguars was unacceptable (Zimmermann 2000). Furthermore, it was clear that many of the opinions of NGOs about protection and development in jaguar habitat were based on scientific and monetary arguments that were not completely understood nor accepted by ranchers. For instance, while cattle ranching was viewed as bad by the conservation community because it promoted jaguar–livestock conflict, ranching was also considered to be one of the most economic uses of land compatible with conservation (Prance and Schaller 1982; Alho *et al.* 1988). Land-use schemes put forth by environmental NGOs, such as ecotourism, while touted as a win–win solution for ranchers and jaguars, were not acceptable to many ranchers (Marchini 2002). The purchase and creation of model ranches by conservation organizations to demonstrate improved husbandry and management practices was viewed as antagonistic and threatening to ranchers who felt that their own views and knowledge were being ignored. In the end ranchers felt that no one was talking or listening to them. They believed that their way of life was being imposed upon, threatened and taken advantage of by government agencies, biologists and conservation NGOs, all with their own agendas.

MODELLING CONFLICT SOLUTIONS

Taking into account the results of such surveys, the Wildlife Conservation Society initiated a rancher outreach programme and restructured its approach to the jaguar–livestock conflict issue, beginning with a series of workshops in Brazil, Venezuela and Mexico. The intention of these workshops was to bring together a cross-section of ranchers with different attitudes, different financial situations and different levels of jaguar–livestock conflict in order to create a model for engaging the ranching community as the first step towards conflict resolution. The workshops were designed so that ranchers had the major voice and the deciding vote in any proposed action. The promise to the ranchers was to take the process to the next step by working with them to implement and fund their *own* ideas, short of killing jaguars, assisting in the killing of jaguars, or translocating jaguars. We also reserved the right to refuse funding if we felt that our offer of assistance was being abused.

Ranchers who attended the workshops said that it was the first time that they had ever been invited by a conservation NGO to participate as speakers, where they were asked to express their knowledge on the problems, as well as dictate possible solutions. While many of the ranchers initially had some level of anger and distrust, such feelings dissipated when they realized that they truly owned the meeting. They became more positive and outspoken, eager to express constructive ideas about what could be done on each of their own ranches to better protect livestock from jaguars. Almost all the ranchers had different ideas or solutions, some of which did not even directly relate to jaguar–livestock conflict. But that did not matter. The objective of creating a process from which a working relationship with the ranching community could be initiated was clearly being achieved.

When ranchers saw that the Wildlife Conservation Society would commit the time and funds necessary to see their ideas through to fruition, they became increasingly cooperative. Even when ranchers did not need outside funding to implement their ideas, most wanted a buy-in on our part as a show of good faith and an indication that we did not consider it their responsibility to shoulder the full burden of jaguar conservation on their land. This was clearly a necessary initial cost for a cooperative working relationship.

The core strategy underlying this approach was that the rancher had to be fully integrated into any decision-making process and that actions had to be taken on both a local and regional level simultaneously. While individual ranches were considered the most basic unit for engagement, initial discussions took place in organized workshops or at meetings of existing rancher organizations, so that the ranching community had the opportunity to influence its members and make collective decisions or suggestions. Only in this way could we work towards achieving lasting changes not only in individual ranch practices, but also in the attitudes and practices across large regions such as the Pantanal of Brazil or the llanos of Venezuela.

THE FUTURE

The fact that we have made little progress with the jaguar–livestock conflict to date, despite knowing how to mitigate at least some of this conflict, is an issue that must be addressed before we can move forward with jaguar conservation. The Wildlife Conservation Society is only at the very beginning of its rancher outreach effort but our initial objective of creating a model of engagement for the ranching community appears successful. Clearly, landowners are more likely to take actions beneficial to wildlife if the information and assistance being delivered is done through continued personal contacts rather than through distant communication (Conover 2002). Future success in specifically addressing or resolving jaguar–livestock conflicts will depend upon our ability to remain on the ground and work with individuals and groups of ranchers to monitor changes in actual conflict, or rancher perception to such conflict. Since all of our actions will be based on ideas by the ranchers themselves, each activity must be worked out on a case-by-case basis to fit possibly unique ecological, social and economic circumstances.

Humans continue to have a very complex relationship with jaguars, a relationship that is often far from tolerant. But this relationship can and must be better managed and restructured if wild jaguar populations are to continue to survive in this world. We can accomplish this goal only when we try to understand more fully how those whom we are trying to influence view the world they live in.

It is imperative that we recognize both sides of the jaguar–livestock conflict if we wish to eliminate or mitigate conflicts between jaguars and people. The world of the rancher is changing as much as that of the jaguar in some respects. Tradition is giving way to some harsh economic realities. The huge extensive landholdings of the past in places such as the Pantanal are being broken up (Fortney 2000) and the cattle industry is in decline in many areas (Crawshaw and Quigley 2002). Still, ranches could play a vital role in jaguar conservation if the involvement of the rancher is encouraged and supported by government agencies and conservation NGOs (Hoogesteijn and Chapman 1997). Unfortunately such encouragement and support has not been forthcoming in the past, and there remains a serious rift in the attitudes and working relationships between all parties involved in jaguar conservation. The responsibility to bridge this rift and move jaguar conservation forward rests on the shoulders of conservation organizations and government agencies, not the ranchers. Only when this is accomplished can we use our knowledge to reshape ideas and influence attitudes so that integrated, acceptable solutions to the jaguar–livestock conflict can be found. Jaguars and people can coexist in this world. But we all must work a lot harder at making this happen.

ACKNOWLEDGEMENTS

The Wildlife Conservation Society's rancher outreach programme is a cooperative effort between the Society and the North of England Zoological Society based at Chester Zoo. The initial funding for the Wildlife Conservation Society Jaguar Conservation Program was provided by Jaguar Cars – North America and is now provided by the Liz Claiborne/Art Ortenberg Foundation. I would like to thank Kathleen Conforti, Howard Quigley and Alexandra Zimmermann for reviewing this chapter and for their efforts in the Wildlife Conservation Society's rancher outreach programme.

18

People and predators in Laikipia District, Kenya

LAURENCE G. FRANK, ROSIE WOODROFFE
AND MORDECAI O. OGADA

INTRODUCTION

In this century, only in Africa do substantial numbers of people and livestock still live alongside sizeable populations of large carnivores. Predators are rarely a threat to humans in modern Africa, but they are a significant source of livestock losses to both commercial and subsistence livestock producers. Killing of predators has been documented for as long as there has been literature (see Homer, *The Iliad*), but a small human population would have had an insignificant effect on total carnivore numbers. However, the press of a very large human population well equipped with firearms and poison has seriously reduced predators even in Africa, a relatively sparsely populated continent (Nowell and Jackson 1996; Woodroffe *et al.* 1997; Mills and Hofer 1998). Few protected areas are large enough to guarantee long-term survival of wide-ranging carnivores (Woodroffe and Ginsberg 1998), as most parks are small and widely separated.

In much of Kenya, wildlife has been eliminated as habitat is converted to cultivation. A growing bushmeat trade has eliminated wildlife from vast regions of southeast Kenya that are unsuitable for agriculture (World Wildlife Fund 2000a). Even the semi-arid northern half of the country, sparsely populated and once rich in wildlife, has been nearly cleared of large mammals by over-grazing, poison and the ubiquitous assault rifle. Outside protected areas, substantial predator populations persist only in the rangelands north of Mount Kenya (particularly Laikipia District), and in the south close to the border with Tanzania.

Laikipia District and the Laikipia Predator Project

The Laikipia Predator Project started in 1997, in an effort to identify the forces that make African predators vulnerable to local extinction, and to

People and Wildlife: Conflict or Coexistence? eds. Rosie Woodroffe, Simon Thirgood and Alan Rabinowitz. Published by Cambridge University Press.

find practical measures to counteract them. Although Laikipia is socio-economically unusual in East Africa, it is an excellent laboratory in which to study the biology of large carnivores outside protected areas, and to find ways of reducing their impact on the human economy.

Laikipia District covers 9700 km^2 of semi-arid bushland. About 35% of the area has been converted to settlement or small-scale urbanization, but the rest is still wildlife habitat. Except for some forest reserves, there are no formally protected areas. Livestock production is the economic base of the district, in the form of both commercial ranching of beef cattle and traditional pastoralism based on goats, sheep and cattle. Both commercial farmers and subsistence pastoralists use traditional livestock husbandry practices: stock are closely herded by day and penned at night in stout thornbush corrals ('bomas': Ogada *et al.* 2003). On most commercial ranches, livestock densities are comparatively low and wildlife is abundant. Our study area encompasses 25 commercial ranches (out of 30 in the district), of which 14 receive non-ranching subsidies in the form of tourism or wealthy owners; the others are largely dependent upon their livestock. We also work on 14 group ranches, communities of pastoralist Mukogodo Masai; both human and livestock densities are higher in these areas, habitat degradation from over-grazing is sometimes severe, and wildlife (particularly lions) are typically more scarce than on commercial ranches (Khaemba *et al.* 2001). Tourism is expanding on both commercial and pastoral lands, and this has involved setting aside land exclusively for wildlife.

Laikipia District supports populations of all the native large carnivore species, most of which are considered globally threatened (IUCN 2002): lions (*Panthera leo*; vulnerable), leopards (*P. pardus*; not listed), cheetahs (*Acinonyx jubatus*; vulnerable), spotted hyaenas (*Crocuta crocuta*; conservation dependent) and striped hyaenas (*Hyaena hyaena*; near threatened) have all persisted, and African wild dogs (*Lycaon pictus*; endangered), which became locally extinct in the early 1980s (Fanshawe *et al.* 1997), recolonized naturally in 2000 and are already well established (Woodroffe 2003). With the exception of wild dogs, all predators can be killed legally if they take livestock (though people who kill big cats are required to report this, and to submit the skins to the Kenya Wildlife Service). Sport hunting is not permitted in Kenya. Predators are killed if they become chronic livestock raiders, but tolerance among commercial ranchers is high; predators are not shot on sight. Although a few ranchers will eliminate a stock-killing lion after the first incident, most ignore low levels of depredation until an individual or group of predators becomes a chronic problem (Frank 1998; Ogada *et al.* 2003). Tourism is growing in importance, and as most tourists to Africa want to see large carnivores, in some areas there is a financial motivation to preserve predators despite livestock depredation (Western and Henry 1979).

METHODS

In 1997, we interviewed the owners or managers of 18 commercial ranches, three group ranches and a sample of eight individual Masai pastoralists on three group ranches (see Frank (1998) for questionnaire). Efforts were concentrated on the commercial ranches because it was apparent that there were very few lions in the communal areas. Respondents were asked over 800 questions about their land, costs and management of their livestock operations, numbers and trends of predators, economic impact of depredation on livestock by predators, how they dealt with predator problems, and what changes in predator populations they would like to see. Two hundred and eighteen pastoralists were interviewed in 46 small groups in 2001 about recent sightings of predators, predators' impact on livestock, and attitudes toward predators (Woodroffe 2001b).

Fieldwork on predator biology commenced in 1998. We have captured and released 103 lions, of which 37 males and 39 females were radio-collared. Collars are monitored weekly from the air and opportunistically from the ground, especially following depredation incidents. We have examined 52 lions shot or poisoned after killing livestock.

Fieldwork also involves daily interactions with ranchers and pastoralists, usually in informal settings, but also in organized gatherings at group ranches or in meetings of the Laikipia Wildlife Forum, which include both ranchers and pastoralists. Formal interviews are inappropriate in these contexts, but these discussions are frank and forthright; we have confidence in our understanding of the varying attitudes toward predators. Moreover, attitudes appear to be relatively homogeneous within both groups of livestock producers.

PATTERNS OF LIVESTOCK DEPREDATION

The different species of predators take livestock in distinctive ways; this influences the methods that can be used to prevent losses. Lions, leopards and hyaenas will take livestock from bomas at night (Fig. 18.1); 72% of cattle kills occur in this way (Frank 1998). Both lions and leopards can enter bomas by leaping over the wall, whereas hyaenas crawl through or under

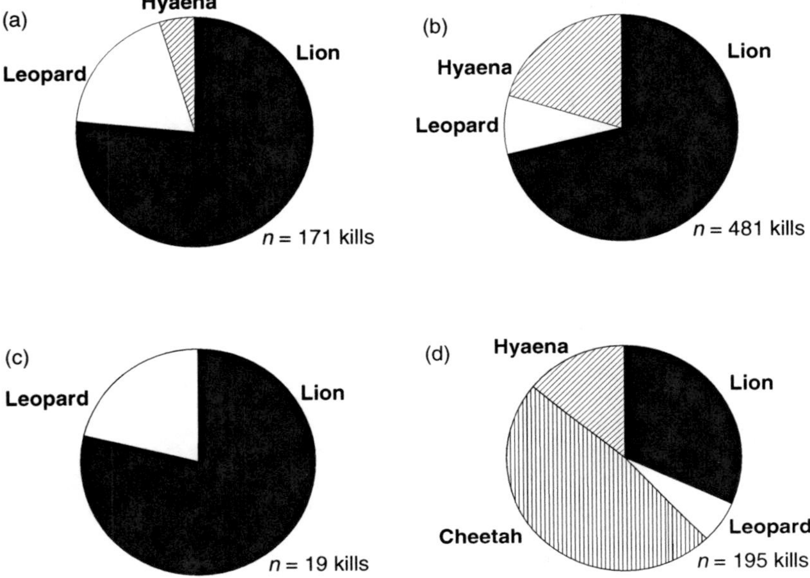

Figure 18.1. Variation in depredation frequency by four predator species in 1999–2000, largely from commercial ranches: proportions of recorded livestock kills attributed to lions, leopards, cheetahs and hyaenas for (a) cattle and (b) sheep and goats by night at bomas (night-time corrals) and (c) cattle and (d) sheep and goats by day away from the boma. Note that these data were gathered before wild dogs recolonized the study area. (Modified from Ogada *et al.* 2003.)

the boma wall. Predators confined in a boma with livestock that cannot escape often kill multiple animals – this helps to explain why attacks on livestock in bomas typically lead to the deaths of more individual animals than when herds are attacked while out grazing (Table 18.1). Cattle panic and break out when lions approach bomas at night. Multiple animals are often killed in such circumstances, and livestock from the scattered herd may also be killed by hyaenas once they are out of the boma. Sixteen per cent of cattle losses to lions involve stray animals inadvertently left outside the boma at night, and only 12% are killed by day (Frank 1998). By contrast, cheetahs and wild dogs are diurnal, and take small stock from herds grazing by day (Woodroffe *et al.* 2005) (Fig. 18.1). Despite the fact that they kill grazing livestock, wild dogs often kill multiple animals when they attack (Table 18.1), presumably because of the high concentration of vulnerable prey.

Table 18.1. *Mean number of livestock killed per attack, when livestock are taken from bomas, at night, and from herds out grazing, by day*

		Mean number killed per attack (range)		
Predator	Livestock	From boma	While grazing	Difference[a]
Lion	Cattle	1.38 (1–6)	1.0	$p = 0.047$
	Sheep and goats	2.06 (1–18)	1.35 (1–5)	$p = 0.087$
Leopard	Cattle	1.07 (1–3)	1.0	$p = 0.71$
	Sheep and goats	2.1 (1–15)	1.5 (1–3)	$p = 0.72$
Cheetah	Sheep and goats	–	1.2 (1–5)	–
Hyaena	Cattle	1.14 (1–2)	–	–
	Sheep and goats	1.43 (1–6)	1.08 (1–2)	$p = 0.029$
Wild dog	Sheep and goats	–	5.9 (1–19)	–

[a] Results of Mann–Whitney *U*-tests.

IMPACTS OF PREDATORS ON LOCAL PEOPLE'S LIVELIHOODS

Each year, carnivores kill approximately 0.8% of cattle and 2.1% of sheep on commercial ranches, and 0.7% of cattle and 1.4% of sheep and goats on pastoralist group ranches (Frank 1998). The slightly lower loss rates in community areas probably reflect both the higher numbers of livestock and lower numbers of predators on these lands. The impact of predation is fairly small in comparison with that of disease (Fig. 18.2); nevertheless, losses to predation are serious, and may have an important impact on the livelihoods of individual pastoralists, and on farm incomes of commercial ranches.

The cost of maintaining stock in the presence of large carnivores can be assessed in two ways. Clearly, the value of livestock killed by predators contributes to overall losses. In addition, costs may also include the infrastructure and staff time required to reduce depredation (e.g. through building bomas and herding livestock). However, while commercial ranches spend considerable money and effort on their herding and security staff, those workers are also required for general husbandry and protection against livestock theft; managers of commercial ranches argued that a total lack of predators would reduce staffing and infrastructure requirements by only 3% (Frank 1998). Hence, the actual cost of maintaining predators reduces essentially to the value of livestock production lost to predators.

Costs are most easily assessed on commercial ranches, which keep systematic records of livestock losses; few group ranches keep records. Since compensation for livestock or crop losses was abandoned due to corruption

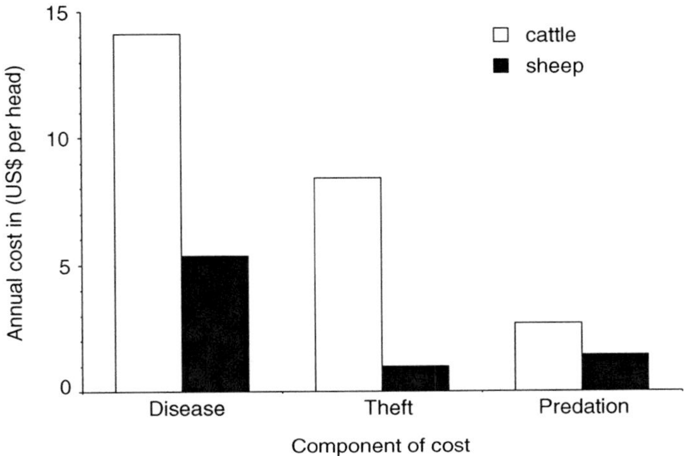

Figure 18.2. Cost of livestock depredation on commercial ranches, in comparison with the costs of losses (and measures to prevent losses) to theft and disease. Costs are calculated in US$, per head of livestock.

(Western and Waithaka, Chapter 22), ranchers have little motivation to misrepresent their losses (indeed, Woodroffe *et al.* (unpubl. data) confirmed the reliability of farmer reports by showing that the number of reported depredation events was closely correlated with the proportion of wild dog scats containing livestock remains). Fig. 18.2 shows the cost of livestock depredation in comparison with the costs of losses (and measures taken to prevent losses) to disease and theft. On commercial ranches, costs of depredation in terms of lost stock amount to about 6% of the cost of raising cattle, 10% of the costs of raising sheep and 11% of the cost of raising camels in Laikipia (Frank 1998). These costs can be converted into approximate costs per predator by comparing them with rough estimates of the number of predators occupying each ranch. As shown in Fig. 18.3, lions are the most costly animals to maintain, costing about US$360 per lion per year, the approximate value of 1 cow, or 9.3 sheep (Frank 1998). Hyaenas, by contrast, are the cheapest to maintain, at about US$35 (about 0.1 cow, or 0.9 sheep) per hyaena per year (Frank 1998). The estimated annual cost of supporting each leopard (US$211) compares well with Mizutani's estimate of US$190 per leopard per year on Lolldaiga Hills ranch in Laikipia (Mizutani 1999).

One group ranch which has a lion population kept written records of livestock loss. Here, depredation losses amounted to US$40 per household per year. While small by Western standards, this amount represents 11% of the average per capita income in Kenya (World Bank 2003). Woodroffe *et al.*

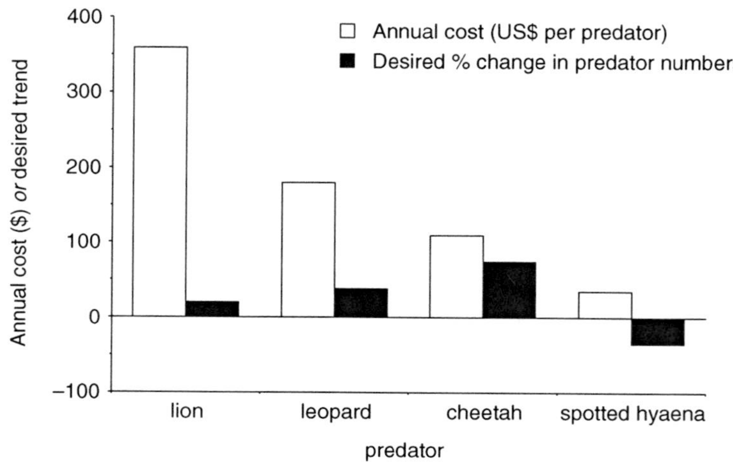

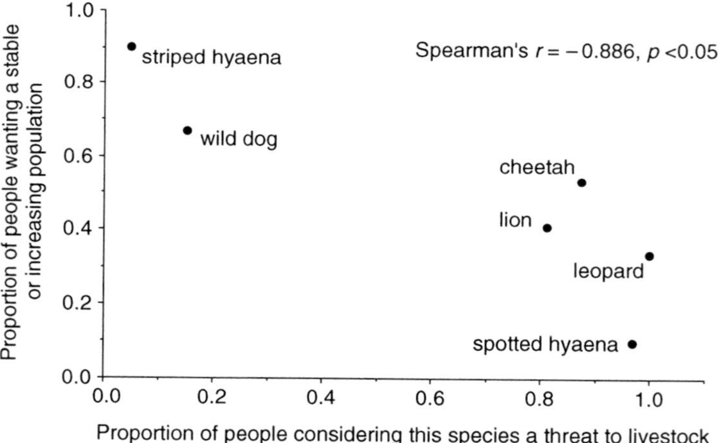

Figure 18.3. Comparison of local people's desired change in predator abundance with assessments of predator impact on (a) commercial ranches (Frank 1998) and (b) pastoralist communities (Woodroffe 2001b). Data from commercial ranches represent the annual cost of depredation by each predator species, calculated from ranch records; data from pastoralists give perceived impacts only.

(unpubl. data) estimated the impact of predation by wild dogs on pastoral lands at less than US$3 per wild dog per year where wild prey remained.

These average figures mask a high variance, and problems may be locally severe. For example, one commercial rancher lost 15 cattle (1% of his herd) and 82 sheep (1.4% of his herd) in a single month. On a recent night in 2004, one Laikipia ranch lost two cows to lions, two goats to hyaenas, two sheep to a cheetah and two sheep to a leopard (A. Mathieson pers. comm.). Wild dog predation is particularly uneven in its distribution. Woodroffe *et al.* (2005) estimated the average rate of wild dog predation as 0.48 attacks per 100 km^2 per year, but this rose to the equivalent of 310 attacks per 100 km^2 per year in the vicinity of the den of the one pack that chronically killed livestock.

Local people's tolerance for predators is not always closely related to the true impact that those predators have on their livelihoods. To assess attitudes, we asked managers/owners of commercial ranches and individual pastoralists if they would prefer to have more, fewer, or the same number of each predator (Frank 1998). Fig. 18.3a) compares the cost of maintaining four large predator species on commercial ranches with the change in predator abundance that ranchers said that they would wish to see over a five-year period. On average, ranchers wished to see population increases of all three big cat species, but wanted a 35% decline in hyaenas, even though they caused the least damage (Frank 1998). Local pastoralists showed a stronger antipathy toward hyaenas, unanimously wishing that there were none. They were also reluctant to tolerate leopards, which they perceive to cause equivalent damage (Fig. 18.3b). Interestingly, however, pastoralists' stated attitudes to predators were greatly improved where they received, or were expecting to receive, income from ecotourism (Fig. 18.4), even though such people were equally likely to experience losses to predators (Woodroffe 2001b). Pastoralists' desire to augment local populations of particular predators was influenced by their perceptions of what foreign tourists would wish to see: hence they were particularly keen to see increases in big cats, but tended not to want spotted hyaenas to increase in number because they did not expect that tourists would wish to see them (Woodroffe 2001b).

IMPACTS OF LOCAL PEOPLE ON PREDATORS

Local people in Laikipia have an unbroken history of coexisting with large carnivores, and compared to livestock producers in most of the world, are remarkably tolerant of them. Only a tiny minority of people would shoot a predator on sight, and two-thirds say that they tolerate some level of loss before attempting to kill offending predators (Frank 1998). When commercial ranchers decide to eliminate a problem animal, they usually track it from a livestock kill, or sit up by a carcass the following night, waiting for the predator (usually lions) to return. Of 27 lions shot in association with attacks on livestock on commercial ranches (including 14 radio-collared adults), we were able to confirm that 26 (96%) had been present at livestock kills as these were shot upon returning to kills, and most had livestock remains in

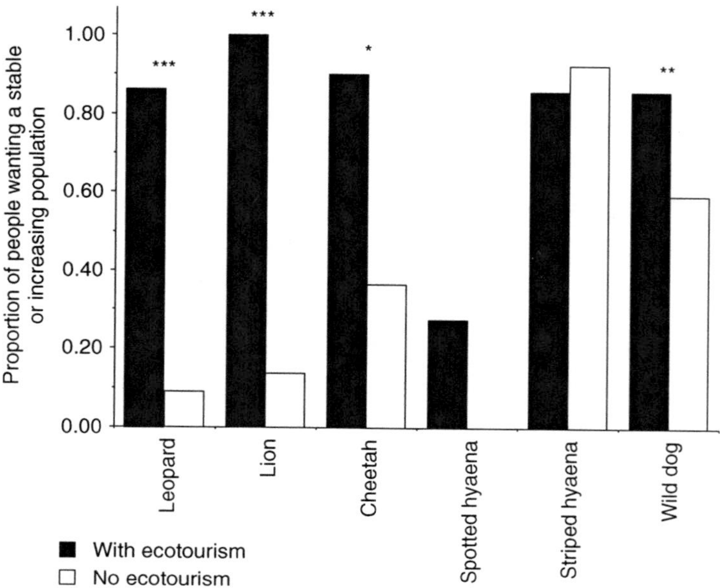

Figure 18.4. Relationships between ecotourism and pastoralists' desired trends in predator numbers. Data give the proportion of people interviewed who wished the predator species listed to increase or remain stable. Asterisks indicate statistically significant differences by χ^2 tests (* $p < 0.05$, ** $p < 0.005$, *** $p < 0.0001$).

their stomachs (Woodroffe and Frank 2005). The number of lions, leopards, cheetahs and hyaenas killed on each ranch was positively correlated with the number of livestock killed by those predators (Ogada *et al.* 2003). This shows that lethal control of carnivores is not indiscriminate, but carried out only in response to depredation and effectively targeted at the individual predators involved.

The situation is quite different on communal lands, however, in which poisoning appears to be on the increase. In the last two years, at least 17 lions, two leopards, and an unknown number of hyaenas and jackals have been poisoned in the communities of Laikipia. A similar trend is evident elsewhere in Kenya: a minimum of 49 lions and many other predators have been speared and poisoned in a 2900-km^2 complex of group ranches in southeast Kenya since 2002 (R. Bonham pers. comm.). Lions were abundant there five years ago, but have since become rare. There has been a surge of similar reports of poisoning and spearing elsewhere in southern Kenya and northern Tanzania (S. Dloniak pers. comm.; C. Packer pers. comm.), suggesting that the problem is rapidly increasing through much of Masailand. It is worth noting that, paradoxically, this may have benefited wild dogs, which are difficult to poison because they rarely scavenge, and appear to favour pastoral lands over commercial ranches where competitors are more abundant (Woodroffe 2001a).

Although lions on commercial ranches are killed only when they kill livestock, lethal control probably does limit their population size. Based on known numbers and home range sizes, lion density is estimated at approximately 5–6 /100 km^2 in the area of Laikipia that is wildlife habitat (L. G. Frank *et al.* unpubl. data), generally lower than that recorded in undisturbed habitats such as Masai Mara (29/100 km^2: Ogutu and Dublin 2002), Serengeti Plains (10/100 km^2: Hanby *et al.* 1995), Ngorongoro Crater (40/100 km^2: Hanby *et al.* 1995) and Kruger (6.5/100 km^2: Mills and Biggs 1993), or even in a sport-hunted population (Selous, 13/100 km^2: Creel and Creel 1997). Lions in Laikipia are well fed and in excellent physical condition: only two of the approximately 140 lions examined has been starving, and these were old, lone individuals (one male, one female) with very worn teeth. We have never seen cub starvation. Of 18 radio-collared lions that died in the course of the study, only one died of natural causes (Woodroffe and Frank 2005). The annual mortality of radio-collared adults, at 19.4%, was substantially higher than that recorded in undisturbed populations such as Serengeti (7–10%: Packer *et al.* 1988), Etosha (3–10%: Orford *et al.* 1988) or the Okavango Delta (5%: Winterbach and Winterbach 2002). A simple model of the Laikipia study population projects an annual decline of approximately 4% (range 12% decline – 3% increase), primarily because of unsustainably high adult mortality (Woodroffe and Frank 2005).

Demographic analysis highlights very strong selection against lions that kill livestock. Sixteen lions that were originally radio-collared returning to livestock kills experienced annual mortality almost four times as high as that of 42 lions collared under other circumstances (49.0%, compared with 12.9%; $\chi^2 = 12.85$, df = 1, $p = 0.0003$). Since farmers had no way of distinguishing radio-collared lions marked under different circumstances, this suggests that some lions were habitual stock-killers; lions originally collared on a livestock kill tend to keep killing stock until they are eliminated (L. G. Frank unpubl. data). As well as experiencing elevated mortality, females originally collared after killing livestock tended to produce fewer cubs than did females with no known history of stock killing at the time of capture (0.231 cubs/female/year, compared with 0.981; $\chi^2 = 4.75$, df = 1, p=0.029), and those cubs were less likely to survive (17% survival to 30 months, compared with 75%; $\chi^2 = 4.75$, df = 1, $p = 0.029$). Hence, while the portion of the lion population not collared as stock-killers (and which rarely killed livestock) was projected to increase at approximately 6% annually, the subpopulation of stock-killers was projected to decline by 46% each year

(Woodroffe and Frank 2005). However, ready availability of livestock, especially on ranches with poorer anti-predator measures, ensures that new lions learn to take stock. If, as seems likely, young lions learn stock-killing behaviour from their mothers, this strong selection may help to explain why most lions in fact kill livestock comparatively rarely.

We have insufficient data to assess the impact of lethal control on populations of leopards, cheetahs and hyaenas. However, spotted hyaenas are virtually absent from some parts of the study area. As hyaenas are slow to recolonize areas where they have been eradicated (Smuts 1978), this almost certainly reflects historic control. The wild dog population in Laikipia is still expanding following natural recolonization in 2000. To date, only two of seven radio-collared wild dogs that have died were killed by people, although several cases of deliberate disturbances of wild dog dens have been recorded (R. Woodroffe unpubl. data).

Human activities also influence predator populations less directly, through their impact on habitat suitability. Ninety-seven percent of 2735 aerial locations of 71 radio-collared lions fell on commercial ranches (Fig. 18.5). The majority of locations from adjoining communal lands were virtually on the boundaries, within the margin of error of the location. This

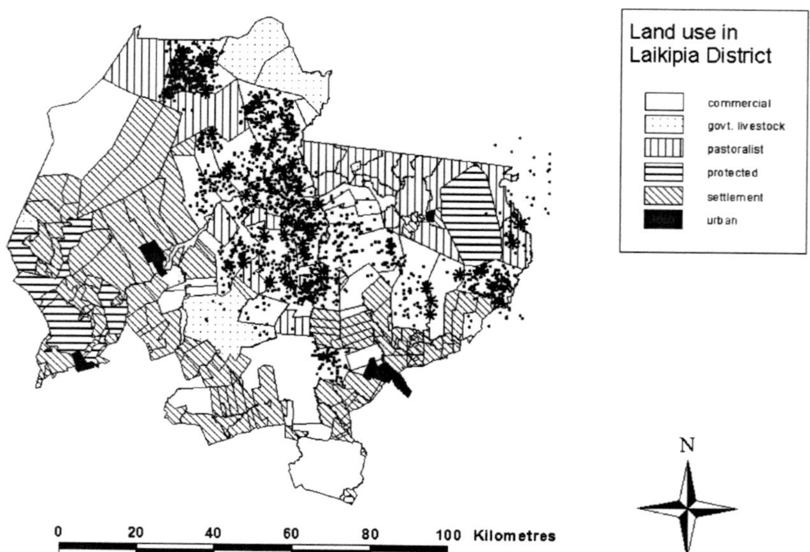

Figure 18.5. Movements of 71 radio-collared lions in relation to patterns of land use, 1998–2004. Black squares indicate locations of lions ($n = 33$ males (1026 fixes) and 38 females (1709 fixes)); stars indicate original capture sites. One female collared in eastern Laikipia moved widely in neighbouring Samburu District.

does not result from the fact that most lions were collared on commerical ranches. Most commerical ranches adjoin communal lands, and lion home ranges are much larger than the ranch sizes (L. G. Frank unpubl. data). The average lion home range encompasses four to six commerical ranches; they move freely between commerical ranches but rarely stray over the boundaries onto communal lands. This suggests that lions strongly prefer the commercial ranches, where human and livestock densities are low, and wild prey densities are relatively high, over adjoining community lands. The low densities of lions on community lands probably reflects this behavioural choice by lions, rather than high levels of lethal control by pastoralists: in 1998–2002 only one collared lion was known to have been poisoned on community lands (but this pattern changed drastically in 2003), whereas 14 were shot on commercial ranches. Interestingly, wild dogs show no such preference for commercial ranches over pastoralist areas (Woodroffe 2003).

ECOLOGICAL CAUSES UNDERLYING THE PROBLEM

As mentioned above, with few exceptions, in Laikipia people kill predators only when predators kill livestock. But predation on livestock depends on a variety of factors including ecological conditions, livestock husbandry and individual behaviour.

During 1999–2000, Laikipia experienced a severe drought. Throughout this period, livestock depredation was minimal; lion mortality was comparatively low (R. Woodroffe and L. G. Frank unpubl. data), and wild dogs to the north presumably fared well enough to produce a crop of dispersers that rapidly recolonized the study area (Woodroffe 2001a). However, at the end of the drought, lion predation on livestock increased markedly, and, as a consequence, lion mortality also rose (Woodroffe and Frank 2005). Saberwal *et al.* (1994) described a similar phenomenon, in which lions around the Gir Forest, in India, increased predation on people following periods of drought. Preliminary analyses suggest that, in Laikipia, drought affects livestock depredation through its impact on the availability of wild prey. Dead and dying wildlife and livestock were abundant during the drought, and predators were presumably well fed. When the rains came and forage recovered, ungulates regained condition and probably became more difficult to capture, leading more lions to turn to stock. Similarly, Laikipia Masai state that hyaena depredation rises with the onset of the rainy seasons (L. G. Frank unpubl. data). Comparison of long-term records of depredation rates on commercial ranches (R. Woodroffe unpubl. data) with data from regular aerial censuses of ungulate prey (Georgiadis and Ojwang'

2001) suggest that losses to lions are more severe where wild prey are scarce. Hemson and Macdonald (2002) described a similar pattern of livestock depredation by lions in dry land areas of Botswana: losses were most severe when prey had migrated away, and least serious when wild prey returned.

Livestock depredation by wild dogs also appears to be influenced by the abundance of wild prey. Woodroffe *et al.* (2005) showed that wild dog predation on livestock occurred almost entirely in areas where wild prey had been very seriously depleted. Interestingly, the threshold prey density to avert wild dog depredation was very low; attacks on livestock were extremely uncommon in pastoralist areas, but severe in a neighbouring area occupied by Pokot people who traditionally hunt wild dogs' natural prey.

TECHNICAL SOLUTIONS TO RESOLVING THE CONFLICT

Effective technology for minimizing livestock depredation has been used in eastern Africa for many centuries. In an 18-months study on nine commercial ranches and one community area, we looked at the efficacy of local livestock management practices, all of which are based on traditional Masai techniques (Ogada *et al.* 2003). This showed that traditional husbandry is a powerful tool for reducing depredation on herds, both at night and by day.

As described above, the vast majority of livestock in Laikipia are confined to bomas at night. The construction of such bomas varies: traditionally, they are built from *Acacia* brush, but enclosures may also be built from stones, wooden posts, or woven branches. Rearing of merino sheep for wool production demands bomas that can be moved every few days (*Acacia* thorns damage the wool, and accumulation of faeces in stationary bomas causes disease problems: G. Powys pers. comm.); hence, these sheep were kept overnight in small, portable bomas made from wire mesh that can be rolled up and moved. Comparison of livestock loss rates experienced at different bomas showed that wire enclosures provided the least protection from predators, with up to five times the depredation rate seen at more traditional bomas (Ogada *et al.* 2003). Recent replacement of rolled mesh with bomas made of portable, inflexible reinforced mesh panels have nearly eliminated losses to predators (G. Powys pers. comm.). Surprisingly, we could detect no effect of the thickness, height or complexity of boma walls on the rate of livestock loss (Ogada *et al.* 2003).

While boma construction did influence predation risk, the level of human activity around the boma had a stronger effect (Fig. 18.6). Lions, leopards and hyaenas were all markedly less likely to attack bomas where large numbers of people were routinely present (Ogada *et al.* 2003). Domestic dogs are also kept at some bomas as a deterrent to both predators and cattle thieves. These dogs are not trained as guards and do not chase

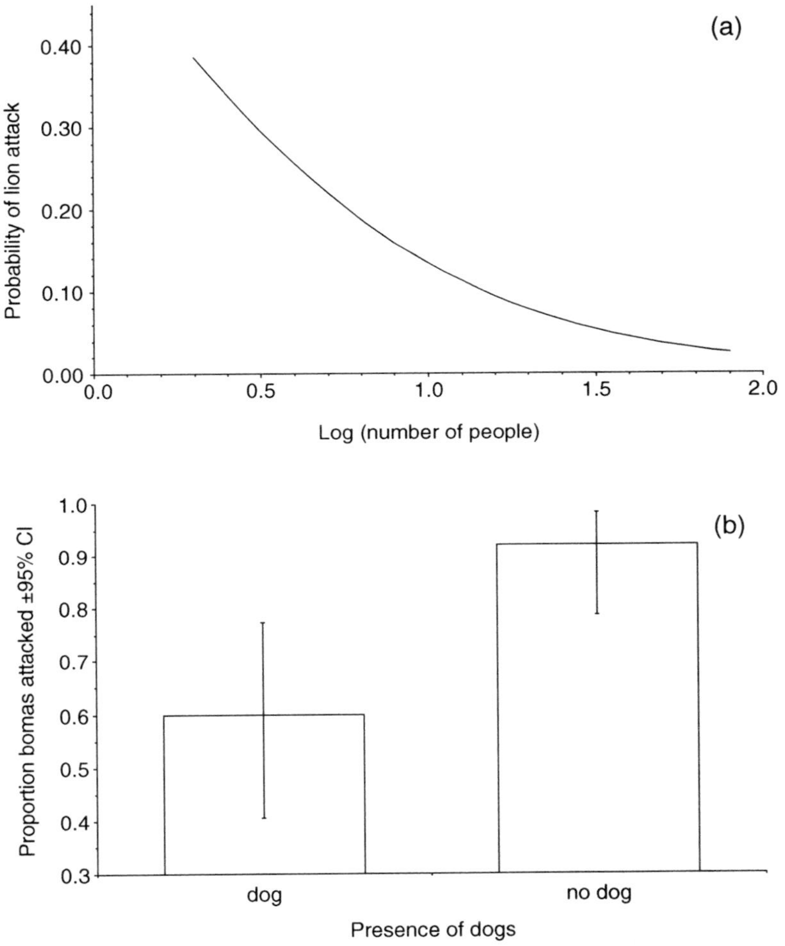

Figure 18.6. Characteristics of bomas (night-time corrals) associated with probabilities of any attack by lions, based on multivariate logistic regression. (a) Effect of log (people); and (b) effect of domestic dogs' presence. Overall $r^2 = 0.28$; effect of log (people) $\chi^2 = 8.36$, $p < 0.005$; effect of dogs' presence $\chi^2 = 8.22$, $p < 0.005$.

predators; rather, they serve to alert people of the presence of predators. Dogs are an effective deterrent against lion attack (Fig. 18.6), although they are less effective in discouraging other predators on commercial ranches. However, pastoralists universally agree that dogs are very effective at warning of hyaena incursions into small stock bomas.

The severity of livestock depredation on herds out grazing by day is also influenced by husbandry practices. Both lion and cheetah attacks were less severe when the number of herders per sheep or goat was higher (Ogada *et al.* 2003). However, there was no such effect for either hyaenas or leopards (and wild dogs had not recolonized the study area when these data were gathered).

ECONOMIC SOLUTIONS TO RESOLVING THE CONFLICT

In today's Kenya, tourism is seen as the only potentially profitable way to recoup the costs of depredation. Compensation was abandoned by the government due to poor infrastructure and corruption (Western and Waithaka, Chapter 22), and sport hunting was banned in 1977. Tourism plays a significant – though not decisive – role in the recovery of wildlife in Laikipia. Several of the commercial ranches gain significant income from tourism, and a number operate smaller ventures such as campsites or camel safaris. However, an equal number of ranches encourage wildlife and tolerate predators out of a conservation ethic not driven by economic considerations. Some of these ranches are owned by wealthy foreigners, but this is by no means the rule. Most encouragingly, several group ranches are developing their own tourism operations. They have built lodges, campsites and 'cultural manyattas' (traditional villages for paying tourists to visit), and have even set aside land for wildlife from which livestock are excluded. Thus far, these efforts have required heavy subsidies from commercial ranches and non-governmental organizations, and it will be some time before local people gain the expertise to raise capital and organize their own ventures. However, the initial successes have stimulated widespread interest and expectations throughout the pastoralist communities of Laikipia, and more ventures are planned.

Tourism, however, is a fragile business; the 2003 terrorism events in Kenya and subsequent warnings by the US and UK governments proved disastrous to tourism (Wallis 2003; see also Walpole and Thouless, Chapter 8). An alternative source of income – not currently an option in Kenya – might come from trophy hunting (Leader-Williams and Hutton, Chapter 9, Lewis and Jackson, Chapter 15). Tanzania has designated 195 000 km^2 for hunting concessions (Leader-Williams, 1993). In 1990, government revenues from hunting licenses alone amounted to some US$4 500 000, compared to $1 900 000 earned from the national parks system (Makombe 1994). Of course, license fees are a small part of what a hunter pays for a safari: in 2004, a one-month safari that includes lion costs well over US$100 000 in Tanzania (Safari Consultants 2004). Note, however, that non-consumptive tourism has flourished since the border with Kenya was reopened in 1986, and by 1998 contributed $570 million

(16% of GDP) to Tanzania (Thirgood *et al.* in press); accurate figures for the current value of consumptive tourism are not available. Also in the early 1990s, the CAMPFIRE programme in Zimbabwe earned $4 000 000 for participating communities from sport hunters, representing about $400 per household (Edwards and Allen 1992; see also Leader Williams and Hutton, Chapter 9, for more details). Hunters pay US$30 000–45 000 for the opportunity to take aged wild male lions and $12 000 to take females from reserves in South Africa (J. Anderson pers. comm.; C. Vermaak pers. comm.). Moreover, trophy hunters do not require scenery to compete with Serengeti or the Okavango, and are often less concerned by politics than are tourists: safari hunting has withstood the political crisis in Zimbabwe, while tourism is essentially extinct (Grobbelaar 2004).

Many Laikipia residents, both ranchers and Masai, have expressed frustration at their inability to offset predator losses through some form of sport hunting (Frank 1998). Every year, about 30 problem lions are killed and left to rot. Given the value of lions in South Africa and Tanzania, these animals might be worth over one million dollars if they were taken by trophy hunters, many times the value of livestock taken annually by predators.

Trophy hunting of predators may prove difficult to administer. Although appealing in theory, the use of sport hunting as a conservation tool is far from simple, especially in the African context (Lewis and Jackson, Chapter 15). To be sustainable, hunting must be carefully monitored and regulated, and proceeds must be distributed in a transparent manner such that all community members benefit. This is difficult to achieve, especially where corruption is widespread. Trophy hunting in South Africa has been disgraced by the widespread 'canned' shooting of captive-reared lions in small enclosures. Traditionally, however, hunters in Kenya policed themselves effectively through the East African Professional Hunters Association, possibly a useful template for the future (Dyer 1996; Herne 2001; Parker and Bleazard 2001). Further, a recent model and accompanying recommendations by Whitman *et al.* (2004), may simplify regulation of trophy hunting by confining it to males that are recognizably above five to six years of age. Achieving sustainability might prove difficult if attempts were made to target trophy hunting at specific 'problem animals' outside the accepted trophy category: bogus claims of depredation incidents would certainly occur, as would arguments between neighbouring communities concerning who should receive proceeds. Since Laikipia's lion population is currently small and largely confined to the commercial ranches, while the most urgent demand for income from wildlife comes from community lands, it would seem prudent to 'test the waters' of trophy hunting by starting with species such as kudu (*Tragelaphus strepsiceros*), zebra (*Equus burchelli*) and leopards, which have fared much better in pastoral areas. This might encourage recovery of both wild prey and lion populations in community lands, eventually allowing lion hunting to be considered in those areas.

CONCLUSIONS: CAN AFRICAN PREDATORS PERSIST OUTSIDE PARKS?

Our results show that conflicts between people and predators are successfully mitigated on those properties that put effort into careful livestock husbandry. Our results also show that traditional methods, perhaps supplemented by modern modifications, can significantly reduce conflict. Even if predators were eliminated from the rangelands of Kenya, some variant of these methods will be necessary as long as traditional cultures maintain cattle raiding as a way of life.

Until the last hundred years, Africans had coexisted with predators out of necessity, because they lacked the technology to eliminate large carnivores. The Laikipia experience shows that modern people can also coexist with predators, if they are willing to make efforts to protect their livestock and tolerate some losses. However, that coexistence is labour-intensive; since herdsmen and farm workers in Kenya are paid only about $30/month, labour costs are low, but they would be prohibitively high in a Western country. Even in parts of Africa, economic expectations have risen to the point that people are no longer willing to tolerate the difficult and uncomfortable life of a pastoralist. In relatively affluent Botswana, for instance, younger people refuse to become herders. As a result, cattle are allowed to wander in the bush untended and are taken by lions. Lion-killing has increased, causing the population to decline (Hemson and Macdonald 2002).

However, there is more to coexistence than economics. Even with the best husbandry, depredation is still a significant economic force throughout the district. Large carnivores persist in Laikipia because the people – of all races – tolerate a level of uncompensated livestock loss that would be unacceptable in the West. For example, in spite of assured compensation, the livestock industry in the Northern Rockies of the USA vigorously resisted the 1995 reintroduction of wolves to Yellowstone National Park (Halfpenny 2003). Between 1996 and 2003, in the Yellowstone ecosystem outside the National Park, Defenders of Wildlife has compensated ranchers for 103 cattle and 568 sheep, an average of 12.9 cattle and 71 sheep per year in an area of 6 790 000 ha (Defenders of Wildlife 2004). By contrast, in 1995 alone, 19 ranches in Laikipia, covering 283 000 ha (4% the size of the

Yellowstone area), lost 202 cattle and 993 sheep to lions, leopards and hyaenas, or a mean of 10.6 cattle and 52.3 sheep per ranch (Frank, 1998). Thus, the average ranch in Laikipia loses nearly as much stock each year to predators as all the ranches in the entire Yellowstone ecosystem lose to wolves, yet few inhabitants of Laikipia share Yellowstone ranchers' vehement opposition to predators, and retributive killing is remarkably uncommon, given predators' impacts.

Many factors no doubt influence the difference in attitudes between the two areas. Ranches that host tourism or are subsidized by wealthy owners are clearly more able to bear the costs of coexisting with wildlife than those that are wholly dependent on livestock income. However, many Laikipia ranchers do not have these advantages, and still encourage robust wildlife populations, including predators. Out of a love of the land, the wildlife and their way of life, these people have stayed in their native country under often difficult political and economic circumstances. Of course, the same can be said of their American counterparts. Moreover, the fathers of current Laikipia ranchers had little more tolerance for depredation losses than do American ranchers today: in 1908 alone, 150 lions were shot on licence in Laikipia District (Playne 1909). However, on Kenyan ranches there has been a sea change in attitudes that is not readily explained by either economics or experience. The difference may be historical. America was settled by pioneers with a strong sense that they were conquering a hostile Nature, in which predators and Native Americans epitomized the forces that had to be overcome in order to civilize the land (Quammen 2003). East Africa was settled in large part by adventurers and big-game hunters who were attracted rather than repelled by wildness (Trzebinski 1988; Herne 2001). Although they had to eliminate wildlife where they created farms and ranches, there were always vast expanses of wilderness into which they safaried to hunt. These people valued wild land, wild animals and traditional peoples in a way that few American pioneers did. The Kenyan settlers may have left a psychological legacy to their descendants that American settlers did not. Further, the loss of Kenyan wildlife has been so fast that each human generation has seen dramatic losses, and may thus be more strongly motivated to reverse that process.

However, as in most of the world, the majority of people must be financially motivated if they are to preserve wildlife. As long as livestock production, either through pastoralism or ranching, remains the primary use of semi-arid African rangeland, some combination of ecotourism and soundly managed sport hunting are probably the only solutions to preserving wildlife on an ecologically meaningful scale.

ACKNOWLEDGEMENTS

We are grateful to the people of Laikipia District who have so enthusiastically shared their information and experience, allowed us to work on their properties and made us welcome in their homes. The work has been sponsored by the Wildlife Conservation Society, the African Wildlife Foundation, Stichting EAWLS Nederland, the National Geographic Society, the US National Cancer Institute, Busch Gardens, the Potreronuevo Fund and the Columbus Zoo.

19

Searching for the coexistence recipe: a case study of conflicts between people and tigers in the Russian Far East

DALE MIQUELLE, IGOR NIKOLAEV, JOHN GOODRICH, BORIS LITVINOV, EVGENY SMIRNOV AND EVGENY SUVOROV

INTRODUCTION

Large carnivores provide the ultimate test of society's willingness to conserve wildlife. They present a unique conservation challenge because first, large carnivores generally require large tracts of land, and second, they can and do kill people and domestic animals. Governments throughout the world are creating protected areas, suggesting that society seems willing to apportion some land for conservation, but whether it is willing to dedicate sufficiently large tracts, and whether it is willing to accept the risk of living in close proximity to large carnivores, are questions yet to be answered. Because human-induced mortality is one of the greatest threats to persistence of carnivore populations worldwide (Woodroffe and Ginsberg 1998), resolving human–carnivore conflicts is key to their survival. Whether a future exists for these most charismatic components of wild ecosystems will largely depend on networks of suitable habitat and intervention programmes that minimize risks to both carnivores and people.

In 1941 Kaplanov (1948) estimated that there were 20–30 Amur tigers (*Panthera tigris altaica*) remaining in the Russian Far East. Harvest of tigers was outlawed in Russia in 1947, and collection of cubs for the world's zoos was sharply curtailed by 1957. Thereafter a slow but apparently steady growth in tiger numbers led to what many believe was a peak population of as many as 600 tigers at the end of the 1980s (Kucherenko 2001). A sharp increase in poaching in the first half of the 1990s (Galster and Vaud Eliot 1999) probably rapidly depressed tiger numbers. In 1996 there were an estimated 330–371 adult (and approximately 100 young) Amur tigers distributed across 156 000 km^2 of habitat in the Russian Far East (Matyushkin *et al.* 1996, 1999).

Presently, only 7% of the remaining tiger habitat is protected as *zapovedniks* (IUCN Category I) or *zakazniks* (IUCN Category IV), and even

People and Wildlife: Conflict or Coexistence? eds. Rosie Woodroffe, Simon Thirgood and Alan Rabinowitz. Published by Cambridge University Press.

best-case scenarios suggest no more than 16% of tiger habitat will come under a protective regime (Miquelle *et al.* 1999a). A conservation strategy dependent solely on protected areas would result in small fragmented subpopulations of tigers scattered across the landscape, a scenario with high extinction risk. A more viable alternative relies on a protected areas network interlaced with multiple-use forest lands (some 84–93% of tiger habitat today), shared between tigers and the approximately 4 million people who live in the region (Miquelle *et al.* 1999a). Survival of Amur tigers will therefore depend largely on whether local people tolerate their presence. Mitigating conflicts, reducing risk to both tigers and people and increasing tolerance of people living with tigers will be fundamental to a successful conservation effort (Miquelle and Smirnov 1999).

In this case study of tigers and people in the Russian Far East, we identify six motives, or situations, that result in human-caused tiger mortality, based on reviews of existing information, as well as new analysis of data from the past 50 years. We focus this discussion on those four motives that represent direct conflicts of interest between tigers and local people. We do not consider those impacts that indirectly influence survivorship of tigers (e.g. habitat loss, road construction, development) or welfare of people (e.g. development restrictions), which are no less important, but require a separate suite of conservation actions beyond the scope of this chapter (Miquelle *et al.* 1999a; Kerley *et al.* 2002). We describe the context in which direct conflicts occur, impacts of each on both tigers and people, and mitigation actions that have been taken to resolve these conflicts. Where possible, we assess effectiveness of these actions.

STUDY AREA

The distribution of Amur tigers in Russia is restricted to Primorski Krai (Province), a region of 165 900 km^2, and the southern portion of Khabarovski Krai (100 450 km^2) (Fig. 19.1). Due to human development elsewhere, approximately 95% of the Amur tigers in Russia remain in the Sikhote-Alin Mountains, a coastal range that parallels the Sea of Japan, from Vladivostok 1000 km north. The remaining 5% of tigers occur in the East Manchurian Mountains in southwest Primorski Krai (Matyushkin *et al.* 1996), and adjacent territories of Jilin and Heilongjiang Provinces, China (Miquelle and Pikunov 2003). These regions also support brown bears (*Ursus arctos*), Himalayan black bears (*U. thibetanus*) and wolves (*Canis lupus*), and ungulate species such as red deer (*Cervus elaphus*), roe deer (*Capreolus capreolus*), sika deer (*Cervus nippon*), musk deer (*Mochus moschiferus*) and wild boar (*Sus scrofa*) (Miquelle *et al.* 1996, 1999b). Tiger habitat in the

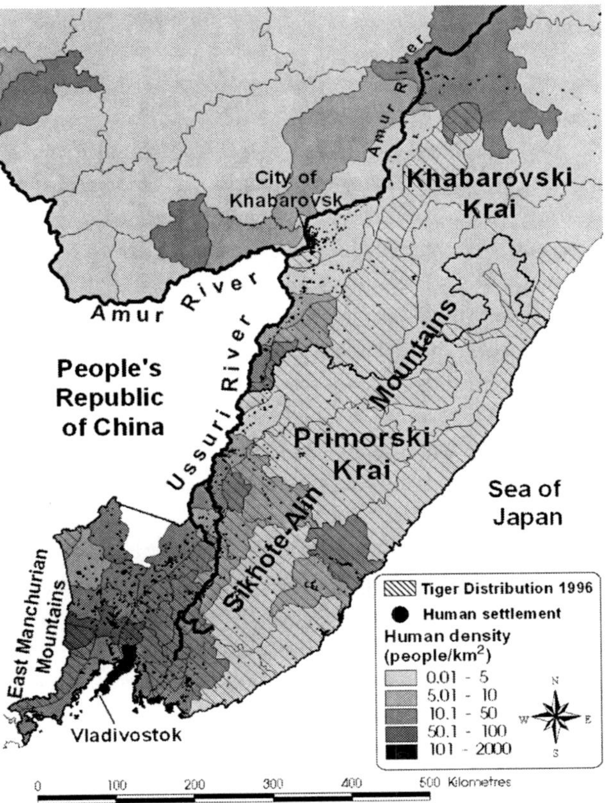

Figure 19.1. Major topographic features, human density (by *raion*, or county) and distribution of tigers in the Russian Far East. (From Matyushkin *et al.* 1996.)

Sikhote-Alin and East Manchurian Mountains is nearly completely forested. The mountains are relatively low (the highest peak is 2004 m), and are covered with a combination of conifer and broad-leaved species. Research on radio-collared tigers is conducted by the Siberian Tiger Project (Goodrich *et al.* 2001), which is centred in Sikhote-Alin State Biosphere Zapovednik, a large (4000-km^2) reserve situated in northeast Primorski Krai, straddling the Sikhote-Alin divide. Human settlements are concentrated around the capital cities of Vladivostok and Khabarovsk, and along the fertile lowlands associated with the Ussuri and Amur Rivers (Fig. 19.1). Nonetheless, small communities are dispersed across the entirety of tiger habitat. People in these small forest communities rely on the fish, wildlife, timber and other natural resources in tiger habitat to provide a means of subsistence and income.

METHODS

Information on tiger mortality comes from official records of permits issued for killing tigers (1985–2001), published data on tiger mortality (Gorokhov 1983; Nikolaev and Yudin 1993), additional unpublished data (collected by I. G. Nikolaev), and information on mortalities of 22 radio-collared animals (Goodrich *et al.* 2000; Goodrich *et al.* unpubl. data). Where more than one incentive for killing a tiger is reported, each death was proportionally allocated. Official records and published data are likely biased towards human-caused deaths because they rely on reports from local informants. Information from collared animals, while probably less biased, may be skewed in the opposite direction because research was centred in a protected area.

Occurrences of tiger attacks on humans were collated from long-term records of the Primorye Hunting Department (up to 1990) and Inspection Tiger (after 1990: Nikolaev and Yudin 1993), and were updated and verified by I. G. Nikolaev (unpubl. data). Attacks were defined as 'provoked' if the person shot at a tiger, if a tiger had been previously wounded by humans, or if a person intentionally or unintentionally approached very close to a tiger. Information on encounters between people and tigers was derived from a survey of local newspapers across Primorski and southern Khabarovski Krai for the period 1992–8 (E. Suvorov, unpubl. data), and from existing literature (e.g. Khramtsov 1995). Data on livestock depredation by tigers was derived from yearly reports for the region surrounding Sikhote-Alin Zapovednik (compiled by E. N. Smirnov), and in Khabarovsk on the basis of responses to a questionnaire (Sukhomirov 2002).

HUMAN-CAUSED MORTALITY OF AMUR TIGERS

Information summarized over the past 50 years (Table 19.1) indicates that human-caused mortality of Amur tigers in the Russian Far East can be categorized as:

(1) Poaching (defined here as intentional killing with intent to profit)
(2) Lethal control with official permit
(3) Self-defence in response to perceived or real threat (including killing of animals that have attacked people and legal and illegal lethal control of animals considered dangerous)
(4) Retaliation for depredation of livestock or other domestic animals
(5) Elimination of 'competitor' by hunters
(6) Accidental killings (mostly vehicle collisions).

Table 19.1. *Mortality factors for adult Amur tigers in the Russian Far East, 1951–2001*

	Percentage of total mortalities			
	Survey of local people[a]	Onsite examinations[b]		Radio-collared tigers[c]
Reason	1951–73 ($n = 49$)	1970–90 ($n = 56$)	1991–2001 ($n = 78$)	1992–2001 ($n = 22$)
Human-caused				
Legal lethal control[d]	(0)	(23.2)	(24.4)	(0)
Poaching[e]	0	0.0	57.7	72.7
Depredation	20	29.5	10.3	4.5
Self-defence	9	20.5	23.1	0.0
Competition for prey	71	1.8	0.0	0.0
Motive unidentified		30.4	1.3	0
Other human-caused[f]		1.8	2.6	4.5
Other causes				
Natural	0	14.3	0.0	18.2
Unknown	0	1.8	5.1	
Total human-caused mortalities	100	83.9	94.9	81.7
Total other causes			5.1	18.2

[a] Gorokhov (1983).
[b] Nikolaev and Yudin (1993) and recent material from Nikolaev, unpubl.; includes confiscated skins.
[c] Goodrich *et al.* unpubl. data.
[d] Permits for legal control (in parentheses) are usually a response to livestock depredations or a potentially dangerous tiger (self-defence). Because this category overlaps with others, it is not included as part of column totals, but only as a percent of total deaths reported.
[e] Includes radio-collared animals who death was categorized as 'suspected poaching'.
[f] Includes accidental killings (vehicular collisions).

Lethal control is initiated almost entirely in response to attacks on humans or livestock and since it overlaps with these two categories, is reported separately (Table 19.1). While accidental killings occur (for example, Nikolaev and Yudin (1993) reported three tigers killed by automobiles) they are rare, probably unimportant to tiger population dynamics, and are not considered further.

Although variation in reporting procedures exists, available data indicate two important trends. First, despite biases in data collection procedures that vary among studies, all evidence indicates that human-caused mortality is responsible for at least 80% of all tiger deaths (Table 19.1). Second, beginning in the 1990s, there was a dramatic shift in the reasons that humans killed tigers. Prior to 1990, tigers were killed for a combination of reasons; hunters killed tigers as competitors, and farmers for livestock depredation, and tigers were not uncommonly killed in self-defence. Based on a survey of local hunters, Gorokhov (1983) reported that the majority of tigers (71%) were killed because they were considered competitors for prey. Nikolaev and Yudin (1993) could not always determine motive for killing, but their examinations of tigers shot and left in the forest (often roadside killings) suggest that tigers were generally viewed as 'bad', probably because of some combination of the fact that they are dangerous and kill prey. Thus, results of Gorokhov (1983) and Nikolaev and Yudin (1993) are probably more similar than the data, as presented, suggest.

From 1972 to 1992 poaching for commercial gain was not reported (Table 19.1): borders were closed during the Soviet era and access to the Asian demand for tiger products was virtually non-existent. Dissolution of the Soviet Union brought an easing in border restrictions and gun laws, and a new, urgent need for village inhabitants to earn income in a collapsed economy with spiralling inflation. Almost instantly tigers turned into a valuable cash crop at a time when there was high demand for tiger parts for traditional Chinese medicines (Mills and Jackson 1994). Data from field examinations and skin confiscations (column 3 in Table 19.1) and from radio-collared animals support the conclusion that the vast majority (58–73%) of deaths were associated with poaching. Those motives that existed prior to the 1990s (self-defence, retaliation for depredations, and elimination of competitors) probably continued, but are masked by the additional commercial value of tiger parts in the black market.

Effect of human-caused mortality on the Amur tiger population

The tiger is considered by some to be a 'resilient' species capable of recovering rapidly from intensive human harvest (Sunquist *et al.* 1999). However, recent analyses suggest that at least some populations may not be so resilient. Smirnov and Miquelle (1999) indicated that recovery of the adult segment of a colonizing Amur tiger subpopulation to 10–15 animals took 20 years, with a modest growth rate of 6%. Reproductive parameters of Amur tigers are not dramatically different from other subspecies (Kerley *et al.* 2002), suggesting that growth rates of other populations may also be slow. Recent modelling suggests that the Amur tiger population can be seriously threatened with extinction if poaching rates exceed 10% of the total population (G. Chapron *et al.* unpubl. data), and available evidence suggests that poaching rates exceeded that level during the 1990s (World Wildlife Fund 2002).

Although poaching for commercial gain has become the primary reason for humans killing tigers over the past decade, this is not a result of 'conflict' between humans and tigers, is not a primary focus of this analysis, and probably masks other continuing conflicts. Since tigers are most often opportunistically shot (e.g. tigers are usually shot during an encounter in the forest, and are not specifically targeted), the incentive for killing probably represents not only the prospect of commercial gain, but a complex mixture of emotions by a poacher, including fear and a sense that tigers are 'bad' because they are dangerous and kill livestock (see below). The long-term data presented here (Table 19.1) suggest that even when commercial gain was not an option, humans were responsible for the majority of tiger deaths. The cumulative impact of these other motives may be critical in determining the fate of small tiger populations. Therefore, efforts to reduce incentives for human-caused tiger mortality other than poaching seem well justified.

LETHAL CONTROL

Hunting of tigers as a game species has been illegal since 1947 in Russia. Permission for lethal control is issued from the appropriate Ministry in Moscow only for animals considered a danger to human life or welfare. Despite the difficulties inherent in obtaining a permit from a governmental agency 8000 km and seven time-zones away, over the 17-year period beginning in 1985, 55 animals have been killed under permit. Nearly all cases are associated with livestock depredation or in defence of human welfare. Aside from the 1986 winter, when 15 tigers were legally killed as a result of severe winter conditions that forced tigers into settlements seeking domestic prey (Nikolaev 1985), yearly kill rates have been consistently low, averaging 2.3 animals per year (3.2 if 1986 is included).

Attempts to reduce lethal control

In 1999, Inspection Tiger created a special Tiger Response Team to address 'problem situations' between tigers and people. The goals of this team are: first, to reduce or eliminate threats and perceived threats caused by tigers to humans (see below), and second, to reduce tiger mortality associated with conflicts with people. While the first priority is to provide safety for local people, many actions are intended to increase survivorship of tigers. The Tiger Response Team was responsible for killing five 'problem tigers' over the four-year period of its existence, resulting in a death rate slightly less than the long-term average (1.75 versus 2.3). However, this analysis does not include the team's impact in reducing tiger mortality due to killing in retaliation and in self-defence (see below). By providing a mechanism by which local citizens can expect a rapid and official response to problem situations, there should be fewer incidents where local people resolve problems 'unofficially.'

SELF-DEFENCE FROM REAL/PERCEIVED THREAT TO HUMAN LIFE

Impact on humans

Of the large predators, tigers are considered the most consistently dangerous species (McDougal 1987; Sillero-Zubiri and Laurenson 2001). Although reports of man-eating Amur tigers were not uncommon across its range in the nineteenth century and the first part of the twentieth century (Prezhewalski 1870, 1923; Baikov 1925), there were no reports of man-eating tigers in the Russian Far East from the 1930s until 1976 (Abramov 1962; Zhivotchenko 1977). Over the most recent 32-year period (1970–2001) there were 51 official reported tiger attacks, with 14 people killed (Fig. 19.2). In eight instances a person escaped an attack uninjured. The probability of being killed by a tiger was identical (27%) for provoked and unprovoked attacks. This attack rate (1.4 attacks per year) and mortality rate (0.4 human deaths per year) is low in comparison to historical rates in Russia and elsewhere (McDougal 1987). In the Sundarbans, where man-killing tigers are most common, Hendrichs (1975) reported 24.3 deaths per year over a 15-year period. Adjusted for area (kills per 1000 km^2 tiger habitat per year) the kill rate in the Sundarbans (6 kills per 1000 km^2 per year) is two orders of magnitude greater than in the Russian Far East (0.01).

While elsewhere repeated attacks on humans by individual tigers are common (Corbett 1944; Hendrichs 1975; McDougal 1987), in the Russian

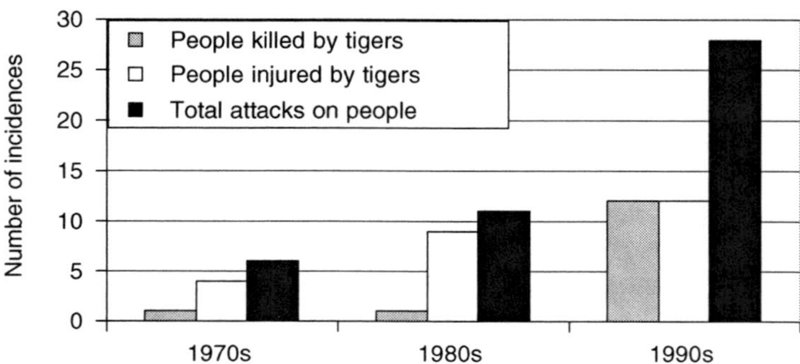

Figure 19.2. Number of attacks and people killed or injured by tigers in the Russian Far East, 1970 to 1999.

Far East there have been only two reported cases in the past 30 years of an individual tiger killing more than one person (in both cases two people were killed). Confirmed man-eaters appear to be rare in the Russian Far East because man-eating tigers are often quickly hunted down and killed and because, with low human densities, man-eaters (often sick or lame) have relatively few opportunities to kill people before succumbing to a natural death in severe winters.

The rate of encounters (as opposed to attacks) between tigers and people provides an indication of the potential and perceived risk to local people. A survey of local newspapers from 1992 to 1998 uncovered 397 articles (66 per year) reporting direct encounters with tigers. The majority of encounters (68%) occurred in tiger habitat (forested areas), but tigers do stray into human-dominated areas where chance of encounter is greater: 15% of encounters were along roads, 11% in transition zones between forest and human settlements (e.g. orchards, dumps, bee-keeping camps), and occasionally (6%) close to or within villages.

The largest percent of encounters (42%) occurred in winter (December–February), and autumn (September – November) (24%), when large numbers of hunters are in the forests. Cumulatively these data indicate that the majority of encounters are between human forest-users (primarily hunters and loggers) and tigers in natural tiger habitat, but excursions by tigers into settled areas do occur.

Low human density within tiger habitat (lower than any other tiger range country) in the Russian Far East is probably an important factor explaining the low attack rate (Fig. 19.1). Nonetheless, villages are scattered throughout the entire range of tigers in the Russian Far East, and the logging, hunting and trapping systems are structured to ensure that nearly all lands with potential yields of timber, game and furs are exploited. Therefore, even in areas with extremely low human densities, hunters and loggers will be scattered across the entirety of tiger habitat, with the result that encounter rates will be relatively high relative to the low densities of both humans and tigers.

Although attacks are rare, the perceived threat by local citizens is considerable, due partially to this high encounter rate. When asked to provide a reason against conserving tigers, respondents in two separate surveys most often cited danger to humans as the primary reason (Zabanova *et al.* 2001; Sukhomirov 2002). The rare but well-publicized appearances of tigers near villages reinforce the perception of danger.

The threat of tigers as perceived by local inhabitants has been reinforced by a significant increase in tiger attacks and human deaths in the early 1990s ($\chi^2 = 17.7$, df $= 2$, $p = 0.0001$) (Fig. 19.2), including all but two of 14 reported deaths since 1970. This increase is coincident with one ecological and two political/economic trends. First, tiger numbers increased consistently through the 1970s and 1980s, probably peaking in the late 1980s (Matyushkin *et al.* 1996; Kucherenko 2001). Political and economic turmoil which brought about the collapse of the Soviet Union in the early 1990s led to an increase in both gun ownership and illegal hunting as greater numbers of villagers entered forests to extract resources for subsistence, resulting in more encounters with tigers. Finally, poaching of tigers apparently reached a peak in the early 1990s (Galster and Vaud Eliot 1999; World Wildlife Fund 2002). The number of provoked attacks consequently increased (18 events in the 1990s compared to 11 from 1970 to 1989) as failed poaching attempts resulted in tiger attacks and human deaths. Thus, an increase in numbers of tigers, an increase in number of people in the forest, and an increase in poaching attempts collectively resulted in an increase in attacks and deaths of humans.

Impacts on tigers

Although some tigers are killed when entering villages, the majority are killed by people with guns in the forest. Fear of attack and a sense of threat no doubt play a role in a hunter's decision to shoot a tiger (Gorokhov 1983). Khramtsov (1995) reported that tigers demonstrated 'exploratory' behaviour (defined as standing motionless, often with tail twitching, and intently watching the person for some time before moving away) in 80% of 120 encounters with people. A motionless tiger staring intently, with tail twitching, could be sufficiently intimidating to elicit a reaction to shoot by any armed person. Although few have attempted to measure the importance of fear and self-defensive reactions as a contributing factor to human-caused mortality, Gorokhov (1983) reported it as one of the three primary motives for killing tigers between 1951 and 1973. It is clear that this sense of fear continues today (Zabanova *et al.* 2001; Sukhomirov 2002), and may be an even more common explanation for killing tigers as more inexperienced recreational and subsistence hunters are entering the forest (V. Solkin pers. comm.).

Attempts to reduce loss of/threat to human life

Given the bureaucratic constraints of obtaining a permit, it is not surprising that local people often prefer to resolve problems without official intervention. As already noted, in 1999 a federally mandated Tiger Response Team was created to address problem situations. To date the team has responded to 73 conflicts in nine different ways (Table 19.2). The most common response (50%) is to investigate and provide 'security' to local people who

Table 19.2. *Responses of the Tiger Response Team to conflicts between Amur tigers and people in the Russian Far East, 1999–2002*

Action	1999	2000	2001	2002	Total
Not investigated	2	3	5	3	13
Surveillance	3	5	14	17	39
Scare tactics	1	4	2		7
Capture and release		1			1
Capture and relocation		1		1	2
Capture, rehabilitation and release			2		2
Removal to captive population			2 (cubs)		2
Killed/died after capture and assessment		2	2	1	5
Killed		2			2
Total	6	18	27	22	73

feel threatened. These situations most often represent a single encounter in which a person or community felt threatened by the presence of a tiger. In these situations the perceived risk is clearly greater than the actual risk, yet, by responding, the team provides an official acknowledgement of the concerns of local people, and helps alleviate the antagonistic relationship between local people and tigers. A steady increase in 'surveillance' responses over four years is indicative of an increasing awareness of the team's existence by local people, and an increasing interest in requesting an official response (Table 19.2). Although difficult to measure, the ability of the Tiger Response Team to reduce the perceived risk may be its most important contribution in lowering human-caused tiger mortality.

Education programmes have attempted to reduce human fear and change local attitudes towards tigers, and publications such as 'Amur tiger: recommendations for human behaviour and domestic animal husbandry in tiger habitat' (produced by Inspection Tiger) attempt to reduce conflicts. Unfortunately, the effectiveness of these outreach programmes in changing local perceptions has not been measured.

RETALIATION KILLING FOR DEPREDATION OF DOMESTIC ANIMALS

Impact on humans

Livestock depredation is the most common source of conflict between large carnivores and humans (Sillero-Zubiri and Laurenson 2001; Rabinowitz, Chapter 17, Frank *et al.*, Chapter 18, Swenson and Andren, Chapter 20). However, in the Russian Far East livestock depredation may not be as important as in other parts of the world. Many villagers retain one to three cows (and to a lesser extent goats and sheep) as a source of milk and meat, but these animals are seldom targeted by tigers because they are normally brought into sheds every night. Larger farms retain sizeable herds of cattle, often close to tiger habitat, but few such enterprises have survived perestroika. Horses are not uncommon, and in winter often range independently, making them a potential target. Nonetheless, the relatively low numbers of livestock and the attentive management of small herds results in relatively few depredation incidents when compared to other parts of the world where depredation by carnivores occurs.

Comprehensive data on livestock depredations across the region are scarce, but E. N. Smirnov has kept records in a 5000-km^2 boundary area surrounding Sikhote-Alin Zapovednik since 1983 (Fig. 19.3). Depredation rates of livestock in this region averaged 3.2 animals per year ± 1.4 (95% confidence interval), ranging from 0 to 10. Extrapolating depredation rates from this study area (which are probably at the high end of the spectrum) across the entire tiger range in the Russian Far East would indicate that approximately 100 livestock per year are killed.

Dogs are the most common domestic animals taken as prey by tigers in the Russian Far East. Dogs are an easy target of tigers entering villages since they are usually chained outside, but most dogs are killed while accompanying hunters in the forest. Hunting dogs are a valuable asset, and loss of a dog can disrupt an entire hunting/trapping season, substantially impacting income. Of the 588 survey respondents in Khabarovski Krai who knew of a depredation event, 55% involved killing of dogs (Sukhomirov 2002). Data from Sikhote-Alin suggest that depredation on dogs is at least as common as on livestock (Fig. 19.3). Unfortunately, there is little that can be done to

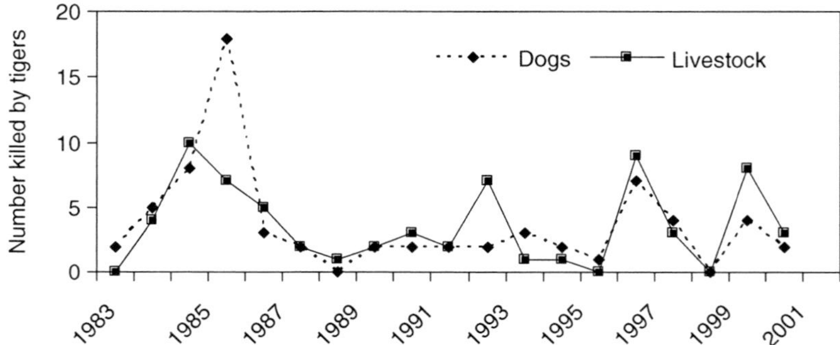

Figure 19.3. Number of dogs and livestock killed by tigers in the 5000-km^2 boundary area of Sikhote-Alin Zapovednik, Russian Far East, 1983–2001.

reduce such losses as long as hunters encourage their dogs to roam freely in tiger habitat.

Impact on tigers

Depredation retaliation accounted for 20–30% of reported tiger mortalities from 1951 to the early 1990s (Table 19.2). Since then depredation retaliation represents a smaller percentage of the total number of tigers killed (Table 19.2), probably reflecting both the increase in poaching rates, and a reduction in depredations as livestock numbers decreased. Gorokhov (1983) cites 'revenge' for loss of dogs as a primary motive for killing tigers. Dogs that run to their owner when pursued by tigers can precipitate encounters between tigers and humans, resulting in either an 'incidental' tiger attack or killing of a tiger in self-defence. Although such events do occur (Nikolaev and Yudin 1993), they are relatively rare, and a relatively unimportant factor affecting tiger population dynamics, but they help fuel the antagonistic attitude of hunters towards tigers (see next section).

Attempts to resolve the depredation problem

Historically, livestock depredations by tigers were covered by state-sponsored insurance. This system evaporated with the political turmoil of the early 1990s. Commercial insurance programmes are available, but premiums are relatively high, and are seldom, if ever, used by local farmers. In 1993 we began an experimental compensation programme in the area surrounding Sikhote-Alin Zapovednik. Compensation payments for tiger depredations were made at a few key farms within the range of radio-collared tigers. Our goals were to increase survivorship of study animals, and reduce antagonistic relationships between farmers and tigers. Antagonism of local farmers towards members of the Siberian Tiger Project appeared to decrease afterwards, and more significantly, we never recorded a loss of a radio-collared tiger that appeared linked to livestock depredation.

Despite the apparent success of this approach, it worked only at the local level, and was dependent largely on personal relationships between research staff and a handful of farmers. More importantly, the system was not sustainable in the long term, requiring continuous international sponsorship, and it provided no incentive for local farmers suffering depredations to improve animal husbandry techniques to reduce losses. While reducing retaliation killing locally, the rate of depredation was not changed, and the approach could not be extended to a larger area.

In an attempt to address these deficiencies, in 1999 the 'Farmers' Fund' was created as a legally registered non-governmental insurance and loan programme for farmers. Farmers paid membership dues that acted as insurance, but could also apply for low-interest loans to improve animal husbandry techniques. Membership fees were dependent on number of animals insured, but ranged from 500 to 4000 rubles (US$16 to 133), while the average value of cattle was approximately US$300 when rates were set. Using the long-term average depredation rate for the model area (4.0 animals per year), we needed to generate a minimum US$1200 per year in membership fees to establish a sustainable programme. Membership fees above this amount could go into a loans programme. A membership base of 20 farmers in the model area would generate sufficient funds to insure sustainability.

In 2000 we publicized in local newspapers and billboards and by word of mouth to recruit members. In 2000 five members joined, in 2001 only four, and in 2002 the number decreased to only three farmers. In total, nine farmers joined for one or more years. Over its first three years the fund collected a total of 34 500 rubles (about US$1150) in membership dues, and paid out US$1182 in compensation for five depredation events (one horse and four cows). Whereas in previous years we knew farmers had killed depredating tigers (including farmers who later joined the fund) we had no records of tigers being killed in retaliation for depredations in the model area.

Due to a lower than average depredation rate, compensation costs only slightly exceeded membership dues, providing a false sense of sustainability. However, running costs and projected increases in depredation rates (to the 20-year average) would run the programme into bankruptcy. Most discouraging was the lack of interest in the programme from local farmers. Despite the fact that all nine members applied for and received low-interest loans, interest in the programme was weak.

We believe there were three interacting reasons for the poor results: (1) there is no cultural tradition of buying insurance in Russia, where formerly the state paid everything, thus the concept is new and unfamiliar to local people; (2) management/advertising of the fund was inadequate; and (3) the risk of depredation loss is too low. Even if cultural and educational barriers were overcome, it appears that for most farmers the risk of depredation is simply too low to justify the cost of insurance. Most farmers joined for one season and abandoned the programme when they incurred no immediate losses. While evidence suggests that a compensation programme does reduce retaliation killings, it appears ineffective to run a privatized insurance programme. Expenses to run such a programme are not high, and could theoretically be absorbed by the government under the auspices of the Federal Program for Conservation of the Amur Tiger.

COMPETITION BETWEEN HUNTERS AND TIGERS FOR UNGULATES

A key parameter of tiger habitat is prey density (Karanth and Stith 1999; Miquelle *et al.* 1999b), and therefore a key component of tiger conservation should be management of ungulate populations. There are over 40 000 registered hunters in Primorski Krai, making them potentially the largest stakeholder group coexisting with tigers. Virtually all unprotected tiger habitat is hunted, so hunter perception of and interaction with tigers is a fundamental component of tiger conservation. Nearly all tigers intentionally killed are shot by hunters, and therefore changing the traditional view of tigers as competitors for ungulates is critically important.

Impact of humans on tigers

Hunters impact tiger populations via prey depletion and direct killing of tigers. An average 8624 licences were issued per year (1998–2000) for all ungulates in Primorski and southern Khabarovski Krais, but actual kill rates were an estimated 250% higher (21 560 animals) than the legal limit (World Wildlife Fund 2002). Although hunting quotas are established conservatively, an estimated 97% of hunters kill more than allowed: an average 1.7–3.5 ungulates on licences issued for single animals. Hunting without a license is even more common (World Wildlife Fund 2002). Therefore illegal harvest has probably been a primary factor depressing ungulate numbers over the past decade, thereby increasing tensions over scarce game between tigers and hunters.

Gorokhov (1983) listed competition for prey as one of the three most important reasons for illegally shooting tigers. Gorokhov (1977) reported that hunters often leave dead tigers in the forest, not even attempting to profit from the killing. Available evidence suggests that the incentive to kill tigers as competitors continues, but now with greater access to Asian black markets, there is the added incentive of economic returns. While it is difficult to separate out these motives clearly, the strong sentiments regularly expressed by many hunters indicate that control of tiger numbers is considered an acceptable and even necessary action.

Impact of tigers on humans

Beginning in the early 1970s there has been concern about the impact of increasing tiger numbers on ungulate populations (Kucherenko 1970, 1993; Gorokhov 1983; Dunishenko 1985). In two recent surveys respondents indicated that the impact of tigers on wild ungulate populations ranked second only to concerns about personal safety as reasons not to protect tigers

(Zabanova *et al.* 2001; Sukhomirov 2002). Although the argument as to whether tigers actually do limit prey population densities is debatable, there is sufficient evidence to suggest that at least in some situations limitation by tigers can occur (Miquelle *et al.* 2005). Uninterested in the scientific debate, hunters view every ungulate killed by tigers as one less ungulate they are able to harvest. The annual offtake of ungulates by tigers (an estimated 20 000–25 000 animals: World Wildlife Fund 2002) is often cited as evidence of the severe impact of tiger predation on ungulate populations. Thus a perception has developed that competition between hunters and tigers is direct and serious, with some suggesting that coexistence is possible only with regulation of the tiger population through controlled harvest (Bragin and Gaponov 1989; Kucherenko 1993).

Responses to the problem

Increasing prey populations, better hunter management, and education are key elements in resolving the hunter–tiger conflict in Russia. Beginning in 1995, new legislature provided opportunities for local people to create non-governmental hunting 'societies' that could obtain and manage hunting lands. Today, in place of a small number of state-controlled operations, there are 102 registered hunting leases in Primorye. For the first time ever, local people have been given responsibility for managing wildlife populations, acquiring in the process some responsibilities for non-game and endangered species as well. While this new arrangement only provides rights to use and manage wildlife, it nonetheless represents a revolutionary change in resource management in Russia. Under the former system, most people believed in maximizing personal consumption of communal (state) properties, including wildlife resources, without concern about sustainability. Now, local people have a vested interest in properly managing a resource that is theirs, and that they depend upon for recreation, income and food.

In 2000, the Wildlife Conservation Society initiated a programme to improve wildlife management through support and education to hunting groups of the region. The intention is to support newly established hunting leases, to increase capacity for self-management and financial independence, to increase ungulate populations, and to improve the relationship between hunters and tigers. Activities within the scope of this project include legal assistance to private leases, increasing stakeholder capacity, increasing financial stability, education programmes, and management initiatives to increase ungulate densities (including improvements in anti-poaching, population management and habitat improvements). It is too early to judge the results of this programme, but in a few experimental leases, increases in ungulate populations have been reported. It remains to be determined whether a grassroots approach to wildlife management will result in higher ungulate densities and improved relations between tigers and hunters.

CONCLUSIONS

Given that densities of both human and tiger populations in the Russian Far East are as low or lower than any other tiger range country, the success or failure of humans and tigers to coexist there may be considered a benchmark case for carnivore conservation. If coexistence is impossible in this setting, prospects for success in other countries, where human densities and their demands on the landscape are magnitudes greater, are dim. The fact that the Russian Far East retains the single largest unfragmented habitat for tigers, and that the Amur tiger population probably represents the largest contiguous population of tigers in the world, provides hope for the future. Intensive investments of time, energy and money on the part of the international conservation community, along with commitment from some key governmental agencies, have apparently secured a stable population of Amur tigers, at least temporarily. While intensive harvest of prey remains problematic, the creation of special anti-poaching teams and tiger response teams have had some success in reducing direct impacts of tigers and people on each other. The success of programmes aimed at increasing prey numbers and reducing the antipathy of hunters towards tigers is yet to be determined. In the absence of adequate government financial support a comprehensive compensation programme for livestock depredation is difficult.

We believe that tigers and people can coexist in the Russian Far East, but active conflict-resolution programmes must be an integral part of the coexistence recipe. Government-sponsored tiger conflict teams, educational programmes and hunter support programmes are important steps towards reducing conflicts to manageable levels. At the same time, it must be recognized that while conflicts can be reduced with proper management and education, it is unlikely that they can be totally eliminated as long as people and tigers are using the same land base. Impacts on people must be reduced to a level that is acceptable, not to the global society as whole, but to the people of the forest communities that incur the cost of living with tigers. At the same time, it is imperative to reduce human-caused mortality to levels that do not threaten viability of the tiger population. Increasing ungulate densities, and improving attitudes of local hunters towards tigers, as well as effective depredation compensation programmes, remain challenges for the future.

ACKNOWLEDGEMENTS

We thank S. A. Zubtsov, head of Inspection Tiger, and I. O. Suslov, Head of the Primorye Hunting Department, for supplying information on conflict situations, and for providing leadership in development of a programme to address conflicts between tigers and people. We thank A. A. Astafiev for administrative, political and logistic support to the Siberian Tiger Project, and the Russian State Committee for Environmental Protection (and later the Ministry of Natural Resources) for providing permits for capture of tigers. Maurice Hornocker and Howard Quigley of the Hornocker Wildlife Institute conceptualized and initiated the Siberian Tiger Project, and we are grateful for their expertise, support and dedication to research and conservation of tigers. The Siberian Tiger Project is funded by The National Fish and Wildlife Foundation's "Save the Tiger Fund", the National Geographic Society, the National Wildlife Federation, Exxon Corporation, the Charles Engelhard Foundation, Disney Wildlife Fund, Turner Foundation, Gary Fink, Richard King Mellon and the Wildlife Conservation Society.

A tale of two countries: large carnivore depredation and compensation schemes in Sweden and Norway

JON E. SWENSON AND HENRIK ANDRÉN

INTRODUCTION

Norway and Sweden share the Scandinavian Peninsula and the people have similar languages, customs and lifestyles. Nevertheless, the situation regarding large carnivores, their depredations and the schemes devised to deal with these problems differ greatly. We view these two countries almost as 'replicates' to give us insight into how different types of livestock husbandry affect depredation levels and to compare compensation schemes. We will concentrate on sheep (*Ovis aries*) and semi-domestic reindeer (*Rangifer tarandus*) and will not consider predation on wild large game.

THE SPECIES OF LARGE CARNIVORES

Sweden and Norway have four species of large carnivores: the grey wolf (*Canis lupus*), wolverine (*Gulo gulo*), brown bear (*Ursus arctos*) and Eurasian lynx (*Lynx lynx*). All have been subject to lethal control since the seventeenth and eighteenth centuries with high state-financed bounties. The official policy was extermination, and this occurred over large areas. After official policies changed, the bounties were removed, and populations of all four species increased in size and distribution. Present policies include maintaining or promoting viable populations, balanced with mitigating losses they cause to farm livestock (primarily sheep), semi-domestic reindeer, hunting dogs, and large game animals (primarily moose (*Alces alces*) and roe deer (*Capreolus capreolus*)).

The wolf originally occurred over all of Scandinavia. During the early nineteenth century about 500 were killed annually, but it was exterminated from Scandinavia in the late 1960s. In 1982 a few wolves were observed in south-central Sweden/Norway. They reproduced in 1983. In 1990 a second

People and Wildlife: Conflict or Coexistence? eds. Rosie Woodroffe, Simon Thirgood and Alan Rabinowitz. Published by Cambridge University Press.

pack formed and since then the population has increased by about 25–30% annually (Wabakken *et al.* 2001). During winter 2000/1 there were 12 wolf packs in south-central Scandinavia (70 000 km^2 or 10% of the peninsula: Aronson *et al.* 2001), isolated from the larger Finnish/Russian population (Ellegren *et al.* 1996; Wabakken *et al.* 2001). At the moment only single dispersing wolves occur within the reindeer husbandry area (Aronson *et al.* 2001). Moose is the main prey of wolves in Scandinavia, but they eat several prey species when available, including roe deer (Olsson *et al.* 1997), as well as sheep and reindeer. Although protected in both countries, the authorities give permission for killing problem wolves.

The bear originally occurred throughout Scandinavia. There were 4000–5000 bears in the mid- nineteenth century, with about 65% in Norway (Swenson *et al.* 1995). In 1996 there were 800–1300, the vast majority in Sweden (Sandegren and Swenson 1997). The Scandinavian brown bear exhibits the highest growth rates ever documented for the species (13–16% annually) (Sæther *et al.* 1998) and dispersing bears from Sweden are recolonizing Norway (Swenson *et al.* 1994, 1995, 1998). It has been hunted with a quota system in Sweden since 1943 and is protected in Norway. Most of the bear's annual energy intake comes from berries in areas without free-ranging sheep, and from sheep in areas with unguarded free-ranging sheep; other important food items are moose calves and ants (Dahle *et al.* 1998).

Wolverines were originally found throughout the boreal forest and mountains in Scandinavia. The 1995–7 populations (without young-of-the-year) were estimated to be 429 (95% confidence interval 290–573) in Sweden, 196 (149–237) in northern Norway and 43 (26–59) in southern Norway (Landa *et al.* 1998a, 2000a). The wolverine is protected in Sweden and hunted on a quota system throughout Norway. Recently, population declines have been observed in both countries. Wolverines prey heavily on sheep and semi-domestic reindeer, but it is a less effective predator on wild reindeer (Landa *et al.* 1997, 2000b).

Lynx numbers have changed greatly during the twentieth century. There were 1400–1800 in winter 2000; 700–800 within the reindeer husbandry area) and 700–1000 outside (Liberg and Glöersen 2000). It is presently decreasing in Sweden, but is recolonizing southern Sweden and increasing outside the reindeer husbandry area, although decreasing within. Numbers in Norway were 300–350 in 2002 and declining (Brøseth *et al.* 2003). Roe deer is the main prey of lynx outside the reindeer husbandry area, as is reindeer within it (Pedersen *et al.* 1999). Sheep are also preyed upon in summer (Odden *et al.* 2002). The lynx is hunted on a quota system in both countries.

THE CONFLICT

Large carnivores cause many conflicts in Scandinavia. However, the most important are depredations on farm livestock, particularly sheep in Norway, and on semi-domestic reindeer in both countries. The killing of hunting dogs by wolves is another serious conflict in both countries, as is predation on moose and roe deer, and fear, especially of wolves.

Depredation of reindeer by large carnivores and the resulting conflict with the indigenous Saami people has a long history. Present reindeer management started during the sixteenth and seventeenth centuries (Jorner *et al.* 1999). The reindeer husbandry area covers 40% and 43% of Sweden and Norway, respectively. Both governments have signed the Convention on Indigenous and Tribal Peoples in Independent States, and have committed themselves to maintain the Saami culture. As reindeer management is an important part of Saami culture, subsidies are paid to maintain it.

THE SCANDINAVIAN PENINSULA

From a European perspective both Sweden (411 820 km^2; 8.9 million people) and Norway (323 877 km^2; 4.5 million people) have low human densities. Norway has a stronger 'district policy' than Sweden, and has maintained a dispersed pattern of rural habitation. Both countries have a long land-use history and very little pristine landscape. The largest protected areas are found in the least productive mountain regions and most protected areas are smaller than the home ranges of these four large carnivores (Sandegren and Swenson 1997; Linnell *et al.* 2001a; Wabakken *et al.* 2001). Therefore, all large carnivores are mainly found outside protected areas in a multi-use landscape (Swenson *et al.* 1994; Landa *et al.* 2000b, Linnell *et al.* 2001a; Wabakken *et al.* 2001).

There is a vast difference between the sheep husbandry practised in Norway and Sweden. In Norway, over 2 million sheep graze on open ranges in mountain and forested habitats. To encourage habitation and economic activity in agriculturally marginal areas, the Norwegian state provides subsidies that vary from US$5.08 to 12.45 per kilogram of meat produced, with the highest subsidies going to farmers with the smallest flocks in the most marginal areas (Aanesland and Holm 2001). In 1998 there were 22 000 farms with sheep, averaging 43 sheep in winter; 76% of the sheep are in flocks <100. In 1998, 100 sheep in winter gave an annual wage of US$9761 (Aanesland and Holm 2001). Most farmers raise sheep only for a second income and usually visit their flocks about once a week during the grazing season. In 1999, 137 339 sheep were reported lost on the grazing range. Sheep-farmers applied for compensation for 50 145 as killed by large carnivores, and the authorities compensated 33109 of these, which means that around 100 000 sheep (1000 per day of the grazing season) die of other reasons. Obviously, this husbandry system only could develop in an essentially predator-free environment, which was the case in Norway. Sheep-farming in Sweden is even more small-scale and is mostly a hobby. In 2001 there were 8100 farms with sheep averaging 26 in winter. The 450 000 sheep in summer in Sweden are almost all kept within fenced pastures on farmsteads, mostly in the south.

In Sweden the reindeer husbandry area is divided into 51 reindeer grazing communities, consisting of several reindeer owners, sometimes organized into enterprises, usually a family. Today there are about 4700 reindeer owners in 960 reindeer management enterprises that have about 220 000 reindeer after slaughter (Jorner *et al.* 1999). The reindeer husbandry area in Norway is divided into six reindeer management units, which in turn are divided into 85 reindeer management districts. There are 577 reindeer management enterprises in Norway today and about 2800 reindeer owners have about 161 000 reindeer after slaughter (Hætta 2002). In 1998 the Swedish state paid US$4.76 million to support the reindeer management and US$2.07 million in subsides for reindeer meat (Jorner *et al.* 1999). In Norway, the support in 2000 was US$2.44 million and subsides for reindeer meat was US$1.22 million (Hætta 2002).

IMPACT OF LARGE CARNIVORES ON PEOPLE

Large carnivores cause tremendously different depredation rates on sheep in Norway and Sweden (Table 20.1). These estimates are only meant for comparison between countries, but at least one value, 16.5 sheep killed per lynx for Norway, agrees rather well with data obtained from radio-tracked lynx (Odden *et al.* 2002). These depredation rates are minimum, because they ignore the category 'unknown predators', which makes up 17% and 12% of the compensated sheep in Norway and Sweden, respectively. The wolverine is the most effective predator (Table 20.1). Wolverines do not kill sheep in Sweden, but sheep are not grazed in wolverine habitats. Individual wolves, lynx and bears in Norway all kill more sheep than those in Sweden (Table 20.1). However, there are many more sheep in Norway. The proportions of available sheep taken per individual carnivore are also much greater in Norway (Table 20.1).

Many reindeer are also lost to large carnivores. In Sweden 46 910 reindeer were slaughtered in 2000, valued at about US$9.03 million and

Table 20.1. *Comparison of depredation rates on sheep by the four large carnivores in counties with these species in Norway and Sweden*

	Norway	Sweden	Ratio N : S
Wolverine (north Norway and Sweden)[a]			
Number of sheep (N: 1999, S: 2000)	359 071	39 191	
Number of wolverines (midpoint of estimates, N: 2000–02, S: 1995–7)	181	429	
Sheep losses compensated (mean 1999–2000)[b]	3 952	0	
Number of sheep killed per wolverine	21.8	0	
Proportion of sheep killed by wolverines	1.10%	0	
Proportion of sheep killed per wolverine	0.006 08%	0	
Wolverine (south Norway)[c]			
Number of sheep (N: 1999)	631 965		
Number of wolverines (midpoint of estimates, N: 2000–2)	64		
Sheep losses compensated (mean 1999–2000)[b]	8 781		
Number of sheep killed per wolverine	137.2		
Proportion of sheep killed by wolverines	1.39%		
Proportion of sheep killed per wolverine	0.021 22%		
Brown bear[d]			
Number of sheep (1995)	1 010 605	84 908	11.9
Number of bears (midpoint of estimates, 1996)	41	1 002	0.04
Sheep losses compensated (mean 1993–5)[b]	1 998	98	20.4
Number of sheep killed per bear	48.7	0.1	487.0
Proportion of sheep killed by bears	0.20%	0.12%	1.7
Proportion of sheep killed per bear	0.004 88%	0.000 12%	40.7
Lynx[e]			
Number of sheep (N: 1999, S: 2001)	1 518 275	103 263	14.7
Number of lynx (midpoint of estimates, 1997–8)	561	1 106	0.51

Table 20.1. (*cont.*)

	Norway	Sweden	Ratio N : S
Sheep losses compensated (mean 1999–2000)[b]	8 542	80	106.8
Number of sheep killed per lynx	15.23	0.07	217.6
Proportion of sheep killed by lynx	0.56%	0.08%	7.0
Proportion of sheep killed per lynx	0.001 00%	0.000 07%	14.3
Grey wolf[f]			
Number of sheep (N: 1999, S: 2000)	389 652	111 800	3.5
Number of wolves (midpoint of estimates, 1998/9)	26	41	0.63
Sheep losses compensated (mean 1999–2000)[b]	654	84	7.8
Number of sheep killed per wolf	25.17	2.05	12.3
Proportion of sheep killed by wolves	0.17%	0.08%	2.1
Proportion of sheep killed per wolf	0.006 46%	0.001 83%	3.5

[a] Counties with reproducing populations of wolverines were: Nord-Trøndelag, Nordland, Troms and Finnmark in northern Norway; Kopparberg (now called Dalarna), Västerbotten, Jämtland, and Norrbotten in Sweden. Population estimates are in spring and exclude cubs.

[b] Including sheep documented as killed by each species and those missing due to suspected predation. The compensated losses include only losses specified as caused by each species by the authorities; no portion of the 'predator unknown' category is included here.

[c] Counties with reproducing populations of wolverines were: Hedmark, Oppland, Sør-Trøndelag, and Møre og Romsdal in southern Norway. Population estimates are in spring and exclude cubs.

[d] Counties reporting depredation on sheep due to brown bear were: Hedmark, Oppland, Sør-Trøndelag, Nord-Trøndelag, Nordland, Troms and Finnmark in Norway; Värmland, Kopparberg (now called Dalarna), Gävleborg, Västernorrland, Jämtland, Västerbotten and Norrbotten in Sweden. Population estimates in spring.

[e] Counties with reproducing populations of lynx were: Aust-Adger, Telemark, Buskerud, Østfold, Askershus, Hedmark, Oppland, Møre og Romsdal, Sør-Trøndelag, Nord-Trøndelag, Nordland, Troms and Finnmark in Norway; Örebro, Värmland, Västmanland, Dalarna, Gävleborg, Västernorrland, Jämtland, Västerbotten and Norrbotten in Sweden.

[f] Counties with stationary pairs or packs of wolves were: Østfold, Askershus, Hedmark, and Oppland in Norway; Västra Götaland, Örebro, Värmland, and Dalarna in Sweden. Population estimates in winter; pairs and packs straddling the border were counted as one-half in each country.

compensation for depredations for about 48 000 reindeer was US$4.76 million. In 2000 about 42 950 reindeer were slaughtered in Norway and the state paid US$4.39 million for 20 033 killed or probably killed by large carnivores. Lynx and wolverine were the most important predators in both countries, but losses to unknown large carnivores were high in Norway (Table 20.2).

Both governments have stated that predator-killed reindeer should be fully compensated. One should expect similar predation rates, because reindeer is the main prey for both lynx and wolverine in the reindeer husbandry areas (Pedersen *et al.* 1999; Persson 2003), but the compensation per large carnivore was somewhat higher in Sweden than in Norway (Table 20.2). However, 71% of the compensation in Norway is for unknown large carnivore species (Table 20.2). If this amount is divided among the large carnivores in the same proportion as the documented large carnivores,

Table 20.2. *The amount of money (in US$) paid to compensate for predator-killed sheep, cattle and reindeer in Norway and Sweden in 2000; the results are presented as total amounts and amounts per predator present, based on the population estimates that were temporally closest to 2000*

	Norway[a]		Sweden	
Species	Total amount	Amount per predator[b]	Total amount	Amount per predator[b]
Sheep and cattle				
Wolverine	3 210 488	11 848	0	0
Brown bear	827 504	20 178	50 410	50
Lynx	2 155 971	3 843	31 478	20
Wolf	178 915	8 947	23 764	475[c]
Unknown	1 319 536		2 737	
Reindeer				
Wolverine	518 224	2 479	1 087 503	3 337
Brown bear	5 874	195	228 697	286
Lynx	613 501	2 455	3 016 912	3 779
Wolf	0	0	42 247	10 561[d]
Unknown	3 137 596		0	

[a] Pro-rated, based on the number of sheep killed by each species and total compensation amount.
[b] No proportion of the 'predator unknown' category is included in the amount per predator.
[c] The population estimate in 1998/9 was about 20 wolves in Norway, 50 wolves in Sweden (Aronson *et al.* 1999).
[d] Only single wolves were found in the reindeer husbandry area in 2000 and the number is uncertain, but Aronson *et al.* (2001) reported four wolves.

Table 20.3. *Comparison of depredation rates on reindeer by large carnivores in the reindeer husbandry area in Norway and in Sweden for 2000*

Subject	Norway	Sweden	Ratio N : S
Wolverine			
Number of reindeer	161 050	220 391	0.73
Number of wolverine (midpoint of estimates)	209	326	0.64
Reindeer losses compensated by authorities[a]	1 685	11 000	0.15
Number of reindeer killed per wolverine	8.06	33.74	0.24
Proportion of reindeer killed by wolverines	1.05%	4.99%	0.21
Proportion of reindeer killed per wolverine	0.0050%	0.0153%	0.33
Brown bear			
Number of reindeer	161 050	220 391	0.73
Number of brown bear (midpoint of estimates)	30	800	0.04
Reindeer losses compensated by authorities[a]	31	2 400	0.01
Number of reindeer killed per brown bear	1.03	3.00	0.34
Proportion of reindeer killed by brown bears	0.02%	1.09%	0.02
Proportion of reindeer killed per brown bear	0.0006%	0.0014%	0.47
Lynx			
Number of reindeer	161 050	220 391	0.73
Number of lynx (midpoint of estimates)	250	800	0.31
Reindeer losses compensated by authorities[a]	1 909	31 000	0.06
Number of reindeer killed per lynx	7.64	38.75	0.20
Proportion of reindeer killed by lynx	1.19%	14.07%	0.08
Proportion of reindeer killed per lynx	0.0047%	0.0176%	0.27

[a] The number of reindeer losses compensated by authorities equals the number documented as killed by each species and those missing due to suspected predation in Norway. The compensated losses include only losses specified as caused by each species by the authorities; no portion of the 'predator unknown' category is included here. In Sweden the number of reindeer compensated by authority is an estimation based on the number of large carnivores in the reindeer husbandry area.

then compensation per lynx and wolverine were actually 2.2 and 2.9 times higher in Norway than in Sweden, respectively.

The estimated number of reindeer killed per lynx and wolverine in Sweden was around 40 per year for both species (Table 20.3), based partly on field studies of lynx and wolverine. A lynx family group in winter kills about 6 reindeer per month (Pedersen *et al.* 1999), or about 36 reindeer per lynx per year. Thus, the estimated number of reindeer lost to lynx in Sweden is realistic.

IMPACT OF PEOPLE ON LARGE CARNIVORES

Quota-regulated hunting is allowed for lynx in both countries, bears in Sweden, and wolverines in Norway. The rationale for this hunting is partly to reduce the potential for depredation. In fact, Sweden, a member of European Union, can only allow hunting of large carnivores as a conflict-mitigating measure.

In Sweden, the authorities have increased quotas for bear to slow the population increase and for lynx to reduce the population size within the reindeer husbandry area. In Norway, hunting has been used to reduce lynx numbers (Brøseth *et al.* 2003) and keep the wolverine population at a level commensurate with acceptable depredation problems. In addition to hunting, the authorities allow the killing of individual depredating large carnivores by state employees (Norway) or regular hunters (both countries). Large carnivores in Norway are generally harvested at much higher rates than in Sweden (Table 20.4). Although the wolverine has been protected in Sweden since 1969, the population is not increasing (Landa *et al.* 2000b). Experimental local boards, usually dominated by sheep-farmers, have administered the hunting of wolverines and lynx in Norway since 1997. Probably no large carnivore population in the world of this small size is subjected to as high a legal offtake as the wolverines in southern Norway (Table 20.4). In fact, based on a population viability analysis, Persson (2003) concluded that continuation of the current levels of harvest in Norway would lead to rapid extinction of the species in large parts of the country.

Bears are hunted in Sweden, with a maximum quota set annually for different hunting areas. In Norway, the bear is protected, but kill permits are issued for depredating bears; more applications for a kill permit are granted than rejected in Norway when the number of killed sheep is 6–10 or more (Hustad 2000). Total mortality of bears due to human causes is just as high, or higher, in Norway, than in Sweden (Table 20.4).

People also kill large carnivores illegally. Although difficult, estimates have been made for illegal hunting of lynx, brown bears, and wolverines in Sweden. Andrén (1999) estimated that 9.9–13.9% of the adult lynx and 5.6% of the young-of-the-year might be killed annually within the reindeer husbandry area, or 1.2–1.6 times the legal kill. Outside the reindeer husbandry area, it was about 9.8% and 15.1%, or about 1.6 times the legal kill. In total, 133–157 lynx might be killed illegally in Sweden annually. Swenson and Sandegren (1999) estimated that about 6.7% of the bears in the reindeer husbandry area were killed annually, or 2.8 times more than the legal kill, and about 1.4% outside the area, or 0.6 times the legal kill. In total, around 40 bears might have been killed illegally each year in Sweden during the period 1984–98. Willebrand *et al.* (1999) estimated that, in the reindeer husbandry area, documented, probable and possible illegal hunting accounted for 17%, 46% and 71%, respectively, of the cases where contact was lost with radio-marked older wolverines. The corresponding values for young-of-the-year were 0%, 6% and 11%, respectively. They could not

Table 20.4. *Levels of human killing of large carnivores in Norway and Sweden (average values for the years indicated)*

Parameter	Norway	Sweden
Wolverine (Sweden and north Norway)[a]		
Population estimate (S: 1995–7 S, N: 1999–2001)	209	429
Harvest quotas, total (% of population), years	33.5 (16%) 00–02	0
Harvest, total (% of population), years	21.5 (10%) 00–02	0
Depredation killings, total (% of population), years	9.5 (4.5%) 00–02	1 (0.2%) 94–00
Wolverine (south Norway)[a]		
Population estimate (1999–2001)	62	
Harvest quotas, total (% of population), years	13.5 (22%) 00–02	
Harvest, total (% of population), years	5.5 (9%) 00–02	
Depredation killings, total (% of population), years	7.5 (12%) 00–02	
Lynx (north Sweden and Norway)		
Population estimate (N: 1995 S: 2000)	545	750
Harvest quotas, total (% of population), years	135 (25%) 95–00	117 (16%) 96–00
Harvest, total (% of population), years	97 (18%) 95–00	63 (8%) 96–00
Depredation killings, total (% of population), years	1.3 (0.2%) 99–02	0
Lynx (south-central Sweden)		
Population estimate (2000)		850
Harvest quotas, total (% of population), years		49.5 (6%) 96–00
Harvest, total (% of population), years		49 (6%) 96–00
Depredation killings, total (% of population), years		0
Brown bear		
Population estimate (1996)	41	1002
Harvest quotas, total (% of population), years	0	60 (6%) 00–02
Harvest, total (% of population), years	0	60 (6%) 00–02
Depredation killings, total (% of population), years	2.7 (6.6%) 00–02	3 (0.3%) 00–02
Wolf		
Population estimate (2000–01)	34	58
Depredation killings, total (% of population), years	4.7 (14%) 00–02	0.3 (0.5%) 00–02

[a] The population estimates for wolverines do not include young-of-the year, but the harvest data do.

conclude to which degree illegal hunting affected the population dynamics of wolverines, but they stated that it was potentially an important factor, and that it might hinder the establishment of wolverines in areas outside their study area.

POLITICAL GOALS

Large carnivore policy has been handled politically in Norway twice, most recently in 1997, and in Sweden once, in 2000. Both parliaments accepted maintaining viable populations of all four species. Beyond this, they have expressed population and distribution goals (Table 20.5). The Swedish goals

Table 20.5. *Management goals for large carnivores, as stated by the parliaments of Norway and Sweden. Specific goals in management plans produced by the wildlife management agencies are not included here*

Species	Norway (1997)	Sweden (2000)
Lynx	Maintain a viable population with continuous distribution, but not allowing establishment in southwestern Norway and some northern coastal areas	A minimum of 300 reproductions per year, i.e. about 1500 lynx; the lynx should mainly be found outside the reindeer husbandry area, but it can occur over all of Sweden
Brown bear	Less than 8–10 adult females will be allowed to establish in each of the five core areas	A minimum of 100 reproductions per year, i.e. about 1000 bears; the distribution of brown bear should be in the northern 2/3 of Sweden
Wolf	Establish at least 8–10 wolf packs *in Scandinavia* outside the reindeer husbandry area	A first minimum level of 20 wolf packs, i.e. about 200 wolves *in Sweden*, and then an evaluation of the consequences; only single wolves should be allowed within the year-round reindeer husbandry area (about 25% of Sweden)
Wolverine	A viable population in the north; establishment of a viable population in the south, with reproduction limited to the core area and quota-limited hunting outside of it	A first minimum level of 90 wolverine reproductions, i.e. 400 wolverines and a more even distribution, and then an evaluation of the consequences; establishment outside the reindeer husbandry area is desirable

Source: Norway: Energi- og miljøkomitéen (1997); Sweden: Regeringen (2000).

are more ambitious, more precise and give minimum goals, whereas the Norwegian goals are modest and vague, with a maximum goal for bears. There is a particularly large difference regarding wolves, with a Norwegian goal of minimum 8–10 packs in southern *Scandinavia*, whereas the first Swedish goal of minimum 20 packs is for southern *Sweden* (Table 20.5).

The Norwegian Parliament has decided that the country will have both viable populations of large carnivores and maintain an active and responsible agricultural production with acceptable economic conditions based on utilizing grazing resources by sheep and reindeer in the mountains and forests (Energi- og miljøkomitéen 1997). As the farmers earn relatively little for keeping sheep, they have little time or incentive to invest in preventative measures, nor does the compensation system require it. In the southern Norway wolverine area, with high levels of sheep losses and conflict, sheep numbers increased fivefold during 1979–94 (Landa *et al.* 1999). Thus, the diverging political goals of having viable populations of large carnivores and viable agriculture in marginal areas, based on extensive sheep grazing, seem to be incompatible (Sagør *et al.* 1997; Landa *et al.* 2000b).

ECOLOGICAL CAUSES UNDERLYING THE PROBLEM

The considerable differences between Norway and Sweden in terms of sheep killed per carnivore clearly indicate that the biological reason for the large problem in Norway and small problem in Sweden is the type of sheep husbandry. There is an abundance of wild prey in both countries.

The conflict within the reindeer husbandry area in both countries differs from many other large carnivore–livestock conflicts because the free-ranging semi-domesticated reindeer are important prey for all the carnivores and the main prey for lynx and wolverine (Pedersen *et al.* 1999; Landa *et al.* 2000b). Thus, the conflict cannot be solved by a complete protection of reindeer, because the predators could not survive.

ECONOMIC AND TECHNICAL SOLUTIONS FOR RESOLVING THE CONFLICT

Economic solutions

In both countries, the basic principle is that the state compensates a livestock owner who suffers depredation caused by protected large carnivores. In 1998, the Norwegian state paid sheep owners US$7.32 million in compensation for 29 700 sheep killed and probably killed by large carnivores and eagles (*Haliaeetus albicilla* and *Aquila chrysaetos*), based on applications for compensation for 51 227 lost sheep. In 2000, the Swedish state paid

US$111 413 to compensate for farm livestock depredations due to large carnivores. The total compensations per large carnivore species and per individual large carnivore are vastly higher in Norway than Sweden (Table 20.2) and the requirements for compensation are quite different. In Norway, a farmer is compensated for losses due to large carnivores if depredation has been documented on or near the grazing area by an inspector. Then the farmer receives payment for all lost animals minus a 'normal loss' without large carnivores, which is about 2% for adults and 5% for lambs. The value per sheep is set annually, and was US$147 for adults and US$121 for lambs in 2000, but can be increased for extra costs the farmer might have had in connection with the depredation and for lost future production value. There is no requirement that the farmer has used specific measures to prevent depredation, other than acting responsibly to avoid or reduce losses. In Sweden, the principle is that the conflict between farm livestock and large carnivores should be solved using preventive measures to protect the livestock. When a depredation occurs, it must be verified by an inspector, as in Norway. However, to obtain compensation, the farmer is required to document that he has used approved and effective protective measures. The state compensates for animals that are killed, injured and missing in each verified depredation. The compensation rates for 2002 were somewhat higher than in Norway (US$329 for rams, US$256 for ewes and US$146 for lambs), but include loss of future production, which is added to the base value in Norway.

As for farm livestock, semi-domesticated reindeer lost to large carnivores should be fully compensated in both countries. The two countries pay similar sums yearly to compensate reindeer owners: US$4.76 million in Sweden and US$4.39 million in Norway. However, the two countries have very different compensation systems. The compensation per reindeer is higher in Norway (US$122 for a calf, US$183 for a yearling, US$305 for an adult male and US$488 for an adult female) than in Sweden (a mean of about US$95 per reindeer).

Technical solutions

In 1998, the Norwegian state paid sheep-owners US$3.29 million for measures to protect their sheep from depredations, which is 20 times more than Sweden paid in the same year. It is a paradox that Norway pays so much more for measures to protect sheep, and yet still suffers much greater depredation rates. The measure receiving the greatest amount of funding was for periodic surveillance, i.e. an increased visitation frequency, with the extra labour funded by the state (Sollie *et al.* 1996). This measure is the most preferred by the sheep-farmers, because it results in the discovery of more carcasses, which results in a more fair compensation payment, it allows depredations to be discovered more quickly, so other protective measures or an attempt to kill the marauder can be initiated quickly, and it strengthens the local economy by increasing summer employment. Neither sheep-farmers nor managers regard it as a method to reduce depredations, which occur mostly at night, but it does have a conflict-reducing effect. A similar measure, with similar results, is 'extraordinary surveillance', which is increased visitation after depredations have begun. The second most used measure, measured in money spent, is an early round-up of sheep where late depredations have occurred or are expected to occur. The state pays for the extra costs of the early round-up and for the extra feed or rent of pastures. This can be an effective method to reduce losses to wolverines and bears, which tend to kill most sheep late in the season. The sheep-raising organizations agree that this is often an effective method, but are against it because they believe the grazing resources should be fully utilized, they do not trust the state to fund this measure over a long time period, and there is often a shortage of adequate pastures. Many other measures also have been tried, such as moving sheep to grazing areas with fewer large carnivores, using collars to protect from lynx or collars with adversive repellents to stop wolverine attack, changing to other forms of agriculture or other races of sheep, using herders and guarding dogs, etc. Scientific evaluations showed that using adversive repellents against wolverines was not effective on a large scale (Landa *et al.* 1998b), that smaller races of sheep were less vulnerable to wolverine attack (Landa *et al.* 1999), and herders and guarding dogs were very effective to prevent losses, although there was some reduction in growth for one race (Krogstad *et al.* 1999). The smaller sheep races are less desirable for slaughter and the farmer receives less per kilogram, and use of herders and guarding dogs is considered to be too expensive for Norwegian conditions. Thus, although there is a myriad of techniques to limit depredation on sheep (Smith *et al.* 2000a, 2000b), there seem to be few quick-and-easy solutions to depredation in the Norwegian system of sheep husbandry. In addition, the agricultural organizations recommend that their members apply for protective measures to improve the chances of discovering depredation quickly rather than those protecting the sheep. They also want to decrease the densities of large carnivores.

In 1998–2000, the Swedish state spent on average US$165 271 annually for protective measures against wolves, bears and lynx. Areas with known depredation problems are prioritized. The most used protective measure is electric fences. The farmer receives money to buy the materials and is given advice about fence construction, but he/she has to construct it. In 1998, 54 electric fences were funded. Smaller amounts are given for early round-up of sheep and increased tending (Ahlqvist 1999).

Since 1996, the compensation to reindeer owners in Sweden has been based on the number of large carnivores within the reindeer grazing community, not on the number of reindeer killed by the large carnivores, because detection rate of large carnivore-killed reindeer, especially reindeer calves, is very low (Bjärvall *et al.* 1990). In this compensation scheme the reindeer owners accept some losses to large carnivores and the Swedish state should give full compensation, not only for the losses but also compensation for the extra work that large carnivores cause. Reindeer owners and county employees survey lynx, wolverine and wolf cooperatively every year (Östergren *et al.* 1998).

Under this new system, if the true losses decrease due to better protection, but the number of large carnivores remains the same, the compensation for large carnivore kills remains the same. However, both lynx and wolverine depend on reindeer for survival. Although the system has been accepted by the reindeer owners, they complain that the compensation is far too low, about US$95 per reindeer killed by large carnivores, compared with about US$311 per reindeer killed by trains or vehicles. The reindeer owners' organization in Sweden (SSR 2002) has estimated at least US$7.07 million in direct production losses due to large carnivores and true costs of having large carnivores at >US$17.6 million, considering the loss in future production and increased herding. This is four to five times higher than the compensation today. Furthermore, the organization proposes a maximum loss of 10% of the reindeer winter population to large carnivore predation. Today it is about 25%. Therefore they propose reducing the populations of all large carnivores. Reducing the lynx population by two-thirds and keeping the other large carnivores at present levels could solve this problem. However, the Swedish management goals for the carnivores calls for larger populations of wolverine, brown bear and wolves in the reindeer husbandry area, although the lynx population should be reduced, but not by two-thirds (Table 20.5).

The compensation for reindeer killed by large carnivores in Norway is based on documented or probable losses of reindeer. The reindeer can either be found killed by a large carnivore or probably killed by a large carnivore. The compensation is the value of slaughtered reindeer, but for a female reindeer an additional value is added for loss of future reproduction.

Reindeer losses are more difficult to mitigate than sheep, because they are the main prey for their predators (lynx and wolverine), they are out all year, and the females are very shy and vulnerable to disturbance during calving, which is a period of high predation. Measures that have been funded in Norway are increased surveillance, and herding during the calving period, and feeding eagles. These measures have not been evaluated critically, but the reindeer owners thought they helped.

CONCLUSIONS

Norway and Sweden represent the two ends of a spectrum in relation to sheep depredation. Basically, the problem was relatively easy to solve in Sweden. Requiring farmers to use technical solutions, primarily effective electric fences, works well, and losses sustained in spite of these measures are compensated. The dual Norwegian political goals of viable populations of large carnivores and partly financing the 'district policy' with subsidies for grazing free-ranging sheep simply does not work. The state invests great amounts in protective measures that are ineffective, although they are popular with the farmers and therefore reduce conflicts. Thus, losses to large carnivores and costs are vastly higher than in Sweden, measured totally or on an individual large carnivore basis. The dual political goals and high losses have lead to much conflict and frustration and are certainly an important reason behind the more modest and unclear population goals for large carnivores and much higher harvest levels in Norway than Sweden. It is hard to imagine how this conflict can be solved in Norway, unless a zonation plan is adopted, with some areas basically without free-ranging sheep and where the large carnivores would be concentrated (see Linnell *et al.*, Chapter 10). However, Sweden has few 'living rural areas' left, compared with Norway. The low loss levels and lower conflict level in Sweden allow for more ambitious population goals there, which further aggravates the loss problem, and conflict level, in Norway, due to increased dispersal of large carnivores to Norway.

Interestingly, losses to large carnivores seem to be more accepted in the reindeer husbandry area in both Sweden and Norway. However, the losses must be at an acceptable level and losses must be fully compensated. The future for large carnivores in the reindeer husbandry areas in both countries is related to these two questions. The compensation per large carnivore in Sweden has changed the discussion from the predation rate on reindeer to a discussion of the compensation per reindeer, which today is lower than the value of a slaughtered reindeer. On the other hand, the compensation system in Norway is based on documented losses, which will underestimate the true losses. Interestingly, the Swedish reindeer owners' organization has quantified both the acceptable level of loss and the amount per lost reindeer for full compensation (SSR 2002). Lynx, wolverine and brown bear can probably coexist with reindeer management by increasing the amount paid for compensation. On the other hand, wolves will cause much more problems and the coexistence of wolves and reindeer management is more doubtful. The acceptance of wolves is not only a matter of money.

The Swedish compensation system requires cooperative surveys for large carnivores by reindeer owners and the authorities. This gives managers better population estimates and reduces the data conflict, because reindeer owners can better accept estimates that they have helped obtain.

The conflict caused by large carnivores goes well beyond the loss of sheep and reindeer. Many rural people in Norway fear the loss of a way of life, loss of their youth to the lure of the cities, loss of incomes from sheep, reindeer, hunting, fishing, etc., loss of outdoor recreation – in short, loss of quality of life. Swedes in rural areas also have these fears, but they are now so few that they do not have much political power, do not have a strong Rural Policy to defend, and do not have the same economic loss as their Norwegian counterparts. The Saami people also fear the loss of their culture and language if they lose the opportunity to raise reindeer.

It is important to point out that most of the people in both countries support the goal of viable populations of large carnivores. The rural people understand this, but feel that the costs of having these large carnivores fall unfairly upon them. Sweden has come farther than Norway in balancing viable populations of large carnivores with conflicting land uses. Norwegian politicians will deal with these problems when deciding a new policy on large carnivores in 2003.

We believe that the experiences from Sweden and Norway are general and even have predictive value. We predict the Norwegian-style sheep husbandry, with free-ranging sheep and high compensation rates for depredation without requirements for effective mitigating factors, will cause high levels of conflict when instituted in areas with large carnivores or when large carnivores come into such areas. This was evident in Slovenia, where the state recently encouraged a similar system in the major bear area, where almost no sheep existed before. The subsequent high depredations and conflict have resulted in a much greater kill of bears (Krystufek and Griffiths 2003). There are also conflicts elsewhere in Europe, where large carnivores are recolonizing areas with similar husbandry and compensation systems (Breitenmoser 1998). We suggest that sheep and large carnivores can only coexist where effective mitigation measures are used and are required for payment of compensation.

ACKNOWLEDGEMENTS

This is a contribution from the ongoing large carnivore research programme in Scandinavia. We would like to thank our many colleagues in these projects for valuable discussions concerning large carnivore problems in Scandinavia. The studies are supported by several sources, but the most important are the Swedish Environmental Protection Agency, the Norwegian Directorate for Nature Management, and WWF-Sweden.

21

Managing wolf–human conflict in the northwestern United States

EDWARD E. BANGS, JOSEPH A. FONTAINE, MICHAEL D. JIMENEZ, THOMAS J. MEIER, ELIZABETH H. BRADLEY, CARTER C. NIEMEYER, DOUGLAS W. SMITH, CURT M. MACK, VAL ASHER AND JOHN K. OAKLEAF

INTRODUCTION

The grey wolf (*Canis lupus*) is the most widely distributed large carnivore in the northern hemisphere (Nowak 1995) and has a reputation for killing livestock and competing with human hunters for wild ungulates (Young 1944; Fritts *et al.* 2003). Wolves rarely threaten human safety, but many people still fear them. In the western USA, widespread extirpation of ungulates by colonizing settlers, wolf depredation on livestock and negative public attitudes towards wolves resulted in extirpation of wolf populations by 1930 (Mech 1970; McIntyre 1995). By 1970, mule deer (*Odocoileus hemionus*), white-tailed deer (*O. virginianus*), elk (*Cervus elaphus*), moose (*Alces alces*) and bighorn sheep (*Ovis canadensis*) populations had been restored throughout the western USA while bison (*Bison bison*) were recovered only in Yellowstone National Park. However, grey wolves were still persecuted. In 1974, grey wolves were protected and managed by the US Fish and Wildlife Service under the federal Endangered Species Act of 1973.

In 1986, the first recorded den in the western USA in over 50 years was established in Glacier National Park by wolves that naturally dispersed from Canada (Ream *et al.* 1989). Restoration of wolves in that region emphasized legal protection and building local public tolerance. Wolves from Canada were reintroduced to central Idaho and Yellowstone National Park in 1995 and 1996 to accelerate restoration (Bangs and Fritts 1996; Fritts *et al.* 1997). The Northern Rocky Mountains wolf population grew from 10 wolves in 1987 to 663 wolves by 2003 (US Fish and Wildlife Service *et al.* 2003) (Fig. 21.1, Table 21.1). Resolving conflicts, both perceived and real, between wolves and livestock remains the dominant social issue for the recovery programme, but perceived competition between hunters and wolves is becoming increasingly controversial (Bangs *et al.* 2001).

People and Wildlife: Conflict or Coexistence? eds. Rosie Woodroffe, Simon Thirgood and Alan Rabinowitz. Published by Cambridge University Press.

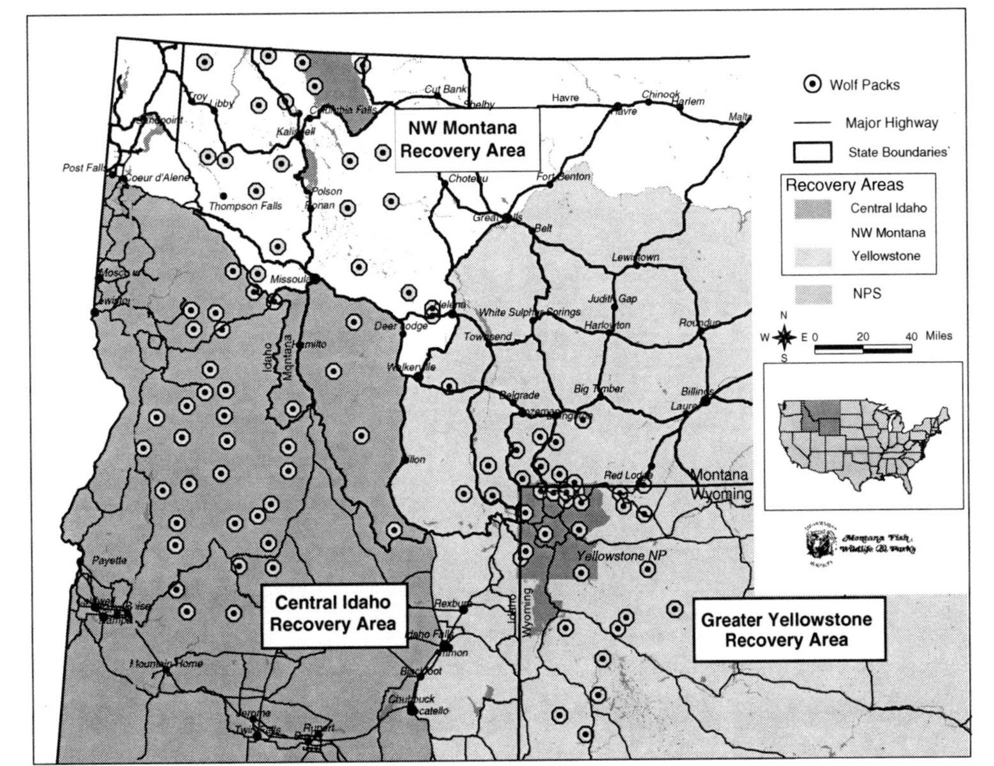

Figure 21.1. Wolf pack distribution in Montana Idaho and Wyoming in 2002.

Table 21.1. *Confirmed wolf depredation, minimum autumn wolf population and wolf removal in Montana, Idaho and Wyoming, 1987–2003*

	1987	1988	1989	1990	1991	1992	1993	1994	1995	1996	1997	1998	1999	2000	2001	2002	2003	Total
Montana																		
Wolf predation[a]																		
Cattle	6	0	3	5	2	1	0	6	3	10	19	10	20	14	12	20	25	156
Sheep	10	0	0	0	2	0	0	0	0	13	41	0	25	7	50	84	86	318
Dogs	0	0	0	1	0	0	0	0	4	1	0	1	2	5	2	5	1	22
Other[b]	0	0	0	0	0	0	0	0	0	0	0	0	0	0	4	5	0	9
Total wolves	10	14	12	33	29	41	55	48	66	70	56	49	74	97	123	183	182	1142
Wolves moved	0	0	4	0	3	0	0	2	8	22	20	0	14	6	17	0	0	96
Wolves killed[c]	4	0	1	1	0	0	0	0	0	5	18	4	19	7	8	26	34	127
Wyoming																		
Wolf predation[a]																		
Cattle	0	0	0	0	0	0	0	0	0	0	2	2	2	3	18	23	34	84
Sheep	0	0	0	0	0	0	0	0	0	0	56	7	0	25	34	0	7	129
Dogs	0	0	0	0	0	0	0	0	0	0	0	3	6	6	2	0	0	17
Other[b]	0	0	0	0	0	0	0	0	0	0	0	0	1	0	0	0	10	11
Total wolves	0	0	0	0	0	0	0	0	21	40	86	112	107	153	189	217	234	1159
Wolves moved	0	0	0	0	0	0	0	0	0	0	1	0	0	0	0	0	0	1
Wolves killed[c]	0	0	0	0	0	0	0	0	0	0	2	3	1	2	4	6	18	36
Idaho																		
Wolf predation[a]																		
Cattle	0	0	0	0	0	0	0	0	0	1	1	9	11	15	10	9	6	62
Sheep	0	0	0	0	0	0	0	0	0	24	29	5	64	48	54	15	118	357
Dogs	0	0	0	0	0	0	0	0	0	1	4	1	7	0	2	4	4	23
Other[b]	0	0	0	0	0	0	0	0	0	0	0	0	0	0	0	0	0	0

Total wolves	0	0	0	0	0	0	0	0	14	42	71	114	156	187	251	263	340	1438
Wolves moved	0	0	0	0	0	0	0	0	0	1	0	3	5	10	1	0	0	20
Wolves killed[c]	0	0	0	0	0	0	0	0	0	1	1	0	3	11	7	14	7	44
Total, three States																		
Wolf predation[a]																		
Cattle	6	0	3	5	2	1	0	6	3	11	22	21	33	32	40	52	65	302
Sheep	10	0	0	0	2	0	0	0	0	37	126	12	89	80	138	99	211	804
Dogs	0	0	0	1	0	0	0	0	4	2	4	5	15	11	6	9	5	62
Other[b]	0	0	0	0	0	0	0	0	0	0	0	0	1	0	4	5	10	20
Total wolves	10	14	12	33	29	41	55	48	101	152	213	275	337	437	563	663	756	3739
Wolves moved	0	0	4	0	3	0	0	2	8	23	21	3	19	16	18	0	0	117
Wolves killed[c]	4	0	1	1	0	0	0	0	0	6	21	7	23	20	19	46	59	207
Northwest Montana Recovery Area																		
Wolf predation[a]																		
Cattle	6	0	3	5	2	1	0	6	3	9	16	9	13	10	8	11		102
Sheep	10	0	0	0	2	0	0	0	0	0	30	0	19	2	5	13		81
Dogs	0	0	0	1	0	0	0	0	3	1	0	0	2	3	1	4		15
Total wolves	10	14	12	33	29	41	55	48	66	70	56	49	63	64	84	109		803
Wolves moved	0	0	4	0	3	0	0	2	2	10	7	0	4	0	5	0		37
Wolves killed	4	0	1	1	0	0	0	0	0	4	14	4	9	4	3	9		53
Yellowstone Recovery Area																		
Wolf predation[a]																		
Cattle									0	0	5	3	4	7	22	37		78
Sheep									0	13	67	7	13	39	117	74		330
Dogs									1	0	0	4	7	8	4	1		25
Total wolves									21	40	86	112	118	177	218	272		1044

Table 21.1. *(cont.)*

	1987	1988	1989	1990	1991	1992	1993	1994	1995	1996	1997	1998	1999	2000	2001	2002	2003	Total
Wolves moved									6	8	14	0	0	6	8	0		42
Wolves killed									0	1	6	3	9	6	9	22		56
Central Idaho Recovery Area																		
Wolf predation[a]																		
Cattle									0	4	1	10	16	15	10	10		66
Sheep									0	24	29	5	57	39	16	15		185
Dogs									0	1	3	1	5	0	1	4		15
Total wolves									14	42	71	114	141	192	261	285		1120
Wolves moved									0	5	0	3	15	10	5	0		38
Wolves killed									0	1	1	0	6	10	7	14		39
Total, three States, three Recovery Areas																		
Wolf predation[a]																		
Cattle	6	0	3	5	2	1	0	6	3	13	22	22	33	32	40	58		246
Sheep	10	0	0	0	2	0	0	0	0	37	126	12	89	80	138	102		596
Dogs	0	0	0	1	0	0	0	0	4	2	3	5	14	11	6	9		55
Total wolves	10	14	12	33	29	41	55	48	101	152	213	275	322	433	563	666		2967
Wolves moved	0	0	4	0	3	0	0	2	8	23	21	3	19	16	18	0		117
Wolves killed	4	0	1	1	0	0	0	0	0	6	21	7	24	20	19	45		148

[a] Numbers of animals confirmed killed by wolves in calendar year.
[b] Includes 1 foal in 1999, 4 llamas in 2001, 5 llamas in 2002 and 10 goats in 2003.
[c] Includes 13 wolves legally shot by ranchers. Others killed in government control efforts.

WOLF HABITAT IN THE US NORTHERN ROCKY MOUNTAINS

The Wolf Recovery Plan for the Northern Rocky Mountains of Montana, Idaho and Wyoming identified preferred wolf habitat as large areas of public land with an adequate year-round supply of wild prey and few livestock (US Fish and Wildlife Service 1987). Based on those criteria northwestern Montana, central Idaho, and the Greater Yellowstone Area, which is mostly in northwestern Wyoming, were recommended for wolf restoration (Fig. 21.1). Each of these areas has a large refugium of National Park or US Forest Service Wilderness, where motorized access and livestock grazing are limited. Mountainous habitat is often undeveloped forested federal public land managed for multiple uses and is typically leased to the adjacent ranches for summer livestock grazing. Lower-elevation valleys are often in private ownership and used for livestock production. Ungulates often summer on higher elevation public lands but winter at lower elevations, so many wolves use private land. The proximity of the Northern Rocky Mountains recovery areas to one another and public land corridors between them, the genetic diversity of reintroduced wolves and the capability of wolves to disperse long distances indicated that genetic diversity is unlikely to be a threat to wolf conservation in the Northern Rocky Mountains (Forbes and Boyd 1997).

Oakleaf (2002) used geographic information system (GIS) modelling to predict preferred wolf habitat in the Northern Rocky Mountains. Higher degree of forest habitat, lower human population density, higher elk density, and lower domestic sheep density were the primary factors related to wolf occupancy. Central Idaho contained the greatest amount of preferred wolf habitat (77 596 km^2), while the Greater Yellowstone Area (45 900 km^2) and northwestern Montana (44 929 km^2) had substantial amounts. However, few believe that any of the recovery areas could conserve an isolated wolf population by itself. Large carnivores are susceptible to extinction even in protected areas because of their large home ranges (Woodroffe and Ginsberg 1998). Wolves must live in and disperse through areas occupied by people and their livestock to ensure their long-term viability in the Northern Rocky Mountains.

NEGATIVE HUMAN IMPACTS ON WOLVES

Humans kill wolves

While wolf populations were severely reduced by human settlement and depletion of ungulates in the late nineteenth century, wolves were extirpated by extensive poisoning and federal and state government-led wolf campaigns. Gipson *et al.* (1998) suggested stories of almost supernatural western wolves that supposedly killed thousands of livestock in the early twentieth century and were all but impossible to kill were largely the result of the political efforts of the Biological Survey (the agency that evolved into the US Fish and Wildlife Service) to create a purpose and funding for itself in Congress. This propaganda campaign capitalized on public hatred of wolves, agricultural development and wolf vulnerability to human-caused mortality, particularly poisoning. By 1930, the western remnants of a once vast southern continental wolf population vanished.

Biologically, wolves are easy to manage and control because of their distinct territories, pack structure, naturally low density and high recruitment potential. Wolves are habitat generalists and can live almost anywhere that has ungulate prey and moderate human-caused mortality. Wolf restoration depended on reducing annual human-caused mortality below 38% (Keith 1983). Unless wolves have been extirpated, restoration can be simply a matter of reducing human-caused mortality to a sustainable level.

Humans cause 85% of adult wolf mortality in the Northern Rocky Mountains (Pletscher *et al.* 1997; Bangs *et al.* 1998). Wolves have been killed illegally; accidentally by vehicles, trains, or traps; by other wolves; by their prey; in accidents (avalanches, drowning); and by other large predators. However, the most common documented cause of radio-collared wolf death is agency control in response to livestock depredation (Bangs *et al.* 1998). Despite the fact livestock depredations are rare, that competition for wild ungulates is minimal, and there have been virtually no restrictions of human recreational or commercial activities to enhance wolf survival and restoration, there remains widespread fear and resentment of wolf restoration. Some people equate wolf recovery with 'outsiders' who would use wolf restoration and federal authority to usurp state rights, control local land uses, confiscate guns, eliminate hunting and depopulate rural areas. Some people have publicly called for widespread wolf shooting and poisoning. Illegal killing is difficult to document but is probably the biggest single cause of adult wolf death. This sporadic illegal killing, including some poisonings with Compound 1080, strychnine and other prohibited chemicals, has not prevented wolf population growth, but it has affected wolf distribution. Wolves have not been able to persist in open prairie habitat in Montana or Wyoming (Fig. 21.1), where they are most vulnerable and least tolerated by people (Bangs *et al.* 1995).

Introduction of diseases and parasites and potential for canid hybridization

Wolves are vulnerable to dog (*Canis familiaris*) diseases and parasites (Brand *et al.* 1995) and can hybridize with dogs and coyotes (*Canis latrans*) (Wayne *et al.* 1995). Humans incidentally transmit dogs' and canid diseases and parasites into wolf populations and habitat modification has greatly increased coyote distribution. However, these factors have not significantly affected wolf population viability in North America.

POSITIVE HUMAN IMPACTS ON WOLVES

Preservation of wolf habitat

In the late 1880s, large areas in the Northern Rocky Mountains were still undeveloped. Many of those lands remained in federal ownership after the western states were created, primarily because they were not productive for agriculture. These public lands were still set aside for commodities such as water, timber, forage and minerals. Most National Parks were originally established because of their unique geologic features or scenic beauty rather than wildlife. On public lands, wildlife and other uses were subservient to production of raw materials. Over the next century, the public's expectations of how these lands should be managed expanded to include outdoor recreation, wildlife conservation and other values associated with nature such as biodiversity, that reflected a national rather than primarily local public perspective (Keiter 1998).

Restoration of wolf prey

The restoration of ungulate populations by state game agencies was one of the most remarkable achievements of wildlife management and without it wolf restoration would be impossible. Native ungulate populations that had been extirpated in the early twentieth century were restored throughout much of their historic range in North America by the 1970s. Ungulate research also provided insights into the ecological role of predators. This new information revealed a more positive image of wolves than the sinister one portrayed in folklore. The ultimate result of this fresh outlook by wildlife professionals, a more informed public and the increasing national concern for a host of environmental issues was that public attitudes about wolves changed dramatically.

Wolf conservation programmes

The grey wolf was protected under the Endangered Species Act in 1974, after it had been deliberately eliminated from nearly 99% of its historic range in the contiguous USA. The Act prohibited all attempts to kill wolves, and mandated that all federal agencies use their authorities to help promote recovery. Changing societal values resulted in enforcement of strict laws with penalties up to one year in jail and a US $100 000 fine for illegally killing wolves. The Endangered Species Act also funnelled federal funding into wolf research projects and public information programmes, and initiated three successful wolf restoration efforts in the Midwest, the Northern Rocky Mountains and Southwest USA (Phillips *et al.* 2004).

NEGATIVE WOLF IMPACTS ON HUMANS

Wolf attacks on humans

Wolves in fairy tales and throughout most popular media, are portrayed as dangerous (Boitani 1995). This perception is a reason that people fear wolves and oppose wolf restoration (Bath 1992; Kellert *et al.* 1996). Wolves have attacked humans but incidents are remarkably rare (Quigley and Herrero, Chapter 3). Wolf attacks are often the result of rabies or some human-caused contributing factor such as dog–wolf conflict or food habituation (Linnell *et al.* 2002). In North America, healthy wild wolves have not been documented to kill anyone since European colonization (McNay 2002). There is no documentation of a wild wolf attacking a person in the Northern Rocky Mountains. Wild wolves kill prey much larger and better able to defend themselves than unarmed humans. Dogs kill a dozen and injure hundreds of thousands of people in North America each year, implying wild wolves choose not to attack people. Documented cases of rabid wolves attacking people in North America are rare; the last human fatality was in Alaska in 1945 (Ritter 1981). Despite some public apprehension, human safety was not a significant issue during wolf restoration efforts, because of accurate information and the fact that any wolf that threatened humans can be legally killed.

Livestock depredations

Wolves can be effective predators and scavengers on livestock (Fritts 1982) and access to livestock and livestock carrion can increase wolf density (Hovens *et al.* 2000; Vos 2000). However, cultures that raise livestock strongly dislike wolves (Lopez 1978; Fritts *et al.* 2003). In today's society, livestock producers typically have the strongest dislike of wolves compared to other segments of society (Boitani 1995; McIntyre 1995; Kellert *et al.* 1996; Williams *et al.* 2002). In the late nineteenth century, as native prey populations diminished, wolf depredation on livestock was perceived as a major problem (McIntyre 1995). While the actual rate and importance of historic wolf depredation will never be known, it was probably exaggerated (Gipson *et al.* 1998). Wolf depredation on livestock is lower than expected given the wolf's effectiveness as a predator and its high exposure to livestock (Fritts *et al.* 1992; Bangs *et al.* 1995; Oakleaf *et al.* 2003). From 1987 to 2002 wolves in the Northern Rocky Mountains were confirmed to have killed 237 cattle (0 bulls, 8 steers, 17 cows, 26 yearlings, 165 calves, 21 unknown; 30% on public land; with an average of 1.3 cattle (range 1–7) killed in 173 depredations events that affected 74–112 cattle producers) and 593 sheep (33 rams, 85 ewes, 66 lambs, 409 unknown; 52% on public land; with an average of 7.0 sheep (range 1–41) killed in 80 depredation events that affected 31–49 sheep producers). Cattle (67%) and sheep (75%) were killed most often from April to September when grazing is most dispersed and young livestock are most common. Small body size makes livestock more generally vulnerable to a wider variety of predators (Fritts 1982), and allowed wolves to kill any sheep. Calves were more susceptible to wolf depredation than adult cattle (Fritts *et al.* 1992; Oakleaf *et al.* 2003). Wolf depredations were dispersed and sporadic and only a few livestock producers incurred multiple losses. While unimportant to the regional livestock industry, wolf depredations could affect the economic viability of a few small ranches, primarily those dependent on remote public land grazing allotments.

We reviewed statistics on livestock numbers, losses, and predation in 2000 to assess the relative importance of wolf depredation to the livestock industry in Northern Rocky Mountains (National Agriculture Stastistics Service 2001a, b). There were 2 210 000 sheep, 9 300 000 cattle and 437 wolves in Montana, Idaho and Wyoming in 2000. Livestock producers there reported that they lost 235 000 cattle and 195 000 sheep from all causes in 2000. Of those losses 82 200 (42%) sheep and 10 300 (4.4%) cattle were reportedly killed by predators. Coyotes were responsible for over 70% of those losses. In 2000, wolves killed 80 sheep and 32 cattle in the Northern Rocky Mountains or 0.04% and 0.01% of all losses, and 0.01% and 0.31% of all predator-caused losses, respectively. Statistics from Montana from 1986–91 (Bangs *et al.* 1995) and Minnesota (Fritts *et al.* 1992) also indicted a very low percentage of wolf-caused loss.

Before wolves were reintroduced livestock producers estimated they annually lost 8340 cattle and 12 993 sheep in the Greater Yellowstone Area and 12 314 cattle and 9 366 sheep in central Idaho (US Fish and Wildlife Service 1994a). The service predicted 100 wolves would kill about 10 to 20 cattle and 50 to 70 sheep per area per year, a small fraction of the livestock losses that were already occurring from all causes before wolves were present (US Fish and Wildlife Service 1994a). Wolf damage was estimated at US $2000–30 000 annually. Despite there being an average of 178 wolves in the Greater Yellowstone Area and 202 wolves in central Idaho from 1998 to 2002, confirmed livestock losses have been fewer than predicted averaging 14 and 12 cattle and 47 and 24 sheep, respectively (US Fish and Wildlife Service *et al.* 2003).

Wolves rarely attack other types of livestock in the Northern Rocky Mountains, with only one horse, ten goats and nine llamas confirmed killed. However, the US Fish and Wildlife Service considers any wolf attack on livestock serious. We relocated problem wolves 117 times and killed 207 to reduce conflicts (Table 21.1). Agency-initiated research with radio-collared livestock indicated that wolf depredation was rare but confirmed losses may be a fraction of actual wolf-caused losses near active dens in densely forested and remote public land grazing allotments. Seven calves might be killed and not documented for every confirmed wolf depredation in the worst-case scenarios (Oakleaf *et al.* 2003). If that ratio is typical, livestock depredations by wolves could still only be significant to a few public land grazing permittees.

Wolf depredation is a rare cause of livestock mortality, but it is inordinately controversial. Nearly every wolf depredation on livestock in the Northern Rocky Mountains becomes a major local and state-wide media story. This high level of publicity exaggerates the actual impacts of wolf depredation. Wolves are routinely discussed at the local, state and federal political level. Several county and state governments have passed 'resolutions' declaring the wolf an 'unacceptable species' and calling for its extirpation, and there has been litigation from livestock interests. Conversely, control by the US Fish and Wildlife Service of depredating wolves generates angry correspondence and litigation from animal rights, anti-public land grazing and pro-wolf advocates.

Wolf attacks on dogs

Wolves infrequently kill dogs and usually do not eat them (Fritts and Paul 1989; Kojola and Kuittinen 2002; Treves *et al.* 2002). To date 57 dogs (10 pet, 11 guard, 19 hunting hounds, and 16 herding, 1 undocumented, average 1.2 dog per depredation, range 1–4 per attack) have been confirmed killed by wolves in the Northern Rocky Mountains. Although Humane Society organizations in each state euthanize thousands more dogs than wolves kill, wolf depredation on dogs is a serious social issue. It is one of the most difficult conflicts that Northern Rocky Mountains biologists address because people form particularly strong emotional bonds with dogs. Depredations near homes also raise fears for human safety and anger over the perceived violation of personal space. Compensation is only provided for herding and guarding dog depredations, but trained hunting dogs can be worth thousands of dollars. Wolves that attack dogs on private land can be legally relocated or killed (US Fish and Wildlife Service *et al.* 2003), but to date none has been because most attacks were isolated incidents in remote areas.

Techniques to reduce wolf conflict with domestic animals

Wolf restoration in the Northern Rocky Mountains has little measurable direct impact on the lives of most people; however, the psychological impact of the programme has been enormous. It is surprising how many local livestock producers whose ancestors founded their ranch remarked that a relative had killed 'the last wolf' in this or that valley, county or state, and how that lone act became a source of pride and remembrance. One hundred years ago, elimination of wolves was seen as a righteous duty and was the symbol that civilization had 'won'. Wolf restoration became the symbol of change from an agricultural heritage and lifestyle to something else that was unfamiliar and undefined.

To moderate the real and psychological effects of wolf restoration, the US Fish and Wildlife Service and its cooperators implemented a wide variety of programmes to minimize the potential for and extent of wolf–human conflict. The Service took a highly publicized position that there was already adequate habitat for wolf recovery and traditional uses of private and public land would not be modified. The Service also stated that depredating wolves would not be the foundation for wolf restoration and chronic depredating wolves would be killed (US Fish and Wildlife Service 1988).

Before wolves were reintroduced, the US Fish and Wildlife Service established regulations that empowered local landowners and livestock producers (US Fish and Wildlife Service 1994b). In 2003 the Service liberalized and expanded that flexibility for problem wolf management to the entire northwestern USA (US Fish and Wildlife Service 2003). Livestock producers are routinely provided radio telemetry receivers so they can locate radio-collared wolves on their property. Landowners can harass wolves in a non-injurious manner at any time. Any wolf seen attacking livestock on private land can be legally shot, and five wolves have been killed. Over a dozen livestock owners obtained US Fish and Wildlife Service permits to shoot wolves seen attacking their livestock on public grazing allotments, but no wolves have been killed. In areas with chronic livestock depredations landowners received permits to shoot wolves on sight. Since the first permits were issued in 1999, eight wolves have been killed, three in 2002 and five in 2003. After a brief training course, over 100 landowners were issued permits and less-than-lethal munitions (12-gauge shotgun cracker shells and rubber bullets), to harass wolves near their livestock or dwellings. Several wolves have been hit but none was seriously injured, and those residents report that wolves are more wary. Biologists have temporarily provided road-killed ungulates to denning wolves when the potential for conflict appeared highest such as after young livestock were turned out on rangeland before ungulates calved. In intensively grazed areas on private land biologists disturbed soon-to-be-active den sites or harassed wolves from rendezvous sites causing them to relocate their pups away from livestock.

Even though the vast majority of livestock producers would never experience wolf depredation, wolf advocates recognized that some losses would occur. Defenders of Wildlife started a private livestock compensation fund in 1987 that has paid ranchers nearly US $275 000 for confirmed and probable damage to livestock and livestock herding and guarding animals caused by wolves (Fischer 1989; Nyhus *et al.*, Chapter 7). Compensation is based on professional field investigations of livestock death routinely conducted by US Department of Agriculture Wildlife Services (Paul and Gipson 1994). The US Fish and Wildlife Service contracts Wildlife Services to investigate reports of wolf damage and uses their findings to determine whether wolf control is warranted (Bangs *et al.* 1995). Defenders of Wildlife also helps livestock producers avoid wolf depredation by cost-sharing guard animals, fencing, fladry, extra livestock surveillance, disposal of livestock carcasses, alternative grazing pastures in lower-risk areas, attempting to purchase and retire public land grazing allotments in areas of chronic conflict, and funding research on non-lethal methods to reduce conflicts (Bangs and Shivik 2001). While these non-lethal efforts reduced conflicts, many were expensive to implement and none has been proven widely effective. Lethal control of chronic depredating wolves is still required.

Competition with human hunters for ungulates

The average adult wolf eats about 5 kg of prey per day. The US Fish and Wildlife Service (1994a) predicted that 100 wolves in central Idaho would kill the equivalent of about 1600 ungulates annually (primarily mule deer and elk) out of a population of nearly 241 000. That level of wolf predation would have little effect on any ungulate population or on the annual hunter harvest of 33 358 ungulates, except for the harvest of female elk that might be reduced by 10% to 15%. The reduced hunter opportunity would result in yearly theoretical economic losses of US $757 000–1 135 000 in hunter benefits (i.e. what hunters thought the loss of that female elk hunting was worth to them) and US $572 000–857 000 in potential reduced hunter expenditures (i.e. what hunters would have spent hunting female elk).

In the Greater Yellowstone Area the Environmental Impact Statement (US Fish and Wildlife Service 1994a) predicted that 100 wolves would kill the equivalent of 1200 ungulates annually, primarily elk, out of nearly 100 000. Few moose, bighorn sheep, mountain goats, bison and antelope would be killed by wolves. A recovered wolf population might reduce elk 5–30% (the higher level only in some small herds), deer 3–19%, moose 7–13% and bison up to 15%. The Environmental Impact Statement predicted that wolf predation would not affect hunter harvest of males but could reduce harvest of female elk, deer and moose for some herds. Wolf predation would cause annual economic losses estimated at US $187 000–465 000 in hunter benefits and US$207 000–414 000 in reduced hunter expenditures.

The issue of how much wolf predation and 'harassment' affects ungulate populations and hunter harvest remains a major public concern. Wolf predation may or may not affect ungulate populations and hunter harvest depending on a wide number of variables (Boyce 1995; Vales and Peek 1995; Kunkel 1997). In anticipation of potential conflict, the regulations for wolf reintroduction allowed for the relocation of wolves if ungulate populations were being significantly impacted. To date, no wolves have been moved because there has been no documented need.

Despite data, including ungulate surveys, several groups with reportedly thousands of members have recently formed around the rumour that ungulate populations have been decimated by wolf predation. This is a powerful psychological issue and thousands of frustrated and angry hunters have the potential to cause significant illegal wolf mortality. State wildlife agencies report that hunters at game check stations routinely ask about 'wolf damage' to ungulate populations even when there are few wolves in that area or ungulate populations and harvests are at historically high levels. It is a foregone conclusion that any declines in ungulate populations or hunter harvest will be attributed solely to wolf predation by wolf opponents and adamantly denied as being wolf-caused by wolf advocates.

To address these public concerns, cooperative research on wolf–ungulate relationships has been continuously initiated and funded by the US Fish and Wildlife Service and other agencies since the 1980s. These often university-led multi-year studies predicted that wildlife managers should anticipate some ungulate population declines and reduced hunter harvest and recommended more intensive monitoring of ungulate population declines to detect changes early (Kunkel 1997). These data continue to be gathered and publicized to better inform the public about the effect of wolf predation on ungulate populations and hunter harvest. However, the effect of wolf predation on ungulate populations and subsequent hunter harvest appears minor and very difficult to detect despite the intensive research conducted to date.

THE SYMBOLISM OF WOLVES TO HUMANS

Perhaps the most interesting aspect and significant effect of wolves is their unusually strong symbolism to humans (Fritts *et al.* 2003). Humans have used wolves as very powerful symbols in many cultures for thousands of years (Lopez 1978; Boitani 1995). Today, wolves have little material effect on people but wolves make many people's lives more interesting. Wolves can positively impact people by their strong symbolism and entertainment value. People can enjoy the opportunity to interact (i.e. hunting, trapping, viewing, photography) with wolves, and wolves can enhance the natural ecological integrity and wildlife diversity of wildlands, enhancing their value to some people.

Economic analysis predicted that wolf restoration, primarily associated with tourism in Yellowstone National Park, would generate up to US$23 000 000 in economic activity in the Northern Rocky Mountains annually (US Fish and Wildlife Service 1994a). In addition, wolf restoration had a potential existence value (i.e., what people thought having wolves in Yellowstone was worth) of US$8 300 000 annually. While specific follow-up studies have not been completed, the trends predicted in the Environmental Impact Statement seem to be occurring (J. Duffield pers. comm.). Most wolves are still fairly wary of people, but high prey density, open habitat and a highway in the northern portion of Yellowstone National Park provides outstanding wolf-viewing opportunities. On 26 June 2002, a Yellowstone National Park naturalist calculated that over 100 000 park visitors had seen wolves since 1995 (R. McIntrye pers. comm.). Several commercial wildlife-viewing tour operators have started since 1995, and wolf-viewing is a cornerstone of their business. Traffic actually became so congested that wildlife-watchers prevented wolves from crossing the highway. Beginning in 2000, National Park 'guards' were hired to direct traffic to protect visitor safety and allow wolves to pass. The powerful symbolism of wolves is still evident and is particularly deeply ingrained in Western culture. Wolves are one of the most popular species in North American wildlife art today. Northern Rocky Mountains wolves, especially those in Yellowstone National Park, have been featured in literally thousands of stories. Wolves have appeared in every form of international, national and local media. The entertainment value of wolves has never been calculated but we suspect it could be worth millions of dollars annually. It appears that whether people love wolves or hate wolves, everyone likes information and gossip about wolves and to share their perspective. Since 1987, we have presented nearly 1000 wolf information programmes to a wide variety of groups. Nearly 60% of the federal Northern Rocky Mountains wolf programme activities involved providing public information (Fritts *et al.* 1995).

Wolf-related legislation or resolutions, universally critical towards wolves and the federal government, are common in Montana, Idaho and Wyoming state legislatures and county governments. Northern Rocky

Mountains wolf recovery staff routinely respond to correspondence generated by members of the US Congress and state Governors. The level of local, national and international interest in wolf restoration was evident when 160 000 public comments from 40 countries were received on the Environmental Impact Statement to reintroduce wolves to central Idaho and Yellowstone National Park (US Fish and Wildlife Service 1994a). That response was one of the largest volumes of public comment received on a planned federal action up to that time. Although the government's legal position has prevailed, since 1994 both wolf opponents and proponents have initiated various wolf-related litigation, some of it going to the US Court of Appeals and Supreme Court level.

CONCLUSIONS

Wolf extermination took place because of negative human values about wildlife in deference to other social values. As a result of those values, government programmes encouraged wildlands to be modified for direct human use; wildlife and indigenous people to be replaced by settlers, crops and livestock; and predators to be exterminated. Society ensured those values and opinions were perpetuated by mythology, popular media, schools and social events. Without large blocks of livestock-free habitat or abundant prey to minimize conflicts, and facing extreme human prejudice and persecution, wolves had no future in the early-twentieth-century American West. Societal values changed, probably fuelled most by urbanization. Society set aside public lands and eventually gave some of them purposes other than commodity production. Hunters established state game agencies to restore ungulate populations but ultimately other species were also valued and conserved. Some predator-control practices such as widespread poisoning became socially unacceptable and were banned. Human values about nature changed and these perspectives were reinforced by scientific research and popular media. 'Charismatic megafauna' benefited from new media technology like television and film, especially wolves because of their large size, familiar dog-like beauty and behaviour, easily anthropomorphized social structure, and inherent interest to people. Wolves also benefited from that fact that few modern people actually had first-hand experience of the real problems associated with living with them (Williams *et al.* 2002).

The Endangered Species Act numerical and distributional population recovery goals established for the grey wolf in the Northern Rocky Mountains were achieved in late 2002 (US Fish and Wildlife Service 1994a; US Fish and Wildlife Service *et al.* 2003). Currently, wolf packs successfully occupy areas where only 20 years ago biologists believed wolves could not exist because the level of conflict would be intolerable (US Fish and Wildlife Service 1988). Successful wolf restoration was inevitable once human values changed, because the biological requirements for a viable wolf population – space and prey – had already been restored. There was enough remote habitat and wild prey so that, with agency management, conflicts between wolves and people were rare. The majority of the public perceived conflicts were rare enough and the presence of wolves valuable enough that local people should tolerate wolves.

Active management of wolf-caused conflicts will be required to maintain public tolerance. (Fritts and Carbyn 1995; Mech 1995; Fritts *et al.* 2003). A viable wolf population can persist in the Northern Rocky Mountains because large areas of suitable habitat are secure in public ownership. Conflict with livestock will remain low because sheep or other highly vulnerable types of livestock are unlikely to return to their former abundance because of global market competition and changing social values about acceptable uses of public land. Ungulate populations will continue to thrive on public and private land and state wildlife management agencies will continue to manage for high population levels for hunting, ensuring an adequate prey base for wolves. Professional wildlife managers should minimize wolf-caused problems to reduce the likelihood of a backlash of public opinion against wolves. Such a backlash could result in widespread vigilantism or public calls for government extermination programmes (Mech 1995). Given some minimal level of secure habitat, wild prey and human tolerance, wolf populations will persist. Wolves will eventually spread to adjacent areas as the changing social values that allowed recovery in the Northern Rocky Mountains, continue to manifest themselves elsewhere.

22

Policies for reducing human–wildlife conflict: a Kenya case study

DAVID WESTERN AND JOHN WAITHAKA

INTRODUCTION

The human–wildlife conflict is a face-off between people and wildlife over space or resources. Typically, conflict involves wildlife that consumes pasture or crops, or attacks domestic stock or even humans – and people who kill wildlife in reprisal (Woodroffe *et al.*, Chapter 1, Thirgood *et al.*, Chapter 2). For humans, the conflict is shrinking as a dwindling proportion of people encounter wildlife. For wildlife, the reverse is true. With between a third and a half of all land transformed and used by humans (Vitousek *et al.* 1997), natural habitats are shrinking. A growing proportion of wildlife competes with people and survives only through conservation measures.

The term 'wildlife' originally referred to large or conspicuous animals. Over the last century, however, the term has come to include a wider variety of species as our sensibilities have broadened from parochial to universal human rights, and recently to the intrinsic value of life as a whole (Nash 1989). The scope of conservation has grown in lockstep, from its origins in sport hunting to wildlife conservation and more recently to conservation of all life forms (Shabecoff 1993). Today most nations are revising their policies and legislation to reflect the global aims of the Convention on Biological Diversity (Heywood 1995; Hempel 1996; Convention on Biological Diversity 2005). Seen in this light, human–wildlife conflict should apply to any species competing with human interests of any sort.

Expanding human–wildlife conflict to human–biodiversity conflict would not be problematic if we valued all species the same. We don't, of course. Nature is untamed and threatening to one person and fascinating and valuable to another (Weston 1999). We see and treat problem locusts (Acrididae) and problem tigers (*Panthera tigris*) differently – largely by ignoring the less charismatic species altogether. Elephants (*Loxodonta*

People and Wildlife: Conflict or Coexistence? eds. Rosie Woodroffe, Simon Thirgood and Alan Rabinowitz. Published by Cambridge University Press.

africana) are endearing creatures to Western animal lovers yet dangerous marauders to African farmers. The less like us a species is, and the more it threatens our interests, the less likely we are to save it. This mixed view of species echoes the deep ambivalence towards nature at the heart of human–wildlife conflict. Treating species according to how they affect us is only natural. Biodiversity policy will therefore need to face the realities of human–wildlife conflict if it is to be publicly acceptable and effective.

Human–biodiversity conflict compounds our ambivalence over wildlife by adding millions of species that we know and care little about. That is not to say they are unimportant. To the contrary, evidence points to our well-being depending on ecological processes driven and maintained by a diversity of life forms (Tillman 2000).

How, then, do we conserve all species when we see them in such disparate ways? How do we develop conservation goals, policies and laws for all life forms, yet confront the conflicts between people and wildlife (often rural and poor) and among people holding differing opinions on the matter? Can we do so using laws and regulations that are democratic, fair and consistent with our larger societal aspirations, yet also practical and affordable?

Here we look at one case study, Kenya, in East Africa, in which both the conservation of biodiversity and resolution of human–wildlife conflict were the focus of policy formulation. Our aim is to look at trends in human–wildlife conflict over the last few decades, explore the implications for conservation and look at policies for minimizing conflict within the overall conservation goal of conserving biodiversity. Our emphasis will be on the policy process and role in human–wildlife conflict, rather than on specific mitigation measures – the latter are discussed extensively in other parts of this book.

KENYA: A CASE STUDY

To trace the historical trends in human–wildlife conflict in Kenya, we examined official government records as far back as possible. All records prior to Independence (in 1963) were consolidated centrally in annual reports (Game Department 1920–63). After 1963 annual consolidation was abandoned, making it difficult to collate records from every field station. We have therefore used the information already compiled, including an elephant database that the Kenya Wildlife Service set up in 1989. Here we focus on elephants, large predators and primates, species for which there are records over several decades and which have been at the centre of human–wildlife conflict (Edwasi 1994).

NUMERICAL AND SPATIAL TRENDS IN HUMAN–WILDLIFE CONFLICT

The longest and most complete official records available are for elephants. Records of the number and locations of elephants killed in conflict with people date from the 1920s. Table 22.1 summarizes the numbers by district. The numbers of elephants killed on control climbed steeply. Control killing spread from a few locations to much of the country between the 1920s and 1960s. Most elephants were shot to open up lands for settlement or as crop-raiders. By the 1960s, many agricultural areas, such as Trans Nzoia and most of Laikipia, had been cleared of elephants and no further incidents of conflict were reported. The annual reports show the same trend, with numbers of all wildlife shot nationally on control rising from 5864 in 1950 to 10 765 in 1962. At least in the public's mind, the conflict continued to intensify after the 1960s, with the number of annual press reports of elephant conflict rising from 10 in the 1970s to 42 in the 1980s to over 96 in the 1990s

The nature of conflict and response by the Game Department varied greatly from district to district. The policy of the Game Department was essentially to solve the problem of conflict in the high-potential agricultural areas by eliminating problem animals altogether. Over 80% of all elephants shot on control between 1950 and 1963 were destroyed in agricultural districts. The scale of such schemes is evident from the control programme the Game Department undertook in Makueni area, where 996 rhinos were killed between 1944 and 1946 to open up 50 000 acres (200 km^2) for settlement (Hunter 1952).

Table 22.1. *Elephants killed on control by districts, prior to Independence in 1963 (the empty cells reflect no reported incidents)*

Area	1920s	1930s	1950s	1960s
Laikipia	60	104	106	
Trans Nzoia			2073	
Mount Kenya		270	360	82
Narok		71	1491	
Tana River			529	943
Kilifi			1237	257
Kwale	12	91	2172	346
Lamu		49	2061	
North Eastern		5	127	1194
Others	2	119	863	2284

CAUSES OF HUMAN-WILDLIFE CONFLICT AND REMEDIAL OPTIONS

The most compelling explanation for the mounting conflict between people and wildlife is the growth of human populations (Yaeger and Miller 1986). The human population in Kenya has risen nearly fivefold since the establishment of National Parks in 1947. Over the last 25 years one-third of Kenya's wildlife has been lost (Grundblatt *et al.* 1995), largely from areas of human settlement. With Kenya's population set to rise to around 55 million at stabilization, the prospect for wildlife looks bleak.

Protected areas are the best way to ensure uncontested space for wildlife. Kenya is fortunate in having 7.6% of its land surface under parks and reserves. Even so, this is proving insufficient, especially for the large migratory herbivores and wide-ranging carnivores. Based on national sample counts, Kenya's protected areas support less than 30% of all wildlife (Western 1989). Species losses likely to occur due to insularization (Soulé *et al.* 1979) and ecological dislocation (Western and Gichohi 1993) further underscore the limitations of parks as a solitary conservation tool. Finally, the compression of large herbivores and carnivores into protected areas threatens the very biodiversity such areas are intended to conserve (Western and Gichohi 1993; Waithaka 1994). All these threats, and the sharp fall in new protected areas in recent decades (Table 22.2), point to the need for additional conservation measures to protect large herbivores and carnivores.

Conservation policy must necessarily look beyond the confines of parks to sustain viable populations of large vertebrates (Harris and Eisenberg 1989). Outside protected areas, the primary incentive to conserve wildlife in the Western world has historically been utilization in some form. Unfortunately, conservation policy in Kenya, like much of the developing world, has intensified the conflict by eroding traditional mitigation methods (Bonner 1993) and denying indigenous occupants any economic use of wildlife (Parker and Bleazard 2001). Democracy, land reforms and changing lifestyles have added to the rising intolerance (Western 2001). Consequently, landowners are increasingly taking action to protect their property where government fails to do so. The ready availability of guns, poisons and other means make it easier than ever to exterminate wildlife, irrespective of legal protection.

Despite the dismal outlook for wildlife, several counter-trends suggest avenues for wildlife policy reform, based on human–wildlife conflict mitigation. Perhaps the most important evidence is that the rate of wildlife loss has slowed greatly since the mid-1980s. For some species such as elephant, rhino (*Diceros bicornis*), eland (*Taurotragus oryx*), buffalo (*Syncerus caffer*) and zebra (*Equus burchelli*), numbers have actually increased in Kenya

Table 22.2. *Protected area set-asides in Kenya*

Decade	Number of parks	Area set aside (km^2)	Cumulative area (km^2)	Number of community sanctuaries	Area set aside	Number of reserves	Area set aside (km^2)	Cumulative area of reserves (km^2)
1940s	4	21 756	21 756	0	0	0	0	0
1950s	2	8 536	30 292	0	0	0	0	0
1960s	7	1 955	32 247	0	0	2	224	224
1970s	4	1 945	34 192	0	0	18	12 088	12 312
1980s	10	3 735	37 927	2	0	7	908	13 220
1990s	1	<1	37 927	30	513	2	684	13 904
Total	28		37 927	32	513	29		13 904

(Grundblatt *et al.* 1995; Githaiga 1998). Githaiga (1998) shows that improvements occur where wildlife has become an economic asset.

Yet another reason for slowdown in wildlife losses in recent years stems from changing national attitudes towards wildlife due to economic transition, formal education and rural-to-urban migration (Western and Manzollilo Nightingale 2002), mirroring the trends that portended the rise of conservation sensibilities in the West (Shabecoff 1993). Global tourism has spawned a US$0.75 billion industry in Kenya that has also raised national consciousness of wildlife.

The softening attitude towards wildlife is, however, deceptive and points to the part played by imbalances in wildlife costs and benefits. Most rural residents in Kenya realize the value of wildlife nationally, yet don't feel they should bear the cost. The exceptions to the hostile rural view occur where conservation programmes have addressed the conflict through economic opportunity, direct conflict reduction using fencing, education and so on (Table 22.3).

Table 22.3. *Attitudes to wildlife among rural communities, relative to wildlife benefits and conflict mitigation measures*

Station	Percentage positive attitude towards wildlife
Areas with tourist benefits	
Narok Station	75
Amboseli National Park	80
Areas with effective electric fencing	
Aberdares National Park	80
Lake Nakuru National Park	64
Areas with education/awareness and problem animal control programmes	
Nairobi Park	50
Kericho, Buret, Bomet	50
Kilgoris-Transmara	50
Transmara	50
Areas with some Kenya Wildlife Service security	
Nasolot/South Turkana	50
Baringo/Koibatek	63
Areas with inadequate interventions	
Kapenguria Station	45
Marsabit National Park	31
Embu Station	40
Nasolot Station	20
Kakamega	4
National average	50

Source: Kenya Wildlife Service (2001).

THE EVOLUTION OF HUMAN–WILDLIFE CONFLICT POLICY

Wildlife policy in Kenya has evolved over the last century in response to the changing conservation goals and shifting relationship between people and wildlife (Simon 1962; Yaeger and Miller 1986; Anderson and Grove 1987; Rosenblum and Willamson 1987; Parker and Bleazard 2001; Western 2002). In general, the early preoccupation with 'game' animals for sport broadened to larger mammals with the rise of National Parks and tourism in the 1940s to 1980s. More recently, the focus expanded to biodiversity (Western 2001) as national conservation policy encompassed all forms of life when Kenya ratified the Convention on Biological Diversity. Human–wildlife conflict was seen largely as a fallout of conserving wildlife and, as such, dealt with reluctantly (Simon 1962). Control measures were confined to the largest and most threatened species, including elephants, rhinos, hippos (*Hippopotamus amphibius*), buffaloes, crocodiles (*Crocodylus niloticus*) and large predators (Game Department 1920–63). The twin goals were to protect big-game animals, and to curb their adverse impact on people. Many species, including bush pig (*Potamochoerus* spp.) and baboons (*Papio* spp.), were classified as vermin to allow landowners to take their own action on species too numerous and problematic for the Game Department to control.

Here, we highlight two advances in human–wildlife conflict policy: the 1977 Wildlife Policy, and 1994 national debate on human–wildlife conflict. DW was involved in both sets of policies and JW in the second. We therefore draw on first-hand accounts of policy formulation as well as available reports.

Wildlife Policy 1977

In 1977, the Kenya government drew up a new wildlife policy based on an ecosystem approach, community involvement (Western 1994) and the growing importance of tourism. The Game Department and Kenya National Parks were amalgamated under the Wildlife Conservation and Management Department (WCMD), specifically to bring wildlife conservation under a single agency and to reconcile conservation with development (Western 2002).

The policy stated that human–wildlife conflict posed a threat not only to wildlife, but also to tourism, wildlife utilization and development. It also recognized for the first time that Kenya's parks were far too small to sustain wildlife numbers and migrations. Wildlife had, therefore, to be justified outside parks as a way to protect ecosystems and the migrations from parks. The biggest impediment lay in the conflict between wildlife and rural communities. The policy saw human–wildlife conflict more as a clash between those who wanted to conserve wildlife and those who wanted it eliminated, rather than 'people versus wildlife' *per se*.

The nub of the new policy lay in mitigating human–wildlife conflict and redressing the imbalance of costs, primarily by enhancing wildlife benefits to landowners. This included not only direct returns, but annual wildlife grazing fees to offset the cost of important migrations from National Parks onto surrounding lands. The policy also gave government the responsibility of guiding conflict-mitigation strategies within the limits of its resources.

The tools in the mitigation arsenal were broad (thunder flashes, fires, dogs, shooting, barriers, translocation, habitat destruction etc) and guided by optimization principles. The Wildlife Conservation and Management Department was to recommend suitable techniques to landowners, where they were authorized to act. Landowners were entitled to kill wildlife that immediately threatened their property, provided they reported the killings and could establish proof of threat. Family, livestock and crops were considered personal property; pasture was not. Landowners were also entitled to claim compensation for property losses. In short, the conflict-mitigation strategy was broad, flexible and based on market forces as a way to devolve action to the lowest appropriate level.

The enabling legislation under the Wildlife (Conservation and Management) Act 1977 (Republic of Kenya 1977) provided a basis for discriminating species according to economic and conservation importance. Schedule 1 included three categories: endangered species, threatened species and species covered by licence. Schedule 2 specified game-bird species covered by licence. Schedule 3 listed vertebrate species subject to legislation. The schedules gave the Wildlife Conservation and Management Department a way to gauge the conservation importance of species in conflict with people and tailor its response. The agency could thus focus its own conflict mitigation on endangered and threatened species – and to a lesser extent valuable game and wildlife species – leaving landowners to deal with the most common and least valuable. The Act was silent, though, on animals of no economic value, whether threatened or not. Plants were ignored altogether and marine and freshwater organisms were covered by the Fisheries Act in which the primary goal was maximum commercial harvest rather than biodiversity conservation.

Unfortunately, the Wildlife Act directly undermined the policy by failing to devolve authority and delegate conflict mitigation. In addition, a hunting ban, issued by government in 1977, and a ban on wildlife produce in 1978, removed important economic incentives as a mitigation tool (Western 2002). Tourism, the only exception, was essentially confined to the

National Parks. In essence, the Act perpetuated protectionist policies and further fuelled human–wildlife conflict as much if not more than growing human activity.

In 1989 the Wildlife (Conservation and Management Act) Amended (Republic of Kenya 1989) rectified some of the shortcomings of the 1977 Act by establishing the Kenya Wildlife Service. This Service, governed by a board of trustees, removed the stifling and corrupt hand of government and allowed all revenues collected to be ploughed directly back into conservation. The deeply corrupted compensation systems for crops and livestock were also eliminated. The Kenya Wildlife Serivce reaffirmed the 1977 Wildlife Policy and drew up a progressive policy framework (Kenya Wildlife Service 1991) to conserve biological diversity nationwide. However, with the hunting and trophy bans still in place, the Kenya Wildlife Service was as constrained as the Wildlife Conservation and Management Department in using economic incentives to mitigate human–wildlife conflict or to redress imbalances. To make matters worse, the Wildlife Service began to shed all non-remunerative activities in order to become self-sufficient by 1997. The upshot was that the Kenya Wildlife Service largely abandoned human–wildlife conflict mitigation in the early 1990s. Human deaths caused by elephants rose steeply from five in 1989, at the time of the ivory ban, to 45 in 1994 (Edwasi 1994). The political fallout brought human–wildlife conflict to national attention (Edwasi 1994) and resulted in a presidential directive that the Kenya Wildlife Service should address the problem immediately.

Human–wildlife conflict policy 1997

In July 1994, the Kenya Wildlife Service launched an independent five-person review of human–wildlife conflict. The review team was charged with soliciting views of communities, landowners, government, the private sector and conservation non-government organizations (NGOs)throughout Kenya on the causes of the conflict and its remedies.

The review (Edwasi 1994) concluded that:

> Wildlife–human conflicts are not just a litany of specific problems, but a whole unacknowledged perspective on reality. Their solution requires a concept of sustainable wildlife management by and for people on their land, not in spite of them.

The public blamed the conflict more on protectionist policies than growing human activity. The Kenya Wildlife Service was seen to manage conflict by avoidance and force applied to poachers, rather than animal marauders. Compensation for human death was low and rarely paid. The Kenya Wildlife Service gave little attention to conflict and lacked the expertise and resources to deal with it. Prevailing laws protected wildlife and prevented the devolution of mitigation measures to affected communities. Centralization of conflict resolution and compensation created dependency and local inaction. Most of all, the review considered the lack of wildlife benefits to rural communities and the heavy costs in lost productivity and livelihoods to be the overriding causes of conflicts.

Based on these findings, the review put forward a number of mitigation measures suggested at public meetings. These can be summarized as follows:

- In keeping with national goals, liberalize wildlife policy in order to raise the benefits and diversify the uses landowners can make of wildlife.
- Custodianship should be retained by the state, but the authority for management should be devolved to landowners as close to the seat of conflict as possible.
- Devolution should be based on clearly prescribed rights and responsibilities of landowners with respect to the Kenya Wildlife Service or the government, and on an equitable cost and benefit-sharing basis.
- Wildlife management should be zoned, with protected areas taking highest priority, pastoral areas next and arable areas least. Human–wildlife conflict mitigation should be prioritized accordingly.
- In keeping with these goals, the Kenya Wildlife Service should assume a role on private and community lands based on advice, supervision, enforcement, building local capacity and arbitration.
- Conflict mitigation should be embedded in a hierarchy of approaches, ranging from large-scale and general strategies based on incentives, to wildlife and land planning, 'scheduling' of species and flexible strategies and tools for actual conflict mitigation. These can vary from physical barriers to the removal or destruction of specific animals in high-conflict areas. 'Scheduling' refers to the conservation status of a species and defines who is responsible for conflict mitigation (the Kenya Wildlife Service in the case of endangered species and landowners in the case of pest species, for example).
- Compensation only for loss of human life, to be operated through an insurance scheme
- Education and public relations.

The findings and recommendations were submitted to the board of the Kenya Wildlife Service, which then released the findings to the media to foster further debate. Four technical studies ('Wildlife Utilization', 'Tourism pricing studies', 'Biodiversity' and 'Land use planning and policies and legal framework') were undertaken with NGO input at the same time, complemented by a national elephant programme looking into human–elephant conflict across Kenya (Waithaka 1997).

A policy evaluation group comprising Kenya Wildlife Service staff and representatives from various constituency groups drew up the policy implications from the public and technical reviews, and prepared a draft policy for national appraisal. The draft policy was also presented at public meetings in the USA and UK to solicit international input. The completed Wildlife Policy 1997 was approved by the Kenya Wildlife Service board and the Minister of Tourism and Wildlife and submitted to the Attorney General's Chambers for Cabinet approval.

The policy framework

The 1997 Wildlife Policy (Kenya Wildlife Service 1997) updated rather than replaced the 1977 Wildlife Policy. Wildlife was explicitly equated with biodiversity to conform with Kenya's obligations under Convention on Biological Diversity and other international conventions. The widening values of biodiversity were recognized and encouraged and the conflict of interests they elicited was to be tackled by enhancing the positive values of biodiversity and reducing the negative through market mechanisms, education, arbitration and direct mitigation using humane methods.

The policy explicitly recognized that government could not conserve everything everywhere, and that land and resources were limited and shrinking. It also recognized that wildlife policies should adhere to national goals – specifically decentralization, liberalization and poverty alleviation. Custodianship of wildlife was to be devolved to the lowest competent level, based on land ownership. User rights were linked to specific responsibilities for conserving wildlife and to competent individuals and associations through zonal associations. The Kenya Wildlife Service was, in collaboration with stakeholders, to draw up the minimum viable conservation area needed to maintain biodiversity in perpetuity and to guide national conservation priorities. The minimum viable conservation area was to link critical biological areas across public and private lands and between protected areas and non-protected.

The mitigation of human–wildlife conflict was essentially to be tackled and devolved through a similar mechanism. The essence lay in prioritizing conservation within the minimum viable conservation area through a hierarchy of land-use plans, conservation and management plans and specific mitigation measures drawn up under the zonal associations. Strategically, the management of an individual species was to be devolved according to the schedule (endangered, threatened, pest, etc.) under which it fell, and the status of that species locally. So, for example, problem elephants as endangered species would be dealt with by the Kenya Wildlife Service, whereas pest rodents should be dealt with by landowners. Land-use planning, zonation and market mechanisms were seen as important ways to minimize conflict. Market solutions should include direct and indirect utilization, compensation, easements, leases, land trusts and management agreements. Direct control of problem animals should include a wide array of mitigation strategies and be cost-effective and humane. An insurance fund was to be established under a public–private board to compensate loss of human life due to conflict.

The devolution called for the regionalization of the Kenya Wildlife Service to scale back its many functions to oversight, enforcement and arbitration of rights and responsibilities, and performance monitoring of landowner and zonal associations. The policy recognized the need for an experimental and adaptive approach, based on transparent science and success indicators for the process of devolution.

POLICY OUTCOME

In 1997 Kenya Wildlife Service began putting the policy into practice through the establishment of eight bioregions, zonal associations, and a Partnership Department to coordinate the activities of these associations. However, conflict mitigation measures began as early as 1994, leading to the establishment of a dedicated Problem Animal Control Unit and a Human–Wildlife Conflict Committee. The Human–Wildlife Conflict Committee comprised the Kenya Wildlife Service, representatives from NGOs and landowner representatives, tasked with identifying and resolving the most extreme and contentious conflict issues. Its activities focussed initially on elephants, larger carnivores and primates.

In 1998, the policy was withdrawn for review following a change of Director at the Kenya Wildlife Service. The review has been set back repeatedly since then, due to regular changes of Director. Nevertheless, much of the policy was implemented from 1994 to 1998 and does give a basis for assessment.

Several success indicators can be used to measure human–wildlife conflict reduction, including social indicators, reduction in human losses and increase in wildlife. The evidence points to significant improvements following the launch of the community-based initiative of the Kenya Wildlife Service and human–wildlife conflict programmes. These include a stabilization or increase in wildlife populations where communities engaged in conservation-for-profit initiatives (Githaiga 1998), improved attitudes towards wildlife as a result of such initiatives (Table 22.3),

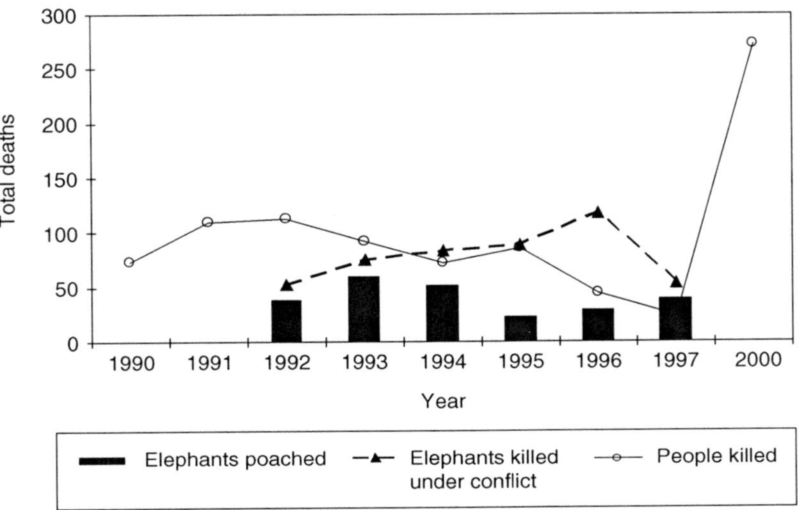

Figure 22.1. Trends in elephant deaths, and wildlife-caused human deaths, in Kenya.

improved crop harvests due to containment fences (Waithaka 1997) and a sharp increase in the number of new community wildlife sanctuaries (Table 22.2), thus reversing the decline in new National Parks and reserves over the previous two decades. In each case, community wildlife scouts have been recruited and trained to protect wildlife in collaboration with the Kenya Wildlife Service.

Yet another measure can be gleaned from efforts by the Kenya Wildlife Service to reduce the escalating deaths of elephants and people in the early 1990s, the highest priority on the national agenda (Edwasi 1994). Poaching had shrunk the national elephant herd from 160 000 in 1970 to around 19 000 in 1989 (Western 2001). A combination of the 1989 international ivory ban and improved Kenya Wildlife Service security cut the poaching rate from around 7000 elephants a year pre-1989 to a maximum of 75 a year after the ban (Fig. 22.1). The resulting spread of elephants from the safety of the parks saw human deaths rise steeply. Reprisal elephant killing surged to the point of becoming the biggest source of elephant mortality (Fig. 22.1).

In designing a strategy to reduce elephant depredations, the Problem Animal Control Unit targeted the main conflict areas identified within the minimum viable conservation area system and endeavoured to minimize the response time for taking action on crop-raiding elephants. A special training programme was established by the Kenya Wildlife Society for community scouts, aimed at improving anti-poaching and security outside as well as inside parks. A concerted effort was also made to construct elephant fences in the worst-hit agricultural areas around Mount Kenya, Abedares, Tsavo and Amboseli National Parks.

The results show that number of elephants killed in conflict (which included animals killed by farmers and those shot on control) rose from 84 in 1994, when the Problem Animal Control Operations began, to 118 in 1996 (Fig. 22.1). The number killed subsequently declined to 54 in 1997, once containment measures began to take effect. The number of human deaths caused by elephants, which had climbed sharply from 12 in 1990 to 45 in 1992, also began to fall steadily from 1994 onwards, to a low of 13 in 1997. It should be borne in mind that the decrease in human deaths due to elephants after 1993 occurred against a rising elephant population during this time, and its steady spread beyond parks.

Unfortunately, the high priority that the Kenya Wildlife Service gave to resolving the human–wildlife conflict in the period 1994 to 1998 was largely abandoned in 1998, when a new Director stated that community programmes detracted attention from protecting parks. Our efforts to obtain more up-to-date data on elephant human mortalities were turned down by the Director on the grounds that the information was too sensitive for public release. The sole figure we could obtain was for 2000, which showed human deaths climbing to 272 from all forms of wildlife, compared with a peak of 113 (1992) for the 1990s. It would appear from field reports that a large majority of the deaths were caused by elephants.

GENERAL CONCLUSIONS

Human–wildlife conflict has long been a cause of wildlife loss, yet has been consistently ignored by wildlife agencies in policy formulation. The Kenya case study gives a historical perspective on the changing goals of conservation over the last century. Several threads relevant to human–wildlife conflict emerge. First, human population has expanded fivefold, leading to intensified human–wildlife conflict. The conflict is sharpest between large herbivores (and carnivores) and rural farmers and herders. Second, conservation goals have continually changed and broadened from largely utilitarian to intrinsic and biological, reflecting expanding societal sensibilities and knowledge. Third, conservation policy has been repeatedly amended to reflect the changing goals. Fourth, conservation policy in Kenya (and much of the developing world) has intensified human–wildlife conflict by excluding rural communities which bear the brunt of conflict from any economic gains from wildlife. Fifth, democratization and liberalization are raising awareness among such communities, leading to less tolerance of threatening species on the one hand, but greater opportunity to use them profitably on the other. Sixth, the wide ambit of modern biodiversity conservation gives more grounds for conflict – both among people who differ over priorities and between those who are differentially affected by conservation.

Despite the mounting conflict with and over wildlife, human–wildlife conflict was ignored as an unwanted side effect of conservation and consequently marginalized in national policy. Even after policies were drafted to redress the conflict in the 1970s, the policies were vitiated by preservation measures that nullified the new policies. Human–wildlife conflict consequently increased sharply and is now one of the gravest threats to Kenya's large wide-ranging herbivores and carnivores. More elephants are now killed due to conflict with people than by poachers. In the mid-1990s, the Kenya Wildlife Service took steps to address the conflict as a central article of policy at the same time as expanding national policy from a narrow wildlife focus to biodiversity as a whole.

Choices are inevitable. But how should a wildlife agency decide how and when to intervene? What should a wildlife agency do about the vast majority of species it can't afford to act on? Where should the obligation of the state end and the rights of its citizens to take their own action begin? What is the dividing line between species considered worthy of custodianship by the state and those too numerous and too unimportant to cover practically?

We have shown that human–wildlife conflict can and should be addressed within national conservation policies, rather than as an unwanted fallout of conservation. Ideally, conflict mitigation should be addressed using a hierarchy of measures from national to local level and economic incentives to conserve species – in addition to direct measures to curb conflict. The broad policy approach to conflict mitigation in Kenya showed a fall in elephant–human conflict, for example, even as the elephant population grew and expanded its range beyond parks.

Incorporating human–wildlife conflict mitigation into national policy calls for a clear sense of the overall national conservation goals and specific strategies for reconciling the positive and negative impact of high-conflict species on different interest groups. We conclude by suggesting a number of tentative principles that help achieve both objectives:

- The need for a national conservation policy that incorporates all life forms.
- Criteria for setting conservation priorities, based on biological, social and economic considerations.
- The identification of a minimum viable conservation network for conserving biodiversity in perpetuity and for setting priorities.

- A conflict-resolution strategy aimed at reducing the impact and increasing the value of high-conflict species.
- A rationalization of species listing criteria based on the degree of threat to humans and species' perceived value.
- A clearly articulated set of rights and responsibilities for devolving and delegating responsibility for human–wildlife conflict mitigation.
- The devolution of conflict resolution to the lowest practical and verifiable level.
- A fair and transparent political dialogue and process for conflict resolution.
- Monitoring and proof of mitigation measures based on transparent and verifiable indicators of success.
- An array of conflict-mitigation measures that can be tailored to specific circumstances according to cost-effective considerations and humane procedures.

ACKNOWLEDGEMENTS

We are grateful to the Kenya Wildlife Service staff for support of our work on human–wildlife conflict during our time working as Director (DW) and Head of Elephant Program (JW) at the Kenya Wildlife Service between 1994 and 1998. We also wish to thank the Wildlife Conservation Society and African Conservation Centre who have both supported our conservation work over the years.

23

An ecology-based policy framework for human–tiger coexistence in India

K. ULLAS KARANTH AND RAJESH GOPAL

INTRODUCTION

Carnivores are in decline across the world for a variety of reasons, among which conflict with humans is the most predominant (Woodroffe 2000). This conflict takes varied forms and involves several carnivore species (see Treves and Karanth (2003) for a recent review). Mitigation of such conflicts should be the most important part of any conservation agenda that strives for continued coexistence of carnivores and humans. Among carnivore taxa, felids in the genus *Panthera* appear to be particularly conflict-prone (Rabinowitz 1986; McDougal 1987; Mishra et al. 2003; Ogada et al. 2003). How conservationists can promote the coexistence of *Panthera* cats and humans in densely populated countries such as India, or can generate potentially useful models for other regions of the world where human population densities and habitat fragmentation levels are relatively lower, but rising rapidly, are urgent problems.

The tiger (*Panthera tigris*) is a felid species of global concern because of its cultural and ecological significance (Jackson 1990; Karanth 2001; Thapar 2002). Over the last three decades, the Indian government has used the tiger as an effective flagship species to protect a wide range of biodiversity. Special tiger reserves have been established in different biomes across India covering mangrove swamps, alluvial grasslands and forests of the deciduous, semi-deciduous and evergreen types. However, India's wild tiger populations are still under serious threat from human impacts such as prey depletion, poaching, habitat loss and fragmentation (Karanth and Stith 1999; Seidensticker *et al.* 1999; Karanth 2003). Such proximate causal factors of tiger decline are ultimately driven by human demographic and resource consumption patterns that fuel expanding economic development.

People and Wildlife: Conflict or Coexistence? eds. Rosie Woodroffe, Simon Thirgood and Alan Rabinowitz. Published by Cambridge University Press.

Tigers are obligate predators of ungulates in the 20–1000-kg body mass category (Sunquist *et al.* 1999; Karanth 2003). Consequently, wherever wild tiger populations survive and interface with landscapes dominated by humans, they pose a threat by preying on livestock, and, less commonly, on people. Although the traditional social ethos in most parts of India is remarkably tolerant of wildlife damage compared with elsewhere (Rangarajan 2001), in conflict situations, local antagonism against tigers often erupts into a serious management problem.

Paradoxically, given the key demographic traits of tigers such as early reproduction, high fecundity, wide-ranging movements and territorial land tenure system (Sunquist 1981; Smith 1993; Chundawat *et al.* 1999; Karanth and Stith 1999; Karanth 2003), conflict with humans becomes an inevitable consequence of successful tiger population recoveries. Because of the tiger's endangered status, mitigation of these conflicts through traditional lethal control has become increasingly problematic.

A tiger management policy that aims to mitigate human–tiger conflicts must necessarily consider situation-specific details of how the needs of humans and tigers clash. How can one address the broader question of coexistence between people and tigers in India? Which of the two competing conservation paradigms (Robinson 1993) – 'preservation' or 'sustainable use' – is more appropriate for effectively addressing this issue? Among the wide array of conflict-mitigation tactics available to managers (Treves and Karanth 2003) – ranging from lethal control to strict preservation – which ones are most relevant to resolving conflicts in specific ecological and social contexts?

We explore the above questions in this case study. Our goal is to generate a policy framework to maintain demographically viable meta-populations of wild tigers across India's conservation landscapes, while trying to keep human–tiger conflicts within socially acceptable limits. In generating this policy framework, we have deliberately avoided delving into the political question of who should manage India's wildlife reserves: the official machinery as at present, or an alternate localized power structure (Kothari *et al.* 1995). We argue that, whoever is responsible for managing human–tiger conflicts, the factors we consider in our analysis are still relevant to promoting coexistence of humans and tigers.

PARADIGM SHIFTS IN TIGER MANAGEMENT

India's geographical area of 3.05 million km^2 supports a human population of over 1 billion. About 70% of this population is rural, with a majority depending on agriculture, animal husbandry and related occupations. Over half the rural people are classified as poor, with typical wage rates being less than a US$1 per day. Because of various cultural and social reasons, rural India supports about 450 million livestock – a large proportion of which are unproductive animals grazed on public lands for the primary purpose of producing dung applied to fertilize crops. Rural domestic energy needs are primarily met by burning wood, animal dung or other biomass. The collection of firewood from forests for sale in nearby villages and towns is a dominant form of rural occupation in many forested parts of the country (Agarwala 1985). All such extractive pressures have been driving the degradation of tiger habitats (Gee 1964; Schaller 1967; Karanth 2001; Madhusudan and Mishra 2003).

Consequent to the rapid expansion of agricultural areas, the remaining tiger habitat has shrunk and fragmented continuously over the last few centuries (Karanth 2001, 2003). In more recent times, economic development projects such as dams for irrigation and power generation, mines, and road and rail transportation have increased levels of fragmentation and facilitated the penetration of forces of modern commerce into relatively remote tiger habitats. At present forests potentially suitable for tigers cover about 300 000 km^2 (Wikramanayake *et al.* 1998) of India's land area, with perhaps half that extent actually harbouring tigers. This potential tiger habitat is patchy and restricted primarily to blocks of forests in the southwestern, central and northeastern parts of the country (Fig. 23.1).

Historically, tigers were widely distributed and viewed as threats to expansion of agriculture and rural livelihood (Karanth 2001; Rangarajan 2001) because of persistent tiger predation on humans and livestock. A compilation by McDougal (1987) showed that tigers killed 798 people in 1877 and 908 people in 1908 in the British-administered provinces of India. Consequently, tigers were simultaneously targeted both as dangerous vermin and desirable trophies (Rangarajan 2001).

The preferred mode of dealing with human–tiger conflicts was lethal control through intensive bounty and sport hunting. The scale of this lethal control effort can be gauged from examples such as the lifetime tallies of tigers hunted by 'sportsmen' like kings of Surguja and Udaipur that exceeded 1000 tigers (Schaller 1967). Even ordinary government officials shot tigers by the thousands. Such 'sportsmen' operated simultaneously with the even more numerous 'native poachers' using an array of devices such as muzzle-loaders, spears, nets, traps, snares and poisons to virtually eradicate wild tigers from large regions of India during the nineteenth and twentieth centuries. Rangarajan (2001) estimates that about 80 000 tigers were killed in India between 1875 and 1925.

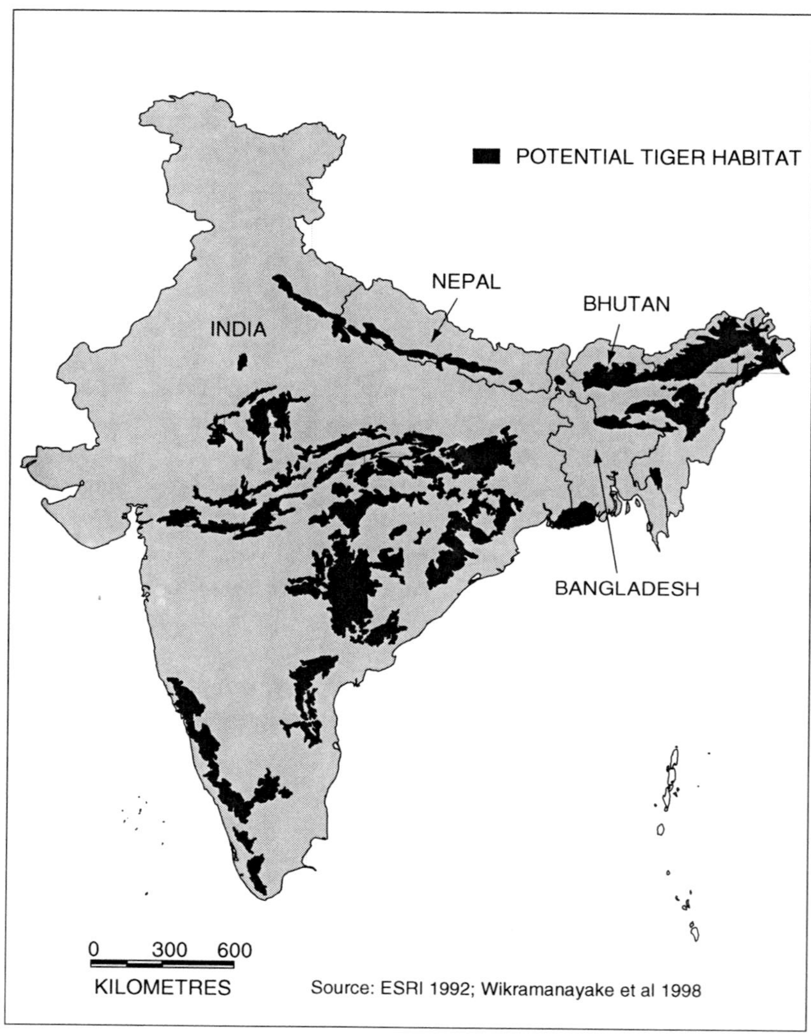

Figure 23.1. Potential tiger habitat in the Indian subcontinent.

By the late 1960s tigers were on the verge of extirpation in the remaining areas (Gee 1964; Schaller 1967). However, a remarkable policy turnaround occurred in the early 1970s, driven by pressure from conservationists. A committed political leadership introduced new initiatives that proscribed tiger hunting and established special protected reserves (Jackson 1990; Karanth 2001, 2003). The core of this new 'preservationist' strategy implemented within protected reserves included active patrolling to deter hunting of tigers and their prey species, as well as habitat recovery measures such as curbs on human-induced forest fires, livestock grazing and harvest of forest products. In some cases, human settlements were also relocated from protected tiger reserves. Effective implementation of such measures in several reserves across the country led to significant recoveries of tiger populations in the 1970s and 1980s (Panwar 1987; Karanth *et al.* 1999; Karanth 2003).

After the 1990s, although the protectionist laws remained in place, their implementation slackened significantly because of several social factors (Karanth 2002, 2003). Driving this change was a shift from earlier 'exclusionary' conservation policies towards more 'inclusive' polices that de-emphasized law enforcement in favour of community development activities around tiger reserves (Mackinnon *et al.* 1999). *Prima facie*, these new approaches appear to have failed either to make conservation gains or to reduce local antagonism towards protected reserves (Karanth 2002; K. U. Karanth unpubl. data).

THE CURRENT STATUS OF HUMAN–TIGER CONFLICT

Although the extent of potential tiger habitat in India is still around 300 000 km^2 (Wikramanayake *et al.* 1998) (Fig. 23.1), the proportion of the habitat that can support adequate reproduction is perhaps only about 10% of this area, and lies mostly within protected nature reserves. Inside such reserves, superior habitat productivity, combined with reduced human impacts, has resulted in ungulate prey attaining high densities of 15–70 animals/km^2. Consequently, tiger densities are also high (5–20 tigers/100 km^2: Karanth and Nichols 1998; Karanth *et al.* 2004). Stochastic simulation models of tiger populations (Kenny *et al.* 1995; Karanth and Stith 1999) show that such clusters of 12–25 breeding tigresses are demographically viable and may produce 10–15% annual 'surpluses'.

The present distribution of tigers in India (Fig. 23.1) comprises several discrete meta-populations, embedded within larger landscape matrices made up of protected reserves, multiple-use forests and agricultural and urban areas (Karanth 2003). The protected reserves are essentially 'sources' for dispersing tigers that may survive for brief periods in the surrounding landscape matrix, before perishing from poaching or prey depletion. However, because of the tiger's habitat specificity (requirements of cover, water, prey) even such transient tigers cannot survive over large parts of the country. As a consequence, conflict with humans is largely restricted to the edges of protected reserves, and some multiple-use forests or plantation crop areas.

Given the pattern of human population densities across India (Guha 2001) and the highly clumped distribution of tiger populations (Wikramanayake *et al.* 1998) (Fig. 23.1), the extent of area of conflict is relatively small. Such conflict zones perhaps cover less than 1% of India's geographical area, and involve an even smaller fraction of its human population. Therefore, in a macro-ecological sense, human–tiger conflict is a relatively localized management problem. However, by its very nature, the conflict poses a serious dilemma for conservationists trying to promote human–tiger coexistence. The conflict assumes the forms of killing of livestock, accidental killing of humans and even persistent predation on humans by tigers, all of which lead to retaliatory killings of tigers.

Tigers readily kill domestic ungulates (Table 23.1). Most such predation takes place inside government-owned forests or common pasturelands,

Table 23.1. *Depredation on livestock by tigers and monetary compensation paid around Kanha tiger reserve, Madhya Pradesh, Central India between 1977 and 2001*

Year (April to March)	Number of cattle kills	Compensation (current US$)
1977–78	4	10
1978–79	7	41
1979–80	21	149
1980–81	26	307
1981–82	26	295
1982–83	22	242
1983–84	61	698
1984–85	80	1 094
1985–86	45	545
1986–87	109	1 419
1987–88	122	1 483
1988–89	107	1 009
1989–90	71	703
1990–91	63	859
1991–92	60	816
1992–93	38	425
1993–94	33	373
1994–95	22	360
1995–96	52	742
1996–97	40	570
1997–98	58	907
1998–99	131	2 234
1999–00	117	2 418
2000–01	129	4 467
Total	1444	22 166

where large numbers of livestock are grazed legally. Usually such livestock kills are not fully consumed by tigers, because herders intervene. In most multiple-use forests, densities of wild prey are low because of hunting and competition with livestock (Madhusudan and Karanth 2002; Madhusudan and Mishra 2003). In such situations tigers may take as much as 12% of the livestock herds annually (Madhusudan 2003).

Tiger depredations lead to retaliation through poisoning of carcasses and other forms of unsanctioned lethal control (Karanth 2003; Madhusudan and Mishra 2003). Organized criminals involved in the illegal trade in tiger body parts (Kumar and Wright 1999) exploit such conflicts by distributing poisons and traps to villagers, and buying tiger body parts from them. As a result, retaliatory killings often escalate into deliberate, market-driven poisoning of both natural and domestic animal kills. Kumar and Wright (1999) reported 123 tiger poaching cases in 44 months between 1994 and 1997. More recent data show that at least 41 tiger-poaching cases were detected during the year 2001. Both these figures are based on law-enforcement records, and total levels of poaching are probably much higher.

In most places, tigers are wary of human beings and avoid encounters. Accidental mauling or killing of humans by tigers is rare, and usually occurs when irate mobs surround tigers that enter human settlements to take livestock. Very rarely, tigers may maul or kill humans they unexpectedly encounter, and the tiger may sometimes eat a part of the cadaver. However, by no means all such encounters lead automatically to persistent predation on humans; hence many incidents may require no further management intervention than compensating the victim's relatives.

In rare cases, individual tigers begin to view human beings as a 'prey species' and persistently stalk them. Such man-eating behaviour, although rare, has been historically documented in several parts of India (McDougal 1987; Daniel 2001). The ecological and social factors that lead to man-eating are unclear, but appear to be influenced by scarcity of natural prey, injuries, transmission of man-eating behaviour from parent to offspring and the lack of effective retaliation following initial attacks on humans (McDougal 1987; Karanth 2001).

Man-eating behaviour is exhibited in an unusually persistent form among the tigers of the Sundarban delta of India and Bangladesh (McDougal 1987; Sanyal 1987; Daniel 2001). In these roadless mangrove forests partially submerged under tidal waters, thousands of people intrude on foot or in boats to collect a variety of forest products, wild honey and fish. Tigers of Sundarban opportunistically kill and eat these people, particularly when they are alone, in small groups or sleeping inside boats at night. Although no scientific data exist, tiger predation on humans in Sundarban

Table 23.2. *Incidental tiger attacks on humans (Kanha tiger reserve, Madhya Pradesh), typical of most areas of India, and records of persistent tiger predation on humans (Sundarban tiger reserve, Bengal) from 1985 to 2001*

	Kanha			Sundarban		
Year (April–March)	Injuries	Deaths	Total attacks	Injuries	Deaths	Total attacks
1985–86	0	0	0	6	32	38
1986–87	0	1	1	6	25	31
1987–88	1	1	2	7	21	28
1988–89	2	5	7	3	14	17
1989–90	1	2	3	2	9	11
1990–91	3	6	9	8	43	51
1991–92	3	3	6	7	38	45
1992–93	1	1	2	5	34	39
1993–94	1	0	1	4	31	35
1994–95	1	1	2	0	5	5
1995–96	1	0	1	0	4	4
1996–97	5	0	5	1	3	4
1997–98	1	0	1	2	5	7
1998–99	0	1	1	2	2	4
1999–00	3	0	3	0	13	13
2000–01	2	1	3	4	15	19

does not appear to be restricted to aberrant individuals as in other parts of tiger range. A substantial proportion of the tiger population appears prone to opportunistic predation on humans.

Sanyal (1987) reported that tigers killed 318 people in Indian Sundarban between 1975 and 1981. Although country-wide records of human fatalities caused by tigers are not available, we provide some data on accidental killings around Kanha reserve in Central India and persistent tiger predation on humans in Indian Sundarban in recent years (Table 23.2). These official figures may be underestimates, because of underreporting of the killing of persons engaged in clandestine activities, and of livestock being grazed illegally.

MITIGATING CONFLICT, FOSTERING COEXISTENCE

Wildlife management in India is carried out under legal provisions of the Wildlife Protection Act of 1972. This law was originally introduced to counter destruction of wildlife occurring in the absence of weak regulations, and is therefore strongly preservationist in its basic thrust. The Act makes it virtually illegal to kill or capture wild animals even when problem animals are involved in severe conflict situations. Only government officials or agents authorized by the Chief Wildlife Warden of the state government can execute such killings or captures. In case of endangered species like tigers, the necessary authorizations can only be issued by the Director-General of Wildlife Preservation in Delhi, based on an application made by the state Chief Wildlife Warden. While admirable in their intent, these strict legal provisions make it very difficult for local wildlife managers to deal effectively with urgent, life-threatening situations of human–tiger conflict.

Following Karanth and Madhusudan (2002), we classify the strategies employed for mitigating human–tiger conflicts into two basic categories: reacting to the conflict in ecological settings where conflict is inevitable, or preventing the conflict by altering the ecological setting itself. In the following analysis, we evaluate the utility of different conflict-mitigation approaches in terms of their value for tiger conservation, technical feasibility and social practicality. We employ the term 'problem tiger' for any animal that is persistently preying on domestic livestock, or has either killed human beings, or is potentially likely to do so immediately.

Lethal control

Killing of problem tigers through shooting, poisoning of livestock kills and less commonly using techniques such as electrocution, snaring and trapping has been the traditional method of conflict mitigation. Killing of 'problem' tigers has been widely accepted and practised by local people in India (Daniel 2001; Karanth 2001). Often, in situations involving tigers cornered by uncontrollable mobs, with the imminent prospect of human deaths, or with injured tigers, lethal control is the only practical option (Karanth and Madhusudan 2002).

However, urban advocacy groups often oppose such tiger killings on the grounds of either conservation or animal welfare. In the case of tigers straying out from small 'source' populations, repeated application of lethal control in the surrounding landscape may eventually lead to population extirpation (e.g. see Woodroffe *et al.*, Chapter 1). Furthermore, in a free-ranging population, it is very difficult to specifically target the individual problem tiger: several tigers may have to be killed before the problem animal gets eliminated. In such situations, local wildlife managers are often unfairly criticized, based on the unrealistic, anthropocentric expectation that only the 'guilty' tiger should have been punished. While lethal control is abhorrent to some conservationists and most animal welfare advocates, unfortunately it is the only practical option open to wildlife managers in many actual cases of human–tiger conflict (see also Treves and Naughton-Treves, Chapter 6).

Capture and removal of 'problem tigers'

Sometimes managers attempt to capture a problem tiger and move it away from the spot of the conflict because this approach has wider social acceptability among conservationists and animal welfare groups. In the case of individual tigers of dispersal age (Smith 1993), moving the animals back to the source population may even be justifiable. However, such translocations are rarely practical, and may not have satisfactory conservation outcomes. In a free-ranging tiger population it is rarely possible to identify the individual problem animal for removal, unless it enters human settlements or is injured. Furthermore, safe chemical capture (or driving away) of tigers is usually rendered difficult because of crowd-control problems, injuries to the animal, lack of technical skills, scarcity of resources and other logistical problems (Karanth and Madhusudan 2002; K. U. Karanth, pers. obs.).

Even after safe capture, the problem tiger has to be permanently housed in captivity or relocated into the source population from which it came or into a new habitat. There are severe constraints on all three options. Wild tigers do not adapt well to life in captivity, and the capacity of Indian zoos to hold tigers is already saturated. Most zoos simply cannot afford to house an ever-increasing number of problem tigers.

Most problem tigers that undergo capture and handling are injured in the process, particularly by losing their canine teeth in steel transport cages that are commonly used. Many are either old or weak animals evicted from their ranges by more vigorous rivals (Smith 1993). Such tigers are unfit for relocation into the wild.

There are several ecological arguments against translocation of even healthy problem tigers into new habitats. First, most such relocations simply result in transfer of the problem to a new location leading to a new situation of conflict, because high-quality tiger habitats devoid of conflict potential are scarce. Second, even after translocation into a large reserve with an adequate prey base, the introduced animal will compete for space and prey with other individuals in the local tiger population. Because tigers are territorial animals (Sunquist 1981; Smith 1993) and their numbers are limited by prey densities (Karanth and Nichols 1998; Karanth 2003; Karanth *et al.* 2004), intraspecific competition is likely to lead to elimination of either the introduced tiger or of another individual from local population. Tiger populations normally go through increased rates of infanticide and mortalities during periods of social instability, which follow natural turnovers of male resident breeders (Sunquist 1981; Smith 1993). Periodic release of new tigers into wild populations may further aggravate such instability, causing more 'problem tigers' to disperse out of the population.

Guarding, barriers and aversive conditioning

A reasonably effective traditional approach involves employing human herders to guard livestock grazing in tiger habitats, wherever such labour is available and inexpensive.

Since most tiger attacks on livestock and humans occur under free-ranging conditions, mechanical barriers like stockades have limited utility. However, barriers made of wooden poles, wire mesh and nylon netting are being used to prevent tigers from entering villages in Sundarban. In the same region, aversive conditioning of tigers using electrified 'human dummies' has been tried out. Another technique used in Sundarban involves attaching backward-facing masks behind the heads of potential victims. The mask is expected to deter tigers, which are behaviourally attuned to avoid frontal attacks on prey. Success has been claimed for both these interesting innovations developed by local wildlife managers (Sanyal 1987). However, their use has been sporadic and irregular with no rigorous experimentation to test their efficacy.

The advanced – and expensive – non-lethal aversive conditioning techniques occasionally used for deterring carnivore attacks in developed countries (Treves and Karanth 2003; Shivik *et al.* 2003; Breitenmoser *et al.*, Chapter 4) do not appear to be very relevant to the technology and resource-scarce social context in which most human–tiger conflict occurs in India.

Compensatory payments

In cases of human predation by tigers, financial aid can never fully compensate the loss suffered by the victim's families. However, prompt delivery of such assistance may help mitigate local hostility towards tigers to some extent (see also Nyhus *et al.*, Chapter 7). Given the relative rarity of tiger attacks on humans (except in Sundarban) and the public pressures that such attacks generate, government schemes for compensating for human lives lost to tigers seem to be working reasonably well.

However, payment of compensation for livestock predation – particularly in multiple-use forests with grazing rights – is problematic (Madhusudan 2003). Livestock compensation schemes fail for a variety of reasons: the massive scale of the problem; the low value of livestock in relation to the expenses involved in getting claims verified; corruption in the official machinery and among claimants; and a general lack of rural financial mechanisms enabling quick transactions (see also Nyhus *et al.*, Chapter 7).

Although compensation schemes of the government (and occasionally non-governmental agencies) have long existed over most parts of India, they do not appear to be highly effective (Karanth and Madhusudan 2002; Madhusudan and Mishra 2003). Furthermore, no systematic evaluation of these schemes appears to have been undertaken during the last 30 years. We note that Indian herders appear to tolerate some degree of carnivore predation as a price they are willing to pay for access to resources in public forests (Sekhar 1998; Madhusudan 2003).

Preventing conflict: relocation of human settlements

All the techniques described earlier are essentially components of an overall 'reactive' mitigation strategy, which tries to deal with the conflict generated by the interspersion of human settlements and tiger habitats. In such settings, conflict is driven by ecological competition because tigers are large, obligate predators (Madhusudan and Mishra 2003).

On the other hand, relocation of human settlements is a proactive strategy that tries to alter the ecological setting, and thus prevent conflict rather than dealing with it after the fact (Karanth 2002, 2003; Karanth and Madhusudan 2002). This strategy has been implemented under the Indian government's wildlife conservation schemes since the early 1970s. As a tool for promoting long-term human–tiger coexistence at the landscape level beyond reserve boundaries, the relocation strategy has several important advantages.

For most tiger populations in India (Fig. 23.1) survival prospects are bleak in the face of escalating habitat fragmentation and resulting conflict with human interests (Karanth 2001, 2003). Relocation of human settlements arrests ongoing conflicts and prevents their escalation. Relocation has been a critical tool in reducing habitat fragmentation and in driving the recovery of many wild tiger populations from the brink of extirpation in several Indian reserves (Karanth *et al.* 1999; Karanth 2002, 2003). When long-term social and economic costs of dealing with perennial human–tiger conflict are considered, relocation appears to be an attractive preventive option.

However, despite their ecological desirability and cost-effectiveness, resettlement projects face many practical hurdles. If the relocation process is not transparent, incentive-driven and fair, it can lead to hardship and resentments (Kothari *et al.* 1995). Scarcity of alternate land, lack of adequate financial resources or other social and cultural factors can also lead to setbacks to resettlement schemes (Karanth *et al.* 1999; Karanth 2002). Relocation schemes are unlikely to work well for large areas under multiple-use forests that support high human population densities, or situations where alternate land is scarce. They are not very relevant for dealing with conflicts that occur at the hard edge between strictly protected tiger reserves and the extensive agricultural landscapes outside. Therefore, the relocation strategy is of primary relevance only for enclaves of human settlements within important tiger habitats, or critical corridors that connect insular tiger populations. Fortunately, in many such situations, there seems to be an incipient local demand for relocations, driven by changing social traditions and economic aspirations of the local people (Karanth 2002). In the light of these ongoing social changes and the escalating costs of delivering economic development and social services to remote settlements, incentive-driven relocation may emerge as a key strategy in India to ensure coexistence of humans and wild tigers at the landscape level.

THE FUTURE OF COEXISTENCE: AN ECOLOGY-BASED POLICY FRAMEWORK

Wildlife managers in India now employ a mix of conflict mitigation strategies. However, this strategic mix is largely *ad hoc* and not guided by clear policy prescriptions based on the aspects of tiger ecology and human social factors discussed earlier. Wildlife managers are severely handicapped by stringent legal requirements, lack of financial resources and technical skills, as well as by social pressures generated locally in conflict situations. A clear policy framework would enable them to avoid *ad hoc* responses and deal with conflict situations much more logically and effectively.

The framework we propose for managing conflict is shaped by the ecological reality that effectively protected breeding tiger populations are restricted to around 1% of India's geographical area and are further losing ground. Clearly, if the societal consensus is that wild tigers must survive in viable numbers in India, it is necessary to ensure that their ecological needs are central to any policy framework for reducing conflict and promoting coexistence.

Therefore, the long-term vision underlying our proposal involves increasing the area that supports demographically viable tiger populations (Karanth and Stith 1999) to at least one-third of the estimated potential tiger habitats in India (about 100 000 km^2). This effort will involve maintaining 50–100 insular or tenuously connected wild tiger populations, each containing 12–50 breeding female ranges. If successful, such a strategy will require the management of tiger populations at average densities of 5–15 animals/100 km^2 with an overall size of 5000–15 000 wild tigers. These populations will naturally produce an annual 'surplus' of 500–1500 dispersing transient tigers that will come into conflict with people in the human-dominated landscapes around them.

We expect that the breeding tiger populations visualized above will be largely confined to protected areas, which will interface with either hard-edge agricultural landscapes or multiple-use forests, both riddled with conflict-prone land uses. Therefore, the above tiger conservation scenario has the potential for substantially escalating the present levels of conflict. Paradoxically, successes in tiger conservation will lead to higher levels of conflict that can only be mitigated by a proactive policy mix of conflict prevention and mitigation, rooted in sound science and practical experience (see Bangs *et al.*, chapter 21). Tactics ranging from lethal control of tigers at one end of the spectrum to relocation of human settlements at the other would have to be part of the mix.

We suggest the contours of an ecology-based policy matrix to guide future management of human–tiger conflict (Table 23.3). Our two-way

Table 23.3. *A policy framework for human–tiger coexistence in India*[a]

	Conflict type		
Land use	Livestock predation	Accidental killing of humans	Persistent predation on humans
Protected area for tiger conservation	Toleration Relocation	Compensation Relocation	Compensation Relocation Lethal control?
Tiger habitat in multiple-use public forest or extractive reserve	Toleration Compensation?	Compensation	Compensation Lethal control
Tiger habitat in privately owned land	Compensation	Compensation Capture–captivity	Compensation Lethal control
Unsuitable tiger habitat in public or private land	Compensation Translocation	Compensation Translocation Lethal control	Compensation Lethal control
Human habitations, livestock enclosures	Compensation Translocation Barriers?	Compensation Translocation Capture–captivity Lethal control	Compensation Lethal control

[a] The management context (each cell in the matrix) is defined by the nature of the conflict (columns) and the priority land use at the site of the conflict (rows). The suggested conflict mitigation tactics (not mutually exclusive) are listed in order of priority within each cell. These tactics include: passive toleration of conflict, financial compensation, tiger capture followed by translocation into the source population or captivity, lethal control of tigers, and incentive-driven relocation of human settlements.

matrix considers the type of conflict and the land-use priority at the site of the conflict, as the basis for management actions listed in order of priority within each cell – a form of zoning (Linnell *et al.*, Chapter 10). This framework prioritizes the need to keep tigers spatially separated from incompatible human land uses at the scale of protected reserves, while at the same time aiming to mitigate conflicts by prioritizing human needs at larger landscape scales. Robinson's (1993) conservation paradigm of establishing 'sustainable landscapes' provides a more appropriate conceptual template for implementing the proposed policy framework than the alternatives of 'sustainable' human use of all tiger habitats favoured by social advocacy groups in India (Kothari *et al.* 1995), or the 'don't kill a single tiger' approach favoured by some votaries of animal rights. We believe that as human populations and resource consumption levels increase, similar policies will be essential for enabling the continued coexistence of all big cat species with human societies in most parts of the world.

ACKNOWLEDGEMENTS

Both the authors would like to acknowledge the facilitation of their work on tigers by the Government of India and several state governments. KUK would also like to acknowledge the long-term support he has received from Wildlife Conservation Society, New York.

24

The future of coexistence: resolving human–wildlife conflicts in a changing world

ROSIE WOODROFFE, SIMON THIRGOOD AND ALAN RABINOWITZ

INTRODUCTION

In this volume, we have sought solutions to the pervasive problem of conflict between people and endangered wildlife. There can be no doubt that human–wildlife conflict has driven global declines of many species (Woodroffe *et al.*, Chapter 1). Equally, there can be no doubt that such animals – even beautiful animals, even endangered animals – can and do cause serious damage to human lives and livelihoods (Thirgood *et al.*, Chapter 2). That much is clear. The question is: what to do about it? The authors of the chapters in this volume have provided a myriad of possible solutions, some of them successful, some of them informative failures, which we hope will help wildlife managers to understand and resolve conflicts. In this concluding chapter, we draw general conclusions from the chapters concerning the conditions underlying human–wildlife conflicts, the most effective means to resolve them, and how we may expect the patterns of conflict to change in future years.

Is coexistence achievable?

For as long as there has been literature, conflicts between people and wildlife have been documented. As early as the eighth century BC, Homer noted how lions 'plunder men's steadings, seizing on their cattle and sturdy sheep, until they too are killed, cut down by the sharp bronze in the men's hands' (*Iliad* V: 548–50). Perhaps the first documentation of a wish to resolve such conflicts came later the same century, when Isaiah prophesied a halcyon future when 'the wolf will live with the lamb, and the leopard lie down with the kid; the calf and the young lion will feed together ... the cow and the bear will be friends ...' (Isaiah 11: 6–7). Is such peaceful coexistence achievable, without the benefit of the divine intervention envisaged by

People and Wildlife: Conflict or Coexistence? eds. Rosie Woodroffe, Simon Thirgood and Alan Rabinowitz. Published by Cambridge University Press.

Isaiah? We doubt that coexistence can ever be so utopian, but many of the studies presented in this book demonstrate that wildlife can live alongside people, livestock and crops, even when their interests compete with those of local people. Perhaps the most striking evidence of this comes from the successful recoveries of threatened wildlife that have been documented in recent years; such successes include the recoveries of large carnivores in Europe and North America described by Swenson and Andrén (Chapter 20), Linnell *et al.* (Chapter 10) and Bangs *et al.* (Chapter 21), of raptors and wild geese in Scotland (Thirgood and Redpath, Chapter 12, Cope *et al.*, Chapter 11), and African wild dogs (*Lycaon pictus*) in Kenya (Frank *et al.*, Chapter 18).

Is there an alternative to coexistence?

We are comforted by our conclusion that coexistence of people and wildlife is – sometimes – possible, since in many areas there is no alternative means of conservation. For example, 80% of the world's African elephants (*Loxodonta africana*) range outside protected areas (Hoare 2000), and the figure for Asian elephants (*Elephas maximus*) may be equally high (Santiapillai and Jackson 1990), creating an urgent need to develop ways to foster coexistence with people. Likewise, Miquelle *et al.* (Chapter 19) comment that, since Amur tigers (*Panthera tigris*) have large territories, and the protected areas within their geographic range are all small, the world's biggest cats can only be conserved by sharing the landscape with local people. Swenson and Andrén (Chapter 20) make exactly the same point concerning wolves (*Canis lupus*), bears (*Ursus arctos*), wolverine (*Gulo gulo*) and lynx (*Lynx lynx*) in Scandinavia, as do Thirgood and Redpath (Chapter 12) for raptors in Europe. If these species cannot be conserved in multiple-use landscapes, there is a very real probability that they cannot be conserved at all.

The alternative to coexistence is partitioning of the landscape into 'human' areas and 'wildlife' areas, an extreme form of zoning discussed by Linnell *et al.* (Chapter 10). Karanth and Gopal (Chapter 23) argue that efforts to promote coexistence of people and tigers in India have largely failed and, instead, argue convincingly in favour of zoning people out of tiger reserves. Even more extreme partition of people and wildlife may be achievable by fencing. While fencing can be effective where it is used to keep wildlife out of small areas that are particularly valuable to people (such as cultivated fields: Osborn and Hill, Chapter 5, or livestock paddocks: Breitenmoser *et al.*, Chapter 4), we caution that fencing may be less useful when it is implemented over larger areas. There are several reasons for this. First, an obvious (but not trivial) limitation is that not all conflict species can be readily excluded from areas by fencing; hen harriers (*Circus cyaneus*)

(Thirgood and Redpath, Chapter 12) and geese (Cope *et al.*, Chapter 11) are examples. If wildlife can cross fences, conflicts will not be resolved. For example, while the Shimba Hills fence discussed by Knickerbocker and Waithaka (Chapter 14) was more or less successful in reducing human fatalities by containing elephants inside the park, baboons' (*Papio* sp.) ability to cross it generated a new problem of crop-raiding at the park border. Moreover, fences which are almost-but-not-quite impenetrable to wildlife can generate occasional problems when animals find a way to cross out of a reserve, but then cannot return (R. Woodroffe *et al.* unpubl. data; Knickerbocker and Waithaka, Chapter 14). Hence Thouless and Sakwa (1995) argued that it might be necessary to shoot elephants that consistently broke fences.

Perhaps the greatest technical argument against fencing is its expense: Knickerbocker and Waithaka (Chapter 14) estimate that maintenance of the Shimba Hills fence costs the Kenya Wildlife Service US\$675/km/year (equivalent of about US\$75 000/year for a perfectly circular 1000-km^2 reserve) – and this does not factor in either the costs of construction, or the contribution of local communities to fence maintenance.

A further problem with fences is that they provide a ready supply of wire which may be used to build snares (Hill *et al.* 2002), often used for illegal hunting. This may be a particularly serious issue where fencing projects lack community support. In many protected areas, snaring is the most common type of illegal hunting (Arcese *et al.* 1995), and poachers will go to great lengths to obtain wire. While fences are certainly not the root cause of snaring problems, they can exacerbate such problems under some circumstances.

Fencing may also have longer-term impacts on wildlife populations. Local declines of wildebeest (*Connochaetes taurinus*) and zebra (*Equus burchelli*) in and around Kruger National Park in South Africa have been convincingly linked to blockage of their migration routes when the park was fenced; unable to move to exploit seasonally available forage, these animals experienced sudden reductions in carrying capacity (Whyte and Joubert 1988; Ben-Shahar 1993). Disruption of migratory routes by veterinary cordon fences in Botswana have also been linked to major declines of wild ungulates in the Kalahari (Williamson and Williamson 1985), which may have long-term effects on the composition of ungulate communities and, hence, on vegetation (Thouless 1998). The long-term demographic and genetic benefits of promoting dispersal between subpopulations of animals are widely acknowledged (Gonzalez *et al.* 1998; Crooks and Sanjayan 2005), and underlie worldwide efforts to promote connectivity (de Lima and Gascon 1999; Webster and Wood 1999; Bruinderink *et al.* 2003; Rouget *et al.* 2003; Freudenberger and Brooker 2004). Decisions about fencing must factor in these ecological costs. However, in some areas of very serious human–wildlife conflict, the conservation benefits of fencing reserves may well outweigh the costs of permanently destroying connectivity with other sites. The long-term costs and benefits of fencing wildlife areas clearly warrant further study.

PATTERNS AND PREDICTORS OF HUMAN–WILDLIFE CONFLICT

Conflicts are not evenly distributed. In virtually every system, some sites are particularly prone to depredation or crop-raiding, while other areas nearby are unaffected (Redpath and Thirgood 1999; Stahl *et al.* 2001; Sitati *et al.* 2003; Treves *et al.* 2004; Woodroffe *et al.* 2005). Curiously, this patchy distribution of conflict may help to engender negative attitudes to wildlife because, while average losses may be low on a regional scale, particular individuals or groups of individuals may be affected severely and often repeatedly. Costs to the affected individual are exacerbated still further, of course, if the animals involved are large or fierce enough to injure or kill people attempting to drive them away. These high individual or local costs can lead to a perception that the risks of conflict are higher than they actually are – simply because the costs are so high if one is unlucky enough to be affected. In the absence of any perceived need to conserve wildlife, these conditions would tend to argue in favour of retributive or pre-emptive killing of 'conflict' species.

The patchy distribution of conflicts may also have positive implications, however. First, it means that conflicts over a wide area might potentially be reduced by dealing with problems in a few small areas. Second, comparison of affected and unaffected areas may reveal the underlying causes of conflicts. Such comparisons can not only point to the management solutions most likely to be effective, but may also allow prediction of when and where future conflicts are likely to occur, so that management can be targeted most effectively. Differences between sites may relate to the management strategies adopted locally (see below). Here, however, we draw attention to several environmental factors, operating at a variety of scales, that may predispose particular areas and systems to conflict.

Opportunity for contact

Human–wildlife conflict can only occur when and where wildlife come into contact with people or their property. For example, Sitati *et al.* (2003) showed that elephant crop-raiding occurred predictably in areas with high crop cover (i.e. where crops are grown), and Treves *et al.* (2004) showed that wolf depredation on livestock occurred in areas where there is extensive pasture (i.e. where cattle live). At first sight such findings may appear trivial: however they can provide extremely valuable quantitative predictors of the extent to which humans can modify a landscape before conflicts ensue. For example, Bangs *et al.* (Chapter 21) showed that recolonizing wolves were unable to persist in areas with high sheep densities – in part because any wolves that did occupy these areas invariably killed livestock and were removed. Analyses of this kind may be extremely valuable for the delineation of management zones (Linnell *et al.*, Chapter 10).

On a smaller spatial scale, proximity to cover or other wildlife habitat is frequently associated with wildlife damage. For example, Naughton-Treves and Treves (Chapter 16) showed that 90% of crop damage occurred within 200 m of the boundary of Kibale National Park, and Stahl *et al.* (2002) showed that 95% of pastures where lynx attacked sheep were within 250 m of forested areas.

Regular contact between wildlife and livestock or crops may not just set the stage for localized conflict – it may also foster the development of 'problem' behaviour such as stock-killing or crop-raiding. For example, Stahl *et al.* (2002) showed that lynx with home ranges containing low densities of sheep never killed sheep, but that some of those exposed to higher sheep densities became habitual sheep-killers, even though wild prey were abundant. Hence, livestock depredation became a localized problem, with mitigation measures on one farm leading to increased predation on neighbouring flocks. Similarly, Woodroffe and Frank (2005) suggested that poor livestock husbandry on one farm essentially 'trained' lions (*Panthera leo*) to become habitual stock-killers, leading to their being eventually shot. It seems that, like the good poets commemorated by Ben Jonson, problem animals are 'made, as well as born'.

Environmental productivity

We note that coexistence has been achieved most readily in areas with comparatively low primary productivity, including areas that are too dry (e.g. Frank *et al.*, Chapter 18) or too cold (e.g. Swenson and Andrén, Chapter 20, Bangs *et al.*, Chapter 21) to support cultivation and, hence, high human densities. This difference is illustrated most starkly by the contrasting case studies of tigers in the boreal forests of the Russian Far East (Miquelle *et al.*, Chapter 19), and the tropical forests of India (Karanth and Gopal, Chapter 23). In Russia, coexistence of people and tigers appears eminently achievable (assuming poaching for the traditional Chinese medicine trade can be controlled): tigers move fairly freely between reserves through areas sustaining low human densities. There are very low levels of livestock depredation, and only sporadic attacks on people; most conflicts concern competition for access to wild ungulates. By contrast, in India human densities are markedly higher, and tigers are almost completely confined to reserves where they persist in small, isolated populations. Livestock depredation is common and attacks on people are a regular occurrence. The difference between the two regions may simply be that, having higher primary productivity, India *can* sustain more people. This creates a circumstance in which competition for space between people and wildlife is severe: hence, where human density is high, reserves are small and hard-edged (Harcourt *et al.* 2001). Some of the most severe human–wildlife conflicts described in this book occur in highly productive areas where reserves abut cultivated areas sustaining high human densities (Knickerbocker and Waithaka, Chapter 14, Naughton Treves and Treves, Chapter 16) – essentially, these are further cases where the opportunities for human–wildlife contact are high. By contrast, where primary production is lower, cultivation rarely possible, and human densities consequently lower, conservation may be a more competitive land use option because the opportunity cost of converting habitat to other uses is low. Walpole and Thouless (Chapter 8) developed this argument in relation to ecotourism as a land use, but we suspect that it can be expanded to a more general explanation of the distribution of conflicts.

Availability of wild prey

Several of the studies presented in this volume have related human–wildlife conflicts to the availability of conflict species' natural prey. Confusingly, however, in some cases abundant wild prey reduces the level of conflict, whereas in others it seems to exacerbate the problem. For example, Frank *et al.* (Chapter 18) linked livestock predation by lions and African wild dogs to depletion of wild prey brought about by drought and hunting, respectively. Likewise Naughton-Treves *et al.* (1998) showed that crop raiding by baboons (*Papio* sp.) and chimpanzees (*Pan troglodytes*) was most severe when natural forest fruits were not available. Miquelle *et al.* (Chapter 19) noted that Amur tiger predation on livestock was most pronounced during a severe winter when wild prey were presumably scarce or inaccessible. Predation on livestock has also been linked to low densities of wild prey in wolves in southern Europe (Meriggi and Lovari 1996) and North America (Mech *et al.* 1988). By contrast, Thirgood and Redpath (Chapter 12) showed that abundant alternative prey (in the form of voles (*Microtus agrestis*) and meadow pipits (*Anthus pratensis*)) increased hen harrier predation on managed red grouse (*Lagopus lagopus*) through a complex trade-off between the numerical and functional responses of harriers to their three most important prey species. Treves *et al.* (2004) and Stahl *et al.* (2002) likewise showed that predation on livestock by wolves and lynx (respectively) was highest where wild prey were abundant. Fruiting trees in the vicinity of cultivated fields may also draw in elephants and trigger crop-raiding (Osborn and Hill, Chapter 5). Hence, while there are clear links between the availability of animals' natural foods and their tendency to feed on crops, livestock, or wildlife valued by people, the relationship seems to be a complex and, currently, a somewhat unpredictable one. This is an area that clearly merits further quantitative study.

Recovering wildlife populations

Many of the wildlife species discussed in this volume continue to experience monotonic declines in abundance. For these species, as Western and Waithaka (Chapter 22) point out, the proportion of people engaged in human–wildlife conflicts is declining, while the proportion of wildlife experiencing such conflicts is increasing. However, some of the species discussed here have responded well to conservation measures, and are now re-expanding their geographic ranges (e.g. geese and hen harriers in Scotland: Cope *et al.*, Chapter 11, Thirgood and Redpath, Chapter 12; African wild dogs in Kenya: Frank *et al.*, Chapter 18; grey wolves in the USA: Bangs *et al.*, Chapter 21); others are projected to do so in future if conservation measures are successful (e.g. tigers in India: Karanth and Gopal, Chapter 23). Such circumstances entail particular challenges for wildlife managers (Mech 1995). Not only do these animals encounter conflicts with people as they recover – they experience conflicts with people who have not encountered them for decades. Such circumstances can be particularly problematic since traditional mitigation measures will usually have been lost. More subtly, the return of problematic wildlife may be perceived as 'a step backwards' in societies that have a strong commitment to 'progress' and economic development. Hence, the environmental conditions that allow population recovery may prompt conflicts more severe than would have been the case had the 'conflict' species been present continuously. Despite this, we advocate zoning recovering populations out of areas only when it is very clear that it will be unsustainable for them to persist there, because the recoveries witnessed in recent years show that wild animals can often survive, without conflict, in unexpected areas. For example, Bangs *et al.* (Chapter 21) note that grey wolves have successfully (and fairly harmlessly) recolonized areas where managers predicted they could not persist. Likewise, one of us (RW) argued against reintroduction of African wild dogs to Laikipia and Samburu Districts, in Kenya, because human and livestock densities were considered too high – only to have the species recolonize naturally the following year and grow to a population of >200 animals with negligible livestock losses (Woodroffe *et al.* 2005).

WHICH MANAGEMENT STRATEGIES AND POLICIES CAN ENCOURAGE COEXISTENCE?

We have noted that certain ecological conditions may predispose particular areas to conflict. A central theme of this book, however, is to understand how management interventions – be they technical measures, economic incentives or policy initiatives – can resolve conflicts and allow sustainable coexistence of people and wildlife. Before we discuss the merits of different approaches, however, we note two general conclusions that have emerged from multiple chapters.

A first general comment is that there is a paucity of well-designed studies demonstrating the effectiveness (or otherwise) of particular approaches. It is not surprising that this should be the case for fairly novel methods such as non-lethal techniques for deterring crop-raiders and livestock-killers (Breitenmoser *et al.*, Chapter 4: Osborn and Hill, Chapter 5). However the technical effectiveness of more traditional methods, such as lethal control, has also been questioned in some cases (Treves and Naughton-Treves, Chapter 6), and clearly needs more study. Interestingly, the authors of all three chapters on the use of economic incentives to resolve conflicts (Nyhus *et al.*, Chapter 7, Walpole and Thouless, Chapter 8, Leader-Williams and Hutton, Chapter 9) drew attention to a need for systematic comparisons of the effectiveness of alternative approaches. Systematic studies would allow managers and local communities to choose among various models that have been shown to be effective elsewhere. In addition, such studies occasionally generate counter-intuitive results that may clarify the seriousness of a conflict (e.g. Naughton Treves and Treves, Chapter 16, showed that two-thirds of 'wildlife' crop damage was in fact caused by livestock), or even fundamentally challenge past management strategies (e.g. Donnelly *et al.* (2003) showed that a badger (*Meles meles*) culling strategy that had been in place for nearly 30 years apparently increased, rather than decreased, impacts on farmers' livelihoods, and Rabinowitz (Chapter 17) showed that pre-emptive attempts to kill jaguars (*Panthera onca*) as pests in Belize in fact contributed to creating livestock predators out of non-problem jaguars). Such studies indicate that 'common sense' is no substitute for systematic study.

While better biological information is often useful, we caution that on its own this is rarely sufficient to resolve conflicts. Reading *et al.* (Chapter 13) dismiss the finding that prairie dogs (*Cynomys ludovicianus*) have little or no detectable impact on livestock production as almost irrelevant in the face of a widespread and systemic perception to the contrary on the part of both ranchers and wildlife managers. Likewise, Bangs *et al.* (Chapter 21) note that, while wolf predation on livestock is infrequent and affects only a small number of farmers, every incident attracts media attention and feeds the perception of a serious impact; similarly Rabinowitz (Chapter 17) discusses how cattle ranchers in the Pantanal continue to blame jaguars for most of their livestock depredation despite data showing that, much of the time, livestock loss was not attributable to jaguars. Finally, Thirgood and Redpath (Chapter 12) comment that their demonstration of a genuine impact of hen harriers on harvests of red grouse may have increased illegal killing of this rare and protected species. It is possible that the failure of such biological insights to advance the resolution of conflicts highlights an even more urgent demand for information from the social sciences on which measures are – or are perceived by local people to be – most effective at resolving conflicts from the stakeholders' points of view. One promising approach involves multi-criteria decision modelling: this was used recently to assess the acceptability to hunters and conservationists of different conflict mitigation techniques in the harrier–grouse example cited above (Redpath *et al.* 2004).

A second pattern to emerge from these studies is that no single solution is effective in all cases – or indeed, in any one case. In the words of one of the chapter authors, 'there is no silver bullet'. For example, Cope *et al.* (Chapter 11) successfully recovered geese in Scotland using a combination of measures including performance payments, diversionary feeding and extensive farmer involvement. At the opposite extreme, Swenson and Andrén (Chapter 20) demonstrate that, in Norway, reliance solely on compensation for livestock losses fails to reduce human conflicts with large carnivores, in part because farmers have no incentives to reduce their losses – exactly as predicted by Nyhus *et al.* (Chapter 7).

Having made these general comments, we now consider the effectiveness of various management interventions in resolving conflicts between people and threatened wildlife.

Technical solutions

Breitenmoser *et al.* (Chapter 4) and Osborn and Hill (Chapter 5) review a multitude of non-lethal technical interventions which may reduce the extent to which wild animals feed upon livestock and crops respectively. We note several general patterns emerging from these and other chapters.

First, traditional approaches to reducing wildlife impacts are often highly successful. This is not surprising – after all, these methods were developed over millennia, when people often had neither the means to eradicate wildlife themselves, nor government institutions to do it for them. Frank *et al.* (Chapter 18) in particular describe how traditional livestock husbandry is effective in reducing depredation by large carnivores in Africa. Likewise, Naughton-Treves and Treves (Chapter 16) comment on how traditional collective land tenure systems spread the risk of crop-raiding more effectively than do modern individualized approaches. Many of these traditional prevention measures – such as remaining in cultivated fields to guard them from crop-raiding elephants – are highly labour-intensive. Hence, they have been abandoned in more developed countries where wage costs for guards (or the opportunity costs of guarding instead of working elsewhere) are higher, and where education produces communities of people unwilling to engage in the somewhat boring (and periodically dangerous) task of guarding fields or livestock. Frank *et al.* (Chapter 18) comment that intensive herding practices have been retained among African pastoralist peoples in part because the tradition of livestock theft has also persisted – hence increased confidence that livestock will not be stolen in developed countries may also have contributed to the loss of traditional husbandry. Finally, of course, many conflicts now occurring in developed countries are in areas where 'conflict' species had been locally extirpated and are now recovering – hence traditional mitigation methods had been abandoned in large part because they were no longer needed. Unfortunately, low-cost alternatives are rarely as effective as guarding: Osborn and Hill (Chapter 5) remark that elephants and primates rapidly habituate to 'empty' threats if they are not backed up with a serious threat from a human guard, and Breitenmoser *et al.* (Chapter 4) likewise comment that guarding dogs are most effective when working together with a shepherd. The conservation and, where appropriate, restoration of traditional farming practices may be a valuable tool in resolving human–wildlife conflicts.

Second, while a host of non-lethal methods are available to dissuade wildlife from consuming crops, livestock and people, there are very few technical solutions to conflicts over access to wild prey. For some species (e.g. hen harriers: Thirgood and Redpath, Chapter 12; prairie dogs: Reading *et al.*, Chapter 13; Amur tigers: Miquelle *et al.*, Chapter 19) this is the principal (or only) basis for conflict with people. Few technical interventions have successfully mitigated such conflicts (although diversionary feeding – used with some success by Redpath *et al.* (2001) on hen harriers and by Gasaway *et al.* (1992) on wolves and bears – may be a promising exception). Hence, resolution of such conflicts is likely to rely primarily on economic incentives, perhaps in combination with some form of lethal control. In this context, we note that the expansion of game farming as a land use, particularly in southern Africa, has been very beneficial for recovery of wild ungulate populations (Lewis and Jackson, Chapter 15), but has generated new conflicts between game farmers and predators which cannot easily be dissuaded from consuming valuable prey (Lindsey 2003).

Third, we note that, paradoxically, occasional lethal control may be a necessary component of conservation strategies in many cases, even when highly endangered species are involved (Treves and Naughton-Treves, Chapter 6). Carefully targeted lethal control may have the potential to reduce serious wildlife impacts on human lives or livelihoods that cannot be resolved in other ways and, perhaps more importantly, may help engender public support by demonstrating managers' willingness to acknowledge the impacts that wildlife may be having on local people. Some authors (Thouless and Sakwa 1995; Woodroffe and Frank 2005) have suggested that occasional targeted lethal control may help to avoid spread of 'conflict' behaviour (such as livestock depredation, crop-raiding or fence breaking) through wildlife populations. The availability of lethal control as a management tool may also promote the acceptability to stakeholders of non-lethal alternatives (e.g. diversionary feeding of hen harriers as part of a management package: Thirgood and Redpath, Chapter 12). It is worth remarking that, if non-lethal techniques are effective enough to reduce conflicts to acceptable levels, then the lethal control 'tool' may remain in the 'toolbox' without being used. Finally, if recovery efforts are successful and threatened wildlife start to colonize unsuitable areas where they have an unsustainably high impact on local people's lives and livelihoods – perhaps areas that would be designated 'no wildlife' zones (Linnell *et al.*, Chapter 10) – lethal control may play a role in removing them.

This certainly does not mean that we advocate a return to the widespread lethal control of 'conflict' species that has made so many of them endangered in the first place (Woodroffe *et al.*, Chapter 1). Moreover, we caution that if 'conflict' behaviour is a predictable response to local conditions (e.g. where cover is available close to agricultural areas and wild prey are scarce), then 'problem animals', once eliminated, are likely to be replaced by other animals that themselves develop the same 'conflict' behaviour sooner or later, perhaps explaining the short-term success of many attempts at lethal control (Treves and Naughton-Treves, Chapter 6). However, we note the important role that targeted lethal control has played in improving public support for some successful conservation plans (e.g. Swenson and Andrén, Chapter 20, Bangs *et al.*, Chapter 21) and the frustration expressed by several authors at not being able to use lethal control when it could be expected to contribute to conflict resolution along the lines described above (e.g. Thirgood and

Redpath, Chapter 12, Karanth and Gopal, Chapter 23), and conclude that lethal control is likely to have a role to play in some conservation programmes. Lethal control should, however, be used only as one of a suite of management interventions that also involve an array of non-lethal measures.

Economic incentives

Some forms of economic incentives to encourage tolerance – in which we include compensation, performance payments, and both consumptive and non-consumptive use of wildlife to generate revenues – show promise and may be important components of future conservation strategies for 'conflict' species. However, systematic studies showing the effectiveness (or otherwise) of these approaches are generally lacking at present. To illustrate the difficulties, we note that Nyhus *et al.* (Chapter 7) suggest that revenue-generating measures such as ecotourism and sport hunting might be more successful than simple compensation, yet Walpole and Thouless (Chapter 8) explain that programmes based on the use of wildlife to generate revenues for conservation are at a competitive disadvantage with ordinary businesses which are expected only to make money, not to support the costs of conservation and address community needs as well. It seems likely that donor funding will be required in many cases to support – or at least initiate – such programmes.

There is a general conclusion that, while compensation may have a role to play in some cases (e.g. Bangs *et al.*, Chapter 21), it has severe limitations, particularly when used in isolation from other methods (Nyhus *et al.*, Chapter 7). Compensation is highly sensitive to corruption (hence its abandonment in many areas, e.g. Kenya: Western and Waithaka, Chapter 22), and can also generate unintended incentives; for example, in Norway farmers argued to have government funds directed towards detection of predator-killed sheep (so that they would receive maximum compensation) rather than towards measures to prevent depredation (Swenson and Andrén, Chapter 20). Nyhus *et al.* (Chapter 7) argue that performance payments may be a more effective measure in some cases, and Cope *et al.*'s (Chapter 11) study of conflicts between geese and farmers in Scotland would tend to support this view. Insurance programmes may be an appealing alternative to donor- or government-funded compensation, in part because (if properly designed) they contain internal incentives to farmers to take measures to reduce the probability of wildlife damage (since premiums rise with risk). However, we note that Miquelle *et al.*'s (Chapter 19) attempt to develop such a programme collapsed largely because the risk of livestock loss to Amur tigers was so low that farmers were not interested in taking part. While compensation and performance payments can reduce the economic burden of coexisting with wildlife, and may help to make conservation programmes more acceptable to the public, they are unlikely, on their own, to be sufficient to engender positive attitudes towards wildlife that are perceived to cause damage (e.g. Reading *et al.*, Chapter 13).

The use of wildlife to generate revenue – simultaneously achieving conservation and local economic development – has, until recently, been the mantra of many conservation organizations. However, there is limited systematic evidence that industries such as ecotourism and sport hunting can reliably drive rural development and conservation at the same time (du Toit *et al.* 2004), and evidence that such programmes can contribute to the resolution of human–wildlife conflict is still more scanty (Walpole and Thouless, Chapter 8, Leader-Williams and Hutton, Chapter 9). While there are a few promising case studies, this approach can certainly not yet be called an unqualified success. A major obstacle is the clarity of the link between conservation activities (such as choosing not to spear the lion which has just killed your cow), and benefits (such as having access to a clinic funded ultimately by tourists who paid to see the lion). As Walpole and Thouless (Chapter 8) point out, there is a strong likelihood that a person would – quite reasonably – kill the lion and still use the clinic. Certainly the effectiveness of such community incentives depends upon the existence of a strong local sense of community – something which is being lost rapidly in most parts of the world (e.g. Knickerbocker and Waithaka, Chapter 14, Naughton-Treves and Treves, Chapter 16). Sport hunting might provide a clearer link between the decision to conserve a conflict species, and revenues generated from that species, simply because the connection is more tangible (if I don't kill that lion now, a rich foreigner will pay US$10 000 next month to shoot it). However, sport hunting involves additional requirements for establishing and enforcing sustainable offtakes which can be difficult to achieve, particularly in community areas (Lewis and Jackson, Chapter 15). Both Walpole and Thouless (Chapter 8) and Leader-Williams and Hutton (Chapter 9) suggest that 'sustainable use' programmes might be more effective at improving local tolerance if the revenues generated were used directly to offset the costs of living with wildlife (e.g. for mitigation measures such as electric fences, or for compensation schemes) rather than for infrastructural improvements such as roads and schools which are, they argue, the responsibility of government to provide.

Legal protection

Laws which limit or prohibit killing of threatened wildlife – such as the US Endangered Species Act (Reading *et al*, Chapter 13, Bangs *et al.*, Chapter 21), Kenya's Wildlife (Conservation and Management) Act (Western and Waithaka, Chapter 22) and India's Wildlife Protection Act (Karanth and Gopal, Chapter 23) – have made a huge contribution to the conservation of endangered species. The recovery of species protected by such legislation, including various European geese (Cope *et al.*, Chapter 11), hen harriers (Thirgood and Redpath, Chapter 12) and grey wolves (Bangs *et al.*, Chapter 21), is testament to the value of legal protection. Of course, legal protection alone is not sufficient to achieve conservation; witness, for example, the large number of countries where African wild dogs receive full legal protection, yet have become extinct (Woodroffe *et al.* 1997). Perhaps of greater concern in the current context, however, is that protectionist laws often seem to exacerbate conflicts. This occurs in several ways.

First, legal protection often creates a feeling of disenfranchisement among local people, which can make them hostile to conservation efforts. Examples are legion. Rabinowitz (Chapter 17) notes that legal protection of jaguars led to resentment among ranchers in Brazil; Linnell *et al.* (Chapter 10) make exactly the same point regarding farmers in Europe. Knickerbocker and Waithaka (Chapter 14) describe resentment among local people towards a government-imposed protected area in Kenya. Bangs *et al.* (Chapter 21) describe how ranchers equated the (government-led) restoration of (legally protected) grey wolves with 'outside' forces that undermined their own autonomy. Perhaps most extreme, Reading *et al.* (Chapter 13) describe the huge amount of effort invested by farming groups and others to avert impending legal protection of prairie dogs. While this opposition does not necessarily argue against legal protection, it does demonstrate that legal protection – where appropriate – may need to be linked with extensive community involvement to avoid people subverting laws. Conservationists themselves are sometimes at fault in this regard. For example, Rabinowitz (Chapter 17) describes how conservation advocates in Brazil exacerbated landowners' hostility to jaguar conservation by assuming (incorrectly) that ranching was incompatible with conservation.

Less stringent legal protection may help to reduce this feeling of disenfranchisement and engender greater public support for conservation of 'conflict' species. Bangs *et al.* (Chapter 21) argue that the legal designation of reintroduced wolves as 'experimental, non-essential' – a clause of the US Endangered Species Act which allowed animals to be legally killed under predetermined circumstances pertaining to human–wildlife conflict – was essential to the success of this recovery effort. The same designation also greatly facilitated restoration of red wolves (*Canis rufus*) in the southern United States (Phillips 1995). Karanth and Gopal (Chapter 23) comment that very strict protection laws in India may prevent lethal control of tigers even in life-threatening situations, which is unlikely to encourage local people to strongly support conservation efforts. Within our own personal experience, we note the extreme difference in attitudes between cattle-farmers in Kenya, whose cattle are undoubtedly killed by lions, but who are permitted to kill lions in active defence of livestock as long as they submit the skins to the Kenya Wildlife Service, and cattle-farmers in Britain whose cattle *might* occasionally acquire TB infection from badgers, but who are not allowed to kill or disturb badgers in any way. In this case it appears that some degree of autonomy greatly encourages positive attitudes towards wildlife.

A second effect of preservationist legislation is that it may constrain the development of creative new management solutions. Western and Waithaka (Chapter 22; echoed by Frank *et al*, Chapter 18) note how the 1977 ban on hunting in Kenya has entirely prevented the development of certain conservation strategies based on sustainable use of wildlife. Similarly, Thirgood and Redpath (Chapter 12) argue that a willingness to permit limited lethal control of hen harriers (which is currently illegal) could substantially increase the land area available to this threatened bird, in part by increasing tolerance of low-density hen harrier populations which cause less damage to game-bird harvests, and in part by allowing more extensive use of diversionary feeding as a mitigation measure.

Where threatened species are strictly protected by law, relaxation of legal protection is highly controversial and likely very difficult to achieve. All of the species discussed here have a history of severe decline due to lethal control – that is how their populations become low enough to require conservation management. Hence, conservation advocacy groups tend to strongly oppose any changes in legislation perceived to allow resurgence of an old threat that had been effectively 'controlled' through past legislation. Extreme controversy surrounds all such initiatives – such as proposals to remove grey wolves from the US Endangered Species Act (Bangs *et al*, Chapter 21), to use lethal control on hen harriers (Thirgood and Redpath, Chapter 12), to shoot problem tigers (Karanth and Gopal, Chapter 23) or allow sport hunting in Kenya (Western and Waithaka, Chapter 22) – which are seen as 'a step backwards', likely to undermine not only conservation of the species in question, but also government's attitude to conservation in general. Unfortunately, with much conservation activity funded by organizations reliant on public donations, interventions that appear unpalatable may be sidelined. Clearly this is an arena which will require a great deal of future attention from conservation managers.

Community involvement

Engagement with local people is clearly a key component of any strategy to resolve human–wildlife conflict. After all, it is the local people who are experiencing the costs of living alongside wildlife and, hence, are those most likely to kill wildlife, legally or illegally. As discussed above, legislation that is perceived to exclude these stakeholders can create hostility and may well lead to reduced tolerance for wildlife and increased (often illegal) lethal control. Several of the studies presented in this volume consider community involvement to have been vital to the success of their programmes (e.g. Cope *et al.*, Chapter 11, Rabinowitz, Chapter 17), or consider lack of such involvement to explain past failures (e.g. Reading *et al.*, Chapter 13, Knickerbocker and Waithaka, Chapter 14).

The extent to which local communities are involved may vary; it can be as little as outreach which acknowledges local people's concerns, or may involve full devolution of management authority to local communities. The latter situation is often viewed as the ultimate goal, particularly among development agencies – however we note that this is not always successful. For example, Swenson and Andrén (Chapter 20) describe how, in Norway, devolution of authority for hunting of wolverines and lynx to local boards led to massive over-exploitation as the boards were dominated by sheep-farming interests. Walpole and Thouless (Chapter 8) caution that devolution of management authority can be potentially dangerous if this is not accompanied by plans for effective management and policing. Hence, Western and Waithaka (Chapter 22) argue that conflict resolution should be devolved to 'the lowest practical and verifiable level'. Devolution of management authority is likely to be helpful in many cases because it helps to dissipate hostility to conservation activities and can also ensure that management plans are tailored to local conditions. However, it is important that devolution, where appropriate, is accompanied by sufficient technical expertise to allow local people to make informed decisions.

While community involvement and empowerment in all aspects of conservation and development plans are often a political ideal for development agencies, we note that experience suggests this is not always the most appropriate approach. For example, Walpole and Thouless (Chapter 8) comment that, while development plans often wish to have ecotourism initiatives entirely community-owned and run, in many cases initiatives are more successful, and provide a more stable income stream to local communities, if they contract with a commercial operator. In a similar example, Lewis and Jackson (Chapter 15) argue that communities can most effectively generate income from sport hunting if local hunters do not simultaneously use the same areas.

ECONOMIC DEVELOPMENT AND CHANGING SOCIETAL VALUES

The studies presented in this volume include examples from a full spectrum of economic development. Somewhat to our surprise, successful conflict resolution seems to be achievable across this continuum. Several general patterns emerge, which will perhaps help us to foresee the kinds of conflicts that managers will be faced with in the future, as human densities rise and economic development proceeds.

Democracy

Western and Waithaka (Chapter 22) note that, as developing countries have become democratized, local people's expectations regarding their own quality of life have increased. Hence, they are likely to become less tolerant of wildlife damage to their livelihoods and property. At the same time, Naughton-Treves and Treves (Chapter 16) note that an increasing focus on individual (rather than community or family) benefits has been associated with the loss of traditional approaches to the sharing and spreading of risk among agriculturalists in Africa. Such societal changes are likely to make conflicts more difficult to resolve.

Urbanization

Several authors have commented that improved national attitudes to wildlife have been associated with urbanization (e.g. Swenson and Andrén, Chapter 20, Bangs *et al.*, Chapter 21, Western and Waithaka, Chapter 22, Karanth and Gopal, Chapter 23). As people become more urbanized, they seem to become more positive towards wildlife (Bandara and Tisdell 2003); of course, they also become more insulated from the problems of actually living with wildlife. Hence, in many countries, both developed and developing, human–wildlife conflict becomes transformed into a social conflict between groups of people with differing views on how wildlife should be managed – a conflict that falls fairly predictably across the urban–rural divide. Urbanization generates wildlife advocacy (e.g. Bangs *et al.*, Chapter 21, Karanth and Gopal, Chapter 23), but the immediate costs of living with wildlife are (or are perceived to be) borne by the rural population (e.g. Swenson and Andrén, Chapter 20), even when governments or donors are contributing to mitigation costs. Conflicts appear particularly acrimonious when the tendency to urbanize is countered by a strong commitment, from rural people, government, or both, to conserve a rural, agricultural heritage. Protection of this heritage underlies conflicts over conservation of prairie dogs in North America's prairies (Reading *et al.*, Chapter 13), and large carnivores in Norway (Swenson and Andrén, Chapter 20). Interestingly in Sweden, where government policy has not attempted to conserve a 'living rural landscape' through agricultural subsidies, farming populations in rural areas have declined and large carnivores have been able to stage an impressive recovery.

CONCLUSIONS

Human–wildlife conflicts are inevitable if we are to share the planet with other species. Such conflicts will only increase as humans encroach on wildlife areas and, potentially, as wildlife repopulate human-dominated landscapes. The studies presented in this volume demonstrate that some conflicts, at least, can be resolved or mitigated, and that coexistence of people and wildlife is achievable under the right circumstances. We need to be continually creative and adaptive in responding to human needs and perceptions, and to changing environmental conditions. However, we must also recognize that living with wildlife has costs, both in terms of the damage caused by wildlife, and the opportunity costs of not converting wildlife areas to other uses. While we support the need to make conservation programmes economically self-sustaining wherever possible, it is important to acknowledge that this will not be achievable in most cases. People in developed countries are accustomed to the idea that society should pay to protect things that they value – ranging from historic buildings to unprofitable industries which nevertheless provide jobs. By extension, then, people may need to accept that the costs of coexisting with wildlife – many of them currently borne by rural people, and in developing countries – may need to be met in large part by funds derived ultimately from the people who benefit most from biodiversity conservation, often in cities, and in developed countries (Balmford and Whitten 2003). While choosing a more expedient option – to not attempt to coexist with wildlife – may seem economically appealing in the short term, we must realize that there can be a more severe price to pay, financially and ecologically, if we choose not to live with wildlife.

References

Aanesland, N. and Holm, O. (2001). *Offentlige tilskuddsordninger for sauenæringen-virkninger på norsk rovdyrpolitikk*. Oslo: WWF-Norway. (In Norwegian)

Abramov, V. K. (1962). On the biology of the Amur tiger, *Panthera tigris longipilis* Fitrzenger, 1868. *Vestn. Ceskhslov. Spolecnosti. Zool.*, **26**(2), 189–202. (In Russian)

Adamic, M. (1994). Evaluation of possibilities for natural spreading of brown bear (*Ursus arctos*) towards the Alps, directions of main migration corridors and disturbances in their functionning. In *Braunbär in den Ländern Alpen–Adria*, ed. M. Adamic. Ljubljana: Ministrstvo R.Slovenije za kmetijstvo, gozdarstvo in prehrano, pp. 145–158.

(1997). The expanding brown bear population of Slovenia: a chance for bear recovery in the southeastern Alps. *International Conference on Bear Research and Management*, **9**, 25–29.

Adesina, A. A., Johnson, D. E. and Heinrichs, E. A. (1994). Rice pests in the Ivory Coast, West Africa: farmers' perceptions and management strategies. *International Journal of Pest Management*, **40**, 293–299.

Agarwala, V. P. (1985). *Forests in India*. New Delhi: Oxford University Press and IBH Publishing Company.

Agrawal, A. (1997). *Community in Conservation: Beyond Enchantment and Disenchantment*. Gainesville, FL: Conservation and Development Forum.

Ahlqvist, I. (1999). Förekomst av skador som björn, varg, lo, järv och kungsörn åstadkommer på annan egendom än ren. *Bilagor till sammanhållen rovdjurspolitik, slutbetånkande av Rovdjursutredningen*, **146**, 97–118. (In Swedish)

Ahmad, A. (1994). Protection of snow leopards through grazier communities – some examples from WWF-Pakistan's projects in the northern areas. In *Proceedings of the 7th International Snow Leopard Symposium*, ed. J. L. Fox and D. Jizeng. Seattle, WA: International Snow Leopard Trust, pp. 265–272.

Albers, H. J. and Grinspoon, E. (1997). A comparison of the enforcement of access restrictions between Xishuangbanna Nature Reserve (China) and Khao Yai National Park (Thailand). *Environmental Conservation*, **24**, 351–362.

Alho, C. J., Lacher, T. E. and Goncalves, H. C. (1988). Environmental degredation in the Pantanal ecosystem. *BioScience*, **38**, 164–171.

Allen, L. R. and Gonzalez, A. (1998). Baiting reduces dingo numbers, changes age structures yet often increases calf losses. *Australian Vertebrate Pest Conference*, **11**, 421–428.

Allen, L. R. and Sparkes, E. C. (2001). The effect of dingo control on sheep and beef cattle in Queensland. *Journal of Applied Ecology*, **38**, 76–87.

Allen, S. H. (1982). Bait consumption and diethylstilbesterol influence on North Dakota red fox reproductive performance. *Wildlife Society Bulletin*, **19**, 307–373.

Almeida, T. de (1990). *Jaguar Hunting in the Mato Grosso and Bolivia*. Long Beach, CA: Safari Press.

Aluma, J., Drennon, C. *et al.* (1989). *Settlement in Forest Reserves, Game Reserves, and National Parks in Uganda*. Kampala: Makerere Institute of Social Research.

Amar, A., Arroyo, B., Redpath, S. M. and Thirgood, S. J. (2004). Habitat variability predicts losses of red grouse to hen harriers. *Journal of Applied Ecology*, **41**, 305–314.

Andelt, W. F. (1992). Effectiveness of livestock guarding dogs for reducing predation on domestic sheep. *Wildlife Society Bulletin*, **20**, 55–62.

(1995). *Livestock Guarding Dogs, Llamas and Donkeys for Reducing Livestock Losses to Predators*. Fort Collins, CO: Colorado State University.

(1996). Carnivores. In *Rangeland Wildlife*, ed. P. R. Krausman. Denver, CO: Society for Range Management, pp. 133–155.

(1999a). *Livestock Guard Dogs, Llamas and Donkeys*. Fort Collins, CO: Colorado State University.

(1999b). Relative effectiveness of guarding-dog breeds to deter predation on domestic sheep in Colorado. *Wildlife Society Bulletin*, **27**, 706–714.

Andelt, W. F. and Gipson, P. S. (1979). Domestic turkey losses to radio-tagged coyotes. *Journal of Wildlife Management*, **43**, 673–679.

Andelt, W. F. and Hopper, S. N. (2000). Livestock guard dogs reduce predation on domestic sheep in Colorado. *Journal of Range Management*, **53**, 259–267. (http://uvalde.tamu.edu/jrm/jrmhome.htm)

Andelt, F. A., Phillips, R. L., Gruver, K. S. and Guthrie, J. W. (1999). Coyote predation on domestic sheep deterred with electronic dog-training collar. *Wildlife Society Bulletin*, **27**, 12–18.

Anderson, D. and Grove, R. (1987). *Conservation in Africa: People, Policies and Practices*. Cambridge, UK: Cambridge University Press.

Andersone, Z. and Ozolins, J. (2000). First results of public involvement in wolf research in Latvia. *Folia Therologica Estonica*, **5**, 7–14.

Andrén, H. (1999). Kvantifiering av illegal jakt på lodjur i Sverige. *Bilagor till Sammanhållen rovdjurspolitik, Slutbetänkande av Rovdjursutredningen*, **146**, 183–190. (In Swedish)

Angst, C. (2001). Procedure to selectively remove stock raiding lynx in Switzerland. *Carnivore Damage Prevention News*, **4**, 8.

Angst, C., Olsson, P. and Breitenmoser, U. (2000). Übergriffe von Luchsen auf Kleinvieh und Gehegetiere in der Schweiz. I: Entwicklung und Verteilung der Schäden. *KORA-Bericht Nr.* 5. (http://www.kora.unibe.ch)

Angst, C., Hagen, S. and Breitenmoser, U. (2002). Übergriffe von Luchsen auf Kleinvieh und Gehegetiere in der Schweiz. II: Massnahmen zum Schutz von Nutztieren. *KORA-Bericht Nr. 10*. (http://www.kora.unibe.ch)

Anonymous (1994). Firing line. *The Weekly Review*, 3–10.

(2000). How to minimize risks of mountain lion attacks. *Outdoor California Magazine* **15**,

(2002a). Follow-up of recommendation No. 82 (2000) concerning brown bear in Slovenia. Report by the Slovenian Government. *Council of Europe*, **T-PVS/Inf(2002) 45**, 1–11.

(2002b). Malaysian state calls for army to kill tigers. *Reuters News Service*, 16 August 2002.

Aranda, M. (2002). Importancia de los peccaries para la conservación del jaguar en Mexico. In *El Jaguar en el nuevo milenio*, ed. R. A. Medellin *et al.* Mexico: Fondo de Cultura Economica, pp. 101–107.

Arcese, P., Hando, J. and Campbell, K. (1995). Historical and present day anti-poaching efforts in Serengeti. In *Serengeti II: Dynamics, Management and Conservation of an Ecosystem*, ed. A. R. E. Sinclair and P. Arcese. Chicago, IL: University of Chicago Press.

Archabald, K. and Naughton-Treves, L. (2001). Tourism revenue-sharing around national parks in Western Uganda: early efforts to identify and reward local communities. *Environmental Conservation*, **28**, 135–149.

Arnold, C. (2001). *Wildlife and Poverty Study: Kunene/Namibia Case Study*. London: Department for International Development.

Arnold, O. (1968). *The Story of Cattle Ranching*. Irving-on-Hudson, NY: Harvey House Inc.

Aronson, Å., Wabakken, P., Sand, H., Steinset, O.-K. and Kojola, I. (1999). *The Wolf in Scandinavia: Status Report of the 1998/9 Winter*. Evenstad, Norway: Høgskolen i Hedmark. (In Swedish with English Summary)

(2001). *The Wolf in Scandinavia: Status Report of the 2000/2001 Winter*. Evenstad, Norway: Høgskolen i Hedmark. (In Swedish with English summary)

Asher, V., Phillips, M., Kunkel, K. *et al.* (2001). Evaluation of electronic aversive conditioning methods for reducing wolf predation on livestock. *8th ITC*, Sun City, South Africa.

Ashley, C. and Roe, D. (1998). *Enhancing Community Involvement in Wildlife Tourism: Issues and Challenges*, Wildlife and Development Series No. 11. London: International Institute for Environment and Development.

Ashley, C., Goodwin, H. and Roe, D. (2001). *Pro-Poor Tourism Strategies: Expanding Opportunities for the Poor*. London: Overseas Development Institute.

Aubert, M. F. A. (1993). Control of rabies in wildlife by depopulation. In *Proceedings of the International Conference on Epidemiology, Control and Prevention of Rabies in Eastern and Southern Africa 1992*, ed. A. King. Lyon: Editions Fondation Marcel Mérieux, pp. 141–145.

Aune, K. E. (1991). Increasing mountain lion populations and human–mountain lion interactions in Montana. In *Mountain Lion–Human Interaction Symposium and Workshop*, ed. C. E. Braun. Denver, CO: Colorado Division of Wildlife, pp. 86–94.

Avery, M. (1989). Experimental evaluation of partial repellent treatment for reducing bird damage to crops. *Journal of Applied Ecology*, **26**, 430–433.

Baikov, N. A. (1925). *The Manchurian Tiger*, Seksia Yestestvoznaniya Otdelnoye Izdaniye Vipusk 1. Harbin: Obschestvo Izuchenia Manchzhurskogo Kraya.

Baker, J. E. (1997). Development of a model system for tourist hunting revenue collection and allocation. *Tourism Management*, **18**, 273–286.

Baker, P., Harris, S. and Webbon, C. (2002). Effect of British hunting ban on fox numbers. *Nature*, **419**, 34.

Balestra, F. A. (1962). The man-eating hyenas of Mlanje. *African Wildlife*, **16**, 25–27.

Balikrishnan, M. and Ndhlovu, D. E. (1992). Wildlife utilization and local people: a case-study in Upper Lupande Game Management Area, Zambia. *Environmental Conservation*, **19**, 135–144.

Ballard, W. B., Ayres, L. A., Krausman, P. R., Reed, D. J. and Fancy, S. G. (1997). Ecology of wolves in relation to a migratory caribou herd in northwest Alaska. *Wildlife Monographs*, **135**, 1–47.

Balmford, A. and Whitten, T. (2003). Who should pay for tropical conservation, and how could the costs be met? *Oryx*, **37**, 238–250.

Balser, D. S. (1964). Management of predator populations with antifertility agents. *Journal of Wildlife Management*, **28**, 352–358.

Bandara, R. and Tisdell, C. (2003). Comparison of rural and urban attitudes to the conservation of Asian elephants in Sri Lanka: empirical evidence. *Biological Conservation*, **110**, 327–342.

Bangs, E. E. and Fritts, S. H. (1996). Reintroducing the gray wolf to central Idaho and Yellowstone National Park. *Wildlife Society Bulletin*, **24**, 402–413.

Bangs, E. E. and Shivik, J. (2001). Managing wolf conflict with livestock in the Northwestern United States. *Carnivore Damage Prevention News*, **3**, 2–5. (http://www.kora.unibe.ch)

Bangs, E. E., Fritts, S. H., Harms, D. A. *et al.* (1995). Control of endangered gray wolves in Montana. In *Ecology and Conservation of Wolves in a Changing World.*, ed. L. N. Carbyn, S. H. Fritts and D. R. Seip. Edmonton, Alberta: Canadian Circumpolar Institute, pp. 127–134.

Bangs, E. E., Fritts, S. H., Fontaine, J. A., *et al.* (1998). Status of gray wolf restoration in Montana, Idaho, and Wyoming. *Wildlife Society Bulletin*, **26**, 785–798.

Bangs, E. E., Fontaine, J., Jimenez, M., *et al.* (2001). Gray wolf restoration in the northwestern United States. *Endangered Species Update*, **18**, 147–152.

Barnes, A. M. (1993). A review of plague and its relevance to prairie dog populations and the black-footed ferret. In *Management of Prairie Dog Complexes for the Reintroduction of the Black-Footed Ferret*, ed. J. Oldemeyer, D. Biggins, B. Miller and R. Crete. Washington, DC: US Fish and Wildlife Service, pp. 28–37.

Barnes, J. I., MacGregor, J. and Weaver, L. C. (2002). Economic efficiency and incentives for change within Namibia's community wildlife use initiatives. *World Development*, **30**, 667–681.

Barnes, R. F. W. (1990). *Famine in the Rain Forest.* New York: Wildlife Conservation International.

(1996). The conflict between humans and elephants in the central African forests. *Mammal Review*, **26**, 67–80.

Barnes, R. F. W., Jensen, K. and Blom, A. (1991). Man determines the distribution of elephants in the rain forests of northeastern Gabon. *African Journal of Ecology*, **29**, 54–63.

Barnes, R. F. W., Azika, S. and Asamoah-Boateng, B. (1995). Timber, cocoa, and crop-raiding elephants: a preliminary study from southern Ghana. *Pachyderm*, **19**, 33–40.

Barnes, R. F. W., Craig, G. C., Dublin, H. T. *et al.* (1998). *African Elephant Database 1998*, Occasional Paper of the IUCN Species Survival Commission No.22. Gland, Switzerland: IUCN.

Baron, D. (2003). *The Beast in the Garden: A Modern Parable of Man and Nature.* New York: W. W. Norton.

Barrett, C. B., Brandon, K., Gibson, C. and Gjertsen, H. (2001). Conserving tropical biodiversity amid weak institutions. *BioScience*, **51**, 497–502.

Barrow, E. and Murphree, M. (2001). Community conservation: from concept to practice. In *African Wildlife and Livelihoods: The Promise and Performance of Community Conservation*, ed. D. Hulme and M. Murphree. Oxford, UK: James Currey, pp. 24–37.

Bateson, P. and Bradshaw, E. L. (1997). Physiological effects of hunting red deer (*Cervus elaphus*). *Proceedings of the Royal Society of London* B, **264**, 1707–1714.

Bath, A. J. (1992). Identification and documentation of public attitudes toward wolf reintroduction in Yellowstone National Park. In *Wolves for Yellowstone? A Report to the United States Congress*, vol. IV, *Research and Analysis*, ed. J. D. Varley and W. G. Brewster. Yellowstone National Park, WY: National Park Service, pp. 2/5–2/30.

(2001). *Human Dimensions in Wolf Management in Savoie and Les Alpes Maritimes, France: Results Targeted toward Designing a More Effective Communication Campaign and Building Better Public Awareness Materials.* Large Carnivore Initiative for Europe. www.large-carnivores-lcie.org

Bath, A. J. and Majic, A. (2001). *Human Dimensions in Wolf Management in Croatia: Understanding Attitudes and Beliefs of Residents in Gorski Kotar, Lika and Dalmatia towards Wolves and Wolf Management.* Large Carnivore Initiative for Europe. Accessible at http://www.large-carnivores-lcie.org

Beard, P. (1963). *The End of the Game.* San Francisco, CA: Chronicle Books.

Beier, P. (1991). Cougar attacks on humans in the United States and Canada. *Wildlife Society Bulletin*, **19**, 403–412.

(2004), Paul Beier Home Page. Accessible at http://nau.edu/pb1

Bell, R. H. V. (1984a). The man–animal interface: an assessment of crop damage and wildlife control. In *Conservation and Wildlife Management in Africa*, ed. R. H. V. Bell and E. McShane-Caluzi. Washington, DC: US Peace Corps Office of Training and Program Support, pp. 387–416.

(1984b). Monitoring of public attitudes. In *Conservation and Wildlife Management in Africa*, ed. R. H. V. Bell and E. McShane-Caluzi. Washington, DC: US Peace Corps Office of Training and Program Support, pp. 442–448.

(1987). Conservation with a human face: conflict and reconciliation in African land use planning. In *Conservation in Africa: People, Policies and Practice*, ed. D. Anderson and R. Grove. Cambridge UK: Cambridge University Press, pp. 79–101.

Bell, W. R. (1921). Death to the rodents. *US Department of Agriculture Yearbook*, 1920, 421–438.

Bennett, E. L. and Robinson, J. G. (2000). Hunting for sustainability: the start of a synthesis. In *Hunting for Sustainability in Tropical Forests*, ed. J. G. Robinson and E. L. Bennett. New York: Columbia University Press, pp. 409–499.

Bennett, J. (1998). Can state regulatory agencies resolve controversial wildlife management issues involving the broad general public? In *Proceedings of The Defenders of Wildlife Restoring the Wolf Conference*, ed. N. Fascione. Washington, DC: Defenders of Wildlife, pp. 85–89.

Ben-Shahar, R. (1993). Does fencing reduce the carrying capacity for populations of large herbivores? *Journal of Tropical Ecology*, **9**, 249–253.

Benson, E. P. (1998). The lord, the ruler: jaguar symbolism in the Americas. In *Icons of Power: Feline Symbolism in the Americas*, ed. N. J. Saunders. London: Routledge, pp. 53–72.

Bensted-Smith, R., Infield, M. Otekat, J. and Thomson-Handler, N. (1995). A review of the multiple-use (resource sharing) programme in Bwindi Impenetrable National Park. Unpublished report. Kampala: CARE-Development through Conservation Project and CARE-International.

Bere, R. M. (1955). The African wild dog. *Oryx*, **3**, 180–182.

Berg, K. A. (1998). The future of the wolf in Minnesota: control, sport or protection? In *Proceedings of The Defenders of Wildlife Restoring the Wolf Conference*, ed. N. Fascione. Washington, DC: Defenders of Wildlife, pp. 40–44.

Berger, D. (1996). The challenge of integrating Maasai tradition with tourism. In *People and Tourism in Fragile Environments*, ed. M. F. Price. Chichester, UK: John Wiley, pp. 175–197.

Berger, J., Stacey, P. B., Bellis, L. and Johnson, M. P. (2001). A mammalian predator–prey imbalance: grizzly bear and wolf extinction affect avian neotropical migrants. *Ecological Applications*, **11**, 947–960.

Bergman, D. L., Huffman, L. E. and Paulson, J. D. (1998). North Dakota's cost-share program for guard animals. In *Proceedings of the 18th Vertebrate Pest Conference*, ed. R. O. Baker and A. C. Crabb. Davis, CA: University of California, pp. 122–125

Bergo, G. (1987). Eagles as predators of livestock. *Fauna Norvegica Series C*, **10**, 95–102.

Bibby, C. J. and Etheridge, B. (1993). Status of the hen harrier in Scotland during 1988. *Bird Study*, **40**, 1–11.

Bingham, J., Foggin, C., Wandeler, A. and Hill, F. (1999). The epidemiology of rabies in Zimbabwe. *Onderstepoort Journal of Veterinary Research*, **66**, 11–23.

Biodiversity Legal Foundation, Sharps, J. C. and Predator Project (1998). Black-tailed prairie dog (*Cynomys ludovicianus*). Unpublished petition. Denver, CO: US Fish and Wildlife Service Region 6.

Biryahwaho, B. (2002). Community perspectives towards management of crop raiding animals: experiences of CARE-DTC with communities living adjacent to Bwindi Impenetrable and Mgahinga Gorilla National Parks, southwest Uganda. In *Human–Wildlife Conflict: Identifying the Problem and Possible Solutions*, Albertine Rift Technical Report Series No. 1. ed. C. M. Hill, F. Osborn and A. J. Plumptre. New York: Wildlife Conservation Society.

Bishop, G. C., Durrheim, D. N., Kloeck, P. E. *et al.* (2002). *Rabies: Guide for the Medical, Veterinary and Allied Professions*. Pretoria: Department of Agriculture.

Bjärvall, A., Franzén, R., Nordkvist, M. and Åhman, G. (1990). *Renar och rovdjur: Rovdjurens effekter på rennäringen*. Stockholm Naturvårdsverkets Förlag. (In Swedish)

Bjerke, T. and Kaltenborn, B. P. (1998). The relationship of ecocentric and anthropocentric motives to attitudes toward large carnivores. *Journal of Environmental Psychology*, **19**, 415–421.

Bjorge, R. R. and Gunson, J. R. (1985). Evaluation of wolf control to reduce cattle predation in Alberta. *Journal of Range Management*, **38**, 483–486.

Black, H. L. and Green, J. S. (1985). Navajo use mixed-breed dogs for management of predators. *Journal of Range Management*, **38**, 11–15. (http://uvalde.tamu.edu/jrm/jrmhome.htm)

Blackwell, B. F., Dolbeer, R. A. and Tyson, L. A. (2000). Lethal control of piscivorous birds at aquaculture facilities in the northeast United States: effects on populations. *North American Journal of Aquaculture*, **6 + 2**, 300–307.

Blair, J. A. S. (1979). *Elephants and Other Animal Pests*, Workshop on Elephant Damage. Trolak, Perak, Malaysia: INPUT, Institute of Land Development.

Blair, J. A. S., Boon, G. and Noor, M. (1979). Conservation or cultivation: the confrontation between the Asian elephant and land development in Peninsular Malaysia. *Land Development Digest*, **2**, 25–58.

Blanco, J. C. (2003). Wolf damage compensation schemes in Spain. *Carnivore Damage Prevention News*, **6**, 7–9.

Blanco, J. C., Reig, S. and Cuesta, L. (1992). Distribution, status and conservation problems of the wolf in Spain. *Biological Conservation*, **60**, 73–80.

Blejwas, K. M., Sacks, B. N., Jaeger, M. M. and McCullough, D. R. (2002). The effectiveness of selective removal of breeding coyotes in reducing sheep predation. *Journal of Wildlife Management*, **66**, 451–462.

Blomley, T. (2003). Natural resource conflict management: the case of Bwindi Impenetrable and Mgahinga Gorilla National Parks, south western Uganda. Unpublished report. Kampala: CARE-Development through Conservation Project and CARE-International.

Bodner, M. (1998). Damage to fish ponds as a result of otter (*Lutra lutra*) predation. *BOKU Reports on Wildlife Research and Game Management*, **14**, 106–117.

Boesch, C. and Boesch-Achermann, H. (2000). *The Chimpanzees of the Taï Forest: Behavioural Ecology and Evolution*. Oxford, UK: Oxford University Press.

Bogges, E. K., Henderson, F. R. and Spaeth, C. E. (1980). *Managing Predator Problems: Practices and Producers for Preventing and Reducing Livestock Losses. Coop. Ect. Serv. Bull.* C-620. Manhatten, KS: Kansas State University.

Boitani, L. (1992). Wolf research and conservation in Italy. *Biological Conservation*, **61**, 125–132.

(1995). Ecological and cultural diversities in the evolution of wolf–human relationships. In *Ecology and Conservation of Wolves in a Changing World*, ed. L. N. Carbyn, S. H. Fritts and D. R. Seip, Edmonton, Alberta: Canadian Circumpolar Institute, pp. 3–11.

(2000). *Action Plan for the Conservation of Wolves in Europe* (Canis lupus), Nature and Environment No. 113. Strasbourg, France: Council of Europe.

Bomford, M. and O'Brien, P. H. (1990). Sonic deterrents in animal damage control: a review of device tests and effectiveness. *Wildlife Society Bulletin*, **18**, 411–422.

Bond, I. (1994). Importance of elephant hunting to CAMPFIRE revenue in Zimbabwe. *Traffic Bulletin*, **14**, 117–119.
(2001). CAMPFIRE and the incentives for institutional change. In *African Wildlife and Livelihoods: The Promise and Performance of Community Conservation*, ed. D. Hulme and M. Murphree. Oxford, UK: James Currey, pp. 227–243.
Bonham, C. D., and Lerwick, A. (1976). Vegetation changes induced by prairie dogs on shortgrass range. *Journal of Range Management*, **29**, 221–225.
Bonner, R. (1993). *At the Hand of Man*. New York: Knopf.
Bonnie, R., McMillan, M. and Wilcove, D. S. (2001). *A Home on the Range: How Economic Incentives Can Save the Threatened Utah Prairie Dog*. Washington, DC: Environmental Defense.
Bookbinder, M. P., Dinerstein, E., Rijal, A., Cauley, H. and Rajouria, A. (1998). Ecotourism's support of biodiversity conservation. *Conservation Biology*, **12**, 1399–1404.
Boonzaier, E. (1996). Local responses to conservation in the Richtersveld National Park, South Africa. *Biodiversity and Conservation*, **5**, 307–314.
Borrini-Feyerabend, G. (1997). *Beyond Fences: Seeking Social Sustainability in Conservation*. Gland Switzerland: IUCN.
Bothma, J. du P. (1971). Food of *Canis mesomelas* in South Africa. *Zoologica Africana*, **6**, 187–193.
(2002). Some economics of wildlife ranching. In *Wildlife Group Symposium on Game Ranching*. South African Veterinary Association, pp. 1–10.
Boulton, A. M., Horrocks, J. A. and Baulu, J. (1996). The Barbados vervet monkey (*Cercopithecus aethiops sabaeus*): changes in population size and crop damage 1980–1994. *International Journal of Primatology*, **17**, 831–844.
Bourne, J. (1994). Protecting livestock with guard donkey. *Agri-fax*, Edmonton, Alberta: Alberta Agriculture.
(2002). Electric fencing for predator control in Alberta. *Carnivore Damage Prevention News*, **5**, 9–10. (http://www.kora.unibe.ch)
Boutin, S. (1990). Predation and moose population dynamics: a critique. *Journal of Wildlife Management*, **56**, 116–127.
Bowland, A. E., Mills, M. G. L. and Lawson, D. (1993). *Predators and Farmers*. Parkview, South Africa: Endangered Wildlife Trust.
Boyce, M. S. (1995). Anticipating consequences of wolves in Yellowstone: model validation. In *Ecology and Conservation of Wolves in a Changing World*, ed. L. N. Carbyn, S. H. Fritts and D. R. Seip. Edmonton, Alberta: Canadian Circumpolar Institute, pp. 199–210.
Boyd, D. K., Paquet, P. C., Donelon, S. *et al.* (1995). Transboundary movements of a recolonizing wolf population in the Rocky Mountains. In *Ecology and Conservation of Wolves in a Changing World*, ed. L. N. Carbyn, S. H. Fritts and D. R. Seip. Edmonton, Alberta: Canadian Circumpolar Institute, pp. 135–140.
Bradley, E. H. (2004). An evaluation of wolf–livestock conflicts and management in the Northwestern United States Ph.D. thesis, University of Montana, Missoula, MT.
Bradley, E. H., Pletscher, D. H., Bangs, E. E. (in press) Evaluating wolf translocation as a non-lethal method to reduce livestock conflicts in the northwestern United States. *Conservation Biology*.

Bragin, A. P. and Gaponov, V. V. (1989). Problems of the Amur tiger. *Okhota i okhotnichie khozyaistvo*, **10**, 12–15. (In Russian)

Brainerd, S. (2003). *Konfliktdempende tiltak i rovviltforvaltningen*. Trondheim, Norway: Norwegian Institute for Nature Research Rapport. (In Norwegian).

Braithwait, J. (1996). *Using Guard Animals to Protect Livestock*. Missouri Department of Conservation.

Brand, C. J., Pybus, M. J., Ballard, W. B. and Peterson, R. O. (1995). Infectious and parasitic diseases of the gray wolf and their potential effects on wolf populations in North America. In *Ecology and Conservation of Wolves in a Changing World* ed. L. N. Carbyn, S. H. Fritts and D. R. Seip. Edmonton, Alberta: Canadian Circumpolar Institute, pp. 419–430.

Brashares, J. S., Arcese, P. and Sam, M. K. (2001). Human demography and reserve size predict wildlife extinction in west Africa. *Proceedings of the Royal Society of London B*, **268**, 2473–2478.

Breitenmoser, U. (1998). Large predators in the Alps: the fall and rise of man's competitors. *Biological Conservation*, **83**, 279–289.

Breitenmoser, U., and Breitenmoser-Würsten, C. (2001). Die ökologischen und anthropogenen Voraussetzungen für die Existenz grosser Beutegreifer in der Kulturlandschaft. *Forest Snow and Landscape Research*, **76**, 23–39.

Breitenmoser, U. and Haller, H. (1993). Patterns of predation by reintroduced European lynx in the Swiss Alps. *Journal of Wildlife Management*, **57**, 135–144.

Breitenmoser, U., Breitenmoser-Würsten, C., Okarma, H. *et al.* (2000). *Action Plan for the Conservation of the Eurasian lynx in Europe* (Lynx lynx). Nature and Environment No. 112. Strasbourg, France: Council of Europe.

British Trust for Ornithology *et al.* (2001). *Red Grouse and Birds of Prey*. Sandy, UK: BTO and 11 others.

Bro, E., Reitz, F., Clobert, J., Migot, P. and Massot, M. (2001). Diagnosing the environmental causes of the decline in grey partridge in France. *Ibis*, **143**, 120–132.

Bromley, (1977). The tiger as a component of biocenoses of the Far East. In *Redkie vidy mlekopitaiuschikh i ikh okhrana*, Materially z Vsesoiuznogo soveschania. Moscow: Nauka, pp. 111–113. (In Russian)

Bromley, C. and Gese, E. M. (2001a). Effects of sterilization on territory fidelity and maintenance, pair bonds, and survival rates of free-ranging coyotes. *Canadian Journal of Zoology*, **79**, 386–392.

(2001b). Surgical sterilization as a method of reducing coyote predation on domestic sheep. *Journal of Wildlife Management*, **65**, 510–519.

Brooker, M. G. and Ridpath, M. G. (1980). The diet of the wedge-tailed eagle in Western Australia. *Australian Wildlife Research*, **7**, 433–452.

Brøseth, H., Odden, J. and Linnell, J. D. C. (2003). *Minimum antall familiegrupper, bestandsestimat og bestandsutvikling for gaupe i Norge i perioden 1996–2002*. Trondheim, Norway: Norwegian Institute for Nature Research Rapport. (In Norwegian)

Brown, D. (1998). *Participatory Biodiversity Conservation: Rethinking the Strategy in the Low Tourist Potential Areas of Tropical Africa*. London: Overseas Development Institute.

Brown, D. and Gonzalez, C. A. L. (2001). *Borderland Jaguars*. Salt Lake City, UT: University of Utah Press.

Brown, W. M. and Parsons, D. R. (2001). Restoring the Mexican gray wolf to the desert southwest. In *Large Mammal Restoration: Ecological and Sociological Challenges in the 21st Century*, ed. D. S. Maehr, R. F. Noss and J. L. Larkin. London: Island Press, pp. 169–186.

Bruinderink, G. G., Van Der Sluis, T., Lammertsma, D., Opdam, P. and Pouwels, R. (2003). Designing a coherent ecological network for large mammals in northwestern Europe. *Conservation Biology*, **17**, 549–557.

Burnett, H. S. (2002). The role and value of hunting. *The Hunting Report*, **22**, 1–4.

Burns, L., Edwards, V., Marsh, J., Soulsby, L. and Winter, M. (2000). *Report of the Committee of Inquiry into Hunting with Dogs in England and Wales*. London: Stationery Office.

Burns, R. J. (1980). Evaluation of conditioned predation aversion for controlling coyote predation. *Journal of Wildlife Management*, **44**, 938–942.

(1983). Coyote predation aversion with lithium chloride: management implications and comments. *Wildlife Society Bulletin*, **11**, 128–133.

(1996). Effectiveness of Vichos non-lethal collars in deterring coyote attacks on sheep. *Vertebrate Pest Conference Proceedings*, **17**, 204–206.

Burns, R. J., Connolly, G. E. and Griffiths, R. E. Jr (1984). Repellent or aversive chemicals in sheep neck collars did not deter coyote attacks. *Vertebrate Pest Conference Proceedings*, **11**, 146–153.

Burns, R. J., Tietjen, H. J. and Connolly, G. E. (1991). Secondary hazard of livestock protection collars to skunks and eagles. *Journal of Wildlife Management*, **55**, 701–704.

Burns, R. J., Zemlicka, D. E. and Savarie, P. J. (1996) Effectiveness of large livestock protection collars against depredating coyotes. *Wildlife Society Bulletin*, **24**, 123–127.

Butler, J. R. A. (2000). The economic costs of wildlife predation on livestock in Gokwe communal land, Zimbabwe. *African Journal of Ecology*, **38**, 23–30.

Campbell, S. C., Smith, A. A., Redpath, S. M. and Thirgood, S. J. (2002). Nest site characteristics and nesting success in red grouse. *Wildlife Biology*, **8**, 169–174.

Campbell, D., Gichohi, H., Mwangi, A., Chege, L. and Sawin, T. (1999). *Interactions between People and Wildlife in SE Kajiado District, Kenya*. Nairobi: Ford Foundation.

Cape Nature Conservation (2001). *Badgers and Beekeepers*. Cape Nature Conservation Information Brochure, 4.

Carter, M. R. (1997). Environment, technology, and the social articulation of risk in West African agriculture. *Economic Development and Cultural Change*, **45**, 557–590.

Casey, D. and Clark, T. W. (1996). *Tales of the Wolf: Fifty-One Stories of Wolf Encounters in the Wild*. Moose, WY: Homestead Publishing.

Causey, D. (2001). What's behind Zambia's closure of hunting? *The Hunting Report*, **21**, 1–3.

Ceballos, G., Chavez, C., Rivera, A., Manterola, C. and Wall, B. (2002). Tamano poblacionaly conservación del jaguar en la Reserva de la Prosfera Calakmul, Campeche, Mexico. In *El Jaguar en el nuevo milenio*, ed. R. A. Medellin *et al.* Mexico: Fondo de Cultura Economica, pp. 403–419.

Ceballos-Lascurain, H. (1993). Ecotourism as a worldwide phenomenon. In *Ecotourism: A Guide for Planners and Managers*, ed. K. Lindberg and D. E. Hawkins. North Bennington, VT: The Ecotourism Society, pp. 12–14.

Chalise, M. K. (2001). Crop raiding by wildlife, especially primates, and indigenous practices for crop protection in Lakuwa Area, East Nepal. *Asian Primates, IUCN/SSC Primate Specialist Group*, **7**, 4–9.

Chapman, C. A. and Onderdonk, D. A. (1998). Forests without primates: primate/plant codependency. *American Journal of Primatology*, **45**, 127–141.

Charudutt, M. (1997). Livestock depredation by large carnivores in the Indian trans-Himalaya: conflict perceptions and conservation prospects. *Environmental Conservation*, **24**, 338–343.

Chaudhary, A. B. and Chakrabarti, K. (1979). The tiger and man: the Sundarbans tiger as viewed by fisherman and other who eke out their livelihood from the area. In *Proceedings of the International Symposium on Tiger*. Government of India, pp. 129–135.

Chauhan, N. P. S., Bargali, H. S. and Akhtar, N.(2002). Human–sloth bear conflicts, causal factors and management implications in Bilaspur Forest Division, Chattishgarh, India. In *Program and Abstracts of Papers, 14th International Congress on Bear Research and Management*, ed. T. Kvam and O. J. Sorensen. Steinkjer, Norway: Nord-Trondelag University College.

Chellam, R. and Johnsingh, A. J. T. (1993). Management of Asiatic lions in the Gir Forest, India. *Symposium of the Zoological Society of London*, **65**, 409–424.

Chester, T. (2004). *Mountain Lion Attacks on Peaople in the US and Canada*. Accessible at http://tchester.org/sgm/lists/lion_attacks.html#summary

Chestin, I. (1999). Brown bear conservation action plan for Asia: Russia. In *Bears: Status Survey and Conservation Action Plan*, ed. C. Servheen, S. Herrero and B. Peyton. Gland, Switzerland: IUCN, pp. 136–143.

Child, G. (1995). *Wildlife and People: The Zimbabwean Success*. Harare: Wisdom, Press.

Chin, R. and Benne, K. D. (1976). General strategies for effecting changes in human systems. In *The Planning of Change, 3rd edn*, ed. W. G. Bennis, K. D. Benne, R. Chin and K. E. Corey. New York: Holt, Rinehart, and Winston, pp. 22–45.

Chiyo, P. I. (2000). Elephant ecology and crop depredation in Kibale National Park, Uganda. M.Sc. thesis, Makerere University, Kampala, Uganda.

Chiyo, P. I., Cochrane, E., Naughton-Treves, L. and Basuta, G. I. (2005). Temporal patterns of crop raiding by elephants at Kibale National Park, Uganda. *African Journal of Ecology*, in press.

Chundawat, R. S., Gogate, N. and Johnsingh, A. J. T. (1999). Tigers in Panna: Preliminary results from an Indian tropical dry forest. In *Riding the Tiger: Tiger Conservation in Human-Dominated Landscapes*, ed. J. Seidensticker, S. Christie and P. Jackson. Cambridge, UK: Cambridge University Press, pp. 123–129.

CITES (1997). Review of the proposals submitted by Botswana, Namibia and Zimbabwe to transfer their national populations of *Loxodonta africana* from CITES Appendix I to Appendix II. Accessible at http://wildnetafrica.co.za/cites/info/iss_002_12.html

Ciucci, P. and Boitani, L. (1998). Wolf and dog depredation on livestock in central Italy. *Wildlife Society Bulletin*, **26**, 504–514.

(2000). Wolves, dogs, livestock and compensation costs: 25 years of Italian experience. In *Beyond 2000: Realities of Global Wolf Restoration Symposium*, Duluth, MN, 23–26 February 2000. Accessible at http://www.wolf.org/wolves/learn/scientific/symposium/abstracts/008.asp

Clark, T. W. (1997). *Averting Extinction: Restructuring the Endangered Species Recovery Process*. New Haven, CT: Yale University Press.

Clark, T. W. and Brunner, R. D. (1997). Making partnerships work in endangered species conservation: an introduction to the decision process. *Endangered Species Update*, **13**, 1–4.

Clark, T. W., and Minta, S. C. (1994). *Greater Yellowstone's Future*. Moose, WY: Homestead Publishing.

Clark, T. W. and Wallace, R. L. (1998). Understanding the human factor in endangered species recovery: an introduction to human social process. *Endangered Species Update*, **15**, 2–9.

(2002). The dynamics of value interactions in endangered species conversation. *Endangered Species Update*, **19**, 95–100.

Clark, T. W., Hinckley, D. and Rich, T. (eds.) (1989). The prairie dog ecosystem: managing for biological diversity. *Montana BLM Wildlife Technical Bulletin*, **2**, 1–55.

Clark, T. W., Begg, R. J. and Lowe, K. W. (2002). Interdisciplinary problem-solving workshops for natural resources professionals. In *The Policy Process: A Practical Guide for Natural Resources Professionals*, ed. T. W. Clark. New Haven, CT: Yale University Press, pp. 173–189.

Clarke, J. (1969). *Man Is the Prey*. New York: Stein and Day.

Clarkson, P. L. and Marley, J. L. (1995). Preventing and managing black and grizzly bear problems in Agricultural and forested areas in North America. *In Proceedings of the 9th (France) International Bear Association*, pp. 306–322.

Clemence, E. (1992). A barking dog. *DogLog*, **3**(3), 3–4.

Collins, A. R., Workman, J. P. and Uresk, D. W. (1984). An economic analysis of black-tailed prairie dog (*Cynomys ludovicianus*) control. *Journal of Range Management*, **37**, 358–61.

Conelly, W. T. (1987). Perception and management of crop pests among subsistence farmers in South Nyanza, Kenya. In *Management of Pests and Pesticides: Farmers' Perceptions and Practices*, ed. Tait, and B. Nampometh. Boulder, CO: Westview Press, pp. 198–209.

Conner, M. M. (1995). Identifying patterns of coyote predation on sheep on a northern California ranch. M.S. thesis, University of California, Berkeley, CA.

Conner, M. M., Jaeger, M. M., Weller, T. J. and McCullough, D. R. (1998). Effect of coyote removal on sheep depredation in northern California. *Journal of Wildlife Management*, **62**, 690–699.

Connolly, G. (1995). Animal damage control research contributions to coyote management. In *Proceedings of the 1995 Joint Fur Resources Workshop*.

Connolly, G. and O'Gara, B. W. (1987) Aerial hunting takes sheep-killing coyotes in western Montana. *Great Plains Wildlife Damage Control Workshop*, **8**, 184–188.

Conover, M. (1984). Comparative effectiveness of Avitrol, exploders and hawk-kites to reduce blackbird damage to corn. *Journal of Wildlife Management*, **48**, 109–116.

(2002). *Resolving Human–Wildlife Conflicts: The Science of Wildlife Damage Management*. Boca Raton, FL: CRC Press.

Conover, M. R. and Decker, D. J. (1991). Wildlife damage to crops: perceptions of agricultural and wildlife professionals in 1957 and 1987. *Wildlife Society Bulletin*, **19**, 46–52.

Conover, M. F. and Kessler, K. K. (1994). Diminished producer participation in an aversive conditioning program to reduce coyote depredation on sheep. *Wildlife Society Bulletin*, **22**, 229–233.

Conover, M. F., Francik, J. G. and Miller, D. E. (1979). An experimental evaluation of aversive conditioning for controlling coyote predation: a critique. *Journal of Wildlife Management*, **43**, 208–211.

Convention on Biological Diversity (2005) http://www.biodiv.org

Cook, S. J., Norris, D. R. and Theberge, J. B. (1999). Spatial dynamics of a migratory wolf population in winter, south-central Ontario (1990–1995). *Canadian Journal of Zoology*, **77**, 1740–1750.

Cope, D. R., Pettifor, R. A., Griffin, L. R. and Rowcliffe, J. M. (2003). Integrating farming and wildlife conservation: the Barnacle Goose Management Scheme. *Biological Conservation*, **110**, 113–122.

Coppinger, C. and Coppinger, L. (1998). Differences in the behaviour of dog breeds. In *Genetics and the Behavior of Domestic Animals*, ed. T. Grandin. London: Academic Press, pp. 167–202.

Coppinger, L. (1992). Getting through that juvenile period. *DogLog* **2**(3–4), 6–12.

Coppinger, L. and Coppinger, R. (1982). Livestock-guarding dogs that wear sheep's clothing. *Smithsonian*, **13**, 65–73.

Coppinger, R. (1992). Can dogs protect livestock against wolves in North America? *DogLog*, **3**, 2–4.

Coppinger, R. and Coppinger, L. (1994). *The Predicament of Flock-Guarding Dogs in the Tatra Mountains, Slovakia*. Amherst, MA: Hampshire College.

(1995). Interaction between livestock guarding dogs and wolves. In *Wolves in a Changing World*, ed. L. N. Carbyn, S. H. Fritts and D. R. Seip. Edmonton, Alberta: Canadian Circumpolar Institute, pp. 523–526.

(2001). *Dogs: A Startling New Understanding of Canine Origin, Behavior and Evolution*. New York: Scribner.

Coppinger, R. and Schneider, R. (1995). Evolution of working dogs. In *The Domestic Dog*, ed. J. Serpell. Cambridge, UK: Cambridge University Press, pp. 21–47.

Coppinger, R., Lorenz, J., Glendinning, J. and Pinardi, P. (1983). Attentiveness of guarding dogs for reducing predation on domestic sheep. *Journal of Range Management*, **36**, 275–279. (http://uvalde.tamu.edu/jrm/jrmhome.htm)

Coppinger, R., Smith, C. and Miller, L. (1985). Observations on why mongrels make effective livestock protecting dogs. *Journal of Range Management*, **38**, 560–561. (http://uvalde.tamu.edu/jrm/jrmhome.htm)

Coppinger, R., Glendinning, J., Torop, E. *et al.* (1987). Degree of behavioural neoteny differentiates canid polymorphs. *Ethology*, **75**, 89–108.

Coppinger, R., Coppinger, L., Langeloh, G., Gettler, L. and Lorenz, J. (1988). A decade of use of livestock guarding dogs. *Vertebrate Pest Conference Proceedings*, **13**, 209–214.

Coppock, D. L., Detling, J. K, Ellis, J. E. and Dyer, M. I. (1983a). Plant–herbivore interactions in a North American mixed-grass prairie. I: Effects of black-tailed prairie dogs on intraseasonal aboveground plant biomass and nutrient dynamics and plant species diversity. *Oecologia*, **56**, 1–9.

(1983b). Plant–herbivore interactions in a North American mixed-grass prairie. II: Responses of bison to modification of vegetation by prairie dogs. *Oecologia*, **56**, 10–15.

Corbett, J. (1944). *The Man Eaters of Kumaon*. London: Oxford University Press.

(1948). *The Man-Eating Leopard of Rudraprayag*. New Delhi: Oxford University Press.

Corbett, L. (1995). *The Dingo in Australia and Asia*. Ithaca, NY: Cornell University Press.

Cote, I. M. and Sutherland, W. J. (1997). The effectiveness of removing predators to protect bird populations. *Conservation Biology*, **11**, 395–405.

Courchamp, F. and Macdonald, D. W. (2001). Crucial importance of pack size in the African wild dog *Lycaon pictus*. *Animal Conservation*, **4**, 169–174.

Courtin, F., Carpenter, T., Paskin, R. and Chomel, B. (2000). Temporal patterns of domestic and wildlife rabies in central Namibia stock-ranching area, 1986–1996. *Preventive Veterinary Medicine*, **43**, 13–28.

Cozza, K., Fico, R., Battistini, M. and Rodgers, E. (1996). The damage–conservation interface illustrated by predation on domestic livestock in central Italy. *Biological Conservation*, **78**, 329–336.

Crawshaw, P. G. and Quigley, H. B. (1991). Jaguar spacing, activity, and habitat use in a seasonally flooded environment in Brazil. *Journal of Zoology*, **223**, 357–370.

(2002). Jaguar and puma feeding habits in the Pantanal of Mato Grosso, Brazil, with implications for their management and conservation. In *El Jaguar en el nuevo milenio*, ed. R. A. Medellin *et al.* Mexico: Fondo de Cultura Economica, pp. 223–227.

Creel, S. and Creel, N. M. (1997). Lion density and population structure in the Selous Game Reserve: evaluation of hunting quotas and offtake. *African Journal of Ecology*, **35**, 83–93.

Crocker-Bedford, D. (1976). Food interactions between Utah prairie dogs and cattle. M.S. thesis, Utah State University, Logan, UT.

Crooks, K. and Soulé, M. (1999). Mesopredator release and avifaunal extinctions in a fragmented system. *Nature*, **400**, 563–566.

Crooks, K. R. and Sanjayan, M. A. (2005). *Connectivity and Conservation*. Cambridge, UK: Cambridge University Press.

Cully, J. F. Jr and Williams, E. S. (2001). Interspecific comparisons of sylvatic plague in prairie dogs. *Journal of Mammalogy*, **82**, 894–905.

Cunningham, A. B. (1996). *People, Parks and Plant Use: Recommendations for Multiple-Use Zones and Development Alternatives around Bwindi Impenetrable National Park, Uganda*, People and Plants Working Paper No. 5. Paris: UNESCO.

Cutlip, S. M. and Center, A. H. (1964). *Effective Public Relations*, 3rd edn. Englewood Cliffs, NJ: Prentice-Hall.

Cutter, S. L. (1996). Vulnerability to environmental hazards. *Progress in Human Geography*, **20**, 529–539.

Czech, B. and Krausman, P. R. (1999). Public opinion on endangered species conservation and policy. *Society and Natural Resources*, **12**, 469–479.

Dahier, T. (2000). Bilan des dommages en 2000. *L'Infoloups*, **8**, 11.

(2002). Bilan des dommages en 2001. *L'Infoloups*, **10**, 11.

Dahle, B., Sørensen, O. J., Wedul, E. H., Swenson, J. E. and Sandegren, F. (1998). The diet of brown bears *Ursus arctos* in central Scandinavia: effect of access to free-ranging domestic sheep *Ovis aries*. *Wildlife Biology*, **4**, 147–158.

Daniel, J. C. (1996). *The Leopard in India: A Natural History*. New Delhi: Natraj Publishers.

(2001). *The Tiger in India: A Natural History*. Dehra Dun, India: Natraj Publishers.

Danz, H. P. (1999). *Cougar!* Athens, OH: Ohio University Press.

Darnton, R. 1985. *The Great Cat Massacre and Other Episodes in French Cultural History*. New York: Vintage.

Davies, R. A. G. (1994). Black eagle predation on rock hyrax and other prey in the Karoo. Ph.D. thesis, University of Pretoria, South Africa.

(1999). The extent, cost and control of livestock predation by eagles with a case study on black eagles in the Karoo. *Journal of Raptor Research*, **33**, 67–72.

(2000). The influence of predation by black eagles on rock hyrax numbers in the Karoo. In *Raptors at Risk*, ed. R. D. Chancellor and B. U. Meyburg. Berlin, Germany: World Working Group for Birds of Prey and Owls, pp. 519–526.

Davis, T. (2002). Division of wildlife to compensate landowners who protect prairie dogs. *Colorado Conservator*, **18**, 9.

Daw, M. and Daw, D. (2001). *The Costs of Wild Geese to Scottish Agriculture: Islay and Loch of Strathbeg Case Studies*. Edinburgh, UK: Scottish Executive Central Research Unit.

De Boer, W. F. and Baquete, D. (1998). Natural resource use, crop damage and attitudes of rural people in the vicinity of the Maputo Elephant Reserve, Mozambique. *Environmental Conservation*, **25**, 208–218.

de Klemm, C. (1996). *Compensation for Damage Caused by Wild Animals*. Strasbourg, France: Council of Europe.

de Lima, M. G. and Gascon, C. (1999). The conservation value of linear forest remnants in central Amazonia. *Biological Conservation*, **91**, 241–247.

deCalesta, D. S. and Cropsey, M. G. (1978). Field test of a coyote proof fence. *Wildlife Society Bulletin*, **6**, 256–259.

Decker, D. J. and Chase, L. C., (1997). Human dimensions of living with wildlife: a management challenge for the twenty-first century. *Wildlife Society Bulletin*, **25**, 788–795.

Decker, D. J. and Purdy, K. G. (1988). Toward a concept of wildlife acceptance capacity in wildlife management. *Wildlife Society Bulletin*, **16**, 53–57.

Defenders of Wildlife (2004). http://www.defenders.org/wildlife/wolf/wolfcomp.pdf

DeLiberto, T. J., Conover, M. R., Gese, E. M. *et al.* (1998). Fertility control in coyotes: is it a potential management tool? *Vertebrate Pest Conference Proceedings*, **18**, 144–149.

Detling, J. K. (1998). Mammalian herbivores: ecosystem-level effects in two grassland national parks. *Wildlife Society Bulletin*, **26**, 438–448.

(2000). Distribution of raptors on heather moorland. *Oikos*, **100**, 15–24.

Deurbrouck, J. and Miller, D. (2001). *Cat Attacks: True Stories and Hard Lessons fron Congar Country*. Seattle, WA: Sasquatch Books.

Dixon, J. A. and Sherman, P. B. (1990). *Economics of Protected Areas: A New Look at Benefits and Costs*. Washington, DC: Island Press.

Dolabella, A. L. (2000). The Brazilian Panatanal: an overview. In *The Pantanal: Understanding and Preserving the World's Largest Wetland*, ed. F. A. Swarts. St Paul, MN: Paragon House, pp. 37–41.

Dolan, C. C. (1999). The National Grasslands and disappearing biodiversity: can the prairie dog save us from an ecological desert? *Environmental Law*, **29**, 213–234.

Dolbeer, R., Holler, N. and Hawthorne, D. (1994). Identification and control of wildlife damage. In *Research and Management Techniques for Wildlife and Habitats*, ed. R. Dolbeer. Bethesda, MD: The Wildlife Society, pp. 474–506.

Donnelly, C. A., Woodroffe, R., Cox, D. R. *et al.* (2003). Impact of localized badger culling on TB incidence in British cattle. *Nature*, **426**, 834–837.

Dorrance, M. J. (1976). Predation losses of sheep in Alberta. *Journal of Range Management*, **29**, 457–460. (http://uvalde.tamu.edu/jrm/jrmhome.htm)

(1982). Predation losses of cattle in Alberta. *Journal of Range Management*, **35**, 690–692. (http://uvalde.tamu.edu/jrm/jrmhome.htm)

Dorrance, M. J. and Bourne, J. (1980). An evaluation of anti coyote electric fencing. *Journal of Range Management*, **33**, 385–387. (http://uvalde.tamu.edu/jrm/jrmhome.htm)

Dorrance, M. J. and Roy, L. D. (1976). Predation losses of domestic sheep in Alberta. *Journal of Range Management*, **29**, 457–460.

(1978). Aversive conditioning tests of black bears in beeyards failed. *Vertebrate Pest Conference Proceedings*, **8**, 251–254.

du Toit, J. T. (2002). Wildlife harvesting guidelines for community-based wildlife management: a southern African perspective. *Biodiversity and Conservation*, **11**, 1403–1416.

du Toit, J. T., Walker, B. H. and Campbell, B. M. (2004). Conserving tropical nature: current challenges for ecologists. *Trends in Ecology and Evolution*, **19**, 12–17.

Dudley, J. P., Mensah-Ntiamoah, A. Y. and Kpelle, D. G. (1992). Forest elephants in a rainforest fragment: preliminary findings from a wildlife conservation project in southern Ghana. *African Journal of Ecology*, **30**, 116–126.

Duguid, J. 1932. *Tiger-Man*. London: Victor Gollancz.

Duke University (2005) *Hippos Newsletter*. http://moray.ml.duke.edu/projects/hippos/Newsletter/NewsFrameSet.html

Dunishenko, Yu. M. (1985). On the problem of tiger conservation in Khahbarovskiy Krai. In *Izuchenie i okhrana redkhikh i ischezaiuschikh vidov zhivotnykh fauny SSSR*. Moscow: Nauka, pp. 62–65. (In Russian)

Dunlap, T. R. (1988). *Saving America's Wildlife*. Princeton, NJ: Princeton University Press.

Dunn, A. (1991). *A Survey of Elephants in Sapo National Park, Liberia*. Gland, Switzerland: WWF International.

Durant, S. M. (2000). Dispersal patterns, social organization and population viability. In *Behaviour and Conservation*, ed. L. M. Gosling and W. J. Sutherland. Cambridge, UK: Cambridge University Press, pp. 25–140.

Dyar, J. A. and Wagner, J. (2003). Uncertainty and species recovery program design. *Journal of Environmental Economics and Management*, **45**, 505–522.

Dyer, A. (1996). *Men for All Seasons: The Hunters and Pioneers*. Agoura, CA: Trophy Room Books.

Ebbinge, B. S. (1991). The impact of hunting on mortality rates and spatial distribution of geese wintering in the western palearctic. *Ardea*, **79**, 197–210.

Eckert, J., Conraths, F. and Tackmann, K. (2000). Echinococcosis: an emerging or re-emerging zoonosis? *International Journal of Parasitology*, **30**, 1283–1294.

Edwards, G. R., Crawley, M. J. and Heard, M. S. (1999). Factors influencing molehill distribution in grassland: implications for controlling the damage caused by molehills. *Journal of Applied Ecology*, **36**, 434–442.

Edwards, S. R. and Allen, C. M. (1992). Sport hunting as sustainable use of wildlife. Unpublished report. of the Sustainable Use of Wildlife Programme. Gland, Switzerland: IUCN.

Edwasi, I. (1994). *Wildlife–Human Conflicts in Kenya*, Report of the Five-Person Review Group. Nairobi: Kenya Wildlife Service.

Ellegren, H., Savolainen, P. and Rosen, B. (1996). The genetical history of an isolated population of the endangered grey wolf *Canis lupus*: a study of nuclear and mitochondrial polymorphisms. *Philosophical Transactions of the Royal Society of London B*, **351**, 1661–1669.

Ellins, S. R. and Catalano, S. M. (1980). Field application of the conditioned taste aversion paradigm to the control of coyote predation on sheep and turkeys. *Behavioral and Neural Biology*, **29**, 532–536.

Ellins, S. R., Catalano, S. M. and Schechinger, S. A. (1977). Conditioned taste aversion: a field application to coyote predation on sheep. *Behavioral Biology*, **20**, 91–95.

Eloff, T. (2002). The economic realities of the game industry in South Africa. In *Sustainable Utilization: Conservation in Practice*, ed. H. Ebedes, B. Reilly, W. Van Hoven and B. Penzhorn. Pretoria: South African Game Ranchers' Organization, pp. 78–86.

Else, J. G. (1991). Nonhuman primates as pests. In *Primate Responses to Environmental Change*, ed. H. O. Box. London: Chapman and Hall, pp. 115–165.

Eltringham, S. K. (1994). *Wildlife Resources and Economic Development*. Chichester, UK: John Wiley.

Emerton, L. (2001). The nature of benefits and the benefits of nature: why wildlife conservation has not economically benefited communities in Africa. In *African Wildlife and Livelihoods: The Promise and Performance of Community Conservation*, ed. D. Hulme and M. Murphree. Oxford, UK: James Currey, pp. 208–226.

Energi- og miljøkomitéen. (1997). *Innstilling fra energi- og miljøkomitéen om rovviltforvaltning*, Innstilling til Stortinget No. 301. Oslo: Energi- og miljøkomitéen. (In Norwegian)

Ernest, H. B. and Boyce, W. M. (2000). DNA identification of mountain lions involved in livestock depredation and public safety incident and investigation. *Vertebrate Pest Conference Proceedings*, **19**,

Errington, P. L. (1946). Predation and vertebrate populations. *Quarterly Review of Biology*, **21**, 221–245.
Estes, J. A. (1996). The influence of large, mobile predators in aquatic food webs: examples from sea otters and kelp forests. In *Aquatic Predators and their Prey*, ed. S. P. R. Greenstreet and M. L. Tasker. Oxford, UK: Blackwell Scientific, pp. 58–64.
Estes, J. A., Tinker, M. T., Williams, T. M. and Doak, D. F. (1998). Killer whale predation on sea otters linking oceanic and nearshore ecosystems. *Science*, **282**, 473–476.
Etheridge, B., Summers, R. W. and Green, R. (1997). The effects of illegal killing and destruction of nests on the population dynamics of hen harriers in Scotland. *Journal of Applied Ecology*, **34**, 1081–1106.
Etling, K. (2001). *Cougar Attacks: Encounters of the Worst Kind*. Gilford, CT: Lyons Press.
Evans, W. (1983). *The Cougar in New Mexico: Biology, Status, Depredation of Livestock and Management Recommendations*. Santa Fe, NM: New Mexico Department of Game and Fish.
Evans Pritchand, E. (1906). *The Criminal Prosecution and Capital Punishment of Animals*. London: Faber and Faber (reprinted 1987).
Eves, J. A. (1999). Impact of badger removal on bovine tuberculosis in east County Offaly. *Irish Veterinary Journal*, **52**, 199–203.
Fanshawe, J. H., Ginsberg, J. R., Sillero-Zubiri, C. and Woodroffe, R. (1997). The status and distribution of remaining wild dog populations. In *The African Wild Dog: Status Survey and Conservation Action Plan*, ed. D. W. Macdonald. Gland, Switzeland: IUCN, pp. 11–57.
Faraizl, S. D. and Stiver, S. J. (1996). A profile of depredating mountain lions. *Vertebrate Pest Conference Proceedings*, **17**, 88–90.
Fawcett, P. H. (1954). *Exploration Fawcett*. London: Companion Book Club.
Fernandez, A. J. G. (1995). Livestock predation in the Venezuelan llanos. *Cat News*, **22**, 14–15.
Ferraro, P. J. and Kiss, A. (2002). Direct payments to conserve biodiversity. *Science*, **298**, 1718–1719.
Fiedler, L. A. (1988). Rodent problems in Africa. In *Rodent Pest Management*, ed. I. Prakash. Boca Raton, FL: CRC Press, pp. 35–65.
Fielding, A., Haworth, P., Morgan, D., Thompson, D. B. A. and Whitfield, D. P. (2003). The impact of golden eagles on a diverse bird of Prey assemblage. In: *Birds of Prey in a Changing Landscape*, ed. D. B. A. Thompson, S. M. Redpath, M. Marquiss, A. Fielding and C. Galbraith. Edinburgh, UK: HMSO, pp. 221–243.
Fischer, F. (2000). *Citizens, Experts, and the Environment: The Politics of Local Knowledge*. Durham, NC: Duke University Press.
Fischer, H. (1989). Restoring the wolf: Defenders launches a compensation fund. *Defender*, **64**, 9, 36.
FitzGibbon, C., Mogaka, H. and Fanshawe, J. (1995). Subsistence hunting in Arabuko-Sokoke Forest, Kenya and its effects on mammal populations. *Conservation Biology*, **9**, 1116–1126.
Fitzwater, W. D. (1972). Barrier fencing in wildlife management. *Vertebrate Pest Conference Proceedings*, **5**, 49–55.

Flack, P. H. (2002). Exotic game: catching up with Texas? *Magnum*, October 2002, 76–80.

Fleck, S. and Herrero, S. (1988). Polar bear conflicts with humans. In *Bear – People Conflicts*, ed. M. Bromley. Yellowknife, NWT: Department of Renewable Resources.

Foeken, D. and Owuor, S. O. (2000). Facts and figures. In *Kenya Coast Handbook: Culture Resources and Development in the East African Littoral*, ed. J. Hoorweg, D. Foeken and R. A. Obudh. London: LIT Verlag, pp. 406–422.

Forbes, G. J. and Theberge, J. B. (1996). Cross-boundary management of Algonquin Park wolves. *Conservation Biology*, **10**, 1091–1097.

Forbes, S. H. and Boyd, D. K. (1997). Genetic structure and migration in native and reintroduced Rocky Mountain wolf populations. *Conservation Biology*, **11**, 1226–1234.

Forrest, S. C. (1988). *Black-Footed Ferret Recovery Plan*. Denver, CO: US Fish and Wildlife Service.

Forrest, S. C., Biggins, D. E., Richardson, L. (1988). Black-footed ferret (*Mustela nigripes*) attributes at Meeteetse, Wyoming, 1981 to 1985. *Journal of Mammalogy*, **69**, 261–276.

Forthman, D. (2000). Experimental application of conditioned taste aversion (CTA) to large carnivores. *Carnivore Damage Prevention News*, **2**, 2–4. (http://www.kora.unibe.ch)

Forthman-Quick, D. L. (1986). Activity budgets and the consumption of human foods in two troops of baboons (*Papio anubis*) at Gilgil, Kenya. In *Primate Ecology and Conservation*, ed. J. G. Else and P. C. Lee. Cambridge, UK: Cambridge University Press, pp. 221–228.

Forthman-Quick, D. L. and Demment, D. (1988). Dynamics of exploitation: differential energetic adaptations of two troops of baboons to recent human contact. In *Ecology and Behaviour of Food-Enhanced Primate Groups*, ed. J. E. Fa and C. H. Southwick. New York: Alan R. Liss, pp. 25–51.

Fortney, R. H. (2000). Cattle grazing and sustainable plant diversity in the Pantanal. In *The Pantanal of Brazil, Bolivia and Paraguay*, ed. F. A. Swarts. Gouldsboro, PA: Hudson MacArthur Publishers, pp. 127–133.

Fourli, M. (1999). *Compensation for Damage Caused by Bears and Wolves in the European Union*. Luxembourg: Office for Official Publications of the European Communities.

Fox, C. H. (2001). Taxpayers say no to killing predators. *Animal Issues*, **32**, 1–2.

Frank, L. G. (1998). *Living with Lions: Carnivore Conservation and Livestock in Laikipia*. Bethesda, MD: Development Alternatives.

Frank, L. G., Simpson, D. and Woodroffe, R. (2003). Foot snares: an effective method for capturing African lions. *Wildlife Society Bulletin*, **39**, 309–314.

Franklin, W. L. and Powell, K. J. (1993). *Guard Llamas*, University Extension No. *PM–1527*. Ames, IA: Iowa State University. Accessible at http://www.extension.iastate.edu/Publications/PM1527.pdf

Freese, C. H. (1996). *The Commercial, Consumptive Use of Wildlife Species: Managing It for the Benefit of Biodiversity Conservation*, WWF Discussion Paper. Washington, DC: WWF United States.

(1998). *Wild Species as Commodities: Managing Markets and Ecosystems for Sustainability*. Washington, DC: Island Press.
Freudenberger, D. and Brooker, L. (2004). Development of the focal species approach for biodiversity conservation in the temperate agricultural zones of Australia. *Biodiversity and Conservation*, **13**, 253–274.
Fritts, S. H. (1982). *Wolf Depredation on Livestock in Minnesota*, Resource Publication No. 145. Washington, DC: US Fish and Wildlife Service.
Fritts, S. H. and Carbyn, L. N. (1995). Population viability, nature reserves, and the outlook for gray wolf conservation in North America. *Restoration Ecology*, **3**, 26–38.
Fritts, S. H. and Paul, W. J. (1989). Interactions of wolves and dogs in Minnesota. *Wildlife Society Bulletin*, **17**, 21–123.
Fritts, S. H., Paul, W. J., Mech, L. D. and Scott, D. P. (1992). *Trends and Management of Wolf–Livestock Conflicts in Minnesota*, Resource Publication No. 181. Washington, DC: US Fish and Wildlife Service.
Fritts, S. H., Bangs, E. E. and Gore, J. F. (1994). The relationship of wolf recovery to habitat conservation and biodiversity in the northwestern United States. *Landscape and Urban Planning*, **28**, 23–32.
Fritts, S. H., Bangs, E. E., Harms, D. R., Brewster, W. G. and Gore, J. F. (1995). Restoring wolves to the northern Rocky Mountains of the United States. In *Ecology and Conservation of Wolves in a Changing World*, ed. L. N. Carbyn, S. H. Fritts and D. R. Seip. Edmonton Alberta: Canadian Circumpolar Institute, pp. 107–126.
Fritts, S. H., Bangs, E. E., Fontaine, M. R. *et al.* (1997). Planning and implementing a reintroduction of wolves to Yellowstone National Park and Central Idaho. *Restoration Ecology*, **5**, 7–27.
Fritts, S. H., Mack, C. M., Smith, D. W. *et al.* (2001). Outcomes of hard and soft releases of reintroduced wolves in central Idaho and the Greater Yellowstone Area. In *Large Mammal Restoration: Ecological and Sociological Challenges in the 21st Century*, ed. D. S. Maehr, R. F. Noss and J. L. Larkin. Washington, DC: Island Press, pp. 125–148.
Fritts, S. H., Stephenson, R. O., Hayes, R. D. and Boitani, L. (2003). Wolves and humans. In *Wolves: Behavior, Ecology, and Conservation*, ed. L. D. Mech and L. Boitani. Chicago, IL: University of Chicago Press, pp. 289–316.
Frost, R. (n.d.). *Possible Answer to Difficult Questions*. Las Cruces, NM: Cooperative Extension Service, New Mexico State University.
Fuller, E. (2000). *Extinct Birds*. Oxford, UK: Oxford University Press.
Fuller, T. K. (1989). Population dynamics of wolves in North-Central Minnesota. *Wildlife Monographs*, **105**, 1–41.
Funston, P. J. (2001). Kalahari transfrontier lion project: Population-ecology and long term monitoring of a free-ranging population in an arid environment.
Gachago, S. and Waithaka, J. (1995). *Human–Elephant Conflict in Kiambu, Murang'a, Kirinyaga, Embu and Meru Districts*. Nairobi: Kenya Wildlife Service.
Galbraith, C., Stroud, D. and Thompson, D. B. A. (2003). Towards resolving raptor–human conflicts. In *Birds of Prey in a Changing Environment*, ed. D. B. A. Thompson, S. M. Redpath, A. H. Fielding, M. Marquiss and C. A. Galbraith. Edinburgh, UK: Stationary Office, pp. 527–536.

Galster, S. R. and Vaud Eliot, K. (1999). Roaring back: anti-poaching strategies from the Russian Far East and the comeback of the Amur tiger. In *Riding the Tiger: Meeting the Needs of People and Wildlife in Asia*, ed. J. Seidensticker, S. Christie and P. Jackson. Cambridge, UK: Cambridge University Press, pp. 230–239.

Game Department (1920–63). *Annual Reports*. Nairobi: Kenya Game Department.

Game Department of Uganda (1924). *Game Department Archives of Uganda*. Kampala: Ugandan Wildlife Authority, Ministry of Tourism and Wildlife, Government of Uganda.

Garshelis, D., Joshi, A. R., Smith, J. L. D. and Rice, C. G. (1999). Sloth bear conservation action plan. In *Bears: Status Survey and Conservation Action Plan*, ed. C. Servheen, S. Herrero and B. Peyton. Gland, Switzerland: IUCN, pp. 225–240.

Gasaway, W., Boertje, R., Grangaard, D. *et al.* (1992). The role of predation in limiting moose at low density in Alaska. *Wildlife Monographs*, **120**, 1–70.

Gates, N., Rich, J. E., Godtel, D. D. and Hulet, C. V. (1978). Development and evaluation of anti coyote electric fencing. *Journal of Range Management*, **31**, 151–153. (http://uvalde.tamu.edu/jrm/jrmhome.htm)

Gede National Museum (2004). *Kipepeo Butterfly Project*. Watamu, Kenya: Gede National Museum. Accessible at http://www.kipepeo.org

Gee, E. P. (1964). *The Wildlife of India*. London: Collins.

Georgiadis, N. and Ojwang', G. (2001). Numbers and distributions of large herbivores in Laikipia, Samburu and Isiolo Districts. In *Laikipia Wildlife Forum*, Nanyuki, Kenya.

Gesicho, A. (1991). *A Survey of the Arabuko Sokoke Elephant Population*. Nairobi: Kenya Wildlife Service Elephant Programme.

Ghiglieri, M. P. (1984). *The Chimpanzees of Kibale Forest: A Field Study of Ecology and Social Structure*. New York: Columbia University Press.

Ghimire, K. B. and Pimbert, M. P. (1997). *Social Change and Conservation*. London: Earthscan.

Giannecchini, J. (1993). Ecotourism: new partners, new relationships. *Conservation Biology*, **7**, 429–432.

Gibson, C. C. and Marks, S. A. (1995). Transforming rural hunters into conservationists: an assessment of community-based wildlife management programs in Africa. *World Development*, **23**, 941–957.

Giles, R. Jr (1989). Wildlife and integrated pest management. *Environmental Management*, **4**, 373–374.

Gillingham, S. and Lee, P. C. (1999). The impact of wildlife related benefits on the conservation attitudes of local people around the Selous Game Reserve, Tanzania. *Environmental Conservation*, **26**, 218–228.

(2003). People and protected areas: a study of local perceptions of wildlife crop-damage conflict in an area bordering the Selous Game Reserve, Tanzania. *Oryx*, **37**, 316–325.

Gipson, P. S. (1975). Efficiency of trapping in capturing offending coyotes. *Wildlife Management*, **39**, 45–47.

Gipson, P. S., Ballard, W. B. and Nowak, R. M. (1998). Famous North America wolves and the credibility of early wildlife literature. *Wildlife Society Bulletin*, **26**, 808–816.

Githaiga, J. (1998). *Recent Population Trends in Kenya*. Nairobi: Kenya Wildlife Service.

Gjertz, I. and Persen, E. (1987). Confrontations between humans and polar bears in Svalbard. *Polar Record*, **34**, 340–347.

Gjertz, I. and Scheie, J. O. (1998). Human casualties and polar bears killed in Svalbard, 1993–1997. *Polar Record*, **34**(191), 347–340.

Glen, A. S. and Short, J. (2000). The control of dingoes in New South Wales in the period 1883–1930 and its likely impact on their distribution and abundance. *Australian Zoologist*, **31**, 432–442.

Glover, P. E. (1968). A report on an ecological survey of the proposed Shimba Hills National Reserve. Unpublished report. Nairobi: Kenya National Parks.

Gniadek, S. J. and Kendall, K. C. (1998). A summary of bear management in Glacier National Park, 1960–1994. *Ursus*, **10**, 155–159.

Goldman, A. (1996). Pest and disease hazards and sustainability in African agriculture. *Experimental Agriculture*, **32**, 199–211.

Goldstein, I. (1991). Spectacled bear predation and feeding behavior on livestock in Venezuela. *Studies on Neotropical Fauna and Environment*, **26**, 231–235.

Gonzalez, A., Lawton, J. H., Gilbert, F. S., Blackburn, T. M. and Evans-Freke, I. (1998). Metapopulation dynamics, abundance, and distribution in a microecosystem. *Science*, **281**, 2045–2047.

Gonzalez-Fernandez, A. J. (1995). Livestock predation in the Venezuelan llanos. *Cat News*, **22**, 14–15.

Goodrich, J. M., Kerley, L. L., Schleyer, B. O. *et al.* (2000). Capture and chemical anesthesia of Amur (Siberian) tigers. *Wildlife Society Bulletin*, **29**, 533–542.

Goodwin, H. J. (1996). In pursuit of ecotourism. *Biodiversity and Conservation*, 5, 277–291.

Goodwin, H. J. and Leader-Williams, N. (2000). Tourism and protected areas: distorting conservation priorities towards charismatic megafauna? In *Has the Panda Had its Day? Priorities for the Conservation of Mammal Diversity*, ed. A. Entwistle and N. Dunstone. Cambridge, UK: Cambridge University Press, pp. 257–275.

Goodwin, H., Kent, I., Parker, K. and Walpole, M. (1998). *Tourism, Conservation and Sustainable Development: Case Studies from Asia and Africa*, Wildlife and Development Series No. 12. London: International Institute for Environment and Development.

Goodwin, H., Johnston, G. and Warburton, C. (2000). *Carnivores and Tourism: The Challenge Ahead*. WWF-UK Report: 1– 26.

Gorokhov, G. F. (1977). The numbers and population structure of the Amur tiger in the south of Sikhote-Alin. In *Redkie vidy mlekopitaiuschikh i ikh okhrana: Materialy 2 Vsesoiuznogo soveschania*. Moscow: Nauka, pp. 119–220. (In Russian)

(1983). The causes of illegal shooting of Amur tigers. In *Redkie vidy mlekopitaiuschikh i ikh okhrana: Materialy 3 Vsesoiuznogo soveschania*. Moscow: Iemezh ran i vto an SSSR, pp. 88–89. (In Russian)

Gossling, S. (1999). Ecotourism: a means to safeguard biodiversity and ecosystem functions? *Ecological Economics*, **29**, 303–320.

Graham, A. D. (1973). *The Gardeners of Eden*. London: Allen and Unwin.

Gray, G. G. (1993). *Wildlife and People: The Human Dimensions of Wildlife Ecology*. Urbana, IL: University of Illinois Press.

Green, J. S. and Woodruff, R. A. (1988). Breed comparison and characteristics of use of livestock guarding dogs. *Journal of Range Management*, **41**, 249–251. (http://uvalde.tamu.edu/jrm/jrmhome.htm)

(1990). *Livestock Guard Dogs: Protecting Sheep from Predators*, Agriculture Information Bulletin No. 588. Washington, DC: US Department of Agriculture.

Green, J. S., Woodruff, R. A. and Tueller, T. T. (1984). Livestock-guarding dogs for predator control: costs, benefits and practicality. *Wildlife Society Bulletin*, **12**, 44–50.

Green, J. S., Woodruff, R. A. and Audelt, W. F. (1994). Do livestock guarding dogs lose their effectiveness over time? *Vertebrate Pest Conference Proceedings*, **16**, 41–44.

Green, R. and Etheridge, B. (1999). Breeding success of the hen harrier in relation to the distribution of grouse moors and the red fox. *Journal of Applied Ecology*, **36**, 472–484.

Greenaway, J. C. (1967). *Extinct and Vanishing Birds of the World*. New York: Dover.

Greenwood, R. J., Sargeant, A. B., Johnson, D. H., Cowardin, L. M. and Shaffer, T. L. (1995). Factors accociated with duck nest success in the prairie pothole region of Canada. *Wildlife Monograph*, **128**, 1–57.

Grey, Z. (1922). *Roping Lions in the Grand Canyon*. New York: Harper.

Griffin, L. R. and Coath, D. C. (2001). *WWT Svalbard Barnacle Goose Project Report 2000–2001*, Wildfowl and Wetlands Trust Internal Report. Slimbridge, UK:

Grobbelaar, C. (2004). Zimbabwe. In *African Hunting Guide*, ed. T. Wieland. Rivonia, South Africa: Future Publishing.

Grundblatt, M., Said, M. Y. and Warugute, P. (1995). *National Rangeland Report: Summary of Population Estimates of Wildlife and Livestock*. Nairobi: Ministry of Planning and National Development, Department of Remote Sensing and Regional Surveys.

Guerrera, W., Sleeman, J. M., Jasper, S. B. *et al.* (2003). Medical survey of the local human population to determine possible health risks to the mountain gorillas of Bwindi Impenetrable Forest National Park, Uganda. *International Journal of Primatology*, **24**, 197–207.

Guggisberg, C. A. W. (1961). *Simba: The Life of the Lion*. Cape Town, South Africa: Howard Timmins.

Guha, S. (2001). *Health and Population in South Asia: From Earliest Times to the Present*. New Delhi: Permanent Black.

Gujadhur, T. (2001). *Joint Venture Options for Communities and Safari Operators in Botswana*, CBNRN Support Programme Occasional Paper No. 6. SNV/IUCN.

Gunther, K. A. (1994). Bear management in Yellowstone National Park, 1960–93. *International Conference on Bear Research and Management*, **9**, 549–560.

Gustavson, C. R., Garcia, J., Hankins, W. G. and Rusiniak, K. W. (1974). Coyote predation control by aversive conditioning. *Science*, **184**, 581–583.

Gustavson, C. R., Jowsey, J. R. and Milligan, D. (1982). A 3-year evaluation of taste aversion coyote control in Saskatchewan. *Journal of Range Management*, **35**, 57–59. (http://uvalde.tamu.edu/jrm/jrmhome.htm)

Gutleb, B. (2001). Experiences of 10 years of damage prevention for brown bears in Austria. *Carnivore Damage Prevention News*, **5**, 9–10. (http://www.kora.unibe.ch)

Haber, G. C. (1996). Biological, conservation, and ethical implications of exploiting and controlling wolves. *Conservation Biology*, **10**, 1068–1081.

Hackel, J. D. (1999). Community conservation and the future of Africa's wildlife. *Conservation Biology*, **13**, 726–734.

Hætta, I. O. 2002. *Ressursregnskap for reindriftsnaeringen*. Alta, Norway: Reindriftsforvaltningen. (In Norwegian)

Haight, R. G. and Mech, L. D. (1997). Computer simulation of vasectomy for wolf control. *Journal of Wildlife Management*, **61**, 1023–1031.

Hakkarainen, H. and Korpimaki, E. (1996) Competitive and predatory interactions among raptors: an observational and experimental study. *Ecology*, **77**, 1134–1142.

Halfpenny, J. C. (2003). *Yellowstone Wolves in the Wild*. Helena, MT: Riverbend Publishing.

Hallowell, I. A. (1926). Bear ceremonialism in the Northern Hemisphere. *American Anthropologist*, **28**, 1–175.

Hamilton, P. H. (1981). *The Leopard* Panthera pardus *and the Cheetah* Acinonyx jubatus *in Kenya*. Report for the US Fish and Wildlife Service, the African Wildlife Leadership Foundation and the Government of Kenya.

Hanby, J. P., Bygott, J. D. and Packer, C. (1995). Ecology, demography and behavior of lions in two contrasting habitats: Ngorongoro Crater and the Serengeti Plains. In *Serengeti II: Research, Management and Conservation of an Ecosystem*, ed. A. Sinclair and P. Arcese. Chicago, IL: University of Chicago Press, pp. 315–331.

Hanks, J., Denshaw, W. D., Smuts, G. L. *et al.* (1981). Management of locally abundant mammals: the South African experience. In *Problems of Managing Locally Abundant Wild Animals*, ed. P. A. Jewell, S. Holt and D. Hart. London: Academic Press, pp. 21–55.

Hanley, N., MacMillan, D., Patterson, I. and Wright, R.E. (2003). Economics and the design of nature conservation policy: a case study of wild goose conservation in Scotland using choice experiments. *Animal Conservation*, **6**, 123–129.

Hansen, I. and Bakken, M. (1999). Livestock-guarding dogs in Norway. I: Interaction. *Journal of Range Management*, **52**, 2–6. (http://uvalde.tamu.edu/jrm/jrmhome.htm)

Hansen, I. and Smith, M. M. (2001). Livestock-guarding dogs in Norway. II: Different working regimes. *Journal of Range Management*, **52**, 312–316. (http://uvalde.tamu.edu/jrm/jrmhome.htm)

Hansen, K. (1992). *Cougar: The American Lion*. Flagstaff, AZ: Northland Publishing.

Hansen, R. M. and Gold, I. K. (1977). Black-tailed prairie dogs, desert cottontails and cattle trophic relations on shortgrass range. *Journal of Range Management*, **30**, 210–214.

Harbo, S. J. Jr and Dean, F. C. (1983). Historical and current perspectives on wolf management in Alaska. In *Wolves in Canada and Alaska: Their Status, Biology and Management*, ed. L. N. Carbyn. Edmonton, Alberta: Canadian Wildlife Service, pp. 51–64.

Harcourt, A. H., Parks, S. A. and Woodroffe, R. (2001). Small reserves face a double jeopardy: small size and high surrounding human density. *Biodiversity and Conservation*, **10**, 1011–1026.

Hardin, G. (1968). The tragedy of the commons. *Science*, **168**, 1243–1248.

Harms, D. R. (1980). Bear management in Yosemite. In *International Conference on Bear Research and Management*, pp. 205–212.

Harris, D. and Eisenberg, J. H. (1989). Enhanced linkages: necessary steps for success in conservation of faunal diversity. In *Conservation for the Twenty-First Century*, ed. D. Western and M. Pearl. Oxford, UK: Oxford University Press, pp. 166–181.

Harwood, J. (2000). Risk assessment and decision analysis in conservation. *Biological Conservation*, **95**, 219–226.

Hatfield, P. G. and Walker, J. W. (1994). *An Evaluation of PRED-X Eartag in Protection of Lambs from Coyote Predation*, Sheep Research Progress Report No. 3. US Department of Agriculture, Agricultural Research Service.

Hawes-Davis, D. (1998). *Varmints*. Missoula, MT: High Plains Films. (video)

Hawkes, R. K. (1991). *Crop and Livestock Losses to Wild Animals in the Bulilimamangwe Natural Resources Management Project Area*. Harare: Centre for Applied Social Sciences, University of Zimbabwe.

Hawthorne, J. (1980). Wildlife damage and control techniques. In *Research and Management Techniques for Wildlife and Habitats*, 3rd edn, ed. E. Schemnitz. Bethesda, MD: The Wildlife Society.

Hayward, G. D., Miquelle, D. G., Smirnov, E. N. and Nations, C. (2002). Monitoring Amur tiger populations: characteristics of track surveys in snow. *Wildlife Society Bulletin*, **30**, 1150–1159.

Hazumi, T. (1999). Asiatic black bear conservation action plan: Japan. In *Bears: Status Survey and Conservation Action Plan*, ed. C. Servheen, S. Herrero and B. Peyton. Gland, Switzerland: IUCN, pp. 207–210

Hearn, R. D. (2002). *The 2000 National Census of Pink-footed Geese and Icelandic Greylag Geese in Britain and Ireland*, Slimbridge, UK: Joint Nature Conservation Committee.

Hefner, R. and Geffen, E. (1999). Group size and home range of the Arabian wolf (*Canis lupus*) in southern Israel. *Journal of Mammalogy*, **80**, 611–619.

Hempel, L. C. (1996). *Environmental Governance*. Washington, DC: Island Press.

Hemson, G. and Macdonald, D. W. (2002). Cattle predation by lions in the Makgadikgadi: some patterns and parameters. In *Lion Conservation Research, Workshop 2: Modelling Conflict*, ed. A. J. Loveridge, T. Lynam and D. W. Macdonald. Oxford, UK: Wildlife Conservation Research Unit, pp. 10–12.

Hendrichs, H. (1975). The status of the tiger *Panthera tigris* (Linne, 1758) in the Sundarbans Mangrove Forest (Bay of Bengal). *Saugetierkundlich Mitteilungen*, **3**, 161–199.

Herne, B. (2001). *White Hunters: The Golden Age of African Safaris*. New York: Henry Holt.

Herrero, S. (1970). Human injury inflicted by grizzly bears. *Science*, **170**, 593–598.

(1985). *Bear Attacks: Their Causes and Avoidance*. Piscataway, NJ: Winchester Press.

(1989). The role of learning in some fatal grizzly bear attacks on people. In *Bear–People Conflicts*, ed. M. Bromley. Yellowknife, NWT: Department of Renewable Resources, pp. 9–14.

Herrero, S. and Fleck, S. (1990). Injury to people inflicted by black, grizzly or polar bears: Recent trends and new insights. *International Conference on Bear Research and Management*, **8**, 25–32.

Herrero, S. and Higgins, A. (1995). Fatal injuries inflicted to people by black bears. In *Proceedings of the 5th Western Black Bear Workshop*, ed. J. Auger and H. L. Black. Provo, UT: Brigham Young University Press, pp. 75–82.

(1998). Field use of capsaicin spray as a bear deterrent. *Ursus*, **10**, 533–537.

(1999). Human injuries inflicted by bears in British Columbia: 1960–. *Ursus*, **11**, 209–218.

(2003). Human injuries inflicted by bears in Alberta: 1960–1998. *Ursus*, **14**, 44–54.

Hewson, R. (1984). Scavenging and predation upon ship and lambs in west Scotland. *Journal of Applied Ecology*, **21**, 843–868.

Heydon, M. J. and Reynolds, J. L. (2000). Demography of rural foxes (*Vulpes vulpes*) in relation to cull intensity in three contrasting regions of Britain. *Journal of Zoology*, **251**, 265–276.

Heywood, V. H. (1995). *Global Biodiversity Assessment*. Cambridge, UK: Cambridge University Press.

Hill, C. M. (1997). Crop-raiding by wild vertebrates: the farmers' perspective in an agricultural community in western Uganda. *International Journal of Pest Management*, **43**, 77–84.

(1998). Conflicting attitudes towards elephants around the Budongo Forest Reserve, Uganda. *Environmental Conservation*, **25**, 244–250.

(2000). Conflict of interest between people and baboons: crop raiding in Uganda. *International Journal of Primatology*, **21**, 299–315.

(in press). People, crops and primates: a conflict of interests. In *Primate Commensalism and Conflict*, ed. J. D. Paterson.

Hill, C. M., Osborn, F. V. and Plumptre, A. J. (2002). *Human–Wildlife Conflict: Identifying the Problem and Possible Solutions*. New York: Wildlife Conservation Society.

Hill, J. and Simper, N. (2002). Evaluation of a high-density polyethylene collar for the prevention of coyote predation on sheep. *Proceedings of the Defenders of Wildlife's Carnivores 2002*, **236**.

Hillman Smith, A. K. K., Merode, E., Nicholas, A., Buts, B. and Ndey, A. (1995). Factors affecting elephant distribution in Garamba National Park and surrounding reserves, Zaire, with a focus on human–elephant conflict. *Pachyderm*, **19**, 39–48.

Hoare, R. E. (1995). Options for the control of elephants in conflict with people. *Pachyderm*, **19**, 54–63.

(1999). Determinants of human–elephant conflict in a land-use mosaic. *Journal of Applied Ecology*, **36**, 689–700.

(2000). African elephants and humans in conflict: the outlook for coexistence. *Oryx*, **34**, 34–38.

(2001). *A Decision Support System (DSS) for Managing Human–Elephant Conflict Situations in Africa*. Nairobi: IUCN African Elephant Specialist Group.

Hoare, R. E. and du Toit, J. (1999). Coexistence between people and elephants in African savannah. *Conservation Biology*, **13**, 633–639.

Hofer, D. (2002). *The Lion's Share of the Hunt: Trophy Hunting and Conservation – A Review of the Legal Eurasian Tourist Trophy Hunting Market and Trophy Trade under CITES*. TRAFFIC Europe.

Hofer, H. and East, M. L. (1993). The commuting system of Serengeti spotted hyaenas: how a predator copes with migratory prey. I: Social organization. *Animal Behaviour*, **46**, 547–557.

Homewood, K., Lambin, E. F., Coast, E. *et al.* (2001). Long-term changes in Serengeti–Mara wildebeest and land cover: pastoralism, population or policies? *Proceedings of the National Academy of Sciences of the USA*, **98**, 12544–12549.

Honey, M. (1999). *Ecotourism and Sustainable Development: Who Owns Paradise?* Washington DC: Island Press.

Hoogesteijn, R. (2002). *A Manual on the Problems of Depredation Caused by Jaguars and Pumas on Cattle Ranches*. New York: Wildlife Conservation Society. Accessible at http://www.savethejaguar.com

Hoogesteijn, R. and Chapman, C. A. (1997). Large ranches as conservation tools in the Venezuelan llanos. *Oryx*, **31**, 274–284.

Hoogesteijn, R. and Mondolfi, E. (1992). *The Jaguar*. Caracas: Armitano Publishers.

Hoogesteijn, R., Hoogesteijn, A. and Mondolfi, E. (1993). Jaguar predation and conservation: cattle mortality caused by felines on three ranches in the Venezuelan llanos. In *Mammals as Predators*, ed. N. Dunstone and M. L. Gorman. London: Zoological Society, pp. 391–406.

Hoogesteijn, R., Boede, E. O. and Mondolfi, E. (2002). Observaciones de la depredación de bovines por jaguares en Venezuela y los programmas Gubernamentales de control. In *El Jaguar en el nuevo milenio*, ed. R. A. Medellin *et al.* Mexico: Fondo de Cultura Economica, pp. 183–199.

Hoogland, J. L. (1995). *The Black-Tailed Prairie Dog: Social Life of a Burrowing Mammal*. Chicago, IL: University of Chicago Press.

(1996). *Cynomys ludovicianus*. *Mammalian Species*, **535**, 1–10.

Hornocker, M. (1992). Learning to live with mountain lions. *National Geographic*, **182**, 52–65.

Horrocks, J. and Baulu, J. (1994). Food competition between vervets (*Cercopithecus aethiops sabaeus*) and farmers in Barbados: implications for management. *Revue d'Ecologie (Terre Vie)*, **49**, 281–294.

Horstman, L. P. and Gunson, J. R. (1982). Black bear predation on livestock in Alberta. *Wildlife Society Bulletin*, **10**, 34–39.

Hötte, M. and Bereznuk, S. (2001). Compensation for livestock kills by tigers and leopards in Russia. *Carnivore Damage Prevention News*, **3**, 6–7.

Hovens, J. P. M., Tungalartuja, K. H., Todgeril, T. and Batdorj, D. (2000). The impact of wolves (*Canis lupus*) on wild ungulates and nomadic livestock in and around the Hustain Nuruu Steppe Reserve, Mongolia. *Lutra*, **43**, 39–50.

Howard, W. E. (1988). Why lions need to be hunted. *Proceedings of the Mountain Lion Workshop*, **3**, 66–68.

Hoyt, J. (1994). *Animals in Peril: How Sustainable Use Is Wiping out the World's Wildlife*. New York: Avery.

Huber, D. and Adamic, M. (1999). Slovenia. In *Bears: Status Survey and Conservation Action Plans*, ed. C. Sevheen, S. Herrero and B. Peyton. Gland, Switzerland: IUCN, pp. 119–122.

Hudson, P. J. (1992). *Grouse in Space and Time.* Fordingbridge, UK: Game Conservancy Trust.

Hudson, P. J., Rizzoli, A., Grenfell, B. T., Heesterbeek, H. and Dobson, A. P. (2002). *The Ecology of Wildlife Diseases.* Oxford, UK: Oxford University Press.

Humphrey, A. and Humphrey, E. (2003). *A Profile of Four Communal Area Conservancies in Namibia.* Windhoek: Ministry of Environment and Tourism.

Hunt, C. (1983). *Deterrents, Aversive Conditioning and Other Practices: An Annotated Bibliography to Aid Bear Management.* National Park Service, USA.

(1985). *Descriptions of Five Promising Deterrent and Repellent Products for Use with Bears.* US Fish and Wildlife Service, Office of Grizzly Bear Recovery Coordinator, Montana.

Hunter, J. A. (1952). *Hunter.* London: Hamish Hamilton.

Hussain, S. (2000). Protecting the snow leopard and enhancing farmers' livelihoods. *Mountain Research and Development,* **20**, 226–231.

(2003). Snow leopards and local livelihoods: managing the emerging conflicts through an insurance scheme. *Carnivore Damage Prevention News,* **6**, 9–11.

Hustad, H. (2000). The issuing of kill permits for brown bears in response to domestic sheep depredation in Norway, 1989–99. Thesis, Agricultural University of Norway, Ås.

Hutton, J. M. (1992). *The CITES Nile Crocodile Project.* Lausanne, Switzerland: CITES Secretariat.

Hutton, J. M. and Leader-Williams, N. (2003). Sustainable use and incentive-driven conservation: realigning human and conservation interests. *Oryx,* **37**, 215–226.

Hutton, J. M. and Webb, G. (2003). Crocodiles: the legal trade snaps back. In *The Trade in Wildlife: Regulation for Conservation,* ed. S. Oldfield. London: Earthscan, pp. 108–120.

Huygens, O. C. and Hayashi, H. (1999). Using electric fences to reduce Asiatic black bear depredation in Nagano prefecture, central Japan. *Wildlife Society Bulletin,* **27**, 959–964.

Idaho Legislative Wolf Oversight Committee (2002). Idaho wolf conservation and management plan. Unpublished report. Boise, ID.

Infield, M. (1988). Attitudes of a rural community towards conservation and a local conservation area in Natal, South Africa. *Biological Conservation,* **45**, 21–46.

Inukai, T. (1935). Damages on people by brown bears. *Syoku-butsu oyo-bi dou-butsu (Plant and Animal),* **1**, 57–64. (In Japanese)

Inukai, T. (1972). Bear damage and bear control in Japan. In *Bears: Their Biology and Management,* ed. S. Herrero. Morges, Switzerland: IUCN, pp.333–

Inverarity, J. D. (1894). Man-eating panther of Basim Berars. *Journal of the Bombay Natural History Society,* **9**, 25–27.

Isenberg, A. C. (2000). *The Destruction of the Bison.* Cambridge, UK: Cambridge University Press.

IUCN (1999). *African Elephant Database.* UNEP Publication.

(2002). *The IUCN Red List of Threatened Species.* Accessible at http://www.redlist.org

(2003). *2003 United Nations List of Protected Areas.* Gland, Switzerland: IUCN.

IUCN/UNEP/WWF (1980). *World Conservation Strategy: Living Resource Conservation for Sustainable Development*. Gland, Switzerland: IUCN/UNEP/WWF.

(1991). *Caring for the Earth: A Strategy for Sustainable Living*. Gland, Switzerland: IUCN/UNEP/WWF.

Jackson, J. J. (1996). An international perspective on trophy hunting. In *Tourist Hunting in Tanzania*, ed. N. Leader-Williams, J. A. Kayera and G. L. Overton. Gland, Switzerland: IUCN, pp. 7–11.

Jackson, P. (1990). *Endangered Species: Tigers*. London: The Apple Press.

Jackson, P. and Nowell, K. (1996). Problems and possible solutions in management of felid predators. *Journal of Wildlife Research*, **1**, 304–314.

Jackson, R., Ahlorn, G., Ale, S. *et al.* (1994). *Reducing Livestock Predation in the Nepalese Himalaya: Case of the Anapurna Conservation Area*. Draft Report. BioSystems Analysis Inc.,

Jagt, C. J. van der, Gujadhur, T. and Bussel, F. van (2000). *Community Benefits through Community Based Natural Resources Management in Botswana*, CBNRM Support Programme Occasional Paper No. 2. IUCN/SNV.

Jenkins, D., Watson, A. and Millar, G. (1964). Predation and red grouse populations. *Journal of Applied Ecology*, **1**, 183–195.

Jenkins, S., Perry, B. and Winkler, W. (1998). Ecology and epidemiology of raccoon rabies. *Reviews of Infectious Diseases*, **10**, 620–625.

Jhala, Y. (2000). Human–wolf conflict in India: beyond 2000. In *Realities of Global Wolf Restoration Symposium*, Duluth, MN, 23–26 February 2000. Accessible at http://www.wolf.org/wolves/learn/scientific/symposium/abstracts/003.asp

Joffe, H. (2003). Risk: from perception to social representation. *British Journal of Social Psychology*, **42**, 55–73.

Jones, J., Kruszon-Moran, D., Wilson, M. *et al.* (2001). *Toxoplasma gondii* infection in the US: seroprevalence and risk factors. *American Journal of Epidemiology*, **154**, 357–365.

Jones, J. M. and Woolf, A. (1983). Relationship between husbandry practices and coyote use of swine in west central Illinois. *Wildlife Society Bulletin*, **11**, 133–135.

Jones, S. (1999). Becoming a pest: prairie dog ecology and the human economy in the Euroamerican West. *Environmental History*, **4**, 531–552.

Jonker, S. A., Parkhusr, J. A., Field, R. and Fuller, T. K. (1998). Black bear depredation on agricultural commodities in Massachusetts. *Wildlife Society Bulletin*, **26**, 318–324.

Jorgensen, C. J. (1979) Bear-sheep interactions, Targhee National Forest. *International Conference on Bear Research and Management*, **5**, 191–200.

Jorgensen, C. J., Conley, R. H., Hamilton, R. J. and Sanders, O. T. (1978). Management of black bear depredation problems. *Proceedings of the Eastern Workshop on Black Bear Management and Research*, **4**, 297–321.

Jorgenson, J. P. (2000). Wildlife conservation and game harvest by Maya hunters in Quintana Roo, Mexico. In *Hunting for Sustainability in Tropical Forests*, ed. J. G. Robinson and E. L. Bennett. New York: Columbia University Press, pp. 251–266.

Jorner, U., Baer, L. A., Karlsson, E. and Danell, Ö. (1999). *Reindeer Husbandry in Sweden*. Umeå, Sweden: Svenska samernas riksförbund.

Kaczensky, P. (1996). *Livestock–Carnivore Conflicts in Europe*. Munich, Germany: Munich Wildlife Society.

Kalema-Zikusoka, G., Kock, R. and Macfie, E. (2002). Scabies in free-ranging mountain gorillas (*Gorilla beringei beringei*) in Bwindi Impenetrable National Park, Uganda. *Veterinary Record*, **150**, 12–15.

Kaltenborn, B. P., Bjerke, T. and Strumse, E. (1998). Diverging attitudes towards predators: do environmental beliefs play a part? *Research in Human Ecology*, **5**, 1–9.

Kaltenborn, B. P., Bjerke, T. and Vittersø, J. (1999). Attitudes towards large carnivores among sheep farmers, wildlife managers, and research biologists in Norway. *Human Dimensions of Wildlife*, **4**, 57–63.

Kangwana, K. (1993). *Elephants and Maasai: Conflict and Conservation in Amboseli, Kenya*. Cambridge, UK: Cambridge University press.

(1995). Human–elephant conflict: the challenge ahead. *Pachyderm*, **19**, 11–14.

Kaplanov, L. G. (1948). Tigers in Sikhote-Alin. In *Tiger, red deer, and moose, Materialy k poznaniyu fauny i flory SSSR, Obschestva Ispytateley Prirody, Novaya seria, Otdel zool.*, **14**, 18–49. (In Russian)

Karami, M. (1992). Nature reserves in the Islamic Republic of Iran. *Species*, **19**, 11.

Karanth, K. U. (2001). *The Way of the Tiger: Natural History and Conservation of the Endangered Big Cat*. Stillwater, MN: Voyageur Press.

(2002). Nagarahole: limits and opportunities in wildlife conservation. In *Making Parks Work: Identifying Key Factors to Implementing Parks in the Tropics*, ed. J. Terborgh, C. van Schaik, L. C. Davenport and M. Rao. Covelo, CA: Island Press, pp. 189–202.

(2003). Tiger ecology and conservation in the Indian subcontinent. *Journal of the Bombay Natural History Society*, **100**, 169–189.

Karanth, K. U. and Madhusudan, M. D. (2002). Mitigating human–wildlife conflicts in Southern Asia. In *Making Parks Work: Identifying Key Factors to Implementing Parks in the Tropics*, ed. J. Terborgh, C. van Schaik, L. C. Davenport and M. Rao. Covelo, CA: Island Press, pp. 250–264.

Karanth, K. U. and Nichols, J. D. (1998). Estimating tiger densities in India from camera trap data using photographic captures and recaptures. *Ecology*, **79**, 2852–2862.

Karanth, K. U. and Stith, B. M. (1999). Prey depletion as a critical determinant of tiger population viability. In *Riding the Tiger: Tiger Conservation in Human-Dominated Landscapes*, ed. J. Seidensticker, S. Christie and P. Jackson. Cambridge, UK: Cambridge University Press, pp. 100–113.

Karanth, K. U., Sunquist, M. E. and Chinnappa, K. M. (1999). Long-term monitoring of tigers: lessons from Nagarahole. In *Riding the Tiger: Tiger Conservation in Human-Dominated Landscapes*, ed. J. Seidensticker, S. Christie and P. Jackson. Cambridge, UK: Cambridge University Press, pp. 114–122.

Karanth, K. U., Nichols, J. D., Kumar, N. S., Link, W. A. and Hines, J. E. (2004). Tigers and their prey: predicting carnivore densities from prey abundance. *Proceedings of the National Academy of Sciences of the USA*, **101**, 4854–4858.

Kayanja, F. and Douglas-Hamilton, I. (1984). The impact of the unexpected on the Uganda national parks. In *National Parks, Conservation and Development: The Role of Protected Areas in Sustaining Society*, ed. J. A. McNeely and K. R. Miller. Washington, DC: Smithsonian Institution Press, pp. 87–92.

Keiter, R. B. (ed.) (1998). *Reclaiming the Native Home of Hope: Community, Ecology and the American West*. Salt Lake City, UT: University of Utah Press.

Keith, L. B. (1983). Population dynamics of wolves. In *Wolves in Canada and Alaska: Their Status, Biology, and Management*, ed. L. N. Carbyn. Report No. 45. Edmonton, Alberta: Canadian Wildlife Service, pp. 66–77.

Kellert, S. R. (1991). Public views of wolf restoration in Michigan. *Transactions of the North American Wildlife and Natural Resources Conference*, **51**, 193–200.

Kellert, S. R., Black, M., Rush, C. R. and Bath, A. J. (1996). Human culture and large carnivore conservation in North America. *Conservation Biology*, **10**, 977–990.

Kenny, J. S., Smith, J. L. D., Starfield, A. M. and McDougal, C. (1995). The long-term effects of tiger poaching on population viability. *Conservation Biology*, **9**, 1113–1127.

Kenward, R. (1977). Predation on released pheasants by goshawks in central Sweden. *Swedish Game Research*, **10**, 79–112.

(1981). Goshawk winter ecology in Swedish pheasant habitats. *Journal of Wildlife Management*, **45**, 397–408.

Kenya Wildlife Service (1991). *A Policy Framework and Development Programme 1991–1996*. Nairobi: Kenya Wildlife Service.

(1997). *Kenya Wildlife Service Wildlife Policy*. Nairobi: Kenya Wildlife Service.

Kerbis-Peterhams, P. (1999). The science of man-eating among lions (*Panthera leo*) with a reconstruction of the natural history of the 'Man-eaters of Tsavo'. *Journal of the East African Wildlife Society*, **90**, 1–40.

Kerley, L. L., Goodrich, J. M., Miquelle, D. G. *et al.* (2002). Effects of roads and human disturbance on Amur tigers. *Conservation Biology*, **16**, 1–12.

(2003). Reproductive parameters of wild female Amur (Siberian) tigers (*Panthera tigris altaica*). *Journal of Mammalogy*, **84**, 288–298.

Khaemba, W. M., Stein, A., Rasch, D., De Leeuw, J. and Georgiadis, N. (2001). Empirically simulated study to compare and validate sampling methods used in aerial surveys of wildlife populations. *African Journal of Ecology*, **39**, 374–382.

Khan, J. A. (1995). Conservation and management of Gir lion sanctuary and national park, Gujarat, India. *Biological Conservation*, **73**, 183–188.

Kharel, F. R. (1997). Agricultural crop and livestock depredation by wildlife in Langtang National Park, Nepal. *Mountain Research and Development*, **17**, 127–134.

Khramtsov, V. S. (1995). Behavior of tigers in encounters with man. *Ecologia*, **3**, 252–254. (In Russian)

Kiiru, W. (1995). Human–elephant interactions around Shimba Hills N. Reserve, Kenya. M.S. thesis, University of Zimbabwe, Harare.

King, D. A. and Stewart, W. P. (1996). Ecotourism and commodification: protecting people and places. *Biodiversity and Conservation*, **5**, 293–305.

King, F. A. and Lee, P. C. (1987). A brief survey of human attitudes to a pest species of primate: *Cercopithecus aethiops*. *Primate Conservation*, **8**, 82–84.

Kleiman, D. G., Reading, R. P., Miller, B. J. *et al.* (2000). The importance of improving evaluation in conservation. *Conservation Biology*, **14**, 1–11.

Klenzendorf, S. A. (1997). Management of brown bears (*Ursus arctos*) in Europe. Thesis, Virginia Polytechnic Institute and State University, Blacksburg, VA.

Knight, J. (ed). (2001). *Natural Enemies: People–Wildlife Conflicts in Anthropological Perspective.* London: Routledge.

(2003). *Waiting for Wolves in Japan.* Oxford, UK: Oxford University Press.

Knowles, C. J. (1988). An evaluation of shooting and habitat alteration for control of black-tailed prairie dogs. In *Proceedings of the 8 Great Plains Wildlife Damage Control Workshop,* Rapid City, SD, 28–30 April 1987 pp. 53–56.

Knowlton, F. F., Gese, E. M. and Jaeger, M. M. (1999). Coyote depredation control: an interface between biology and management. *Journal of Range Management,* **52**, 398–412. (http://uvalde.tamu.edu/jrm/jrmhome.htm)

Koehler, A. E., Marsh, R. E. and Salmon, T. P. (1990). Frightening methods and devices/stimuli to prevent mammal damage: a review. *Vertebrate Pest Conference Proceedings,* **14**, 168–173.

Kojola, I. and Kuittinen, J. (2002). Wolf attacks on dogs in Finland. *Wildlife Society Bulletin,* **30**, 498–501.

Korpimaki, E. and Norrdahl, K. (1998). Experimental reduction of predators reverses the crash phase of small rodent cycles. *Ecology,* **79**, 2448–2455.

Kothari, A. (1996). Is joint management of protected areas possible and desirable? In *People and Protected Areas: Towards Participatory Conservation in India,* ed. A. Kothari, N. Singh and S. Suri. New Dehli: Sage Publications, pp. 17–49.

Kothari, A., Suri, S. and Singh, N. (1995). People and protected areas: rethinking conservation in India. *Ecologist,* **25**, 188–194.

Kotliar, N. B., Miller, B. and Reading, R. P. (in press). Black-tailed prairie dogs as keystone species. In *Prairie Dog Conservation,* ed. J. L. Hoogland. Washington, DC: Island Press.

Krange, O. and Skogen, K. (2001). Naturen i Stor-Elvdal, ulven og det sosiale landskapet: en kortrapport fra prosjektet Konfliktlinjer i utmarka. *Norwegian Social Research Temahefte,* **1**, 1–31.

Krebs, C. J., Boutin, S., Boonstra, R. *et al.* (1995). Impact of food and predation on the snowshoe hare cycle. *Science,* **269**, 1112–1115.

Krebs, J. R., Anderson, R., Clutton-Brock, T. *et al.* (1997). *Bovine Tuberculosis in Cattle and Badgers: An Independent Scientific Review.* London: HMSO.

Krishke, H., Lyamuya, V. and Ndunguru, I. F. (1996). The development of community-based conservation around the Selous Game Reserve. In *Community-Based Conservation in Tanzania,* ed. N. Leader-Williams, J. A. Kayera and G. L. Overton. Gland, Switzerland: IUCN, pp. 75–83.

Krogstad, S., Christiansen, F., Smith, M. *et al.* (1999). *Protective Measures against Depredation on Sheep: Sheep-Herding and Use of Livestock Guarding Dogs in Lierne,* Annual Report Phase II, 1998. Trondheim, Norway: Norwegian Institute for Nature Research. (In Norwegian with English summary)

Kruuk, H. (1972). *The Spotted Hyena: A Study of Predation and Social Behavior.* Chicago, IL: University of Chicago Press.

(1980). *The Effects of Large Carnivores on Livestock and Animal Husbandry in Marsabit District, Kenya,* United Nations Environmental Program, Man and Biosphere, Integrated Project in Arid Lands Technical Report No. E-4. Nairobi: UNEP/MAB.

(2002). *Hunter and Hunted: Relationships between Carnivores and People.* Cambridge, UK: Cambridge University Press.

Krystufek, B. and Griffiths, H. I. (2003). Anatomy of a human: bear conflict: case study from Slovenia 1999–2000. In *Living with Bears: A Large European Carnivore in a Shrinking World*, ed. B. Krystufek, B. Flasman and H. I. Griffiths. Ljubljana, Slovenia: Ecological Forum of the Liberal Democracy of Slovenia, pp. 127–153.

Kucherenko, S. P. (1970). The Amur tiger (present distribution and numbers). *Okhota i okhotnichie khozyaistvo*, 2, 20–23. (In Russian)

(1993). The price of tiger conservation. *Okhota i okhotnichie khozyaistvo*, 2, 16–19. (In Russian)

(2001). Amur tigers at the turn of the century. *Okhota i okhotnichie khozyaistvo*, 4, 20–24. (In Russian)

Kumar, A. and Wright, B. (1999). Combating tiger poaching and illegal wildlife trade in India. In *Riding the Tiger: Tiger Conservation in Human-Dominated Landscapes*, ed. J. Seidensticker, S. Christie and P. Jackson. Cambridge, UK: Cambridge University Press, pp. 243–251.

Kumar, S. (2001). Compensation policies complicate wolf depredation conflicts. *International Wolf*, **11**, 8–9.

Kunkel, K. E. (1997). Predation by wolves and other large carnivores in northwestern Montana and southeastern British Columbia. Ph.D. thesis, University of Montana, Missoula, MT.

Kvam, T. (1996). *Bestandsestimat for gaupe 1995–96*. Trondheim, Norway: Norwegian Insitute for Nature Research.

Lahiri-Choudhury, D. (1993). Problems with wild elephant translocation. *Oryx*, **27**, 53–55.

Lahm, S. A. (1994). A nation wide survey of crop-raiding by elephants and other species in Gabon. *Pachyderm*, **21**, 69–77.

Laikipia Wildlife Forum (2004). http://www.//laikipia.org/

Lamb, B. L., Reading, R. P. and Andelt, W. F. (in press). Public attitudes and perceptions toward black-tailed prairie dogs. In *Prairie Dog Conservation*, ed. J. L. Hoogland. Washington, DC: Island Press.

Landa, A. and Tømmerås, B. A. (1996). Do volatile repellents reduce wolverine *Gulo gulo* predation on sheep? *Wildlife Biology*, **2**, 219–226.

(1997). A test of aversive agents on wolverines. *Journal of Wildlife Management*, **61**, 510–516.

Landa, A., Strand, O., Swenson, J. E., and Skogland, T. (1997). Wolverines and their prey in southern Norway. *Canadian Journal of Zoology*, **75**, 1292–1299.

Landa, A., Franzén, R., Bø, T. *et al.* (1998a). Active wolverine *Gulo gulo* dens as a minimum population estimator in Scandinavia. *Wildlife Biology*, **4**, 159–168.

Landa, A., Krogstad, S, Tømmerås, B. Å., and Tufto, J. (1998b). Do volatile repellents reduce wolverine *Gulo gulo* predation on sheep? Results of a large-scale experiment. *Wildlife Biology*, **4**, 111–118.

Landa, A., Strand, O., Linnell, J. D. C. and Skogland, T. (1998c). Home range sizes and altitude selection for arctic foxes and wolverines in an alpine environment. *Canadian Journal of Zoology*, **76**, 448–457.

Landa, A., Gudvangen, K., Swenson, J. and Roskaft, E. (1999). Factors associated with wolverine predation on domestic sheep. *Journal of Applied Ecology*, **36**, 963–973.

Landa, A., Tufto, J., Andersen, R. and Persson, J. (2000a). *Reanalyse av aktive ynglehi hos jerv som bestandsestimator basert på nye data om alder for første yngling*. Unpublished report. Trondheim, Norway: Norwegian Institute for Nature Research. (In Norwegian)

Landa, A., Linnell, J. D. C., Lindén, M. *et al.* (2000b). Conservation of Scandinavian wolverines in ecological and political landscapes. In *Mustelids in a Modern World: Management and Conservation Aspects of Small Carnivore–Human Interactions*, ed. H. I. Griffiths. Leiden, Netherlands: Backhuys, pp. 1–20.

Landa, A., Lindén, M. and Kojola, I. (2000c). *Action Plan for the Conservation of Wolverines in Europe* (Gulo gulo). Nature and Environment Report No. 115. Strasbourg, France: Council of Europe.

Landry, J.-M. (1998). *The Use of Guard Dogs in the Swiss Alps: A First Analysis*, KORA Report No. 2. Muri, Switzerland: Coordinated Research Projects for the Protection and Management of Carnivores in Switzerland. Accessible at http://www.kora.unibe.ch

(2000). Testing livestock guard donkeys in the Swiss Alps. *Carnivore Damage Prevention News*, **1**, 6–7. (http://www.kora.unibe.ch)

(2001). The guard dog: protecting livestock and large carnivores. In *Wildlife, Land, and People: Priorities for the Twenty-First Century*, ed. R. Field, R. J. Warren, H. Okarma and P. R. Sievert. Bethesda, MD: The Wildlife Society, pp. 209–121.

Large Carnivore Initiative for Europe (2002). Core group position statement on the use of hunting, and lethal control, as means of managing large carnivore populations. Accessible at http://www.large-carnivores-lcie.org

Larsson, K. and Forslund, P. (1992). Population dynamics of the barnacle goose *Branta leucopsis* in the Baltic area: density-dependent effects on reproduction. *Journal of Animal Ecology*, **63**, 954–962.

Lasswell, H. D. and McDougal, M. S. (1992). *Jurisprudence for a Free Society*. The Hague: Kluwer Law International.

Laws, R., Parker, I. and Johnstone, R. (1975). *Elephants and Their Habitats: The Ecology of Elephants in North Bunyoro, Uganda*. Oxford, UK: Clarendon Press.

Le Fevre, A. M., Johnston, W. T., Bourne, F. J. *et al.* (2003). Changes in badger setts over the first three years of the randomized badger culling trial. Poster at Society for Veterinary Epidemiology and Preventive Medicine Conference. Accessible at http://www.svepm.org.uk/Posters2003/poster_files/LeFevre.pdf

Leader-Williams, N. (2000). The effects of a century of policy and legal change upon wildlife conservation and utilization in Tanzania. In *Conservation of Wildlife by Sustainable Use*, ed. H. H. T. Prins, J. G. Grootenhuis and T. T. Dolan. Boston, MA: Kluwer, pp. 219–245.

Leader-Williams, N. and Dublin, H. T. (2000). Charismatic megafauna as 'flagship species'. In *Has the Panda Had its Day? Priorities for the Conservation of Mammalian Diversity*, ed. A. Entwistle and N. Dunstone. Cambridge, UK: Cambridge University Press, pp. 53–81.

Leader-Williams, N. and Tibanyenda, R. K. (1996). *The Live Bird Trade in Tanzania*. Gland, Switzerland: IUCN.

Leader-Williams, N., Albon, S. D. and Berry, P. S. M. (1990a). Illegal exploitation of black rhinoceros and elephant populations: patterns of decline, law enforcement and patrol effort in Luangwa Valley, Zambia. *Journal of Applied Ecology*, **27**, 1055–1087.

Leader-Williams, N., Harrison, J. and Green, M. J. B. (1990b). Designing protected areas to conserve natural resources. *Science Progress*, **74**, 189–204.

Leader-Williams, N., Kayera, J. A. and Overton, G. L. (1996a). *Tourist Hunting in Tanzania*. Gland, Switzerland: IUCN.

(1996b). *Community-Based Conservation in Tanzania*. Gland, Switzerland: IUCN.

Leader-Williams, N., Smith, R. J and Walpole, M. J. (2001). Elephant hunting and conservation. *Science*, **293**, 2203.

Leader-Williams, N., Oldfield, T. E. E., Smith, R. J. and Walpole, M. J. (2002). Science, conservation and foxhunting. *Nature*, **419**, 878.

Lee, R. J. (2000). Impact of subsistence hunting in North Sulawesi, Indonesia, and conservation options. In *Hunting for Sustainability in Tropical Forests*, ed. J. G. Robinson and E. L. Bennett. New York: Columbia University Press, pp. 455–472.

LeFranc, M. N., Moss, M. B., Patnode, K. A. and Sugg, W. C. (1987). *The Inter-agency Grizzly Bear Committee: Grizzly Bear Compendium*. Missoula, MT: Office of the Grizzly Bear Recovery Coordinator.

Lehner, P. N. (1987). Repellents and conditioned avoidance. In *Protecting Livestock from Coyotes*, ed. J. S. Green. Dubois, ID: US Department of Agriculture, pp. 56–61.

Leopold, A. S. and Wolfe, T. O. (1970). Food habits of nesting wedge-tailed eagles in Southeastern Australia. *CSIRO Wildlife Research*, **15**, 1–17.

Lévi-Strauss, C. (1964). *The Raw and the Cooked: Introduction to a Science of Mythology*. New York: Harper.

Levin, M. (2002). How to prevent damage from large predators with electric fences. *Carnivore Damage Prevention News*, **5**, 5–8. (http://www.kora.unibe.ch)

Lewis, D. M. (1998). Procedures and expectations of the Conservation Bullet Certification. Unpublished report.

(1999). *Comparative Study of Factors Influencing ADMADE Success*. USAID report. Lusaka: USAID.

(2001). *Wildlife Enterprise and Management Approaches in Sichifulo GMA: A Case Study Analysis*. Lusaka: CONASA.

Lewis, D. M. and Alpert, P. (1997). Trophy hunting and wildlife conservation in Zambia. *Conservation Biology*, **11**, 59–68.

Lewis, D. M. and Phiri, A. (1998). Wildlife snaring: an indicator of community response to a community-based conservation project. *Oryx*, **32**, 111–121.

Lewis, D. M. and Tembo, N. (1999). *Improving Food Security to Reduce Illegal Hunting of Wildlife*. Lusaka: ADMADE.

(2000). Non-conventional approaches to wildlife management in an African landscape. In *Pretoria Game Producers Symposium*, pp. 1–14.

Liberg, O. and Glöersen, G. (2000). *Rapport fran lo- och varginventeringen 2000*. Spånga, Sweden: Svenska Jägareforbundet. (In Swedish)

Licht, D. S. (1997). *Ecology and Economics of the Great Plains*. Lincoln, NE: University of Nebraska Press.

Linden, H. and Wikman, M. (1983). Goshawk predation on Tetraonids: availability of prey and diet of the predator in the breeding season. *Journal of Animal Ecology*, **52**, 953–968.

Lindsey, P. A. (2003). Conserving wild dogs (*Lycaon pictus*) outside state protected areas in South Africa: ecological, sociological and economic determinants of success. Ph.D. thesis, University of Pretoria, South Africa.

Linhart, S. B., Roberts, J. D., Shumake, S. A. and Johnson, R. (1976). Avoidance of prey by captive coyotes punished with electric shock. *Vertebrate Pest Conference Proceedings*, **7**, 302–306.

Linhart, S. B., Roberts, J. D. and Dasch, G. J. (1982). Electric fencing reduces coyote predation on pastured sheep. *Journal of Range Management*, **53**, 276–281. (http://uvalde.tamu.edu/jrm/jrmhome.htm)

Linhart, S. B., Sterner, R. T., Dasch, G. J. and Theade, J. W. (1984). Efficacy of light and sound stimuli for reducing coyote predation upon pastured sheep. *Protection Ecology*, **6**, 75–84.

Linhart, S. B., Dasch, G. J., Johnson, R. R., Roberts, J. D. and Packham, C. J. (1992). Electronic frightening devices for reducing coyote predation on domestic sheep: efficacy under range conditions and operational use. *Vertebrate Pest Conference Proceedings*, **15**, 386–392.

Linnell, J. D. C. (2000). Taste aversive conditioning: a comment. *Carnivore Damage Prevention News*, **2**, 4. (http://www.kora.unibe.ch)

Linnell, J. D. C. and Bjerke, T. (2002). Frykten for ulven: en tverrfaglig utredning. *Norwegian Institute for Nature Research Oppdragsmelding*, **722**, 1–109.

Linnell, J. D. C. and Brøseth, H. (2003). Compensation for large carnivore depradation of domestic sheep 1994–2001. *Carnivore Damage Prevention News*, **6**, 11–13.

Linnell, J. D. C., Smith, M. E., Odden, J., Kaczensky, P. and Swenson, J. E. (1996). Strategies for the reduction of carnivore–livestock conflicts: a review. *Norwegian Institute for Nature Research Oppdragsmelding*, **443**, 1–118. (http://nidaros.nina.no/Publikasjoner/Rapporter/opm%20550.pdf)

Linnell, J. D. C., Aanes, R., Swenson, J. E., Odden, J. and Smith, M. E. (1997). Translocation of carnivores as a method for managing problem animals: a review. *Biodiversity and Conservation*, **6**, 1245–1257.

Linnell, J. D. C., Odden, J., Smith, M. E., Aanes, R. and Swenson, J. E. (1999). Large carnivores that kill livestock: do problem individuals really exist? *Wildlife Society Bulletin*, **27**, 698–705.

Linnell, J. D. C., Andersen, R., Kvam, T. *et al.* (2001a). Home range size and choice of management strategy for lynx in Scandinavia. *Environmental Management*, **27**, 869–879.

Linnell, J. D. C., Swenson, J. E. and Andersen, R. (2001b) Predators and people: conservation of large carnivores is possible at high human densities if management policy is favorable. *Animal Conservation*, **4**, 345–349.

Linnell, J. D. C., Løe, J., Okarma, H. *et al.* (2002). The fear of wolves: a review of wolf attacks on humans. *Norwegian Institute for Nature Research Oppdragsmelding*, **731**, 1–65.

Litoroh, M. (1997). *Shimba Hills Elephant Count*. Nairobi: Kenya Wildlife Service.

Liverman, D. (1990). Vulnerability to global environmental change. In *Understanding Global Environmental Change*, ed. R. E. Kasperson. Worcester, MA: The Earth Transformed Program, pp. 27–44.

Logan, K. A. and Sweanor, L. L. (2001) *Desert Puma: Evolutionary Ecology and Conservation of an Enduring Carnivore*. Washington, DC: Island Press.

London, J. (1913). *White Fang*. New York: MacMillan.

Lonely Planet (2004). *Lonely Planet Online*. Accessible at http://www.lonelyplanet.com

Lopez, B. H. (1978). *Of Wolves and Men*. New York: Scribner.

Lorenz, J. R. (1985). *Introducing Livestock-Guarding Dogs*. Extension Circular No. 1224. Corvallis, OR: Oregon State University Extension Service.

Lorenz, J. R. and Coppinger, L. (1986). *Raising and Training a Livestock-Guarding Dog*, Extension Circular No. 1238. Corvallis, OR: Oregon State University Extension Service.

Lorenz, J. R., Coppinger, R. and Sutherland, M. R. (1986). Causes and economic effects of mortality in livestock guarding dogs. *Journal of Range Management*, **39**, 293–295. (http://uvalde.tamu.edu/jrm/jrmhome.htm)

Luce, B. (2001). *An Umbrella, Multi-State Approach for the Conservation and Management of the Black-Tailed Prairie Dog,* Cynomys ludovicianus, *in the United States*. Cheyenne, WY: Black-Tailed Prairie Dog Conservation Team.

Luce, R. J. (2003). *A Multi-State Conservation Plan for the Black-Tailed Prairie Dog,* Cynomys ludovicianus, *in the United States: An Addendum to the Black-Tailed Prairie Dog Conservation Assessment and Strategy*. Sierra Vista, AZ: Prairie Dog Conservation Team.

Lukarevsky, V. (2002). Saving the central Asian leopard in Turkmenistan. *Russian Conservation News*, **28**, 25–26. (http://www.kora.unibe.ch)

(2003). Saving the central Asian leopard in Turkmenistan. *Carnivore Damage Prevention News*, **6**, 13–15.

Lynch, O. J. and Alcorn, J. B. (1994). Tenurial rights and community-based conservation. In *Natural Connections: Perspectives in Community-Based Conservation*, ed. D. Western and R. M. Wright. Washington, DC: Island Press, pp. 373–392.

Macdonald, D. (1980). *Rabies and Wildlife*. Oxford, UK: Oxford University Press.

Macdonald, D. and Sillero-Zubiri, C. (2004). Dramatis personae. In *Biology and Conservation of Wild Canids*, ed. D. Macdonald and C. Sillero-Zubiri. Oxford, UK: Oxford University Press, pp. 3–36.

Mace, R. D. and Waller, J. S. (1996). Grizzly bear distribution and human conflicts in Jewel Basin Hiking Area, Swan Mountains, Montana. *Wildlife Society Bulletin*, **24**, 461–467.

(1998). Demography and population trend of grizzly bears in the Swan Mountains, Montana. *Conservation Biology*, **12**, 1005–1016.

MacKenzie, J. M. (1988). *The Empire of Nature: Hunting, Conservation and British Imperialism*. Manchester, UK: Manchester University Press.

Mackinnon, K., Mishra, H. and Mott, J. (1999). Reconciling the needs of conservation and local communities: Global Environmental Facility support for tiger conservation in India. In *Riding the Tiger: Tiger Conservation in Human-Dominated Landscapes*, ed. J. Seidensticker, S. Christie and P. Jackson. Cambridge, UK: Cambridge University Press, pp. 307–315.

Madhusudan, M. D. (2003). Living amidst large wildlife: livestock and crop depredation by large mammals in the interior villages of Bhadra Tiger Reserve, southern India. *Environmental Management*, **31**, 460–475.
Madhusudan, M. D. and Karanth, K. U. (2002). Local hunting and the conservation of large mammals in India. *Ambio*, **3**, 49–54.
Madhusudan, M. D. and Mishra, C. (2003). Why big, fierce animals are threatened: conserving large mammals in densely populated landscapes. In *Battles over Nature: Science and the Politics of Conservation*, ed. V. Saberwal and M. Rangarajan. New Delhi: Permanent Black, pp. 31–55.
Madzou, Y. C. (1999). *Situation conflictuelle des éléphants à Bomassa*. New York: Wildlife Conservation Society.
Makin, J. (1968). The soils in the country around Shimba Hills Settlement, Kikoneni and Jombo Mountain. Unpublished report. Nairobi: Kenya Soil Survey Unit, Ministry of Agriculture.
Makombe, K. (ed.) (1994). *Sharing the Land: Wildlife, People and Development in Africa*. IUCN/ROSA Environmental Issues Series No. 1. Harare: IUCN/ROSA.
Makombo, J. (2003). Responding to the challenge: how protected areas can best provide benefits beyond boundaries. A case study of Bwindi Impenetrable National Park in western Uganda. Unpublished report. Kampala: Uganda Wildlife Authority.
Manfredo, M. and Dayer, A. (2004). Concepts for exploring the social aspects of human–wildlife conflict in a global context. *Human Dimensions of Wildlife Management*.
Manfredo, M. J., Zinn, H. C., Sikorowski, L. and Jones, J. (1998) Public acceptance of mountain lion management: a case study of Denver, Colorado, and nearby foothill areas. *Wildlife Society Bulletin*, **26**, 964–970.
Manosa, S. (1994). Goshawk diet in a Mediterranian area of northeastern Spain. *Journal of Raptor Research*, **28**, 84–92.
(2002). Conflict between gamebird hunting and raptors in Europe. Accessible at http://www.uclm.es/irec/reghab/informes_3.htm
Maples, W. R., Maples, W. K., Greenhood, W. F. and Walek, M. L. (1976). Adaptations of crop-raiding baboons in Kenya. *American Journal of Physical Anthropology*, **45**, 309–316.
Marchini, S. (2002). Local public opinion about environment and socio-Economic development in the Pantanal. Unpublished report. Wildlife Conservation Society.
Marcstrom, V., Kenward, R. and Engren, E. (1998). The impact of predation on boreal tetraonids during vole cycles. *Journal of Animal Ecology*, **57**, 859–872.
Marker, L. L. (2000a). Donkeys protecting livestock in Namibia. *Carnivore Damage Prevention News*, **2**, 7–8. (http://www.kora.unibe.ch)
(2000b). Livestock guarding dogs. Unpublished panel report.
Marker, L. L., Dickman, A. J., Mills, M. G. L. and MacDonald, D. W. (2003a). Aspects of the management of cheetahs, *Acinonyx jubatus jubatus*, trapped on Namibian farmland. *Biological Conservation*, **114**, 000–000.
(2003b). Demography of the Namibian cheetah, *Acinonyx jubatus jubatus*. *Biological Conservation*, **114**, 000–000.

Marker, L. L., Mills, M. G. L. and MacDonald, D. W. (2003c). Factors influencing perceptions of conflict and tolerance towards cheetahs on Namibian farmlands. *Conservation Biology*, **17**, 1290–1298.

Marker-Kraus L. (1994). The Namibian free-ranging cheetah. *Environmental Conservation*, **21**, 369–370.

Marker-Kraus L., Kraus, D., Barnett, D. and Hurlbut, S. (1996). *Cheetah Survival on Namibian Farmlands*. Windhoek: Cheetah Conservation Fund.

Markham, D. (1995). *Guard Llamas*. Kalispell, MT: International Llama Association. Accessible at http://www.internationalllama.org

Marquiss, M., Madders, M., Irvine, J. and Carss, D. (2002). *The Impact of White-Tailed Eagles on Sheep Farming on Mull*, SEERAD Report No. ITE/004/99. Edinburgh, UK: Scottish Executive Environment and Rural Affairs Department.

Marquiss, M., Madders, M. and Carss, D. (2003). White-tailed eagles and lambs. In: *Birds of Prey in a Changing Landscape*, ed. D. B. A. Thompson, S. M. Redpath, A. Fielding, M. Marquiss and C. Galbraith. Edinburgh, UK: HMSO, pp. 471–480.

Marsh, R. E. (1984). Ground squirrels, prairie dogs, and marmots as pests on rangeland. In *Proceedings of the Conference for Organization and Practice of Vertebrate Pest Control*, Fernherst, UK, 30 August–3 September 1982, pp. 195–208.

Maryland (1997). *A New Conservation Strategy for the Namibian Cheetah* (Acinonyx jubatus), Study by the 1997 Problem Solving Team, Graduate Program in Sustainable Development and Conservation Biology. College Park, MD: University of Maryland.

Mascarenhas, A. (1971). Agricultural vermin in Tanzania. In *Studies in East African Geography and Development*, ed. S. H. Ominde. London: Heinemann, pp. 259–267.

Mason, J. (1989). Avoidance of methiocarb-poisoned apples by red-winged blackbirds. *Journal of Wildlife Management*, **53**, 836–840.

Mason, J. R., Shivik, J. A. and Fall, M. W. (2001). Chemical repellents and other aversive strategies in predation management. *Endangered Species Update*, **18**, 175–181.

Matchett, M. R. and O'Gara, B. W. (1987). Methods of controlling golden eagle predation on domestic sheep in Southwestern Montana. *Journal of Raptor Research*, **21**, 85–94.

Matsuzawa, T., Hasegawa, Y., Gotoh, J. and Wada, K. (1983). One-trial long-lasting food-aversion learning in wild Japanese monkeys (*Macaca fuscata*). *Behavioral and Neural Biology*, **39**, 155–159.

Matyushkin, E. N., Pikunov, D. G., Dunishenko, Y. M. (1996). *Numbers, Distribution, and Habitat Status of the Amur Tiger in the Russian Far East*. Final report to the USAID Russian Far East Environmental Policy and Technology Project.

(1999). Distribution and numbers of Amur tigers in the Russian Far East in the mid-1990s. In *Rare Mammal Species of Russia and Neighboring Territories*, ed. A. A. Aristova. Moscow: Russian Academy of Sciences Therological Society, pp. 242–271. (In Russian)

Maveneke, T. (1996). The elephant's importance to CBCD: the CAMPFIRE example. In *Rural Development and Conservation in Africa*. African Resources Trust, pp. 31–33.

McDougal, C. (1987). The man-eating tiger in geographical and historical perspective. In *Tigers of the World: The Biology, Biopolitics, Management and Conservation of an Endangered Species*, ed. R. L. Tilson and U. S. Seal. Park Ridge, NJ: Noyes Publications, pp. 435–448.

(1999). Tiger attacks on people in Nepal. *Cat News*, **30**, 9–10.

McDougal, C., Cotton, M., Barlow, A., Kumal, S., and Tamang, D. B. (2001). Tigers claim more human victims in Nepal. *Cat News*, **35**, 2–3.

McIntyre, R. (ed.) (1995). *War against the Wolf: America's Campaign to Exterminate the Wolf.* Stillwater, MN: Voyaguer Press.

McLellan, B. N. (1990). Relationships between human industrial activity and grizzly bears. *International Conference on Bear Research and Management*, **8**, 57–64.

McLeod, W. T. (ed.) (1982), *The New Collins Concise English dictionary*. Glasgow, UK: Collins.

McNay, M. E. (2002). Wolf–human interactions in Alaska and Canada: a review of the case history. *Wildlife Society Bulletin*, **30**, 831–843.

McNeely, J. (1988). *Economics and Biological Diversity: Developing and Using Economic Incentives to Conserve Biological Resources.* Gland, Switzerland: IUCN.

(1989). Protected areas and human ecology: how national parks can contribute to sustaining societies. In *Conservation for the Twenty-First Century*, ed. D. Western and M. Pearl. Oxford, UK: Oxford University Press, pp. 150–157.

(ed.) (1993). *Parks for Life: Report of the 4th World Congress on National Parks and Protected Areas.* Gland, Switzerland: IUCN.

McNeely, J. A. and Miller, K. R. (eds.) (1984). *National Parks, Conservation and Development: The Role of Protected Areas in Sustaining Society.* Washington, DC: Smithsonian Institution Press.

McNeilage, A., Plumptre, A. J., Brock-Doyle, A. and Vedder, A. (2001). Bwindi Impenetrable National Park, Uganda: gorilla census 1997. *Oryx*, **3**, 39–47.

Meadows, L. E. and Knowlton, F. F. (2000). Efficacy of guard llamas to reduce canine predation on domestic sheep. *Wildlife Society Bulletin*, **28**, 614–622.

Mech, L. D. (1970). *The Wolf: Ecology and Behavior of an Endangered Species.* Garden City, NY: Natural History Press.

(1995). The challenge and opportunity of recovering wolf populations. *Conservation Biology*, **9**, 270–278.

Mech, L. D. and Nelson, M. E. (2000). Do wolves affect white-tailed buck harvest in northeastern Minnesota? *Journal of Wildlife Management*, **64**, 129–136.

Mech, L. D., Fritts, S. H. and Paul, W. J. (1988). Relationship between winter severity and wolf depredations on domestic animals in Minnesota. *Wildlife Society Bulletin*, **16**, 269–272.

Mech, L. D., Harper, E. K., Meier, T. J., and Paul, W. J. (2000). Assessing factors that may predispose Minnesota farms to wolf depredations on cattle. *Wildlife Society Bulletin*, **28**, 623–629.

Medellin, R. A., Equihua, C., Chetiewicz, C. *et al.* (eds.) (2002). *El Jaguar en el nuevo milenio.* Mexico: Fondo de Cultura Economica.

Mehta, J. N. and Kellert, S. R. (1998). Local attitudes towards community-based conservation policy and programmes in Nepal: a case study in the Makalu-Barun Conservation Area. *Environmental Conservation*, **25**, 320–33.

Meier, K. J. (1993). *Politics and the Bureaucracy: Policymaking in the Fourth Branch of Government*. Belmont, CA: Wadsworth.

Menon, V., Sukumar, R. and Kumar, A. (1998). *A God in Distress: Threats of Poaching and the Ivory Trade to the Asian Elephant in India*. New Delhi: Wildlife Protection Society of India.

Meriggi, A. and Lovari, S. (1996). A review of wolf predation in southern Europe: does the wolf prefer wild prey to livestock? *Journal of Applied Ecology*, **33**, 1561–1571.

Merriam, C. H. (1902). The prairie dog of the Great Plains. In *Yearbook of the United States Department of Agriculture 1901*. Washington, DC: US Government Printing Office, pp. 257–270.

Mertens, A., Promberger, C. and Gheorge, P. (2002). Testing and implementing the use of electric fences for night corrals in Romania. *Carnivore Damage Prevention News*, **5**, 2–5. (http://www.kora.unibe.ch)

Messier, F. (1985). Solitary living and extraterritorial movements of wolves in relation to social ststus and prey abundance. *Canadian Journal of Zoology*, **63**, 239–245.

Messmer, T. (2000). The emergence of human–wildlife conflict management: turning challenges into opportunities. *International Biodeterioration and Biodegradation*, **45**, 97–102.

Miller, B. and Reading, R. P. (2002). The black-tailed prairie dog: threats to survival and a plan for conservation. *Wild Earth*, **12**, 46–55.

Miller, B., Wemmer, C., Biggins, D. and Reading, R. (1990). A proposal to conserve black-footed ferrets and the prairie dog ecosystem. *Environmental Management*, **14**, 763–769.

Miller, B., Ceballos, G. and Reading, R. P. (1994). The prairie dog and biotic diversity. *Conservation Biology*, **8**, 677–681.

Miller, B., Reading, R. P. and Forrest, S. (1996). *Prairie Night: Black-Footed Ferrets and the Recovery of Endangered Species*. Washington, DC: Smithsonian Institution Press.

Miller, D. J. and Jackson, R. (1994). Livestock and snow leopards, making room for cometing users on the Tibetan Plateau. In *Proceedings of the 7th International Snow Leopard Symposium*, ed. J. L. Fox and D. Jizeng. Seattle, WA: International Snow Leopard Trust.

Miller, L. E. (2002). *Eat or Be Eaten: Predator Sensitive Foraging among Primates*. Cambridge, UK: Cambridge University Press.

Miller, S. and Cully, J. F. Jr (2001). Conservation of black-tailed prairie dogs (*Cynomys ludovicianus*). *Journal of Mammalogy*, **82**, 889–893.

Mills, J. A. and Jackson, P. (1994). *Killed for a Cure: A Review of the World-Wide Trade in Tiger Bone*. Cambridge, UK: TRAFFIC International.

Mills, M. G. L. (1990). *Kalahari Hyaenas: Comparative Behavioural Ecology of Two Species*. London: Unwin Hyman.

Mills, M. G. L. and Biggs, H. C. (1993). Prey apportionment and related ecological relationships between large carnivores in Kruger National Park. *Symposia of the Zoological Society of London*, **65**, 253–268.

Mills, M. G. L. and Hes, L. (1997). *The Complete Book of Southern African Mammals*. Cape Town, South Africa: Struik Publishers.

Mills, M. G. L. and Hofer, H. (1998). *Hyaenas: Status Survey and Conservation Action Plan*. Gland, Switzerland: IUCN.

Milner-Gulland, E. J. and Leader-Williams, N. (1992). A model of incentives for the illegal exploitation of black rhinos and elephants: poaching pays in Luangwa Valley, Zambia. *Journal of Applied Ecology*, **29**, 388–401.

Milner-Gulland, E. J. and Mace, R. (1998). *Conservation of Biological Resources.* Oxford, UK: Blackwell.

Minnesota Department of Natural Resources (2001) *Minnesota Wolf Management Plan.* Grand Rapids, MN: Division of Wildlife in collaboration with the Minnesota Department of Agriculture.

(2003). Minnesota wolf management plan. Unpublished report from Minnesota Department of Natural Resources.

Miquelle, D. G. and Pikunov, D. G. (2003). Status of the Amur tiger and Far Eastern leopard. In *The Russian Far East: A Reference Guide for Conservation and Development*, ed. J. P. Newell. McKinleyville, CA: Daniel and Daniel, pp. 106–109.

Miquelle, D. G. and Smirnov, E. N. (1999). People and tigers in the Russian Far East: searching for the 'coexistence recipe'. In *Riding the Tiger: Tiger Conservation in Human-Dominated Landscapes*, ed. J. Seidensticker, S. Christie and P. Jackson. Cambridge, UK: Cambridge University Press, pp. 290–293.

Miquelle, D. G., Smirnov, E. N., Quigley, H. G. *et al.* (1996). Food habits of Amur tigers in Sikhote-Alin Zapovednik and the Russian Far East, and implications for conservation. *Journal of Wildlife Research*, **1**, 138–147.

Miquelle, D. G., Smirnov, E. N., Merrill, W. T. *et al.* (1999a). Hierarchical spatial analysis of Amur tiger relationships to habitat and prey. In *Riding the Tiger: Tiger Conservation in Human-Dominated Landscapes*, ed. J. Seidensticker, S. Christie and P. Jackson. Cambridge, UK: Cambridge University Press, pp. 71–99.

Miquelle, D. G., Stevens, P.A., Smirnov, E. N., *et al.* (2005). Competitive exclusion, functional redundancy, and conservation implications: tigers and wolves in the Russian Far East. In *Large Carnivores and the Conservation of Biodiversity*, ed. J. Ray, J. Berger, K. H. Redford and R. Stereck. Washington, DC: Island Press, pp. 179–207.

Miquelle, D. G., Merrill, W. T., Dunishenko, Y. M. (1999b). A habitat protection plan for the Amur tiger: developing political and ecological criteria for a viable land-use plan. In *Riding the Tiger: Tiger Conservation in Human-Dominated Landscapes*, ed. J. Seidensticker, S. Christie and P. Jackson. Cambridge, UK: Cambridge University Press, pp. 273–295.

Mishra, C. (1997). Livestock predation by large carnivores in the Indian Trans-Himalaya: conflict perceptions and conservation prospects. *Environmental Conservation*, **24**, 338–343.

Mishra, C., Madhusudan, M. D., Allen, P. and McCarthy, T. (2003). The role of incentive schemes in conserving the snow leopard, *Uncia uncia. Conservation Biology*, **17**, 1512–1520.

Mishra, H. R. (1984). A delicate balance: tigers, rhinoceros, tourists and park management vs. the needs of local people in the Royal Chitwan National Park. In *National Parks, Conservation and Development: The Role of Protected Areas in Sustaining Society*, ed. J. A. McNeely and K. R. Miller. Washington, DC: Smithsonian Institution Press, pp. 197–205.

Misra, M. (2003). Evolution, impact and effectiveness of domestic wildlife bans in India. In *The Trade in Wildlife: Regulation for Conservation*, ed. S. Oldfield. London: Earthscan, pp. 78–85.

Mizutani, F. (1993). Home range of leopards and their impact on livestock on Kenyan ranches. *Symposia of Zoological Society of London*, **65**, 425–439.

(1999). Impact of leopards on a working ranch in Laikipia, Kenya. *African Journal of Ecology*, **37**, 211–225.

Mnene, R. (1992). *Interaction between People and Wildlife around Shimba Hills National Reserve and Mwaluganje Forest*. Nairobi: Kenya Wildlife Service.

Monaghan, P. and Wood-Gush, D. (1990). *Managing the Behaviour of Animals*. New York: Chapman and Hall.

Mondolfi, E. and Hoogesteijn, R. (1986). Notes on the biology and status of the small wild cats in Veenzuela. In *Cats of the World: Biology, Conservation and Management*, ed. S. D. Miller and D. D. Everett. Washington, DC: National Wildlife Federation, pp. 125–146.

Montana Fish, Wildlife and Parks (2003). Draft Montana gray wolf conservation and management plan. Montana Fish Wildlife and Parks Unpublished report.

Moss, A. H. (1903). Forest panther attack. *Journal of the Bombay Natural History Society*, **15**, 516.

Moss, C. J. (2001). The demography of an African elephant (*Loxodonta africana*) population in Amboseli, Kenya. *Journal of Zoology*, **255**, 145–156.

Moxey, A., White, B. and Ozanne, A. (1999). Efficient contract design for agrienvironmental policy. *Journal of Agricultural Economics*, **50**, 187–202.

Mubalama, L. (1996). *An Assessment of Crop Damage by Large Mammals in the Reserve de Faune à Okapis in the Ituri Forest, Zaire: With Special Emphasis on Elephants*. Canterbury, UK: Institut Zairois pour la Conservation de la Nature.

Mubalama, L. and Hart, J. A. (1995). *An Assessment of Crop Damage: Damage by Large Mammals in the Reserve de Faune à Okapis, Ituri Forest, Zaire: with a Special Emphasis on Elephants*. Epulu: Institut Zairois pour la Conservation de la Nature.

Muchapondwa, E. (n.d.). *Risk Management through Community-based Wildlife Conservation and Wildlife Damage Insurance*. Göteborg, Sweden: School of Economics and Commercial Law, Göteborg University.

Mueller, L. (1985). Cougar attack. *Outdoor Life*, **175**, 108–111.

Mugisha, S. (1994). *Land cover/use around Kibale National Park*. Kampala: MUIENR and RS/GIS Laboratory.

Munn, L. C. (1993). Effects of prairie dogs on physical and chemical properties of soils. In *Management of Prairie Dog Complexes for the Reintroduction of the Black-Footed Ferret*, ed. J. L. Oldemeyer, D. E. Biggins and B. J. Miller. Washington, DC: US Department of the Interior, pp. 11–17.

Murombedzi, J. (1992). *Decentralization or Recentralization? Implementing CAMPFIRE in the Omay Communal Lands of the Nayminyami District*. Harare: Centre for Applied Social Sciences.

(2001). Committees, rights, costs and benefits: natural resource stewardship and community benefits in Zimbabwe's CAMPFIRE programme. In *African Wildlife and Livelihoods: The Promise and Performance of Community Conservation*, ed. D. Hulme and M. Murphree. Oxford, UK: James Currey, pp. 244–255.

Murphree, M. (1993). *Communities as Resource Management Institutions*, Gatekeeper Series No. 36. London: International Institute for Environment and Development.

(1995). *The Lesson from Mahenye: Rural Poverty, Democracy and Wildlife Conservation*, Wildlife and Development Series No. 1. London: International Institute for Environment and Development.

(2001). Community, council and client: a case study in ecotourism development from Mahenye, Zimbabwe. In *African Wildlife and Livelihoods: The Promise and Performance of Community Conservation*, ed. D. Hulme and M. Murphree. Oxford, UK: James Currey, pp. 177–194.

Muruthi, P., Stanley-Price, M. R., Soorae, P., Moss, C. and Lanjouw, A. (2000). Conservation of large mammals in Africa: what lessons and challenges for the future? In *Has the Panda Had Its Day? Priorities for the Conservation of Mammalian Diversity*, ed. A. Entwistle and N. Dunstone. Cambridge, UK: Cambridge University Press.

Musgrove, A. J., Pollitt, M. S., Hall, C. *et al.* (2001). *The Wetland Bird Survey 1999–2000: Wildfowl and Wader Counts*. Slimbridge, UK: British Trust for Ornithology/Wildfowl and Wetlands Trust/Royal Society for the Protection of Birds/Joint Nature Conservation Committee.

Musiani, M. and Visalberghi, E. (2001). Effectiveness of fladry on wolves in captivity. *Wildlife Society Bulletin*, **29**, 91–98.

Musiani, M., Mamo, C., Boitani, L. *et al.* (2003). Wolf depredation trends and the use of fladry barriers to project livestock in Western North America. *Conservation Biology*, **17**, 1538–1547.

Musters, C. J. M., Kruk, M., Graaf, H. J. and Terkeurs, W. J. (2001). Breeding birds as a farm product. *Conservation Biology*, **15**, 363–369.

Mwathe, K. M. (1992). A preliminary report on elephant crop damage in areas bordering Shimba Hills National Reserve. Unpublished report. Nairobi: Kenya Wildlife Service Elephant Program.

Nash, R. F. (1989). *The Rights of Nature: A History of Environmental Ethics*. Madison, WI: University of Wisconsin Press.

Nass, R. D. and Theade, J. (1988). Electric fences for reducing sheep losses to predators. *Journal of Range Management*, **41**, 251–252. (http://uvalde.tamu.edu/jrm/jrmhome.htm)

Nass, R. D., Lynch, G. and Theade, J. (1984). Circumstances associated with predation rates on sheep and goats. *Journal of Range Management*, **37**, 423–426. (http://uvalde.tamu.edu/jrm/jrmhome.htm)

National Agriculture Statistics Service (2001a). *Cattle and Calf Loss in 2000*. Washington, DC: Agricultural Board, US Department of Agriculture.

(2001b). *Sheep and Lamb Loss in 2000*. Washington, DC: Agricultural Board, US Department of Agriculture.

National Environment Action Plan Secretariat (1995). *National Environmental Action Plan for Uganda*. Kampala: Ministry of Natural Resources.

National Goose Forum (1997a). NGF3/97 Development of a National Policy Framework. Accessible at http://www.scotland.gov.uk/nationalgooseforum/MEETING1/NGF03_97.pdf

(1997b). NGF2/97 Terms of reference and membership. Accessible at http://www.scotland.gov.uk/nationalgooseforum/MEETING1/NGF02_97.pdf

(1998a). NGF10/98 Review of management techniques and habitat creation. Accessible at http://www.scotland.gov.uk/nationalgooseforum/MEETING3/NGF10_98.pdf

(1998b). NGF14/98 Goose management in other european countries. Accessible at http://www.scotland.gov.uk/nationalgooseforum/MEETING4/NGF14_98.pdf

(1998c). NGF9/98 Population viability analyses: theoretical basis. Accessible at http://www.scotland.gov.uk/nationalgooseforum/MEETING3/NGF09_98.pdf

(1998d). NGF16/98 The viability of goose populations wintering in Scotland: a population viability analysis approach. Accessible at http://www.scotland.gov.uk/nationalgooseforum/MEETING4/NGF16_98.pdf

(1998e). NGF19/98 Management arrangements for geese. Accessible at http://www.scotland.gov.uk/nationalgooseforum/MEETING5/NGF19_98.pdf

(1999). NGF4/99 Options for the delivery of future goose management schemes. Accessible at http://www.scotland.gov.uk/nationalgooseforum/MEETING6/NGF04_99.pdf

(2000). Policy report and recommendations of the National Goose Forum. Accessible at http://www.scotland.gov.uk/nationalgooseforum/ngf-00.asp

National Goose Management Review Group (2000). NGMRG report on proposals submitted by Local Goose Management Groups. Accessible at http://www.scotland.gov.uk/library3/environment/ngmrg-00.asp

National Trust (1993). *The Conservation and Management of Red Deer in the West Country*. Unpublished report to the Council of the National Trust by the Deer Working Party.

National Wildlife Federation (1998). *Petition for Rule Listing the Black-Tailed Prairie Dog (*Cynomys ludovicianus*) as Threatened throughout its Range.* Denver, CO: US Fish Wildlife Service Region 6.

Naughton-Treves, L. (1996). Uneasy neighbors: wildlife and farmers around Kibale National Park, Uganda. Ph.D. thesis, University of Florida, Gainesville, FL.

(1997). Farming the forest edge: vulnerable places and people around Kibale National Park, Uganda. *Geographical Review*, **87**, 27–46.

(1998). Predicting patterns of crop damage by wildlife around Kibale National Park, Uganda. *Conservation Biology*, **12**, 156–168.

(1999). Whose animals? A history of property rights to wildlife in Toro, western Uganda. *Land Degradation and Development*, **10**, 311–328.

(2001). Farmers, wildlife and the forest fringe. In *African Rain Forest Ecology and Conservation*, ed. A. Weber, L. White, A. Vedder and L. Naughton-Treves. New Haven, CT: Yale University Press, pp. 369–284.

Naughton-Treves, L., Treves, A., Chapman, C. and Wrangham, R. (1998). Temporal patterns of crop-raiding by primates: linking food availability in croplands and adjacent forest. *Journal of Applied Ecology*, **35**, 596–606.

Naughton-Treves, L., Rose, R. and Treves, A. (1999). *The Social Dimension of Human–Elephant Conflict in Africa: A Literature Review and Case Studies from Uganda and Cameroon*. Gland, Switzerland: IUCN African Elephant Specialist Group.

(2000). The spatial and social dimensions of human–elephant conflict in Africa: case studies from Uganda and Cameroon. Accessible at http://www.iucn.org/themes/ssc/sgs/afesg/hectf/pdfs/hecugcarev.pdf

Naughton-Treves, L., Mena, J. L., Treves, A., Alvarez, N. and Radeloff, V. C. (2003a). Wildlife survival beyond park boundaries: the impact of swidden agriculture and hunting on mammals in Tambopata, Peru. *Conservation Biology*, **17**, 1106–1117.

Naughton-Treves, L., Grossberg, R. and Treves, A. (2003b). Paying for tolerance: the impact of depredation and compensation payments on rural citizens' attitudes toward wolves. *Conservation Biology*, **17**, 1500–1511.

Nchanji, A. C. and Lawson, D. P. (1998). A survey of elephant crop damage around the Banyang-Mbo wildlife sanctuary, 1993–1996. Yaoundé: Cameroon Biodiversity Project and Wildlife Conservation Society.

Nelson, R. (1997). *Heart and Blood: Living Together with Deer in North America*. New York: Knopf.

Nemtzov, S. C. (2003). A short-lived wolf depredation compensation program in Israel. *Carnivore Damage Prevention News*, **6**, 16–17.

Nepal, S. K. and Weber, K. E. (1993). *Struggle for Existence: Park–People Conflict in the Royal Chitwan National Park*. Bangkok: Asian Institute of Technology.

New, T. R. (1994). Butterfly ranching: sustainable use of insects and sustainable benefits to habitats. *Oryx*, **28**, 169–172.

New Mexico Department of Game and Fish (2005). Home pages. Accessible at http://www.gmshs.state.nm.us

Newby, F. and Brown, R. (1958). A new approach to predator management in Montana. *Montana Wildlife*, **8**, 22–27.

Newmark, W. D. (1995). Extinction of mammal populations in western North American national parks. *Conservation Biology*, **9**, 512–526.

(1996). Insularization of Tanzanian parks and the local extinction of large mammals. *Conservation Biology*, **10**, 1549–1556.

Newmark, W. D., Leonard, N. L., Sariko, H. I. and Gamassa, D.-G. M. (1993). Conservation attitudes of local people living adjacent to five protected areas in Tanzania. *Biological Conservation*, **63**, 177–83.

Newmark, W. D., Manyanza, D. N., Gamassa, D. and Sariko, H. (1994). The conflict between wildlife and local people living adjacent to protected areas in Tanzania: human density as a predictor. *Conservation Biology*, **8**, 249–255.

Newton, I. (1979). *Population Ecology of Raptors*. London: Poyser.

(1998). *Population Limitation in Birds*. London: Academic Press.

Nicanor, N. (2001). *Practical Strategies for Pro-Poor Tourism: NACOBTA the Namibian Case Study*, PRo-Poor Tourism Working Paper No. 4. London: Centre for Responsible Tourism/International Institute for Environment and Development/Overseas Development Institute

Nielsen, O. K. (1999). Gyrfalcon predation on ptarmigan: numerical and functional responses. *Journal of Animal Ecology*, **68**, 1034–1050.

Niemeyer, X. Y. Z. (1995). Control of endangered gray wolves in Montana. In *Ecology and Conservation of Wolves in a Changing World*, ed. L. N. Carbyn, S. H. Fritts, and D. R. Seip. Edmonton, Alberta: Canadian Circumpolar Institute, pp. 127–134.

(1998). Status of gray wolf restoration in Montana, Idaho, and Wyoming. *Wildlife Society Bulletin*, **26**, 785–798.

Nikolaev, I. G. (1985). Last winter's loss of tigers. *Okhota i okhotnichie khozyaistvo*, **9**, 18–19. (In Russian)

Nikolaev, I. G. and Yudin, V. G. (1993). Tiger and man in conflict situations. *Bull. Mosk. Obschestva Ispytateley Priorody. Otd. Biol. V.*, **98**, 23–26.

Nolte, D., Farley, J., Campbell, D., Epple, G. and Mason, J. (1993). Potential repellents prevent mountain beaver damage. *Pesticide Science*, **12**, 624–626.

Norton-Griffiths, M. and Southey, C. (1995). The opportunity costs of biodiversity conservation in Kenya. *Environmental Economics*, **12**, 125–139.

Nowack, R. (1991). *Walker's Mammals of the World*. Baltimore, MD: Johns Hopkins University Press.

Nowak, R. (1995). Another look at wolf taxomony. In *Ecology and Conservation of Wolves in a Changing World*. ed. L. N. Carbyn, S. H. Fritts, and D. R. Seip. Edmonton, Alberta: Canadian Circumpolar Institute, pp. 375–397.

Nowell, K. and Jackson, P. (1996). *Wild Cats: Status Survey and Conservation Action Plan*. Cambridge, UK: Burlington Press.

Nunez, R., Miller, B. and Lindzey, F. (2002). Ecology of jaguars in the Chamela-Cuixmala Biosphere Reserve, Jalisco, Mexico. In *El Jaguar en el nuevo milenio*, ed. R. A. Medellin *et al.* Mexico: Fondo de Cultura Economica, pp. 107–127.

Nybakk, K., Kjelvik, O., Kvam, T., Overskaug, K. and Sunde, P. (2002). Mortality of semi-domestic reindeer *Rangifer tarandus* in central Norway. *Wildlife Biology*, **8**, 63–68.

Nyhus, P., Tilson, R. and Sumianto (2000). Crop-raiding elephants and conservation implications at Way Kambas National Park, Sumatra Indonesia. *Oryx*, **34**, 262–274.

Nyhus, P., Fischer, H., Madden, F. and Osofsky, S. (2003). Taking the bite out of wildlife damage: the challenges of wildlife compensation schemes. *Conservation Biology*, **4**, 37–40.

Oakleaf, J. K. (2002). Wolf–cattle interactions and habitat selection by recolonizing wolves in the northwestern United States. M. S. thesis, University of Idaho, Moscow, Idaho.

Oakleaf, J. K., Mack, C. and Murray, D. L. (2003). Effects of wolves on livestock calf survival and movements in central Idaho. *Journal of Wildlife Management*, **67**, 299–306.

Oates, J. F. (1999). *Myth and Reality in the Rainforest: How Conservation Strategies are Failing in West Africa*. Berkeley, CA: University of California Press.

O'Connell-Rodwell, C., Rodwell, T., Rival, L. and Hart, L. (2000). Living with the modern conservation paradigm: can agricultural communities co-exist with elephants? *Biological Conservation*, **93**, 381–391.

Odden, J., Linnell, J. D. C., Moa, P. F. *et al.* (2002). Lynx depredation on domestic sheep in Norway. *Journal of Wildlife Management*, **66**, 98–105.

Oerke, E. C., Dehne, H. W., Schonbeck, F. and Weber, A. (1995). *Crop Production and Crop Protection: Estimated Losses in Major Food and Cash Crops*. Amsterdam, Netherlands: Elsevier.

Ogada, M. O., Woodroffe, R., Oguge, N. and Frank, L. G. (2003). Limiting depredation by African carnivores: the role of livestock husbandry. *Conservation Biology*, **17**, 1521–1530.

Ogutu, J. O. and Dublin, H. T. (2002). Demography of lions in relation to prey and habitat in the Maasai Mara National Reserve, Kenya. *African Journal of Ecology*, **40**, 120–129.

Okarma, H. and Jedrzejewski, W. (1997). Livetrapping wolves with nets. *Wildlife Society Bulletin*, **25**, 78–82.

Okwemba, A. (2004). Proposal to reduce parks' sizes. *Daily Nation*, 18 March, 2.

Oldfield, T. E. E., Smith, R. J., Harrop, S. R. and Leader-Williams, N. (2003). Field sports and conservation in the United Kingdom. *Nature*, **423**, 531–533.

Oli, M. K. (1991). The ecology and conservation of the snow leopard (*Panthera uncia*) in the Annapurna Conservation Area, Nepal. M.Phil. thesis, University of Edinburgh, Edinburgh, UK.

Oli, M. K., Taylor, I. R. and Rogers, M. E. (1994). Snow leopard predation of livestock: an assessment of local perceptions in the Annapurna Conservation Area, Nepal. *Biological Conservation*, **68**, 63–68.

Oliveira, T. G. de. (1992). Ecology and conservation of neotropical felids. M.S. thesis, University of Florida, Gainesville, FL.

Olsen, J. W. (1985). Prehistoric dogs in mainland East Asia. In *Origins of the Domestic dog: The Fossil Record*, ed. S. J. Olsen. Tucson, AZ: University of Arizona Press, pp. 47–70.

Olsen, L. (1991). Compensation: giving a break to ranchers and bears. *Western Wildlands*, Spring, 25–29.

Olsson, O., Wirtberg, J., Andersson, M. and Wirtberg, I. (1997). Wolf predation on moose and roe-deer in south-central Scandinavia. *Wildlife Biology*, **3**, 13–25.

O'Meilia, M. E., Knopf, F. L. and Lewis, J. C. (1982). Some consequences of competition between prairie dogs and beef cattle. *Journal of Range Management*, **35**, 580–585.

Orejuela, J. and Jorgenson, J. (1999). Spectacle bear conservation action plan: Colombia. In *Bears: Status Survey and Conservation Action Plan*, ed. C. Servheen, S. Herrero and B. Peyton. Gland, Switzerland: IUCN, pp. 179–181.

Orford, H. J. L., Perrin, M. R. and Berry, H. H. (1988). Contraception, reproduction and demography of free-ranging Etosha lions. *Journal of Zoology*, **216**, 717–733.

Osborn, F. V. (1998). The ecology of crop-raiding elephants in Zimbabwe. Ph.D. thesis, University of Cambridge, Cambridge, UK.

(2002a). Capsicum oleoresin as an elephant repellent: field trials in the communal lands of Zimbabwe. *Journal of Wildlife Management*, **66**, 674–677.

(2002b). Elephant-induced change in woody vegetation and its impact on elephant movements out of a protected area in Zimbabwe. *Pachyderm*, **33**, 50–57.

Osborn, F. V. and Parker, G. E. (2002). Community-based methods to reduce crop loss to elephants: experiments in the communal lands of Zimbabwe. *Pachyderm*, **33**, 32–38.

(2003). Towards an integrated approach for reducing the conflict between elephants and people: a review of current research. *Oryx*, **37**, 80–84.

Osborn, F. V. and Welford, L. (1997). *Living with Elephants: A Manual for Wildlife Managers in the SADC Region.* Liloagwe, Malawi: Southern African Development Community/Natural Resources Management Programme.

Osmaston, H. A. (1959). *Working Plan for the Kibale and Itwara Central Forest Reserves, Toro District, W. Province, Uganda*. Kampala: Forest Department, Uganda Protectorate.

Östergren, A., Bergström, M.-R., Attergaard, H., From, J. and Mellquist, H. (1998). *Wolverine, Lynx and Wolf in the Reindeer Husbandry Area of Sweden: Results from the 1998 Inventory*. Umeå, Sweden: Länsstyrelsen i Västerbotten, Meddelande 3. (In Swedish with English summary)

Ostrom, E., Burger, J., Field, C. B., Norgaard, R. B. and Policansky, D. (1999). Sustainability: revisiting the commons. *Science*, **284**, 278–282.

Ottichilo, W. K. (2001). Population trends of resident wildebeest (*Connochaetes taurinus hecki* (Neumann)) and factors influencing them in the Masai Mara ecosystem, Kenya. *Biological Conservation*, **97**, 271–282.

Outwater, A. (1996). *Water: A Natural History*. New York: Basic Books.

Owen, M. (1971). The selection of feeding site by white-fronted geese in winter. *Journal of Applied Ecology*, **8**, 905–917.

(1977). The role of wildfowl refuges on agricultural land in lessening the conflict between farmers and geese in Britain. *Biological Conservation*, **11**, 209–222.

(1980). The role of refuges in wildfowl management. In *Bird Problems in Agriculture*, ed. E. N. Wright, I. R. Inglis and C. J. Feare. London: British Crop Protection Council, pp. 144–156.

(1990). The damage-conservation interface illustrated by geese. *Ibis*, **132**, 238–252.

Owen, M. and Black, J. M (1991). Geese and their future fortunes. *Ibis*, **133** (suppl. 1), 28–35.

Owen, M. and Norderhaug, M. (1977). Population dynamics of barnacle geese breeding in Svalbard. *Ornis Scandinavia*, **8**, 161–174.

Owen, M., Black, J. M., Agger, M. K. and Campbell, C. R. G. (1987). The use of the Solway Firth, Britain, by barnacle geese *Branta leucopsis* Bechst. in relation to refuge establishment and increases in numbers. *Biological Conservation*, **39**, 63–81.

Packer, C., Herbst, L., Pusey, A. E. *et al.* (1988). Reproductive success of lions. In *Reproductive Success*, ed. T. H. Clutton-Brock. Chicago, IL: University of Chicago Press.

Palomares, F. and Caro, T. (1999). Interspecific killing among mammalian carnivores. *American Naturalist*, **153**, 492–508.

Palomares, F., Gaona, P., Ferreras, P. and Delibes, M. (1995). Positive effects on game species of top predators by controlling smaller predator populations: an example with lynx, mongooses and rabbits. *Conservation Biology*, **9**, 295–305.

Panwar, H. S. (1987). Project Tiger: The reserves, the tigers and their future. In *Tigers of the World: The Biology, Biopolitics, Management and Conservation of an Endangered Species*, ed. R. L. Tilson and U. S. Seal. Park Ridge, NJ: Noyes Publications, pp. 110–117.

Parker, I. and Bleazard, S. (2001). *An Impossible Dream*. Elgin, UK: Librario Press.

Parker, I. S. C. and Graham, A. D. (1989a). Elephant decline. I: Downward trends in African elephant distribution and numbers. *International Journal of Environmental Studies*, **34**, 287–330.

(1989b). Man, elephants and competition. *Symposium of the Zoological Society of London*, **61**, 241–252.

Patterson, I. J. (1991). Conflict between geese and agriculture: does goose grazing cause damage to crops? *Ardea*, **79**, 179–186.

Patterson, I. J. and Cosgrove, P. J. (1997). Monitoring of a goose management scheme for the Strathbeg area. Unpublished report. Scottish Natural Heritage Commissioned Grampian Area No. GR/97/006.

Patterson, I. J. and Fuchs, R. M. E. (2001). The use of nitrogen fertilizer on alternative grassland feeding refuges for pink-footed geese in spring. *Journal of Applied Ecology*, **38**, 637–646.

Patterson, I. J., Abdul Jalil, S. and East, M. L. (1989). Damage to winter cereals by greylag and pink-footed geese in north-east Scotland. *Journal of Applied Ecology*, **26**, 879–895.

Patterson, J. H. (1907). *The Maneaters of Tsavo*. London: Fontana.

Patton, D. L. H. and Frame, J. (1981). The effect of grazing in winter by wild geese on improved grasslands in West Scotland. *Journal of Applied Ecology*, **18**, 311–325.

Paul, W. J. and P. S. Gipson. (1994). Wolves. In *Prevention and Control of Wildlife Damage*. Lincoln, NE: University of Nebraska Press, pp. 123–129.

Pearl, M. C. (1994). Local initiatives and the rewards of biodiversity conservation: Crater Mountain Wildlife Management Area, Papua New Guinea. In *Natural Connections: Perspectives in Community-Based Conservation*, ed. D. Western and R. M. Wright. Washington, DC: Island Press, pp. 193–214.

Pearson, E. W. and Caroline, M. (1981). Predator control in relation to livestock losses in central Texas. *Journal of Range Management*, **34**, 435–441. (http://uvalde.tamu.edu/jrm/jrmhome.htm)

Pedersen, V., Linnell, J. D. C., Andersen, R. *et al.* (1999). Winter lynx predation on semi-domestic reindeer in northern Sweden. *Wildlife Biology*, **5**, 203–212.

Percival, S. M. (1993). The effects of reseeding, fertiliser application and disturbance on the use of grassland by barnacle geese, and the implications for refuge management. *Journal of Applied Ecology*, **30**, 437–443.

Percival, S. M. and Houston, D. C. (1992). The effect of winter grazing by barnacle geese on grassland yields on Islay. *Journal of Applied Ecology*, **29**, 35–40.

Percival, S. M., Halpin, Y. and Houston, D. C. (1997). Managing the distribution of barnacle geese on Islay, Scotland, through deliberate human disturbance. *Biological Conservation*, **82**, 273–277.

Persson, J. (2003). Population ecology of Scandinavian wolverines. Ph.D. thesis, Swedish University of Agricultural Sciences, Umeå, Sweden.

Peterson, R. (1999) Wolf–moose interactions on Isle Royale: the end of natural regulation? *Ecological Applications*, **9**, 10–16.

Pettifor, R. A., Black, J. M., Owen, M., Rowcliffe, J. M. and Patterson, D. (1998). Growth of the Svalbard barnacle goose *Branta leucopsis* winter population 1958–1996: an initial review of temporal demographic changes. *Norsk Polarinstitutt Skrifter*, **200**, 147–164.

Peyton, B. (1999). Spectacled bear conservation action plan. In *Bears: Status Survey and Conservation Action Plan*, ed. C. Servheen, S. Herrero and B. Peyton. Gland, Switzerland: IUCN, pp. 157–198.

Pfeifer, W. K. and Goos, W. W. (1982). Guard dogs and gas exploders as coyote depredation control tools in North Dakota. *Vertebrate Pest Conference Proceedings*, **10**, 55–61.

Phillips, M. (1995). Conserving the red wolf. *Canid News*, **3**, 13–17.
Phillips, M. K., Bangs, E. E., Mech, L. D., Kelly, B. T. and Fazio, B., (2004). Living alongside canids: lessons from the extermination and recovery of red and grey wolves in the contiguous United States. In *The Biology and Conservation of Wild Canids*, ed. D. MacDonald and C. Sillero. Oxford, UK: Oxford University Press, pp. 297–309.
Phillips, R. L. and Blom, F. S. (1988). Distribution and magnitude of eagle/livestock conflicts in the United States. *Vertebrate Pest Conference Proceedings*, **13**, 241–244.
Picozzi, N. (1978). Dispersion, breeding and prey of the hen harrier in Glen Dye. *Ibis*, **120**, 489–509.
Pierce, B. M., Bleich, V. C., Wehausen, J. D. and Bowyer, R. T. (1999). Migratory patterns of mountain lions: implications for social regulation and conservation. *Journal of Mammalogy*, **80**, 986–992.
Pitman, M. R. P. L., de Oliveira, T. G., de Paula, R. C. and Indrusiak, C. (2002). *Manual de identificacao, prevencao e controle de predacao por carnivos*. Brasilia: IBAMA.
Pitt, J. (1988). *Des chiens 'montagne des Pyrénées' pour la protection des troupeaux ovins en région Rhône Alpes*. Paris: Institut technique de l'élevage ovin et caprin.
Playne, S. (1909). *East Africa: Its History, People, Commerce, Industries and Resources*. London: Unwin Bros.
Pletscher, D. H., Ream, R. R., Boyd, D. K., Fairchild, M. W. and Kunkel, K. E. (1997). Population dynamics of a recolonizing wolf population. *Journal of Wildlife Management*, **61**, 459–465.
Plumptre, A. J. and Bizumuremyi, J. B. (1996). *Ungulates and Hunting in the Parc National des Volcans, Rwanda*. New York: Wildlife Conservation Society.
Polisar, J., Maxit, I., Scognamillo, D. *et al.* (2003). Jaguars, pumas, their prey base and cattle ranching: ecological interpretations of a management problem. *Biological Conservation*, **109**, 297–310.
Pollard, J. (1966). *Wolves and Were-Wolves*. London: Robert Hale.
Porter, P. and Sheppard, E. S. (1998). *A World of Difference*. New York: Guilford Press.
Porter, P. W. (1976). *Agricultural Development and Agricultural Vermin in Tanzania*. Boston, MA: American Association for the Advancement of Science.
(1979). *Food and Development in the Semi-Arid Zone of East Africa*. Syracuse, NY: Syracuse University.
Porter, P. W, and Sheppard, E.S. (1998). *A World of Difference*. New York: Euilford Press.
Potts, G. R. (1998). Global dispersion of nesting hen harriers: implications for grouse moors in the UK. *Ibis*, **140**, 76–88.
Poulle, M.-L., Dahier, T., de Beaufort, R. and Durand, C. (2000). *Conservation du loup en France*, Programme Life-Nature, rapport final 1997–1999.
Prairie Dog Coalition (2002). Prairie dog summit forms coalition. *Prairie Dog Tales*, **1**, 1.
Prance, G. T. and Schaller, G. B. (1982). Preliminary study of some vegetation types of the Pantanal, Mato Grosso, Brazil. *Brittonia*, **34**, 228–251.

Predator Conservation Alliance (2001). *Restoring the Prairie Dog Ecosystem of the Great Plains: Learning from the Past to Ensure the Prairie Dog's Future.* Bozeman, MT: Predator Conservation Alliance.

Prezhewalski, N. M. (1870). *Travelling in Ussuriisky Krai, 1867–1869.* St Petersburg: Tipografiya N. Nekludova.

Price Waterhouse (1996). The trophy hunting industry: an African perspective. In *Tourist Hunting in Tanzania*, ed. N. Leader-Williams, J. A. Kayera and G. L. Overton. Gland, Switzerland: IUCN, pp. 9–11.

Priston, N. E. C. (2001). Assessment of crop damage by *Macaca ochreata brunnescens* in Southeast Sulawesis: a farmer's perspective. Undergraduate thesis, University of Cambridge, Cambridge, UK.

Quammen, D. (2003). *Monster of God: The Man-Eating Predator in the Jungles of History and the Mind.* New York: W. W. Norton.

Quigley, H. and Crawshaw, P. G. (1992). A conservation plan for the jaguar *Panthera onca* in the Pantanal region of Brazil. *Biological Conservation*, **61**, 149–157.

Rabinowitz, A. (1986). Jaguar predation on domestic livestock in Belize. *Wildlife Society Bulletin*, **14**, 170–174.

(1995). Jaguar conflict and conservation, a strategy for the future. In *Integrating People and Wildlife for a Sustainable Future*, ed. J. A. Bissonette and P. R. Krausman. Bethesda, MD: The Wildlife Society, pp. 394–397.

(1999). The present status of jaguars (*Panthera onca*) in the Southwestern United States. *Southwestern Naturalist*, **44**, 96–100.

(2000). *Jaguar: One Man's Struggle to Establish the World's First Jaguar Preserve.* Washington, DC: Island Press.

Rabinowitz, A. and Nottingham, B. G. (1986). Ecology and behavior of the jaguar (*Panthera onca*) in Belize, Central America. *Journal of Zoology*, **210**, 149–159.

Rajpurohit, K. S. (1999). Child lifting: wolves in Hazaribagh, India. *Ambio*, **28**, 162–166.

Rajpurohit, K. S. and Krausman, P. R. (2000). Human–sloth-bear conflicts in Madhya Pradesh, India. *Wildlife Society Bulletin*, **28**, 393–399.

Rangarajan, M. (2001). *India's Wildlife History: An Introduction.* New Delhi: Permanent Black.

Rasker, R. and Freese, C. (1995). Wildlife in the marketplace: opportunities and problems. In *Wildlife Conservation Policy*, ed. V. Geist and I. McTaggert-Cowan. Calgary, Alberta: pp. 177–204.

Rasmussen, G. (1999). Livestock predation by the painted hunting dog in a cattle ranching region of Zimbabwe: a case study. *Biological Conservation*, **88**, 133–139.

Ratnaswamy, M. J., Warren, R. J., Kramer, M. T. and Adam, M. D. (1997). Comparisons of lethal and nonlethal techniques to reduce raccoon depredation of sea turtle nests. *Journal of Wildlife Management*, **61**, 368–376.

Raveneau, A. and Daveze, J. (1994). *Le Livre de l'âne.* Paris: Editions Rustica.

Ray, J. C., Berger, J., Redford, K. H. and Steneck, R. (in press). *Large Carnivores and Biodiversity: Does Saving One Conserve the Other?* Washington, DC: Island Press.

Raynor, R., Strachan, R. and McClellan, A. (1996). The Loch of Strathbeg Goose Management Scheme Part 1. A final report on the scheme for the period 1994 to 1996. Unpublished report. Scottish Natural Heritage, North-East Area No. NE/96/037.

Reading, R. P. (1993). Toward an endangered species reintroduction paradigm: a case study of the black-footed ferret. Ph.D. thesis, Yale University, New Haven, CT.

Reading, R. P. and Kellert, S. R. (1993). Attitudes toward a proposed black-footed ferret (*Mustela nigripes*) reintroduction. *Conservation Biology*, **7**, 569–580.

Reading, R. P., Miller, B. J. and Kellert, S. R. (1999). Values and attitudes toward prairie dogs. *Anthrozoös*, **12**, 43–52.

Ream, R. R., Fairchild, M. W., Boyd, D. K. and Blakesley, A. J., (1989). First wolf den in western United States in recent history. *Northwest Naturalist*, **70**, 39–40.

Redpath, S. (1991). The impact of hen harriers on red grouse breeding success. *Journal of Applied Ecology*, **28**, 659–671.

Redpath, S. and Thirgood, S. (1997). *Birds of Prey and Red Grouse*. London: HMSO.

(1999). Numerical and functional responses in generalist predators: hen harriers and peregrines on Scottish grouse moors. *Journal of Animal Ecology*, **68**, 879–892.

(2003). The impact of hen harrier predation on red grouse: linking models with field data. In *Birds of Prey in a Changing Landscape*, ed. D. B. A. Thompson, S. M. Redpath, A. Fielding, M. Marquiss and C. Galbraith. Edinburgh, UK: Stationery Office, pp. 499–511.

Redpath, S. M., Thirgood, S. J. and Leckie, F. M. (2001). Does supplementary feeding reduce harrier predation on red grouse? *Journal of Applied Ecology*, **38**, 1157–1168.

Redpath, S. M., Thirgood, S. J. and Clarke, R. (2002). Field vole abundance and hen harrier diet and breeding in Scotland. *Ibis*, **144**, E130–E138.

Redpath, S., Arroyo, B., Leckie, F. *et al.* (2004). Using decision modelling to resolve human–wildlife conflicts: a raptor–grouse case study. *Conservation Biology*, **18**, 350–359.

Reeve, R. (2000). *Eselenkei Community Conservation Project*. Nairobi: International Fund for Animal Welfare.

Regeringen (2000). *Sammanhållen Rovdjurspolitik*, Regeringens proposition No. 2000/01:57. Stockholm: Regeringen. (In Swedish)

Reid, D. G. and Gong, J. (1999). Giant panda conservation action plan. In *Bears: Status Survey and Conservation Action Plan*, ed. C. Servheen, S. Herrero and B. Peyton. Gland, Switzerland: IUCN, pp. 241–254.

Reif, V., Tornberg, R., Jungell, S. and Korpimaki, E. (2001). Diet variation of common buzzards in Finland supports the alternative prey hypothesis. *Ecography*, **24**, 267–274.

Reiger, J. F. (1986). *American Sportsmen and the Origins of Conservation*. Norman, OK: University of Oklahoma Press.

Reiter, D. K., Brunson, M. W. and Schmidt, R. H. (1999). Public attitudes toward wildlife damage management and policy. *Wildlife Society Bulletin*, **27**, 746–758.

Republic of Kenya (1977). *The Wildlife (Conservation and Management) Act*. Nairobi: Government Printer.

(1989). *The Wildlife (Conservation and Management) Act Amended*. Nairobi: Government Printer.

Reynolds, J. C. and Tapper, S. C. (1996). Control of mammalian predators in game management and conservation. *Mammal Review*, **26**, 127–156.

Reynolds, J. C., Goddard, H. N. and Brockless, M. H. (1993). The impact of local fox (*Vulpes vulpes*) removal on fox populations at two sites in southern England. *Gibier Faune Sauvage*, **10**, 319–334.

Rigg, R. (2001). *Livestock Guarding Dogs: Their Current Use World Wide*, IUCN/SSC Canid Specialist Group Occasional Paper No. 1. Accessible at http://www.canids.org/occasionalpapers/

Rijksen, H. D. and Meijaard, E. (1999). *Our Vanishing Relative: The Status of Wild Orang Utans at the Close of the Twentieth Century*. Dordrecht, Netherlands: Kluwer.

Rilling, S., Kelly, L., Lindquist, H., Scully, K. and Moriarty, D. (2002). Effectiveness of a fladry in captive wolves. *Proceedings of the Defenders of Wildlife's Carnivores 2002*, 253.

Risley, E. H. (1966). An urgent plea for a national park to be created in the Shimba Hills. *Africana*, **9**, 6–8.

Ritter, D. G. (1981). Rabies in Alaskan furbearers: a review. In *6th North American Furbearer Conference*, Fairbanks, AK, 10–11 April 1991.

Robel, R. J., Dayton, A. D., Henderson, R. R., Meduna, R. L. and Spaeth, C. W. (1981). Relationship between husbandry methods and sheep losses to canine predators. *Journal of Wildlife Management*, **45**, 894–911.

Robertson, P. A., Park, K. J. and Barton, A. F. (2001). Loss of heather moorland in the Scottish uplands: the role of red grouse management. *Wildlife Biology*, **7**, 11–16.

Robertson, S. A. and Luke, W. R. W. (1993). *Coast Forest Survey*, Nairobi: National Museums of Kenya and World Wildlife Fund.

Robinson, J. G. (1993). Limits to caring: sustainable living and the loss of biodiversity. *Conservation Biology*, **7**, 20–28.

Rodgers, W. A. and Lobo, W. D. (1982). Elephant control and legal ivory exploitation, 1920–1976. *Tanganyika Notes and Records*, **85**, 25–54.

Roe, D., Leader-Williams, N. and Dalal-Clayton, D. B. (1997). *Take Only Photographs, Leave Only Footprints: The Environmental Impacts of Wildlife Tourism*. London: International Institute for Environment and Development.

Roe, D., Grieg-Ryan, M. and Schalken, W. (2001). *Getting the Lion's Share from Tourism: Private Sector Community Partnerships in Namibia*. London: International Institute for Environment and Development.

Rollins, K. and Briggs, H. C. (1996). Moral hazard, externalities, and compensation for crop damages from wildlife. *Journal of Environmental Economics and Management*, **31**, 368–386.

Rondeau, D. and Bulte, E. H. (2003). *Compensation for Wildlife Damage: Habitat Conversion, Species Preservation and Local Welfare*, REPA Working Paper No. 2003–01, revised 7 April 2004. Victoria, British Columbia: Department of Economics, University of Victoria.

Roosevelt, T. (1926). *Through the Brazilian Wilderness*. New York: Scribner.

Roper, T. J., Findlay, S. R., Lups, P. and Sheperdson, J. (1995). Damage by badgers *Meles meles* to wheat *Triticum vulgare* and barlery *Hordeum sativum* crops. *Journal of Applied Ecology*, **32**, 720–726.

Rosenblum, M. and Williamson, D. (1987). *Squandering Eden*. Orlando, FL: Harcourt Brace Jovanovich.

Røskaft, E., Bjerke, T., Kaltenborn, B. P. and Linnell, J. D. C. (2003). Patterns of self reported fear towards large carnivores among the Norwegian public. *Evolution and Human Behaviour*, **24**, 184–198.

Ross, P. (1998). *Crocodiles: Status Survey and Conservation Action Plan*. Gland, Switzerland: IUCN.

Ross, P. I., Jalkotzy, M. G. and Gunson, J. R. (1996). The quota system of cougar harvest management in Alberta. *Wildlife Society Bulletin*, **24**, 490–494.

Rouget, M., Cowling, R. M., Pressey, R. L. and Richardson, D. M. (2003). Identifying spatial components of ecological and evolutionary processes for regional conservation planning in the Cape Floristic Region, South Africa. *Diversity and Distributions*, **9**, 191–210.

Rousselot, M.-C. and Pitt, J. (1999). *Guide pratique du chien de protection*. Paris: Institut technique de l'élevage ovin et caprin.

Rowcliffe, J. M., Pettifor, R. A. and Mitchell, C. R. (2000). *Icelandic Population of the Greylag Goose (*Anser anser*): The Collation and Statistical Analysis of Data and Population Viability Analyses*. Edinburgh, UK: Scottish Natural Heritage.

Royal Society for the Protection of Birds (1996). *Wild Geese and Agriculture in Scotland*, RSPB response to Scottish Office Discussion Paper. Edinburgh, UK: RSPB.

Royal Society for the Protection of Birds and British Association for Shorting and Conservation (1998). NGF12/98 Geese and local economies in Scotland: a report to the National Goose Forum by RSPB and BASC. Accessible at http://www.scotland.gov.uk/nationalgooseforum/MEETING4/NGF12_98.pdf.

Rudacille, D. (1998). Activism for animals. In *Encyclopedia of Animal Rights and Animal Welfare*, ed. M. Beckoff and C. A. Meaney. Westport, CT: Greenwood Press, pp. 1–3.

Runte, A. (1987). *National Parks: The American Experience*. Lincoln, NE: University of Nebraska Press.

Saberwal, V. K., Gibbs, J. P., Chellam, R. and Johnsingh, A. J. T. (1994). Lion–human conflict in the Gir Forest, India. *Conservation Biology*, **8**, 501–507.

Sacks, B. N. (1996). Ecology and behavior of coyotes in relation to depredation and control on a California sheep ranch. M.S. thesis, University California, Berkeley, CA.

Sacks, B. N. and Neale, J. C. C. (2002). Foraging strategy of generalist predator toward a special prey: coyote predation on sheep. *Ecological Applications*, **12**, 299–306.

Sacks, B. N., Blejwas, K. M. and Jaeger, M. M. (1999a). Relative vulnerability of coyotes to removal methods on a northern California ranch. *Journal of Wildlife Management*, **63**, 939–949.

Sacks, B. N., Jaeger, M. M., Neale, J. C. C. and McCullough, D. R. (1999b). Territoriality and breeding status of coyotes relative to sheep predation. *Journal of Wildlife Management*, **63**, 593–605.

Saenz, J. C. and Carrillo, E. (2002). Jaguares depredators de Ganado en Costa Rica: Un problema sin solution? In *El Jaguar en el nuevo milenio*, ed. R. A. Medellin *et al.* Mexico: Fondo de Cultura Economica, pp. 127–139.

Sæther, B. E., Engen, S., Swenson, J. E, Bakke, Ø. and Sandegren, F. (1998). Assessing the viability of Scandinavian brown bear, *Ursus arctos*, populations: the effects of uncertain parameter estimates. *Oikos*, **83**, 403–416.

Safari Consultants (2004). http://www.safariconsultants.com.

Safety in Bear Country Society (2001). *Staying Safe in Bear Country*. Available at http://www.magiclantern.ca. at (video)

Sagør, J. T., Swenson, J. E., and Roskaft, E. (1997). Compatibility of brown bear *Ursus arctos* and free-ranging sheep in Norway. *Biological Conservation*, **81**, 91–95.

Saj, T., Sicotte, P. and Paterson, J. (2001). The conflict between vervet monkeys and farmers at the forest edge in Entebbe, Uganda. *African Journal of Ecology*, **39**, 195–199.

Sandegren, F. and Swenson, J. (1997). *Björnen: viltet, ekologin och människan*. Stockholm: Svenska Jägareförbundet. (In Swedish)

Sandell, M. (1989). The mating tactics and spacing behaviour of solitary carnivores. In *Carnivore Behavior, Ecology and Evolution*, ed. J. L. Gittleman. Ithaca, NY: Cornell University Press, pp. 164–182.

Sanderson, E. W., Redford, K. H., Chetkiewicz, C. B. *et al.* (2002). Planning to save a species: the jaguar as a model. *Conservation Biology*, **16**, 58–72.

Santiapillai, C. and Jackson, P. (1990). *The Asian Elephant: An Action Plan for its Conservation*. Gland, Switzerland: IUCN.

Sanyal, P. (1987). Managing the man-eaters in the Sundarbans tiger reserve of India: a case study. In *Tigers of the World: The Biology, Biopolitics, Management and Conservation of an Endangered Species*, ed. R. L. Tilson and U. S. Seal. Park Ridge, NJ: Noyes Publications, pp. 427–434.

Sathyakumar, S. (1999). Asiatic black bear conservation action plan: India. In *Bears: Status Survey and Conservation Action Plan*, ed. C. Servheen, S. Herrero and B. Peyton. Gland, Switzerland: IUCN, pp. 202–206.

Saunders, G., McIlroy, J., Berghout, M. *et al.* (2002). The effects of induced sterility on the territorial behaviour and survival of foxes. *Journal of Applied Ecology*, **39**, 56–66.

Saunders, N. J. 1989. *People of the Jaguar: The Living Spirit of Ancient America*. London: Souvenir Press.

(1991). *The Cult of the Cat*. London: Thames and Hudson.

Schaller, G. B. (1967). *The Deer and the Tiger*. Chicago, IL: University of Chicago Press.

(1983). Mammals and their biomass on a brazilian ranch. *Arquivos de Zoologia, São Paulo*, **31**, 1–36.

Schaller, G. B. and Crawshaw, P. (1980). Movement patterns of jaguar. *Biotropica*, **12**, 161–168.

Schaller, G. B., Jinchu, H., Wenshi, P. and Jing, Z. (1985). *The Giant Pandas of Wolong*. Chicago, IL: University of Chicago Press.

Schiaffino, K., Malmierca, L. and Perovic, P. G. (2002). Depredacion de credos domesticos por jaguar en un area rural vecina a un parque nacional en el noreste de Argentina. In *El Jaguar en el nuevo milenio*, ed. R. A. Medellin *et al.* Mexico: Fondo de Cultura Economica, pp. 251–265.

Schusler, T. M. and Decker, D. J. (2002). Engaging local communities in wildlife management area planning: an evaluation of the Lake Onterio Islands search conference. *Wildlife Society Bulletin*, **30**, 1226–1237.

Schwartz, M. W. (1999). Choosing the appropriate scale of reserves for conservation. *Annual Review of Ecology and Systematics*, **30**, 83–108.

Scognamillo, D., Maxit, I., Sunquist, M. and Farrell, L. (2002). Ecologia del jaguar y el problema de la depredación de ganado en un hato de Los Llanos venezolanos. In *El Jaguar en el nuevo milenio*, ed. R. A. Medellin *et al.* Mexico: Fondo de Cultura Economica, pp. 139–151.

Scott, J. C. (1976). *The Moral Economy of the Peasant*. New Haven, CT: Yale University Press.

Scott, J. P. and Fuller, J. L. (1965). *Genetics and the Social Behaviour of the Dog*. Chicago, IL: University of Chicago Press.

Scottish Executive Envirionment and Rural Affairs Department (2002). *Agriculture Facts and Figures 2002*. Accessible at http://www.scottishexecutive.gov.uk/library5/agri/afaf-00.asp

Scottish Natural Heritage (1998). *Good Practice for Grouse Moor Management*. Edinburgh, UK: SNH.

(2002). *Facts and Figures 2001/02*. Perth, UK: SNH.

Scottish Office Agriculture, Environment and Fisheries Department (1996). *Wild Geese and Agriculture in Scotland*, a discussion paper. Edinburgh, UK: Scottish Office.

Scottish Raptor Study Groups (1997). The illegal persecution of raptors in Scotland. *Scottish Birds*, **19**, 65–85.

Scrivner, J. H., Howard, W. E., Murphy, A. H. and Hays, J. R. (1985). Sheep losses to predators on a California range (1973–1983). *Journal of Range Management*, **38**, 418–421. (http://uvalde.tamu.edu/jrm/jrmhome.htm)

Seal, U. S., Thorne, E. T., Bogan, M. A. and Anderson, S. H. (1989). *Conservation Biology and the Black-Footed Ferret*. New Haven, CT: Yale University Press.

Segerson, K. (1988). Uncertainty and incentives for non-point pollution control. *Journal of Environmental Economics and Management*, **15**, 87–98.

Seidensticker, J., Sunquist, M. E. and McDougal, C. (1990). Leopards living at the edge of the Royal Chitwan National Park, Nepal. In *Conservation in Developing Countries: Problems and Prospects*, ed. J. C. Daniel and J. S. Serrao. Bombay, India: Oxford University Press, pp. 415–423.

Seidensticker, J., Christie, S. and Jackson, P. (1999). Preface. In *Riding the Tiger: Tiger Conservation in Human-Dominated Landscapes*, ed. J. Seidensticker, S. Christie and P. Jackson. Cambridge, UK: Cambridge University Press, pp. xv–xix.

Sekhar, N. U. (1998). Crop and livestock depredation caused by wild animals in protected areas: the case of Sariska Tiger Reserve, Rajasthan, India. *Environmental Conservation*, **25**, 160–171.

Serpell, J. and Jagoe, J. A. (1995). Early experience and the development of behavior. In *The Domestic Dog*, ed. J. Serpell. Cambridge, UK: Cambridge University Press, pp. 79–102.

Servheen, C. (1999). Sun bear conservation action plan. In *Bears: Status Survey and Conservation Action Plan*, ed. C. Servheen, S. Herrero and B. Peyton. Gland, Switzerland: IUCN, pp. 219–224

Servheen, C., Herrero, S. and Peyton, B. (eds.) (1999). *Bears: Status Survey and Conservation Action Plan*. Gland, Switzerland: IUCN.

Shabecoff, P. (1993). *A Fierce Green Fire: The American Environmental Movement*. New York: Hill and Wang.

Shafi, M. M. and Khokhar, A. R. (1986). Some observations on wild boar (*Sus scrofa*) and its control in sugar cane areas of Punjab, Pakistan. *Journal of the Bombay Natural History Society*, **83**, 63–67.

Shaw, H. G., Woolsey, N. G., Wegge, J. R. and Day, R. L. J. (1988). *Factors affecting Mountain Lion Densities and Cattle Depredation in Arizona*. Arizona Game and Fish Department.

Shelton, M. (1973). Some myths concerning the coyote as a livestock predator. *BioScience*, **23**, 719–720.

Shivik, J. A. (2001). The other tools for wolf management. *Wolf*, **19**, 3–7.

Shivik, J. A. and Gruver, K. S. (2002). Animal attendance at coyote trap sites in Texas. *Wildlife Society Bulletin*, **30**, 502–557.

Shivik, J. A. and Martin, D. J. (2001). Aversive and disruptive stimulus applications for managing predation. *Proceeding of the Wildlife Damage Management Conference*, **9**, 111–119.

Shivik, J. A., Asher, V., Bradley, L. *et al.* (2002). Electronic aversive conditioning for managing wolf predation. *Vertebrate Pest Conference Proceedings*, **20**,

Shivik, J. A., Treves, A. and Callahan, M. (2003). Non-lethal techniques: primary and secondary repellents for managing predation. *Conservation Biology*, **17**, 1531–1537.

Siex, K. and Struhsaker, T. (1999). Colobus monkeys and coconuts: a study of perceived human–wildlife conflicts. *Journal of Applied Ecology*, **36**, 1009–1020.

Sillero-Zubiri, C. and Laurenson, M. K. (2001). Interactions between carnivores and local communities: conflict or co-existence? In *Carnivore Conservation*, ed. J. Gittleman, S. M. Funk, D. W. MacDonald and R. K. Wayne. Cambridge, UK: Cambridge University Press, pp. 283–312.

Sillero-Zubiri, C., Reynolds, J. and Novaro, A. J. (2004). Management and control of wild canids alongside people. In *Biology and Conservation of Wild Canids*, ed. D. W. Macdonald and C. Sillero-Zubiri. Oxford, UK: Oxford University Press, pp. 107–122.

Sim, I. M. W., Gibbons, D. W., Bainbridge, I. P. and Mattingley, W. A. (2001). Status of the hen harrier in the UK in 1998. *Bird Study*, **48**, 341–354

Simon, H. A. (1983). *Reason in Human Affairs*. Stanford, CA: Stanford University Press.

Simon, N. (1962). *Between the Sunlight and the Thunder*. London: Collins.

Sinclair, A. R. E. (1995). Equilibria in plant–herbivore interactions. In *Serengeti II: Dynamics, Management and Conservation of an Ecosystem*, ed. A. R. E. Sinclair and P. Arcese. Chicago, IL: Chicago University Press, pp. 91–113.

Singer, D. J. (1916). *Big Game Fields of America: North and South*. London: Hodder and Stoughton.

Siriri, M. N. (2002). Knowledge and attitudes of the communities bordering Bwindi and Mgahinga National Parks in south western Uganda, December 2001. Unpublished report. Kampala: CARE-Development Through Conservation Project and CARE-International.

Sitati, N. W., Walpole, M. J., Smith, R. J. and Leader-Williams, N. (2003). Predicting spatial aspects of human–elephant conflict. *Journal of Applied Ecology*, **40**, 667–677.

Skogen, K. (2001). Who's afraid of the big, bad wolf? Young peoples respones to the conflicts over large carnivores in eastern Norway. *Rural Sociology*, **66**, 203–226.

(2003). Adapting adaptive management to a cultural understanding of land use conflicts.. *Society and Natural Resources*, **16**, 435–450.

Skogen, K. and Haaland, H. (2001). En ulvehistorie fra Østfold: samarbeid og konflikter mellom forvaltning, forskning og lokalbefolkning. *Norwegian Institute for Nature Research Fagrapport*, **52**, 1–51.

Skogen, K., Haaland, H., Brainerd, S. and Hustad, H. (2003). Lokale syn på rovvilt og rovviltforvaltning. En undersøkelse i fire kommuner: Aurskog-Høland, Lesja, Lierne og Porsanger. *Norwegian Institute for Nature Research Fagrapport*, **70**, 1–30.

Skuja, A. (2002). Lion–human conflicts in Tanzania. Ph.D. thesis, University of Wisconsin–Madison, Madison, WI.

Smirnov, E. N. and Miquelle, D. G. (1999). Population dynamics of the Amur tiger in Sikhote-Alin State Biosphere Reserve. In *Riding the Tiger: Tiger Conservation in Human–Dominated Landscapes*, ed. J. Seidensticker, S. Christie and P. Jackson. Cambridge, UK: Cambridge University Press, pp. 61–70.

Smith, A. A., Redpath, S. M., Campbell, S. C. and Thirgood, S. J. (2001). Meadow pipits, red grouse and the habitat characteristics of managed grouse moors. *Journal of Applied Ecology*, **38**, 390–400.

Smith, D. H., Ralls, K., Davenport, B., Adams, B. and Maldonado, J. E. (2001). Canine assistants for conservationists. *Science*, 291–435.

Smith, D. W., Peterson, R. O. and Honston, D. B. (2003). Yellowstone after wolves. *BioScience*, **53**, 330–340.

Smith, J. L. D. (1993). The role of dispersal in structuring the Chitwan tiger population. *Behaviour*, **124**, 165–195.

Smith, M. E. (1984)., Repellents and deterrents for brack and grizzly bears. M.Sc. thesis, University of Montana.

Smith, M. E., Linnell J. D. C., Odden J., and Swenson, J. E. (2000a). Methods for reducing livestock losses to predators. A: Livestock guardian animals. *Acta Agriculturae Scandinavica*, **50**, 279–290.

(2000b). Methods for reducing livestock losses to predators. B: Aversive conditioning, deterrents and repellents. *Acta Agriculturae Scandinavica*, **50**, 304–315.

Smith, N. J. H. (1976). Spotted cats and the Amazon skin trade. *Oryx*, **13**, 362–271.

Smuts, G. L. (1978). Effects of population reduction on the travels and reproduction of lions in Kruger National Park. *Carnivore*, **1**, 61–72.

Solberg, E. J., Sand, H., Linnell, J. D. C. *et al.* (2003). Store rovdyrs innvirkning på hjorteviltet i Norge: økologisk prosesser og konsekvenser for jaktuttak og jaktutøvelse. *Norwegian Institute for Nature Research Fagrapport*, **63**, 1–78. (In Norwegian)

Sollie, P. V., Finset, P., Jaren, V. *et al.* (1996). *Forebyggende tiltak mot rovviltskader i landbruket*. Trondheim, Norway: Directorate for Nature Management. (In Norwegian)

Songorwa, A. N. (1999). Community-based wildlife management (CWM) in Tanzania: are the communities interested? *World Development*, **27**, 2061–2079.

Soulé, M. E. and Terborgh, J. (1999). *Continental Conservation: Scientific Foundations of Regional Reserve Networks*. Washington, DC: Island Press.

Soulé, M. E., Wilcox, B. A. and Holtby, C. (1979). Benign neglect: a model of faunal collapse in the game reserves of East Africa. *Biological Conservation*, **95**, 259–272.

Southwood, T. (1977). The relevance of population dynamic theory of pest status. In *origins of Pest, Parasite, Disease and Weed Problems*, ed. J. M. Cherret and G. R. Sagar. Oxford, UK: Blackwell, pp. 36–54.

Sowerby, A. de C. (1923). *The Naturalist in Manchuria*. Tientsin.

Sprague, D. (2002). Monkeys in the backyard: encroaching wildlife and rural communities in Japan. In *Primates Face to Face*, ed. A. Fuentes and L. Wolfe. Cambridge, UK: Cambridge University Press, pp. 254–272.

Sri Lanka Wildlife Conservation Society (2000). *The Sri Lankan Elephant*. Accessible at http://www.benthic.com/sri_lanka/issues.htm

SSR (2002). Swedish reindeer owners' organization. Accessible at http://www.sapmi.se/ssr

Stahl, P. and Vandel, J. M. (2001). Factors influencing lynx depredation on sheep in France: problem individuals and habitat. *Carnivore Damage Prevention News*, **4**, 6–8.

Stahl, P., Vandel, J. M., Herrenschmidt, V. and Migot, P. (2001a). Predation on livestock by an expanding reintroduced lynx population: long-term trend and spatial variability. *Journal of Applied Ecology*, **38**, 674–687.

(2001b). The effect of removing lynx in reducing attacks on sheep in the Jura. *Biological Conservation*, **101**, 15–22.

Stahl, P., Vandel, J. M., Ruette, S. *et al.* (2002). Factors affecting lynx predation on sheep in the French Jura. *Journal of Applied Ecology*, **39**, 204–216.

Steinset, O. K., Fremming, O. R. and Wabakken, P. (1996). *Protection Collars on Lambs as Preventive Activity against Lynx Predation in Stange, Hedmark*. Koppang, Norway: Department of Forestry and Wilderness Management.

Stellflug, J. N., Leathers, C. W. and Green, J. S. (1984). Antifertility effect of busulfan and DL-6-(N-2-pipecolinomethyl)-5-hydroxy-indane maleate (PMHI) in coyotes (*Canis latrans*). *Theriogenology*, **22**, 533–543.

Stirling, I. and Guravich, D. (1988). *Polar Bears*. Ann Arbor, MI: University of Michigan Press.

Stirling, I., Jonkel, C., Smith, P., Robertson, P. and Cross, D. (1977). *The Ecology of the Polar Bear along the Western Coast of Hudson Bay*, Canadian Wildlife Service Occasional Paper No. 33.

Stowell, L. R. and Willging, R. C. (1992). Bear damage to agriculture in Wisconsin. *Proceedings of the Eastern Wildlife Damage Control Conference*, **5**, 96–104.

Strickland, D. (1999). Algonquin Park struggles with 'fearless' wolves. *Wolf!* **17**, 7–9.

Struhsaker, T. T. (1997). *Ecology of an African Rain Forest*. Gainesville, FL: University Press of Florida.

Strum, S. C. (1994). Prospects for management of primate pests. *Revue d'Ecologie (Terre et Vie)*, **49**, 295–306.

Strum, S. C. and Southwick, C. (1986). Translocation of primates. In *Primates: The Road to Self-Sustaining Populations*, ed. K. Benirschke. New York: Springer-Verlag.

Sukhomirov, G. (2002). *A Survey of Social Perspectives on Conservation of Tigers and their Habitat*. Khabarovsk, Russia: Khabarovsk State Technical University Press. (In Russian)

Sukumar, R. (1989). *The Asian Elephant: Ecology and Management*. Cambridge, UK: Cambridge University Press.

(1990). Ecology of the Asian elephant in southern India. II: Feeding habits and crop raiding patterns. *Journal of Tropical Ecology*, **6**, 33–53.

(1991). The management of large mammals in relation to male strategies and conflict with people. *Biological Conservation*, **55**, 93–102.

Suminski, H. R. (1982). Mountain lion predation on domestic livestock in Nevada. *Vertebrate Pest Conference Proceedings*, **10**, 62–66.

Sunde, P., Overskaug, K., and Kvam, T. (1998). Culling of lynxes *Lynx lynx* related to livestock predation in a heterogeneous landscape. *Wildlife Biology*, **4**, 169–175.

Sunquist, M. E. (1981). Social organization of tigers (*Panthera tigris*) in Royal Chitawan National Park, Nepal. *Smithsonian Contributions to Zoology*, **336**, 1–98.

Sunquist, M. E. and Sunquist, F. (2002). *Wild Cats of the World*. Chicago, IL: University of Chicago Press.

Sunquist, M. E., Karanth, K. U. and Sunquist, F. (1999). Ecology, behaviour, and resilience of the tiger and its conservation needs. In *Riding the Tiger: Tiger Conservation in Human-Dominated Landscapes*, ed. J. Seidensticker, S. Christie and P. Jackson. Cambridge, UK: Cambridge University Press, pp. 5–18.

Sutherland, G. D., Harestad, A. S., Price, K. and Lertzman, K. P. (2000). Scaling of natal dispersal distances in terrestrial birds and mammals. *Conservation Ecology*, **4** (http://www.consecol.org/vol4/iss1/art 16)

Sutton, W. R., Larson, D. M. and Jarvis, L. S. (2002). A new approach to contingent valuation for assessing the costs of living with wildlife in developing countries. In *World Congress of Environmental and Resource Economists*, Monterrey, CA, June 28

Svensson, L., Ahlqvist, I. and Kjellander, P. (1998). *Elstängsel som förebyggande åtgärd mot bjärnskador på bikupor*. Viltskade center, Grimsö Forskningsstation.

Swanson, T. (1994). *International Regulation of Extinction*. London: Macmillan.

Swenson, J. and Sandegren, F. (1999). Misstänkt illegal björnjakt i Sverige. In *Bilagor till Sammanhållen rovdjurspolitik: Slutbetänkande av Rovdjursutredningen*. Statens offentliga utredningar, pp. 201–206. Stockholm: (In Swedish)

Swenson, J. E., Sandegren, F., Bjärvall, A. *et al.* (1994). Size, trend, distribution and conservation of the brown bear *Ursus arctos* population in Sweden. *Biological Conservation*, **70**, 9–17.

Swenson, J. E., Wabakken, P., Sandegren, F., *et al.* (1995). The near extinction and recovery of brown bears in Scandinavia in relation to the bear management policies of Norway and Sweden. *Wildlife Biology*, **1**, 11–25.

Swenson, J., Sandegren, F., Heim, M. *et al.* (1996). Is the Scandinavian brown bear dangerous? *Norwegian Institute for Nature Research Oppdragsmelding*, **404**, 1–26. (In Norwegian)

Swenson, J. E., Sandegren, F. and Söderberg, A. (1998). Geographic expansion of an increasing brown bear population: evidence for presaturation dispersal. *Journal of Animal Ecology*, **67**, 819–826.

Swenson, J. E., Gerstl, N., Dahle, B. and Zedrosser, A. (2000). *Action Plan for the Conservation of the Brown bear (Ursus arctos) in Europe*. Nature and Environment No. 114. Strasbourg, France: Council of Europe.

Talbot, L. M. (1963). The wildebeest in western Maasailand. *Wildlife Monographs*, **12**, 1–88.

Tapper, S. C. (1999). *A Question of Balance: Game Animals and Their Role in the British Countryside*. Fordingbridge, UK: Game Conservancy Trust.

Tapper, S. C., Potts, G. R. and Brockless, M. H. (1996). The effect of an experimental reduction in predation pressure on the breeding success and population density of grey partridges. *Journal of Applied Ecology*, **33**, 965–978.

Tapscott, B. (1997). *Guidelines for Using Donkeys as Guard Animals with Sheep*. Toronto: Ontario Ministry of Agriculture and Food. Accessible at http://www.gov.on.ca/OMAFRA/english/livestock/sheep/facts/donkey2.htm

Taylor, R. (1999). *A Review of Problem Elephant Policies and Management Options in Southern Africa*. Nairobi: HEC Task Force, IUCN AfESG.

Taylor, R. G., Workman, J. P. and Bowns, J. E. (1979). The economics of sheep predation in southwestern Utah. *Journal of Range Management*, **32**, 317–321.

Tchamba, M. N. (1995). The problem elephants of Kaele: a challenge for elephant conservation in Northern Cameroon. *Pachyderm*, **19**, 26–31.

(1996). History and present status of the human/elephant conflict issue in the Waza-Cogone region, Cameroon, West Africa. *Biological Conservation*, **75**, 35–41.

Tchamba, M. N., Bauer, H. and De Iongh, H. H. (1995). Applications of VHF-radio and satellite telemetry techniques on elephants in northern Cameroon. *African Journal of Ecology*, **33**, 335–346.

Terborgh, J., Lopez, L., Nunez, P. *et al.* (2002). Ecological meltdown in predator-free forest fragments. *Science*, **294**, 1923.

Thapar, V. (2002). *Cult of the Tiger*. New Delhi: Oxford University Press.

Tharme, A. P., Green, R. E., Baines, D., Bainbridge, I. P. and O'Brien, M. (2001). The effect of management for red grouse shooting on the population density of breeding birds on heather-dominated moorland. *Journal of Applied Ecology*, **38**, 439–457.

Thiel, R. P. and Ream, R. R. (1992). Status of the grey wolf in the lower 48 States to 1992. In *Ecology and Conservation of Wolves in a Changing World*, ed. L. N. Carbyn, S. H. Fritts and D. R. Seip. Edmonton, Alberta: Canadian Circumpolar Institute, pp. 59–62.

Thiollay, J. M. (1989). Area requirements for the conservation of rain forest raptors and game birds in French Guiana. *Conservation Biology*, **3**, 128–137.

Thirgood, S. and Redpath, S. (1997). Red grouse and their predators. *Nature*, **390**, 547.

Thirgood, S., Redpath, S., Newton, I. and Hudson, P. (2000a). Raptors and red grouse: conservation conflicts and management solutions. *Conservation Biology*, **14**, 95–104.

Thirgood, S. J., Redpath, S., Rothery, P. and Aebischer, N. (2000b). Raptor predation and population limitation in red grouse. *Journal of Animal Ecology*, **69**, 504–516.

Thirgood, S. J., Redpath, S., Haydon, D. *et al.* (2000c). Habitat loss and raptor predation: disentangling causes of red grouse declines. *Proceedings of the Royal Society of London B*, **267**, 651–656.

Thirgood, S., Redpath, S., Campbell, S. and Smith, A. (2002). Do habitat characteristics influence predation on red grouse? *Journal of Applied Ecology*, **39**, 217–225.

Thirgood, S. J., Redpath, S. M. and Graham, I. (2003). What determines the foraging distribution of raptors on heather moorland? *Oikos*, **100**, 15–24.

Thirgood, S., Polasky, S., Mlingwa, C. *et al.* (in press). Financing conservation in the Serengeti Ecosystem. In *Serengeti III: Biodiversity and Biocomplexity in a Human-Influenced Ecosystem*, ed. A. R. E. Sinclair, C. Packer, M. Coughenour, K. Galvin and S. Mduma. Chicago, IL: University of Chicago Press.

Thomas, H. (2002). Managing and controlling the quarry species: deer. Unpublished report to Westminster Hearings on Hunting with Hounds.

Thompson, B. C. (1978). Fence-crossing behavior exhibited by coyotes. *Wildlife Society Bulletin*, **6**, 14–17.

(1979). Evaluation of wire fences for coyote control. *Journal of Range Management*, **32**, 457–461. (http://uvalde.tamu.edu/jrm/jrmhome.htm)

Thompson, D. B. A., MacDonald, A. J., Marsden, J. H. and Galbraith, C. A. (1995). Upland heather moorland in the UK: a review of international importance, vegetation change and objectives for conservation. *Biological Conservation*, **71**, 163–178.

Thompson, J. G. (1993). Addressing the human dimensions of wolf reintroduction: an example using estimates of livestock depredation and costs of compensation. *Society and Natural Resources*, **6**, 165–179.

Thompson, M. and Homewood, K. (2002). Entrepreneurs, elites and exclusion in Maasailand: trends in wildlife conservation and pastoralist development. *Human Ecology*, **30**, 107–138.

Thompson, P. C. (1984). The use of buffer zones in dingo control. *Journal of Agriculture, Western Australia*, **25**, 32–34.

(1986). The effectiveness of aerial baiting for the control of dingoes in North-Western Australia. *Australian Wildlife Research*, **13**, 165–176.

Thomsen, J. B., Edwards, S. R. and Mulliken, T. A. (1992). *Perceptions, Conservation and Mangement of Wild Birds in Trade.* Cambridge, UK: TRAFFIC International.

Thouless, C. R. (1994). Conflict between humans and elephants in Sri Lanka. Unpublished report. GEF. Oxford, UK: EDG.

(1995). Long-distance movements of elephants in northern Kenya. *African Journal of Ecology*, **33**, 321–334.

(1998). Large mammals inside and outside protected areas in the Kalahari. *Transactions of the Royal Society of South Africa*, **53**, 245–255.

Thouless, C. R. and Sakwa, J. (1995). Shocking elephants: fences and crop raiders in Laikipia District, Kenya. *Biological Conservation*, **72**, 99–107.

Tietenberg, T. (1996). *Environmental and Natural Resource Economics*. New York: HarperCollins.

Tigner, J. A. and Larson, G. E. (1977). Sheep losses on selected ranches in southern Wyoming. *Journal of Range Management*, **30**, 244–252. (http://uvalde.tamu.edu/jrm/jrmhome.htm)

Till, J. A. and Knowlton, F. F. (1983). Efficacy of denning in alleviating coyote depredations upon domestic sheep. *Journal of Wildlife Management*, **47**, 1018–1025.

Tillman, D. (2000). Causes, consequences and ethics of biodiversity. *Science*, **405**, 208–211.

Timm, R. M. and Connolly, G. E. (2001). Sheep-killing coyotes a continuing dilemma for ranchers. *California Agriculture*, **55**(6), 26–31.

Todd, A. W. and Keith, L. B. (1976). Responses of coyote to winter reductions in agricultural carrion. *Alberta Wildlife Technical Bulletin*, **5**, 1–32.

Tompa, F. S. (1983). Problem wolf management in British Columbia: conflict and program evaluation. In *Wolves in Canada and Alaska: Their Status, Biology and Management*, ed. L. N. Carbyn. Edmonton, Alberta: Canadian Wildlife Service, pp. 112–119.

Tornberg, R. (2001). Pattern of goshawk predation on four forest grouse species in Finland. *Wildlife Biology*, **7**, 245–256.

Torres, S. G. (1997). *Mountain Lion Alert*. Helena, MT: Falcon Press.

Torres, S. G., Mansfield, T. M., Foley, J. E., Lupo, T. and Brinkhaus, A. (1996). Mountain lion and human activity in California: testing speculations. *Wildlife Society Bulletin*, **24**, 457–460.

Treves, A. (2002). Wolf justice: managing human–carnivore conflict in the 21st century. *Wolf Print*, **13**, 6–9.

Treves, A. and Karanth, K. U. (2003). Human–carnivore conflict and perspectives on carnivore management worldwide. *Conservation Biology*, **17**: 1491–1499.

Treves, A. and Naughton-Treves, L. (1999). Risk and opportunity for humans coexisting with large carnivores. *Journal of Human Evolution*, **36**, 275–282.

Treves, A., Jurewicz, R., Naughton-Treves, L. *et al.* (2002). Wolf depredation on domestic animals in Wisconsin, 1976–2000. *Wildlife Society Bulletin*, **30**, 231–241.

Treves, A., Naughton-Treves, L., Harper, E. K. *et al.* (2004). Predicting human–carnivore conflict: a spatial model derived from 25 years of data on wolf predation on livestock. *Conservation Biology*, **18**, 114–125.

Trzebinski, E. (1988). *The Kenya Pioneers*. New York: W. W. Norton.

Turnbull-Kemp, P. (1967). *The Leopard*. Cape Town, South Africa: Howard Timmins.

Turyahikayo-Rugyema, B. (1974). *The History of the Bakiga in Southwestern Uganda and Northern Rwanda, 1500–1930*. Ann Arbor, MI: University of Michigan Press.

US Fish and Wildlife Service (1987). *Northern Rocky Mountain Wolf Recovery Plan*. Denver, CO: US Fish and Wildlife Service.

(1988). *Interim Wolf Control Plan: Northern Rocky Mountains of Montana and Wyoming*. Denver, CO: US Fish and Wildlife Service.

(1994a). *The Reintroduction of Gray Wolves to Yellowstone National Park and Central Idaho*, Final Environmental Impact Statement. Helena, MT: US Fish and Wildlife Service.

(1994b). Establishment of a nonessential experimental population of gray wolves in Yellowstone National Park in Wyoming, Idaho, and Montana and central Idaho and southwestern Montana. Final Rule, Nov. 22. *Federal Register*, **59**(224), 60252–60281.

(1999). *90-Day Finding for a Petition to List the Black-Tailed Prairie Dog as Threatened*. Washington, DC: US Fish and Wildlife Service.

(2000). Endangered and threatened wildlife and plants: 12-month finding for a petition to list the black-tailed prairie dog as threatened. *Federal Register*, **65**(24), 5476–5488.

(2001a). *Draft Environmental Impact Statement: Light Goose Management*. Washington, DC: US Fish and Wildlife Service.

(2001b). Endangered and threatened wildlife and plants: annual notice of findings on recycled petitions. *Federal Register*, **66**(5), 1295–1300.

(2002). Endangered and threatened wildlife and plants: review of species that are candidates or proposed for listing as endangered or threatened; Annual notice of findings on recycled petitions; Annual description of progress on listing actions. *Federal Register*, **67**(114), 40657–40679.

(2003a). Endangered and threatened wildlife and plants; final rule to reclassify and remove the gray wolf from the list of endangered and threatened wildlife in portions of the conterminous United States; establishment of two special regulations for threatened gray wolves. *Federal Register*, **68**(62), 15804–15875.

(2003b). Final rule to reclassify and remove the gray wolf from the list of endangered and threatened wildlife in portions of the conterminous United States; Establishment of two special regulations for threatened gray wolves; Final and proposed rules. *Federal Register*, **50 CFR Part 17 RIN 1018-AF20**, 1–73.

(2003c). *Final Revised Recovery Plan for the Southern Sea Otter* (Enhydra lutris nereis). Portland, OR: US Fish and Wildlife Service.

US Fish and Wildlife Service, Nez Perce Tribe, National Park Service and US Department of Agriculture Wildlife Services (2003). *Rocky Mountain Wolf Recovery 2002 Annual Report*, ed. T. Meier. Helena, MT: US Fish and Wildlife Service. Accessible at http://westerngraywolf.fws.gov/

United Nations Environmental Programme (2001). *East Africa Database and Atlas Project.* Nairobi: National Museums of Kenya.

Uresk, D. W. (1984). Black-tailed prairie dog food habits and forage relationships in western South Dakota. *Journal of Range Management*, **37**, 325–329.

Uresk, D. W. and Paulson, D. B. (1988). Estimated carrying capacity for cattle competing with prairie dogs and forage utilization in western South Dakota. In *Symposium on Management of Amphibians, Reptiles, and Small Mammals in North America*, Flagstaff, AZ, 19–21 July 1988, pp. 387–390.

Urquart, K. A. and McKenrick, I. J. (2003). Survey of permanent wound tracts in the carcasses of culled wild red deer in Scotland. *Veterinary Record*, **152**, 497–501.

US Department of Agriculture (2005). Wildlife data. Accessible at http://www.aphis.usda.gov/ws/tblfrontpage.html

US Federal Register (2000). *Register*, July 28, pp. 46391–46398.

Vales, D. J. and Peek, J. M. (1995). Projecting the potential effects of wolf predation on elk and mule deer in the east front portion of the northwest Montana wolf recovery area. In *Ecology and Conservation of Wolves in a Changing World*, ed. L. N. Carbyn, S. H. Fritts and D. R. Seip. Edmonton, Alberta: Canadian Circumpolar Institute, pp. 211–221.

Valkama, J., Korpimaki, E., Arroyo, B. *et al.* (2005). Birds of prey as limiting factors of gamebird populations in Europe: a review. *Biological Reviews*, **80**, 170–203.

Van der Walt, J. (2002). Proliferation of game ranches. *Game and Hunting* **8**(10), 7.

Van Eerden, M. R. (1990). The solution of goose damage problems in The Netherlands, with special reference to compensation schemes. *Ibis*, **132**, 253–261.

van Oosten, V. (2000). *The Conflict between Primates and the Human Population in a Protected Area in North Cameroon.* Centre d'Etude de l'Environnement et du

Developpement au Cameroun (CEDC), Centre des Etudes de l'Environnement Université de Leiden (CML) and Organisation Neerlandaise de Developpement (SNV).

Van Pelt, W. E. (1999). *The Black-Tailed Prairie Dog Conservation Assessment and Strategy*. Phoenix, AZ: Arizona Game and Fish Department.

Van Vuren, D. (1998). Mammalian dispersal and reserve design. In *Behavioral Ecology and Conservation Biology*, ed. T. M. Caro. Oxford, UK: Oxford University Press, pp. 369–393.

Vandel, J.-M., Stahl, P., Durand, C., Balestra, L. and Raymond, J. (2001). Des chiens de protection contre le lynx. *Faune sauvage*, **254**, 22–27.

VanDruff, L. W., Bolen, E. G. and San Julian, G. J. (1994). Management of urban wildlife. In *Research and Management Techniques for Wildlife and Habitats*, ed. T. A. Bookout. 5th edn, Bethesda, MD: The Wildlife Society, pp. 507–530.

Veeramani, A. (1996). Man–wildlife conflict: cattle lifting and human casualties in Kerala. *Indian Forester*, **122**, 897–902.

Vickery, J. A. and Gill, J. A. (1999). Managing grassland for wild geese in Britain: a review. *Biological Conservation*, **89**, 93–106.

Vickery, J. A. and Summers, R. W. (1992). Cost-effectiveness of scaring brent geese *Branta b. bernicla* from fields of arable crops by a human bird scarer. *Crop Protection*, **11**, 480–484.

Vickery, J. A., Sutherland, W. J. and Lane, S. J. (1994). The management of grass pasture for brent geese. *Journal of Applied Ecology*, **31**, 282–290.

Vinuela, J. and Arroyo, B. (2002). Gamebird hunting and biodiversity conservation: synthesis, recommendations and future research priorities. Accessible at www.uclm.es/irec/reghab/inicio.html

Vinuela, J. and Villafuerte, R. (2003). Predators and rabbits in Spain: a key conflict for European raptor conservation. In *Birds of Prey in a Changing Landscape*, ed. D. B. A. Thompson, S. M. Redpath, A. H. Fielding, M. Marquiss and C. A. Galbraith. Edinburgh, UK: Stationery Office, pp. 511–526.

Virtanen, P. (2003). Local management of global values: community-based wildlife management in Zimbabwe and Zambia. *Society and Natural Resources*, **16**, 179–190.

Vitousek, P. M., Ehrlich, P. R., Ehrlich, A. A. and Matson, P. A. (1986). Human appropriation of the products of photosynthesis. *BioScience*, **36**, 368–373.

Vitousek, P. M., Mooney, H. A., Lubchenco, J. and Melillo, J. M. (1997). Human domination of earth's ecosystems. *Science*, **277**, 494–498.

Volpi, G., Boitani, L., Callaghan, C. *et al.* (2002). Anti-wolf barriers to manage captive and wild wolves and to protect livestock. *Abstract Wildlife Society 9th Annual Conference*, Bismarck, ND, 24–28 September.

Vos, J. (2000). Food habits and livestock depredation of two Iberian wolf packs (*Canis lupus*) in the north of Portugal. *Journal of Zoology*, **251**, 457–462.

Vosburgh, T. C. and Irby, L. R. (1998). Effects of recreational shooting on prairie dog colonies. *Journal of Wildlife Management*, **62**, 363–372.

Wabakken, P., Sand, H., Liberg and, O. and Bjärvall, A. (2001). The recovery, distribution and population dynamics of wolves on the Scandinavian Peninsula. *Canadian Journal of Zoology*, **79**, 710–725.

Wade, D. (1982). The use of fences for predator damage control. *Vertebrate Pest Conference Proceedings*, **10**, 24–33.

Wagner, K. K. and Conover, M. R. (1999) Effect of preventive coyote hunting on sheep losses to coyote predation. *Journal of Wildlife Management*, **63**, 600–612.

Wagner, K. K., Schmidt, R. H. and Conover, M. R. (1997). Compensation programs for wildlife damage in North America. *Wildlife Society Bulletin*, **25**, 312–319.

Waithaka, J. (1993). The effects of different elephant densities on biodiversity. *Pachyderm*, **16**.

(1994). The ecological role of elephants in restructuring plant and animal communities in different eco-climatic zones in Kenya and their impacts on land-use patterns. Ph. D. thesis, Kenyatta University, Nairobi, Kenya.

(1997). Management of elephants in Kenya: what have we learnt so far? *Pachyderm*, **24**, 33–36.

Waithaka, J. and Mwathe, K. M. (1995). *Report on Crop and Property Losses in Shimba Hills National Reserve.* Nairobi: Kenya Wildlife Service Elephant Program.

Wakajummah, J. O. (2000). Population dynamics. In *Kenya Coast Handbook: Culture Resources and Development in the East African Littoral*, ed., J. Hoorweg, D. Foeken and R. A. Obudh. London: LIT Verlag, pp. 55–81.

Wakeley, J. S. and Mitchell, R. C. (1981). Blackbird damage to ripening field corn in Pennsylvania. *Wildlife Society Bulletin*, **9**, 52–55.

Waller, J. and Reynolds, V. (2001). Limb injuries resulting from snares and traps in chimpanzees (*Pan troglodytes schweinfurthii*) of the Budongo Forest, Uganda. *Primates*, **42**, 135–139.

Wallis, W. (2003). Terror takes toll on Kenya's tourism industry. *Financial Times*, 8 December 2003.

Walpole, M. J. (2001). Feeding dragons in Komodo National Park: a tourism tool with conservation complications. *Animal Conservation*, **4**, 67–74.

Walpole, M. J. and Goodwin, H. J. (2000). Local economic impacts of dragon tourism in Indonesia. *Annals of Tourism Research*, **27**, 559–576.

(2001). Local perceptions of ecotourism and conservation around Komodo National Park, Indonesia. *Environmental Conservation*, **28**, 160–166.

Walpole, M. J. and Leader-Williams, N. (2001). Masai Mara tourism reveals partnership benefits. *Nature*, **413**, 771.

(2002). Ecotourism and flagship species in conservation. *Biodiversity and Conservation*, **11**, 543–547.

Walpole, M. J., Goodwin, H. J. and Ward, K. G. R. (2001). Pricing policy for tourism in protected areas: lessons from Komodo National Park, Indonesia. *Conservation Biology*, **15**, 218–227.

Walpole, M. J., Karanja, G. G., Sitati, N. W. and Leader-Williams, N. (2003). *Wildlife and People: Conflict and Conservation in Masai Mara, Kenya*, Wildlife and Development Series No. 14. London: International Institute for Environment and Development.

Walton, L. R., Cluff, H. D., Paquet, P. C. and Ramsay, M. A. (2001). Movement patterns of barren-ground wolves in the central Canadian arctic. *Journal of Mammalogy*, **82**, 867–876.

Wang, Y. (1999). Asiatic black bear conservation action plan: Taiwan. In *Bears: Status survey and Conservation Action Plan*, ed. C. Servheen, S. Herrero and B. Peyton. Gland, Switzerland: IUCN, pp. 213–215.

Warbington, M. C. (2000). Predator control with dogs and llamas. *North American Veterinary Conference Proceedings*, **14**, 237–239.
Warren, J. T. (1998). Conservation biology and agroecology in Europe. *Conservation Biology*, **12**, 499–500.
Washington Department of Fish and Wildlife (2003). *Game Management Plan 2003–2009*. Washington Department of Fish and Wildlife Report.
Watkin, J. R. (2002). *The Evolution of Ecotourism in East Africa: From an Idea to an Industry*, Wildlife and Development Series No. 15. London: International Institute for Environment and Development.
Watson, D. (1977). *The Hen Harrier*. London: Poyser.
Watson, J. (1997). *The Golden Eagle*. London: Poyser.
Watson, M. and Thirgood, S. J. (2001). Could translocation aid hen harrier conservation in the UK? *Animal Conservation*, **4**, 37–43.
Wayne, R. K., Lehman, N. and Fuller, T. K. (1995). Conservation genetics of the gray wolf. In *Ecology and Conservation of Wolves in a Changing World*, ed. L. N. Carbyn, S. H. Fritts and D. R. Seip. Edmonton, Alberta: Canadian Circumpolar Institute, pp. 399–407.
Weber, J.-M. (2000). Wolf return in Switzerland: a project to solve conflicts. *Carnivore Damage Prevention News*, **2**, 8–9. (http://www.kora.unibe.ch)
Weber, W. and Rabinowitz, A. (1996). A global perspective on large carnivore conservation. *Conservation Biology*, **10**, 1046–1054.
Webster, D. and Wood, T. (1999). Walking a wildlife highway from Yellowstone to the Yukon. *Smithsonian*, **30**, 58.
Wells, M. P. (1992). Biodiversity conservation, affluence and poverty: mismatched costs and benefits and efforts to remedy them. *Ambio*, **21**, 237–243.
(1993). Neglect of biological riches: the economics of nature tourism in Nepal. *Biodiversity and Conservation*, **2**, 445–464.
Wells, M. and Brandon, K. (1992). *People and Parks: Linking Protected Area Management with Local Communities*. Washington, DC: World Bank.
Wells, M., Guggenheim, S., Khan, A., Wardojo, W. and Jepson, P. (1999). *Investing in Biodiversity: A Review of Indonesia's Integrated Conservation and Development Projects*. Washington, DC: World Bank.
Weltzin, J. F., Dowhower, S. L. and Heitschmidt, R. K. (1997a). Prairie dog effects on plant community structure in southern mixed-grass prairie. *Southwestern Naturalist*, **42**, 251–258.
Weltzin, J. F., Archer, S. and Heitshmidt, R. K. (1997b). Small mammal regulation of vegetation structure in a temperate savanna. *Ecology*, **78**, 751–763.
Western, D. (1989). Conservation without parks: conservation in the rural landscape. In *Conservation for the Twenty-First Century*, ed. D. Western and M. Pearl. Oxford, UK: Oxford University Press, pp. 159–165.
(1994). Ecosystem conservation and rural development: the case of Amboseli. In *Natural Connections: Perspectives in Community-Based Conservation*, ed. D. Western and R. M. Wright. Washington, DC: Island Press, pp. 15–52.
(1997). *In the Dust of Kilimanjaro*. Washington, DC: Island Press.
(2001). Conservation in a human-dominated world. *Issues in Science and Technology*, **16**, 53–60.
(2002). *In the Dust of Kilimanjaro*, 2nd edn. Washington, DC: Shearwater.

Western, D. and Gichohi, H. (1993). Segregation effects and the impoverishment of savanna ecosystems: the case for ecosystem viability analysis. *African Journal of Ecology*, **31**, 268–281.

Western, D. and Henry, W. (1979). Economics and conservation in third world national parks. *BioScience*, **29**, 414–418.

Western, D. and Manzollilo Nightingale, D. (2002). Environmental change and the vulnerability of pastoralists to drought: a case study of the Maasai in Amboseli, Kenya. In *Africa Environmental Outlook Report*. Stevenage, UK: Earthprint, pp. 31–50.

Western, D. and Pearl, M. (eds.) (1989). *Conservation for the Twenty-First Century*. Oxford, UK: Oxford University Press.

Western, D. and Wright, R. M. (eds.) (1994). *Natural Connections: Perspectives in Community-Based Conservation*. Washington, DC: Island Press.

Weston, A. (1999). *An Invitation to Environmental Philosophy*. Oxford, UK: Oxford University Press.

White, D., Kendall, K. C. and Picton, H. D. (1999). Potential energetic effects of mountain climbers on foraging grizzly bears. *Wildlife Society Bulletin*, **27**, 146–151.

White, L. J. T., Tutin, C. E. G. and Fernandez, M. (1993). Group composition and diet of forest elephants (*Loxodonta africana cyclotis* Matschie 1900), in the Lope Reserve, Gabon. *African Journal of Ecology*, **31**(3), 181–199.

White, P. C. L. and Whiting, S. J. (2000). Public attitudes towards badger culling to control bovine tuberculosis in cattle. *Veterinary Record*, **147**, 179–184.

Whitman, K., Starfield, A. M., Quadling, H. S, and Packer, C. (2004). Sustainable trophy hunting of African lions. *Nature*, **428**, 175–178.

Whyte, I. (1993). The movement patterns of elephant in the Kruger National Park in response to culling and environmental stimuli. *Pachyderm*, **16**, 72–80.

Whyte, I. J. and Joubert, S. C. J. (1988). Blue wildebeest population trends in the Kruger National Park and the effects of fencing. *South African Journal of Wildlife Research*, **18**, 78–87.

Wick, P. (1995). Minimizing bear–sheep conflicts through herding techniques. *International Conference on Bear Research and Management*, **9**, 367–373.

(1998). *Le Chien de protection sur troupeau ovin: utilisation et méthode de mise en place*. Blois, France: *ARTUS*.

Wickens, P. (1996). Conflict between Cape (South African) fur seals and line-fishing operations. *Wildlife Research*, **23**, 109–117.

Wieland, T. *African Hunting Guide*. Rivonia, South Africa: Future Publishing.

Wikramanayake, E. D., Dinerstein, E., Robinson, J. G. *et al.* (1998). An ecology-based method for defining priorities for large mammal conservation: the tiger as case study. *Conservation Biology*, **12**, 865–878.

Wilcove, D. (1999). *The Condor's Shadow*. New York: W. H. Freeman.

Wilcox, R. (1992). Cattle and environment in the Pantanal of Mato Grosso, Brazil, 1870–1970. *Agricultural History*, **66**, 232–256.

Wild, R. G. and Mutebi, J. (1996). *Conservation through Community Use of Plant Resources*, People and Plants Working Paper No. 5. Paris: UNESCO.

Wilkie, D. S. and Carpenter, J. F. (1999). The potential role of safari hunting as a source of revenue for protected areas in the Congo Basin. *Oryx*, **33**, 339–345.

Willebrand, T., Lindén, M., Persson, J. and Segerström, P. (1999). Överlevnad och dödsorsaker hos märkta järvar i Sarek. *Bilagor till Sammanhållen rovdjurspolitik, Slutbetänkande av Rovdjursutredningen*. 146, 191–199. (In Swedish)

Williams, C. K., Ericsson, G. and Heberlein, T. A. (2002). A quantitative summary of attitudes toward wolves and their reintroduction (1972–2000). *Wildlife Society Bulletin*, **30**, 575–584.

Williamson, D. and Williamson, J. (1985). Botswana's fences and the depletion of Kalahari wildlife. *Parks*, **10**, 5–7.

Wilson, M. A. (1997). The wolf in Yellowstone: science, symbol, or politics? Deconstructing the conflict between environmentalism and wise use. *Society and Natural Resources*, **10**, 453–468.

Windberg, L. A., Knowlton, F. F., Ebbert, S. M. and Kelly, B. T. (1997). Aspects of coyote predation on Angora goats. *Journal of Range Management*, **50**, 226–230. (http://uvalde.tamu.edu/jrm/jrmhome.htm)

Winterbach, H. and Winterbach, C. W. (2002). Okavango Delta lions: ecology, home range and population dynamics. In *Lion Conservation Research Workshop 2: Modelling Conflict*, ed. A. J. Loveridge, T. Lynam and D. W. Macdonald. Oxford, UK: Wildlife Conservation Research Unit, pp. 83–84.

Wisconsin Department of Natural Resources (1999). Wisconsin wolf management plan. Unpublished report. Wisconsin Department of Natural Resources.

(2002). *Guidelines for Conducting Depredation Control on Wolves in Wisconsin following Federal Reclassification to Threatened Status*. Madison, WI: WDNR.

Wise, S. M. (2000). *Rattling the Cage: Toward Legal Rights for Animals*. Cambridge, MA: Perseus Books.

Wittemyer, G. (2001). The elephant population of Samburu and Buffalo Springs National Reserves, Kenya. *African Journal of Ecology*, **39**, 357–365.

Woodroffe, R. (2000). Predators and people: using human densities to interpret declines of large carnivores. *Animal Conservation*, **3**, 165–173.

(2001a). *African Wild Dogs and African People: Conservation through Coexistence*, First Annual Report of the Samburu–Laikipia Wild Dog Project. Davis, CA: University of California–Davis.

(2001b). Strategies for carnivore conservation: lessons from contemporary extinctions. In *Carnivore Conservation*, ed. J. L. Gittleman, S. Funk, D. W. Macdonald and R. K. Wayne. Cambridge, UK: Cambridge University Press, pp. 61–92.

(2003). *African Wild Dogs and African People: Conservation through Coexistence*, Third Annual Report of the Samburu–Laikipia Wild Dog Project. Davis, CA: University of California–Davis.

Woodroffe, R. and Frank, L. G. (2005). Lethal control of African lions (*Panthera leo*): local and regional population impacts. *Animal Conservation*, **8**, 91–98.

Woodroffe, R. and Ginsberg, J. R. (1998). Edge effects and the extinction of populations inside protected areas. *Science*, **280**, 2126–2128.

(2000). Ranging behaviour and vulnerability to extinction in carnivores. In *Behaviour and Conservation*, ed. L. M. Gosling and W. J. Sutherland. Cambridge, UK: Cambridge University Press, pp. 125–140.

Woodroffe, R., Ginsberg, J., Macdonald, D. W. and IUCN/SSC Canid Specialist Group (1997). *The African Wild Dog: Status Survey and Conservation Action Plan*. Gland, Switzerland: IUCN.

Woodroffe, R., Frost, S. D. W. and Clifton-Hadley, R. S. (1999). Attempts to control tuberculosis in cattle by culling infected badgers: constraints imposed by live test sensitivity. *Journal of Applied Ecology*, **36**, 494–501.

Woodroffe, R., Bourne, F. J., Donnelly, C. A., *et al.* (2002). Towards a sustainable policy to control TB in cattle. In *Conservation and Conflict: Mammals and Farming in Britain*, ed. W. Manley. London: Linnean Society.

Woodroffe, R., Lindsey, P. A., Romañach, S. S., Stein, A., Ranah, S. M. K. O. (2005). Livestock predation by endangered African wild dogs (*Lycaon pictus*) in northern Kenya. *Biological Conservation*, **124**, 225–234.

World Bank (2003). Kenya at a glance. Accessible at http://www.worldbank.org/data/countrydata/aag/ken_aag.pdf

World Commission on the Environment and Development (1987). *Our Common Future*. Oxford, UK: Oxford University Press.

World Health Organization (1998). *World Survey of Rabies*. Geneva, Switzevland: WHO.

World Parks Congress (2003). Recommendation 4: Building comprehensive and effective protected area systems. Accessible at http://www.iucn.org/wpc2003/pdfs/outputs/recommendations/approved/english/html/r04.htm

World Resources Institute (1994). *World Resources 1994–95: A Guide to the Global Environment*. Oxford, UK: Oxford University Press.

World Tourism Organization (2002). *Yearbook of Tourism Statistics*. Madrid: WTO.

World Travel and Tourism Environment Research Centre (1993). *World Travel and Tourism Environment Review 1993*. Oxford, UK: WTTERC.

World Wildlife Fund (2000a). *Food for Thought: The Utilization of Wild Meat in Eastern and Southern Africa*.

(2000b). *Tourism and Carnivores: The Challenge Ahead*. Godalming, UK: WWF-UK.

(2002). *An Analysis of the Effectiveness of the Amur Tiger Anti-Poaching Brigades in the Russian Far East*, WWF Project No. RU0005.08. Vladivostok, Russia: WWF.

(2004). World Wildlife Fund online news reports. Accessible at http://www.worldwildlife.org

Wu, J. and Babcock, B. A. (1996). Contract design for the purchase of environmental goods from agriculture. *American Journal of Agricultural Economics*, **78**, 935–945.

Wunder, M. B. (1997). *Of Elephants and Men: Crop Destruction, CAMPFIRE, and Wildlife Management in the Zambezi Valley, Zimbabwe*. Ann Arbor, MI: University of Michigan Press.

Wydeven, A. P., Treves, A., Brost, B. and Wiedenhoeft, J. (2004). Characteristics of wolf packs in Wisconsin: identification of traits influencing depredation. In *People and Predators: From Conflict to Coexistence*, ed. N. Fascione, A. Delach and M. Smith. Washington, DC: Island Press, pp. 28–50.

Wyoming Game and Fish Department (2003). *Final Draft Wyoming Gray Wolf Management Plan*. Wyoming Game and Fish Department Report, 1–43.

Yaeger, R. and Miller, N. N. (1986). *Wildlife, Wild Death.* Albany, NY: State University of New York Press.

Yamazaki, K. (2002). Conflicts between Japanese black bears and human beings in the Okutama Mts., Central Japan. In *14th International Congress on bear research and Management: Program and Abstracts of Papers.* Steinkjer, Norway: Nord-Trondelag University College, p. 80.

Yellowstone National Park. *Wildlife Society Bulletin,* **24**, 402–413.

Yodsis, P. (2001). Must top predators be culled for the sake of fisheries? *Trends in Ecology and Evolution,* **16**, 78–84.

Yom-Tov, Y., Ashkenazi, S. and Viner, O. (1995). Cattle predation by the golden jackal *Canis aureus* in the Golan Heights, Israel. *Biological Conservation,* **73**, 19–22.

Young, S. P. (1944). The wolves of North America. Part 1. Their history, life habits, economic status, and control. In *The Wolves of North America,* ed. S. P. Young and E. A. Goldman. New York: Dover, pp. 1–385.

Young, S. P. and Goldman, E. A. (eds.) (1944). *The Wolves of North America.* New York: Dover.

(1946). *The Puma: Mysterious American Cat.* Washington, DC: American Wildlife Institute.

Yudelman, M., Ratta, A. and Nygaard, D. (1998). *Pest Management and Food Production.* Washington, DC: International Food Policy Research Institute.

Zemlicka, D. E. (1995). Seasonal variation in the behavior of sterile and nonsterile coyotes. M.S. thesis, Utah State University, Logan, UT.

Zhang, L. and Wang, N. (2003). An initial study on habitat conservation of Asian elephant (*Elephas maximus*), with a focus on human elephant conflict in Simao, China. *Biological Conservation,* **112**, 453–459.

Zhivotchenko, V. I. (1977). A man-eating tiger in Primorskiy Krai. *Priroda,* **3**, 123–124. (In Russian)

Zimmermann, A. (2000). Jaguar–rancher conflict in the north Pantanal of Brazil. M.Sc. thesis, Durrell Institute for Conservation and Ecology, University of Kent, UK.

Zubanova, S. D., Leshnevskaya, L. A., Panchenko, P. P. *et al.* (2001). *The Relationship of Local People and Tigers in the Sikhote–Alin Mountains: A Sociological Study.* Vladivostok, Russia: World Wildlife Fund.

Zube, C. H. and Busch, M. L. (1990). Park–people relationships: an international review. *Landscape and Urban Planning,* **19**, 117–131.

Index

Printed in the United Kingdom by
Lightning Source UK Ltd., Milton Keynes
139192UK00001B/19/P

9 780521 532037